Plant Biotechnology: Principles and Future Prospects

Plant Biotechnology: Principles and Future Prospects

Editor: Ava Freeman

CALLISTO REFERENCE

www.callistoreference.com

Callisto Reference,
118-35 Queens Blvd., Suite 400,
Forest Hills, NY 11375, USA

Visit us on the World Wide Web at:
www.callistoreference.com

ISBN: 978-1-64116-198-5 (Hardback)

Cataloging-in-Publication Data

Plant biotechnology : principles and future prospects / edited by Ava Freeman.
    p. cm.
Includes bibliographical references and index.
ISBN 978-1-64116-198-5
1. Plant biotechnology. 2. Agricultural biotechnology. 3. Transgenic plants. I. Freeman, Ava.
SB106.B56 P53 2019
631.523 3--dc23

Table of Contents

Preface

It is often said that books are a boon to mankind. They document every progress and pass on the knowledge from one generation to the other. They play a crucial role in our lives. Thus I was both excited and nervous while editing this book. I was pleased by the thought of being able to make a mark but I was also nervous to do it right because the future of students depends upon it. Hence, I took a few months to research further into the discipline, revise my knowledge and also explore some more aspects. Post this process, I begun with the editing of this book.

Agricultural biotechnology falls under the domain of agricultural science. It uses the techniques of genetic engineering in order to modify crops to exhibit desired traits of growth rate, resistance to diseases and pests, size of crops, etc. The practice of traditional crossbreeding is used to mate two compatible species in order to create a new plant variety with all the desirable traits of the parents. Chromosomes in a crop can also be modified to induce polyploidy. This potentially influences the size and fertility of the crop. Besides these, genome editing, RNA interference and protoplast fusion can also be employed for plant engineering. This book elucidates the concepts and innovative models around prospective developments with respect to plant biotechnology. It presents researches and studies performed by experts across the globe. The extensive content of this book provides the readers with a thorough understanding of the subject.

I thank my publisher with all my heart for considering me worthy of this unparalleled opportunity and for showing unwavering faith in my skills. I would also like to thank the editorial team who worked closely with me at every step and contributed immensely towards the successful completion of this book. Last but not the least, I wish to thank my friends and colleagues for their support.

Editor

Display of whole proteins on inner and outer surfaces of grapevine fanleaf virus-like particles

Lorène Belval[1,2], Caroline Hemmer[1,2], Claude Sauter[3], Catherine Reinbold[1], Jean-Daniel Fauny[3], François Berthold[2], Léa Ackerer[1,2,4], Corinne Schmitt-Keichinger[2], Olivier Lemaire[1], Gérard Demangeat[1,*] and Christophe Ritzenthaler[2,*]

[1]SVQV, INRA, Université de Strasbourg, Colmar, France
[2]Institut de Biologie Moléculaire des Plantes CNRS-UPR 2357, associée à l'Université de Strasbourg, CNRS, Strasbourg, France
[3]Institut de Biologie Moléculaire et Cellulaire du CNRS, UPR 9002, Architecture et Réactivité de l'ARN, Université de Strasbourg, Strasbourg, France
[4]Institut Français de la Vigne et du Vin, Domaine de l'Espiguette, Le Grau-du-Roi, France

*Correspondence (C. Ritzenthaler
G. Demangeat

e-mails ritzenth@unistra.fr and gerard.
demangeat@colmar.inra.fr)

Keywords: virus like particles, nanoparticles, virus, nepovirus, nanocarrier.

Summary

Virus-like particles (VLPs) derived from nonenveloped viruses result from the self-assembly of capsid proteins (CPs). They generally show similar structural features to viral particles but are noninfectious and their inner cavity and outer surface can potentially be adapted to serve as nanocarriers of great biotechnological interest. While a VLP outer surface is generally amenable to chemical or genetic modifications, encaging a cargo within particles can be more complex and is often limited to small molecules or peptides. Examples where both inner cavity and outer surface have been used to simultaneously encapsulate and expose entire proteins remain scarce. Here, we describe the production of spherical VLPs exposing fluorescent proteins at either their outer surface or inner cavity as a result of the self-assembly of a single genetically modified viral structural protein, the CP of grapevine fanleaf virus (GFLV). We found that the N- and C-terminal ends of the GFLV CP allow the genetic fusion of proteins as large as 27 kDa and the plant-based production of nucleic acid-free VLPs. Remarkably, expression of N- or C-terminal CP fusions resulted in the production of VLPs with recombinant proteins exposed to either the inner cavity or the outer surface, respectively, while coexpression of both fusion proteins led to the formation hybrid VLP, although rather inefficiently. Such properties are rather unique for a single viral structural protein and open new potential avenues for the design of safe and versatile nanocarriers, particularly for the targeted delivery of bioactive molecules.

Introduction

From a structural and biotechnological standpoint, virions can be viewed as self-assembled nanometre-scale cages or viral nanoparticles (VNPs) containing the viral genome. Genetic engineering has allowed the development of nucleic acid-free VNPs also named virus-like particles (VLPs) that can be regarded as a subclass of VNPs. VNP-based nanomaterials have been adapted through chemical reactions or genetic engineering to gain new functionalities. First, their external surfaces can be modified or decorated with other molecules, in which case VNPs can act as biocompatible nanocarriers for antigen presentation, immunomodulation, customized targeting, etc., as exemplified by the plant-infecting cowpea mosaic virus (CPMV, Chatterji et al., 2004; Lewis et al., 2006; Sainsbury et al., 2010; Steinmetz et al., 2011), tobacco mosaic virus (TMV, Alonso et al., 2013) or cowpea chlorotic mottle virus (CCMV, Suci et al., 2007). Second, as self-assembled cages, the inner cavity of VNPs can be used to encapsulate or encage a variety of active molecules, including pharmaceuticals, image enhancers and nucleic acids (Arcangeli et al., 2014; Bruckman et al., 2013; Mueller et al., 2011; Shriver et al., 2013). By acting on both external surface and inner cavity, VNPs can be adapted theoretically at will and are therefore regarded as extremely versatile tools with great potentials in medicine, as enzyme nanocarriers or even as novel biomaterials (for review, see Reference Alonso et al., 2013; Cardinale et al., 2012; Pokorski and Steinmetz, 2010).

Despite the interesting traits offered by VNPs, the possibilities to modify them are often limited due to numerous constraints. For instance, whether derived from attenuated or killed animal viruses, bacteriophages or plant viruses, VNPs may contain a viral genome and may therefore be considered as hazardous infectious entities. In this respect, VLPs are highly advantageous over VNPs, because they are noninfectious and a multitude of viral structural proteins have been adapted to this end (for review see Reference Chen and Lai, 2013; Liu et al., 2012).

In general, while the external surface of VLPs can easily accommodate some modifications, the encapsulation capacity of VLPs is often restricted to only small molecules such as ions, oligonucleotides, pharmaceuticals or small peptides (for review, see Reference Glasgow and Tullman-Ercek, 2014; Li and Wang, 2014; Saunders and Lomonossoff, 2013). To date, relatively few examples of encapsulation of foreign cargo proteins have been reported. These are limited to VLPs derived from viruses with spherical architecture, involves at least two separate protein components or nucleic acid linkers and is rather restrictive in stoichiometry of encapsulated proteins (for review, see Reference Lee, 2016; Lua, 2014; Sainsbury et al., 2014). To our knowledge, VLP made from a single structural protein compatible with the simultaneous inner and outer exposure of entire proteins via genetic fusions has not yet been described.

Grapevine fanleaf virus (GFLV) is the main causal agent of fanleaf degeneration, probably the most detrimental viral disease on the emblematic crop grapevine. GFLV belongs to the genus *Nepovirus* in the family Secoviridae (order *Picornavirales*) that also includes the *Comovirus* CPMV (Sanfaçon *et al.*, 2009). GFLV, which has a bipartite positive-sense RNA genome, is a nonenveloped icosahedral virus of approximately 30 nm in diameter, comprising 60 identical CP subunits of 56 kDa, arranged in a pseudo T = 3 symmetry, whose structure has been resolved at 2.7 Å resolution (Schellenberger *et al.*, 2011b). Particles have been observed in transgenic plants expressing GFLV CP coding sequence, suggesting that the CP of GFLV is able to self-assemble into VLPs (Barbier *et al.*, 1997). Here, we show that the CP of GFLV is a highly versatile protein of biotechnological interest, compatible with the simultaneous encapsulation and exposure of large proteins through the genetic fusion to the CP N- or C-terminal end.

Results

GFLV coat protein self-assembles into virus-like particles

To address the ability of GFLV CP to produce VLPs *in planta*, the sequence encoding the CP from GFLV isolate F13 was introduced into the pEAQ-*HT*-DEST1 binary vector (Sainsbury *et al.*, 2009) and used for transient expression in *Nicotiana benthamiana* leaves by agro-infiltration (Figure 1). Samples were analysed by double-antibody sandwich ELISA (DAS-ELISA) at 7 days postagro-infiltration (dpa) using GFLV-F13-infected *N. benthamiana* leaves at 14 days postinoculation as a positive control and pEAQ-*HT*-DEST1-driven TagRFP (TR, Merzlyak *et al.*, 2007) agro-infiltrated leaves at 7 dpa and leaves from healthy plants as negative controls. A strong positive signal was detected in CP-expressing and GFLV-infected samples but not in extracts from TR-infiltrated or healthy leaf material (Figure 2a). To test the ability of transiently expressed CP to self-assemble into VLPs, the

same leaf extracts were further analysed by transmission electron microscopy (TEM) after immunocapture with GFLV-specific polyclonal antibodies (see experimental procedures). Observation of negatively stained material (Figure 2b) revealed the presence of icosahedral particles of about 30 nm in diameter in CP-expressing samples but not in TR-expressing or healthy negative controls (Figure 2b). Although not very abundant on grids, CP-derived icosahedral particles were very similar to GFLV particles observed under equivalent conditions (Figure 2b). This indicates that GFLV CP is able to self-assemble into VLPs upon transient expression in *N. benthamiana*.

N- and C-terminal CP fusion proteins assemble into VLPs

Analysis of the GFLV atomic structure (Schellenberger *et al.*, 2011b) reveals that the GFLV CP amino-terminal residue Gly_1 and the three carboxy-terminal residues Phe_{502}, Pro_{503} and Val_{504} do not contribute to the quaternary interactions of the virus capsid and are exposed at the inner and outer surfaces of the GFLV particle, respectively (Figure 3a and 3b). In this respect, both extremities could be expected compatible with the addition of extra residues without interfering with the capacity of the CP to form a capsid. To test this hypothesis, N- or C-terminal CP fusions to TR were produced and, respectively, named TRCP and CPTR hereafter (Figure 1). Both fusions included a Gly_3-Ser-Gly_3 linker peptide (Figure 1 and Figure S1) to maintain flexibility between the CP and TR domains (Zilian and Maiss, 2011) and were transiently expressed in *N. benthamiana* leaves. Samples were analysed by epifluorescence macroscopy for TR expression at 5 dpa (Figure S2), and 2 days later by DAS-ELISA for CP expression (Figure 3c) and TEM for VLPs (Figure 3d). While TR fluorescence was observed in all samples (Figure S2) suggesting

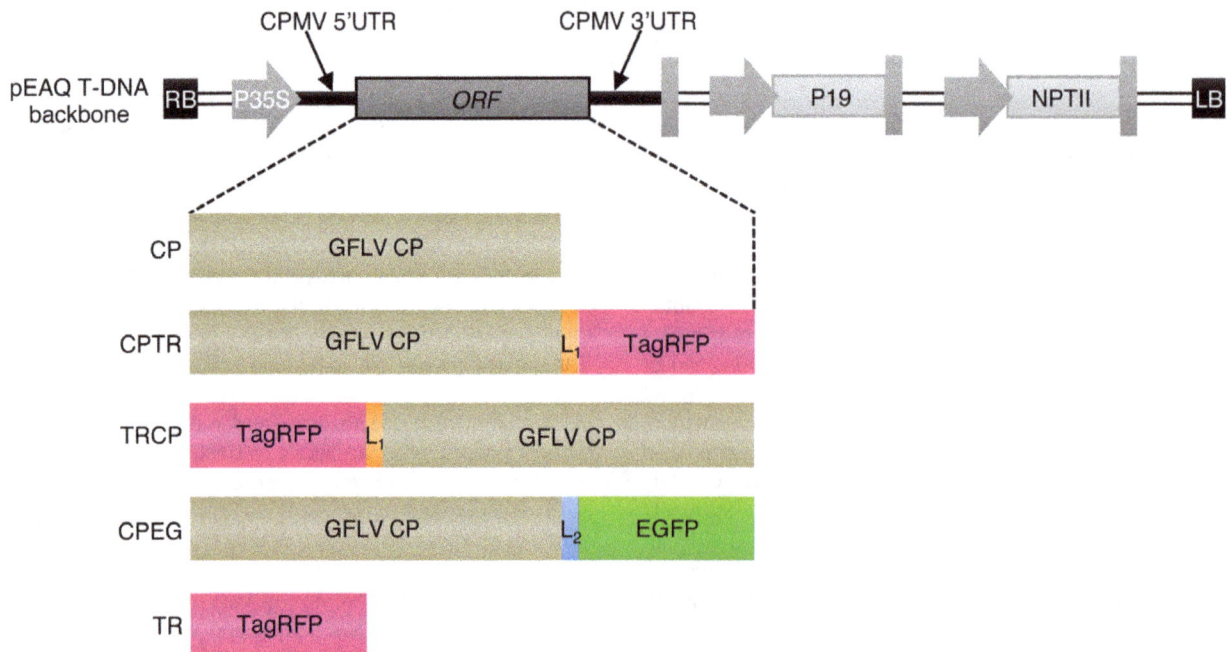

Figure 1 Schematic representation of pEAQ vector T-DNA region and constructs derived thereof. The backbone of the T-DNA region, extending from the right border (RB) to the left border (LB), is represented in grey shades. Open-reading frame (ORF) of interest is flanked by sequences of cowpea mosaic virus untranslated regions (CPMV UTRs) under the control of cauliflower mosaic virus 35S promoter (P35S). Native GFLV CPs as well as its TR- and EG-tagged variants (CPTR, TRCP, CPEG) or TagRFP alone (TR) were introduced into the pEAQ vector as schematically indicated. L1 corresponds to the 7-amino-acid Gly_3-Ser-Gly_3 linker sequence. L2 corresponds to the 15-amino-acid linker sequence resulting from Gateway recombination. Complete amino acid sequences of the expressed proteins are provided in Figure S1. P19: tombusvirus P19 silencing suppressor gene. NPTII: neomycin phosphotransferase II gene.

(a)

(b)

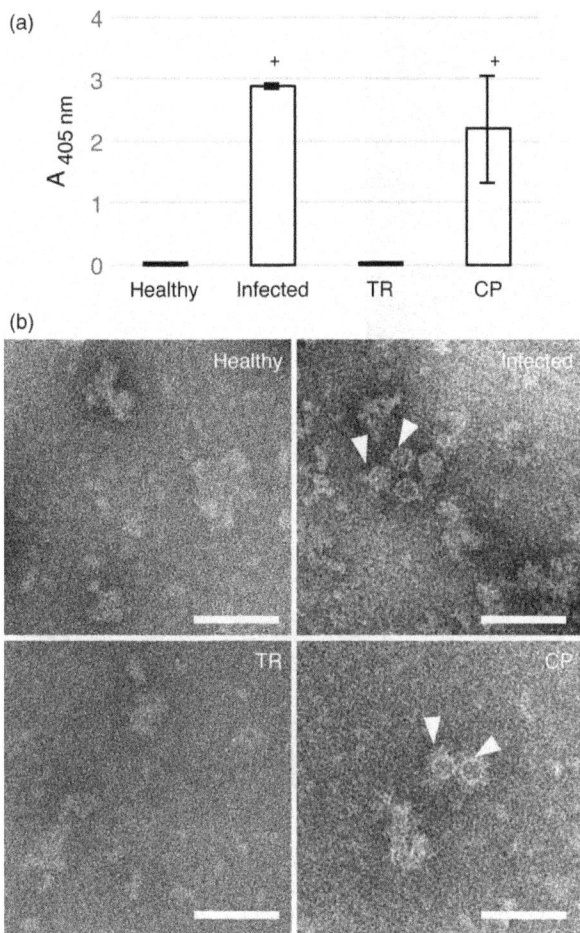

Figure 2 Transient expression of GFLV CP in *N. benthamiana* leaves leads to VLP production. (a) Expression of GFLV CP in *N. benthamiana* leaves at 7 dpa (TR and CP) or at 14 days of infection was determined by DAS-ELISA using anti-GFLV antibodies for detection. Bars represent the mean absorbance obtained with three different leaves for each condition. Error bars correspond to 95% confidence intervals. Samples were considered positive (+) when $A_{405\ nm}$ exceeded the healthy control sample mean value by at least a factor of three. (b) ISEM micrographs resulting from observations performed on the same extracts were then analysed by DAS-ELISA. Particles of approximately 30 nm (arrowheads) were detected only in GFLV-infected and CP-expressing samples. Scale bars: 100 nm.

the expression of the different proteins, CP was detected only in CPTR and TRCP crude extracts by DAS-ELISA (Figure 3c), which correlated with the presence of VLPs in TEM (Figure 3d). These results suggest that GFLV CP retains its capacity to form VLPs upon fusion of its N- or C-terminal end to TR.

To confirm our results and to gain insights into the biochemical properties of VLPs, large-scale production of VLPs in *N. benthamiana* leaves was carried out followed by their purification using standard GFLV purification procedure in the absence of protease inhibitors (see experimental procedures). In parallel, GFLV was purified from infected *Chenopodium quinoa* leaves. After linear sucrose gradient, a pink band was observed in the TRCP gradient (Figure 3Sa). Two millilitres of sucrose gradient fractions was collected and those enriched in VLPs identified by semiquantitative DAS-ELISA. While *bona fide* GFLV particles sedimented essentially towards the bottom of the gradient in fractions 8—10, CP-, CPTR- and TRCP-derived

particles distributed to the lighter fractions 3—5, 4—6 and 6—8, respectively (Figure S3b), well in agreement with a previous report indicating that empty GFLV particles show lower sedimentation coefficient than RNA-containing virions (Quacquarelli *et al.*, 1976). Enriched fractions were further pooled and processed for final concentration by ultracentrifugation. Remarkably, pink pellets were observed in both TRCP and CPTR samples (Figure 4a). The final concentration of purified material (Figure 4b) was determined by quantitative DAS-ELISA using purified GFLV as a standard. Determined yields ranged from 386 to 445 µg GFLV particles equivalent per kg of fresh leaves for the three VLP types, which is in the same order of magnitude as GFLV purification yields from infected *N. benthamiana* (Schellenberger *et al.*, 2011a).

To assess their quality and purity, purified samples were analysed by Coomassie blue staining after SDS-PAGE (Figure 4c), immunoblotting using anti-GFLV or anti-TR antibodies (Figure 4d and 4e) and mass spectrometry (Figure S4). For Coomassie blue staining, 6 mg of GFLV particles equivalent of each sample was loaded on SDS-denaturing gel. In line with the purification of VLPs, one major protein of about 57 kDa co-migrating with the CP of GFLV (calculated mass 56 kDa) was present in purified samples from CP-expressing leaves (Figure 4c, bands 1 and 2). For CPTR and TRCP samples, profiles were more complex with three major proteins of approximate molecular masses of 87, 73 and 57 kDa being detected in both samples (Figure 4c, bands 3—5 for CPTR and 6—8 for TRCP), but in inverse proportions. For TRCP, the largest product was the most abundant and the smallest the least abundant (approximately 69%, 24% and 7% respective abundance), whereas for CPTR the proportions were as follows: 2% for band 3, 35% for band 4 and 63% for band 5. Upon immunoblotting with anti-GFLV antibodies, the shortest product present in the CPTR sample (Figure 4c, b and 5) was clearly detected (Figure 4d), strongly suggesting that band 5 corresponds to the CP of GFLV and probably represents a processing product of CPTR. In the TRCP sample, the largest product (Figure 4c, b and 6) immunoreacted clearly with anti-GFLV antibodies (Figure 4d). Considering this band is about the expected size of TRCP (calculated mass 82.8 kDa), our results suggest that the full-length TRCP is the principal protein present in the purified TRCP sample. Accordingly, band 6 gave also a strong signal upon immunodetection with anti-TR antibodies (Figure 4e). Anti-TR antibodies immunoreacted also but weakly with the largest product present in the CPTR sample (band 3) and with the 73 kDa truncated products observed in CPTR and TRCP samples (Figure 4e, bands 4 and 7). Our results suggest that TRCP and CPTR have the capacity to self-assemble into VLPs *in planta*. They also reveal clear differences in the capacity of these VLPs to accommodate fusion proteins with TRCP-derived VLPs being far less prone to degradation or truncation than CPTR-derived ones.

To gain further insights into the composition of the purified products, Coomassie-stained bands were subjected to mass spectrometry leading to the identification of peptides covering nearly the entire CP for each band analysed (Figure 4c and Figure S4). Peptides corresponding to TR were only observed for bands 3, 4, 6 and 7. Nearly full coverage of the CPTR or TRCP proteins was strictly restricted to bands 3 and 6. The 73 kDa products corresponding to bands 4 and 7 showed only partial coverage of the TR and thus represent a truncated version of CPTR or TRCP, possibly due to proteolytic degradation during the purification process performed without protease inhibitors. Our

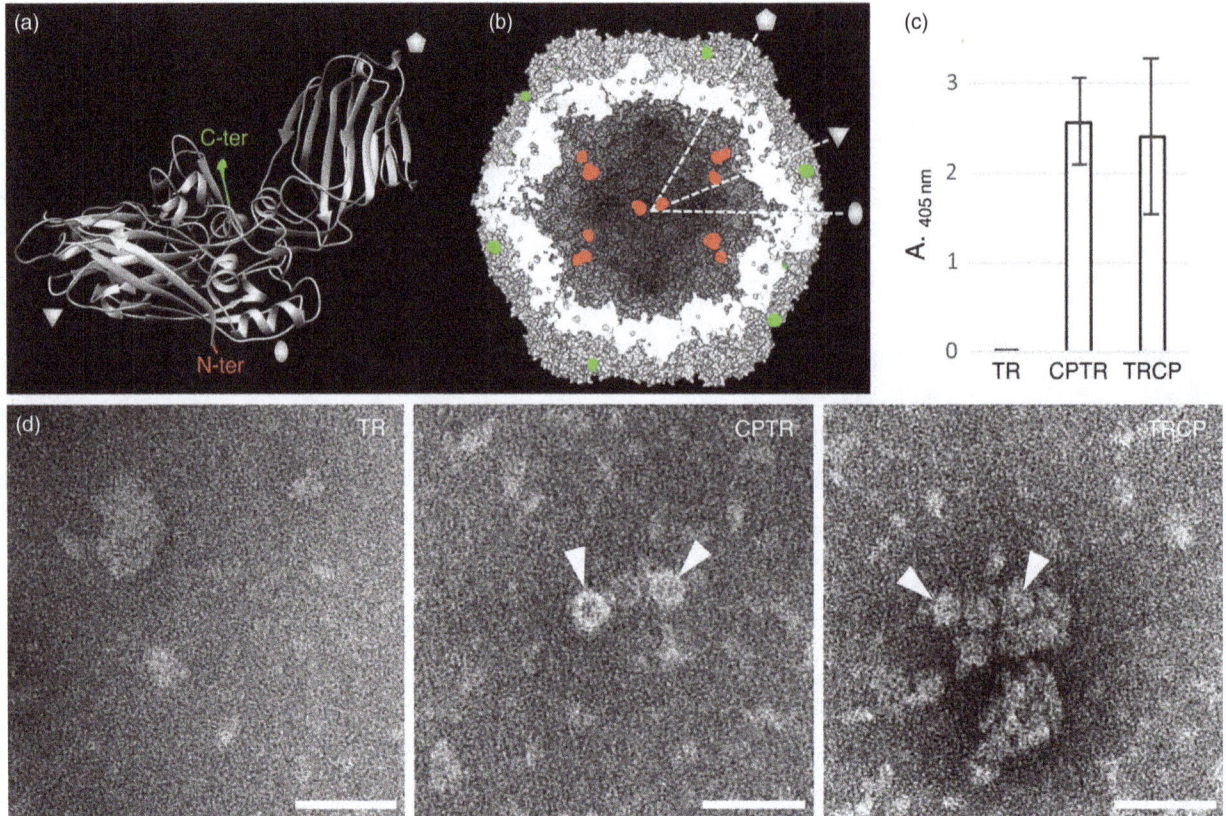

Figure 3 Fusion of TagRFP to the N- or C-terminal end of GFLV CP is compatible with VLP formation. (a) Ribbon diagram view of a GFLV CP subunit and (b) surface-shaded cross section of a particle according to the 3 Å resolution crystal structure (PDB code 4V5T, (Schellenberger *et al.*, 2011b). Positions of the CP N- and C-termini are indicated in red and green, respectively. The pentagon, triangle and oval symbolize the fivefold, threefold and twofold icosahedral symmetry axes, respectively. Cross-sectioned residues appear in white. (c) Detection of GFLV CP in TR-, CPTR- or TRCP-expressing *N. benthamiana* crude leaf extracts at 7 dpa. Bars represent the mean absorbance obtained with three different leaves for each condition. Error bars correspond to 95% confidence intervals. (d) ISEM micrographs resulting from observations performed on the same extracts than analysed by DAS-ELISA. Arrowheads point to VLPs trapped by anti-GFLV antibodies in CPTR- and TRCP-expressing leaf extracts. Scale bars: 100 nm.

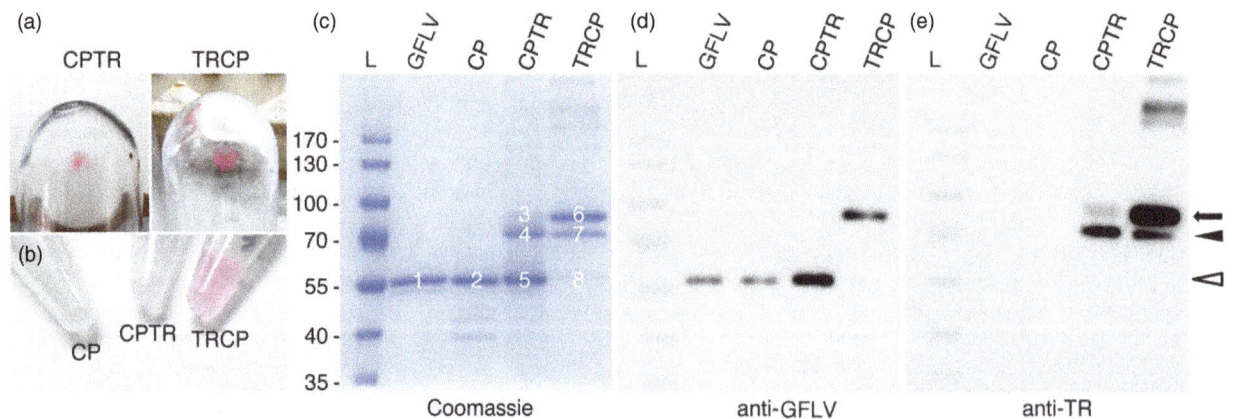

Figure 4 Recombinant VLPs can be purified from CP-, CPTR- and TRCP-expressing leaves. (a) Pink pellets resulting from CPTR (left panel) and TRCP (right panel) purifications after final ultracentrifugation. (b) Purified CP, CPTR and TRCP in solution. Note the pink colour of CPTR and TRCP samples. (c) Coomassie blue-stained gel of GFLV-, CP-, CPTR- and TRCP-purified particles after SDS-PAGE. Six micrograms of GFLV particles equivalent was separated in each lane. Major bands in the gel are numbered from 1 to 8. (d and e) Corresponding Western blotting analyses of GFLV, CP, CPTR and TRCP samples using anti-GFLV (d) or anti-TR antibodies (e). About 0.05 μg of GFLV particles equivalent was used in each lane. White arrowhead indicates bands with expected size for CP. Arrow points to expected size for full-length TRCP or CPTR fusions. Black arrowhead points to major TRCP or CPTR truncated products. L: molecular mass markers. Masses (kDa) are indicated in the left.

results demonstrate that the CPTR or TRCP full-length chimeric proteins can be purified following standard virus purification procedures likely reflecting their capacity to self-assemble into VLPs. They also reveal that the CPTR fusion is more labile than TRCP, possibly as a consequence of the predicted orientation of TR towards the VLP inner or outer surface.

N- and C-terminal CP fusions are oriented towards the interior and exterior of VLPs, respectively

To confirm the presence of VLPs and the different orientation shown by TR in CPTR and TRCP samples, negative staining and immunosorbent electron microscopy (ISEM) analyses were performed. As expected, negative staining revealed the presence of numerous VLPs in all samples (Figure 5d, 5g and 5j) that clearly resemble GFLV particles (Figure 5a). Under these conditions, the interiors of CP and CPTR particles appeared electron dense (Figure 5d and 5g) similar to those of GFLV particles (Figure 5a), whereas those of TRCP particles appeared rather electron lucent (Figure 5j), likely reflecting the increased inner density of particles and decreased penetrability to heavy metals linked to the orientation of TR inside TRCP VLPs. To verify the topology of VLPs, decoration assays were performed with anti-GFLV (Figure 5b, 5e, 5h and 5k) or anti-TR antibodies (Figure 5c, 5f, 5i and 5l). While all purified particles reacted to anti-GFLV antibodies as expected (Figure 5b, 5e, 5h and 5k), only CPTR particles were decorated with anti-TR antibodies (Figure 5i), despite the significantly greater proportion of full-length protein present in TRCP versus CPTR particles (Figure 4). The observed differences in accessibility to TR antibodies confirm the exposure of TR at the outer surface of CPTR-derived VLPs and the encaging of TR inside the particles in TRCP-derived VLPs. Perhaps most importantly, our results also clearly show that GFLV CP can accommodate the fusion of foreign proteins as large as fluorescent proteins (FP) that represent approximately half the mass of the CP without losing its capacity to self-assemble into VLPs.

Hybrid VLPs can be produced

In view of our results, we next examined the capacity of GFLV CP to form hybrid VLPs upon coexpression of N- and C-terminal CP fusions. To do so, EGFP (named EG hereafter) was fused to the CP C-terminus as indicated in Figure 1 (construct CPEG). As before, agro-infiltrated *N. benthamiana* leaves were used for expression assays and purification of VLPs. CPEG-only-expressing leaves were used as negative control and compared to leaves coexpressing CPEG and TRCP (CPEG + TRCP). In compliance with our previous results, CPEG VLPs could be purified and located to the same linear sucrose gradient fractions as CPTR VLPs (Figure S3b). Coexpressed CPEG and TRCP also enabled the purification of DAS-ELISA immunoreactive material cosedimenting with CPEG VLPs (Figure S3b). ISEM analysis confirmed the presence of VLPs in both CPEG and CPEG + TRCP samples that clearly immunoreacted with both anti-GFLV and anti-EG antibodies (Figure 6), well in agreement with the predicted exposure of EG towards the external surface of VLPs. Considering TR is inaccessible to antibodies in ISEM upon fusion to the CP N-terminus, we further assessed the presence of EG and TR by fluorescence imaging of VLPs separated by native agarose gel electrophoresis (Figure 7). Under such conditions, distinct bands with specific migration profiles were detected for TRCP VLPs (Figure 7a), CPEG VLPs (Figure 7b) and CPEG + TRCP VLPs (Figure 7a and b) that emitted in both green and red channels as expected for hybrid particles (Figure 7a and 7b).

To confirm the production of *bona fide* hybrid VLPs and hence the presence of particles that emit simultaneously in green and red and should therefore appear yellow, purified samples were further processed for single-particle microscopy. In this manner, TRCP VLPs appeared as numerous individual red spots (Figure 7c). Similarly, distinct green spots likely corresponding to CPEG VLPs were also detected but in lower density (Figure 7d), likely reflecting the low abundance of full-length protein in purified CPEG VLPs samples. Importantly, the observation of a mix of separately purified CPEG and TRCP VLPs revealed individual red- or green-only VLPs (Figure 7e). Although not abundant probably as a consequence of proteolysis of EG molecules on the outer surface of VLPs, yellow spots likely corresponding to hybrid VLPs were detected only in CPEG + TRCP-derived samples (Figure 7f). Altogether, our results demonstrate that GFLV CP is compatible with the production of hybrid VLPs in which foreign proteins as large as FPs can be exposed at the outer surface and encaged inside individual VLPs when N- and C-terminal fusions to CP are coexpressed.

GFLV CP-derived VLPs are nucleic acid free

To examine the composition of VLPs, native agarose gel electrophoresis was performed and gel-stained with Coomassie blue or ethidium bromide for protein (Figure 8a) or nucleic acid content (Figure 8b). Upon protein staining, the migration profiles of CP, CPTR and TRCP VLPs as well as purified GFLV differed significantly, probably as a consequence of differences in net charge, density and mass of the various particles. Under UV illumination, a clear signal was observed with purified virus but not with CP-derived VLPs (Figure 8a). We attributed the faint bands observed with CPTR and TRCP VLPs (Figure 8b, arrowheads) to the slight TR excitation under UV light (Merzlyak et al., 2007) and the use of filters unable to fully discriminate TR and nucleic acid spectra rather than to the presence of nucleic acids. Upon spectrophotometer analysis, only purified virus led to high $A_{260/280}$ values (>1.6) indicative of the presence of nucleic acids, whereas those measured for the different VLPs ranged from 0.89 to 1.07 (Figure 8c). Our results suggest that VLPs are, within the limits of our detection methods, nucleic acid free. Similar results were obtained with CPMV-derived VLPs (Hesketh et al., 2015; Saunders et al., 2009).

Discussion

We have shown that GFLV CP when expressed transiently in plants is able to form VLPs that can be purified using standard nepovirus purification procedures. Such VLPs look similar to native GFLV particles and appear nucleic acid free. Altogether, our results indicate that GFLV CP has the capacity to self-assemble into VLPs in an RNA-independent manner as already suggested by the presence of empty GFLV particles in samples from infected plants (Quacquarelli et al., 1976) and in CP-expressing transgenic plants (Barbier et al., 1997). This capacity is probably a general feature of nepovirus-encoded CPs considering empty particles are commonly observed upon the expression of nepoviral CPs in transgenic plants or insect cells (Barbier et al., 1997; Bertioli et al., 1991; Seitsonen et al., 2008; Singh et al., 1995) or upon virus purification (Lai-Kee-Him et al., 2013). Most remarkable is the ability of GFLV CP to maintain its capacity to self-assemble into VLPs upon genetic fusion to proteins as large as FP that represent about half the size of the CP (256 residues including linker sequence for EG vs 504 for CP).

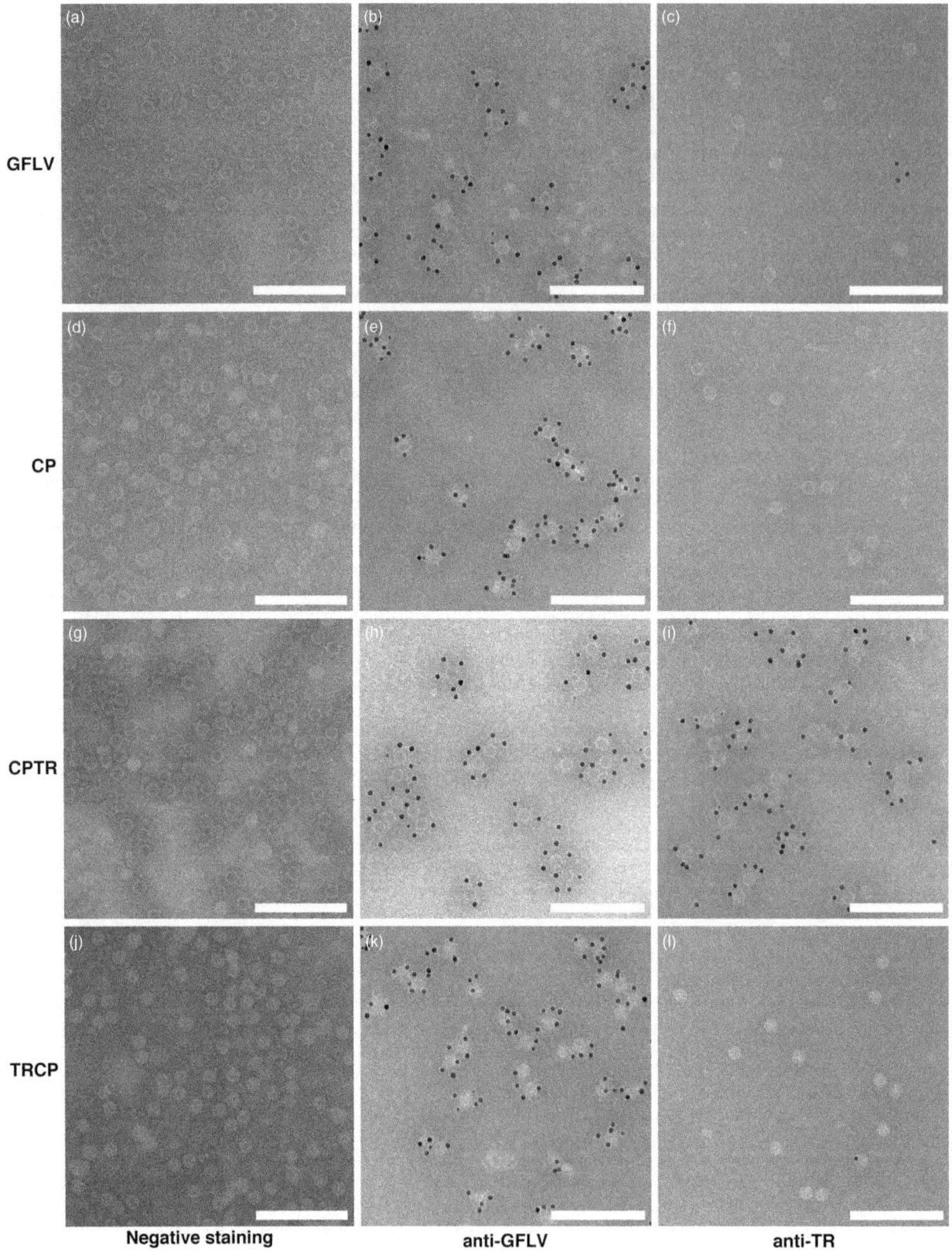

Figure 5 Proteins fused to the N- or C-terminal end of GFLV CP are encaged or exposed at the outer surface of VLPs, respectively. Electron micrographs of purified GFLV (a, b, c), CP VLPs (d, e, f), CPTR VLPs (g, h, i) and TRCP VLPs (j, k, l). Samples were processed for negative staining only (a, d, g, j) or for ISEM using anti-GFLV (b, e, h, k) or anti-TR (c, f, i, l) antibodies and antibodies conjugated to 10-nm colloidal gold particles for labelling. Scale bars: 200 nm.

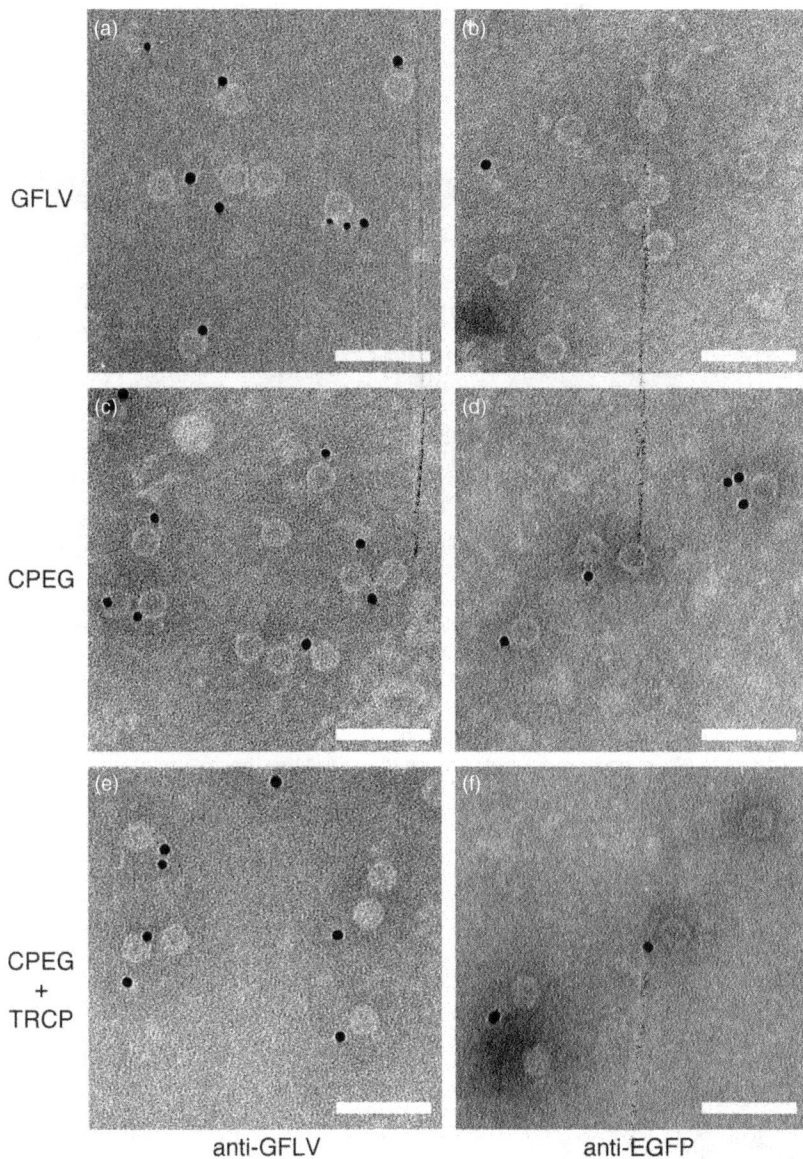

Figure 6 VLPs can be purified from *N. benthamiana* leaves coexpressing CPEG and TRCP. Transmission electron micrographs of purified GFLV (a, b), CPEG VLPs (c, d) and CPEG + TRCP VLPs (e, f) after immunogold labelling. Samples were processed for ISEM using anti-GFLV (a, c, e) or anti-EGFP (b, d, f) antibodies and secondary antibodies conjugated to 10-nm colloidal gold for decoration. Scale bars: 100 nm.

While CPs of numerous viruses accommodate genetic fusions and retain their ability to self-assemble into VLPs, those are commonly restricted to peptides as exemplified with simian virus 40 (Takahashi *et al.*, 2008), bacteriophage P22 (Servid *et al.*, 2013) or CCMV (Brumfield *et al.*, 2004; Hassani-Mehraban *et al.*, 2015) rather than full-length proteins. Another common limitation observed with VLPs is linked to sites of insertion of foreign proteins not necessarily located at the extremities but often exposed within loops of the CP, as it is the case for hepatitis B virus (Kratz *et al.*, 1999; Peyret *et al.*, 2015) or CPMV (Porta *et al.*, 2003), requiring the inserted proteins to be fused to both their extremities. From this viewpoint, GFLV CP appears rather special among the realm of VLP-compatible CPs from animal, bacterial and plant viruses.

Certainly the most remarkable of findings is the differential orientation of the foreign proteins upon fusion to the N- or C-terminal end of GFLV CP, resulting in either encaging or exposure of FP to the particle outer surface, respectively. While exposure of proteins of interest at the VLPs outer surface is commonly achieved, encaging of proteins is less frequent and currently only demonstrated for few icosahedral viruses such as CCMV in plant

(Minten *et al.*, 2009) or the well-established example of human papillomavirus (Schiller and Lowy, 2012). To our best knowledge, identification of a single CP having the capacity to (i) self-assemble into nucleic acid-free VLPs, (ii) accommodate N- and C-terminal genetic fusions and (iii) expose and encage large recombinant proteins is unique and highlights the versatility of the CP of GFLV as a nanocarrier. Despite the low primary sequence conservation among CP of nepoviruses (Sanfaçon *et al.*, 2009), it is likely that the properties established here for the CP of GLFV extend to most if not all CPs from the genus *Nepovirus*. Hence, the reported atomic structures of tobacco ringspot virus (Chandrasekar and Johnson, 1998) and blackcurrant reversion virus (Seitsonen *et al.*, 2008) and the backbone model of arabis mosaic virus (Lai-Kee-Him *et al.*, 2013) all show CP N-terminal ends facing the interior of the particles and CP C-terminal residues exposed at the capsid outer surface.

GFLV atomic structure (Figure 3a, b; Schellenberger *et al.*, 2011b) reveals that the N- and C-terminal ends of all 60 CP subunits are surface accessible and should therefore accommodate fusion to proteins, at least theoretically. This seems in disagreement with our experimental data, in particular those

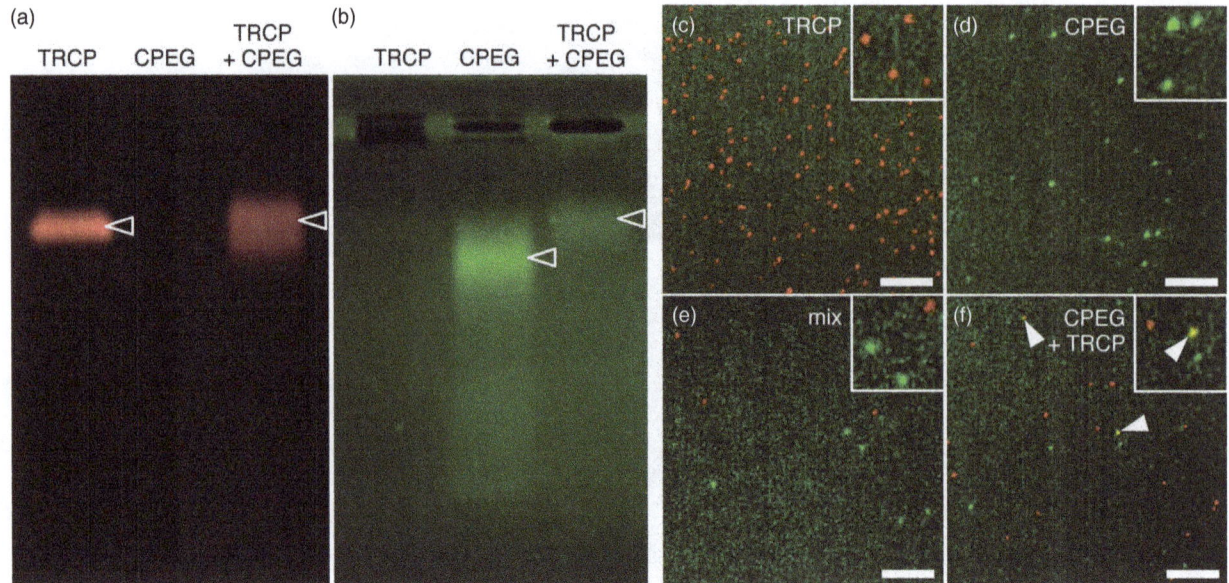

Figure 7 Hybrid VLPs are produced upon coexpression of CPEG and TRCP. (a, b) Fluorescence imaging of TRCP, CPEG or CPEG + TRCP VLPs separated by native agarose gel electrophoresis. Imaging was performed sequentially using a G:box imaging system, first at $\lambda_{ex}480$—540 nm for excitation and $\lambda_{em}590$—660 nm for emission to detect TR (a), then at $\lambda_{ex}450$—485 nm Schellenberger et al., 2011b and $\lambda_{em}510$—540 nm to detect EG (b). Fluorescent VLPs in the gel are indicated by empty arrowheads. (c to f) Single-particle microscopy images of purified TRCP VLPs (c), CPEG VLPs (d), mixed TRCP and CPEG VLPs at 1 : 1 ratio (e) and coexpressed CPEG + TRCP VLPs (f). Epifluorescence imaging was performed sequentially at $\lambda_{ex}455$—495 nm to $\lambda_{em}505$—555 nm to detect EG and at $\lambda_{ex}532.5$—557.5 to $\lambda_{em}570$—640 nm to detect TR. White arrowheads point at hybrid VLPs. Scale bars: 5 μm.

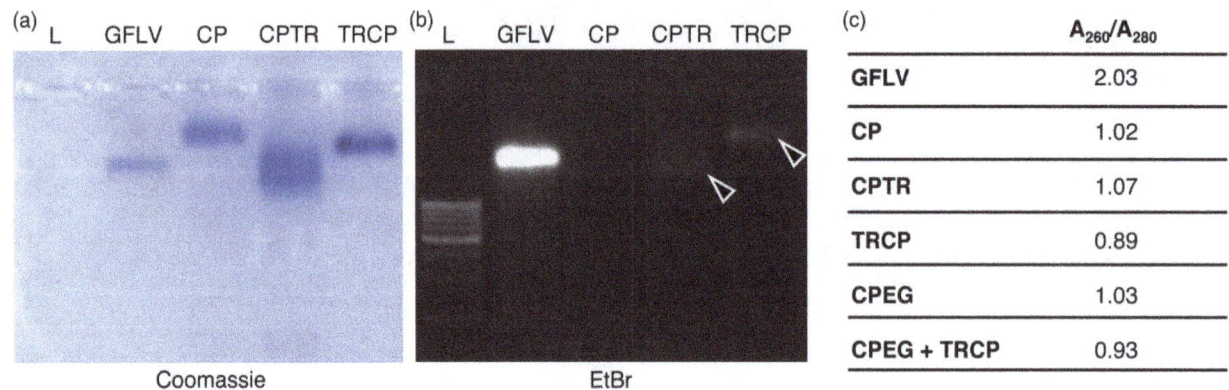

	A_{260}/A_{280}
GFLV	2.03
CP	1.02
CPTR	1.07
TRCP	0.89
CPEG	1.03
CPEG + TRCP	0.93

Figure 8 Purified VLPs are nucleic acid free. (a and b) Purified GFLV and CP, CPTR and TRCP VLPs separated by native agarose gel electrophoresis after Coomassie blue (a) and ethidium bromide staining (b). Arrowheads point to bands corresponding to crosstalk. (c) A_{260}/A_{280} ratio of purified particles.

concerning proteins exposed at the outer surface of VLPs that represent only a minor proportion in purified particles (Figure 4). The CP of tomato black ring virus (TBRV) was shown to undergo proteolytic processing resulting in the loss of the 9 C-terminal amino acids (Demangeat et al., 1992). Although the CP of GFLV does not seem to undergo similar processing (Schellenberger et al., 2011b), the addition of residues may promote the cleavage of additional C-terminal amino acids as already reported for CMPV-derived VLP (Montague et al., 2011). Flexibility provided by the linker peptide may also contribute to increased fragility of the exposed recombinant proteins and their truncation. Probably the most likely explanations reside in the suboptimal transient expression system used and the purification procedure performed in the absence of protease inhibitors. This could clearly also account for the discrepancy in the proportion of full-length recombinant proteins vs truncated products present in purified

VLPs (Figure 4) as FPs encaged into VLPs are likely to be less prone to proteolysis than those exposed at the outer surface. Future work will aim at optimizing the VLP production system before multiple biotechnological applications can be foreseen. This includes reducing protein truncation in particular of outer surface-exposed proteins which remain highly sensitive to proteolysis, improving VLPs purification yield and exploring whether a GFLV CP with both ends fused to foreign proteins still forms VLPs.

Finally, GFLV CP-derived VLPs may also provide new perspectives to explore fundamental mechanisms such as tubule-guided movement in plants (Amari et al., 2010; Ritzenthaler and Hofmann, 2007) and virus transmission from plant to plant by nematode vectors (Marmonier et al., 2010; Schellenberger et al., 2010). Indeed, the production of VLPs in planta in which FPs are encaged should prove to be a unique and powerful tool to track particles by fluorescence microscopy approaches, in particular, during intracellular and cell-to-cell movement of virus through plasmodesmata and also during the nematode feeding process.

Experimental procedures

Construction of binary plasmids

Coding sequences for GFLV-F13 CP, TagRFP and EGFP were amplified by PCR using Phusion high-fidelity DNA polymerase according to the manufacturer's instructions (New England Biolabs, Ipswich, MA; Thermo Fisher Scientific, Villebon sur Yvette, France) using pVec$_{Acc65I}$2ABC (Schellenberger *et al.*, 2010), pTagRFP-C (Evrogen, Moscow, Russia) and pEGFP-N1 (Clontech, Palo Alto, CA) as templates, respectively. The translational fusions TRCP and CPTR, corresponding, respectively, to N- or C-terminal fusions of GFLV CP with TagRFP, were obtained by overlapping PCRs (Ho *et al.*, 1989) using above-described PCR products as templates and overlapping primers encoding the Gly$_3$-Ser-Gly$_3$ peptide linker sequence. The *att*B-flanked CP, TR, CPTR and TRCP PCR products were cloned by Gateway recombination into the pDONR/Zeo donor vector (Invitrogen, Carlsbad,CA) and further recombined into the pEAQ-*HT*-DEST1 binary plasmid (Sainsbury *et al.*, 2009). For CPEG, in which the C-terminus of the CP is fused to EG, a pDONR/Zeo entry vector containing the CP coding sequence devoid of stop codon was used for recombination in a homemade Gateway expression vector deriving from the pEAQ-*HT*-DEST1 (Sainsbury *et al.*, 2009) vector by the introduction of the EG encoding sequence downstream of the *att*R2 recombination site. Recombination resulted in the introduction of the DPAFLYKVVRSFGPA linker peptide between CP C-terminal residue and EG (Figure 1 and Figure S1). All the primers used for cloning are available upon request.

Plant material, virus infection and VLP production

C. quinoa and *N. benthamiana* plants were grown in chambers at 22/18 °C (day/night). GFLV infectious crude sap derived from pMV13 + pVec$_{Acc65I}$2ABC-infected material (Schellenberger *et al.*, 2010) was used to mechanically inoculate three-week-old *C. quinoa*. Plants were harvested 14 days postinoculation and used for virus purification. For mechanical inoculations of *N. benthamiana*, three-week-old plants were inoculated with 300 ng of purified GFLV per plant. VLPs were produced by transient expression via agro-infiltration of *N. benthamiana* leaves. Binary plasmids were introduced by electroporation into *Agrobacterium tumefaciens* strain GV3101 (pMP90). Cultures were grown to stable phase in Luria-Bertani media with appropriate antibiotics, pelleted and resuspended in sterile water, alone or in a 1 : 1 ratio for coexpression, to a final absorbance of 0.5 at 600 nm. Suspensions were infiltrated into four-week-old *N. benthamiana* leaves with 2-mL plastic syringes. Healthy, infected and agro-infiltrated *N. benthamiana* plants were maintained in a growth chamber set at 14-h light/10-h dark photoperiod (4800 lx) with a temperature setting of 21/18 °C (day/night) for 7 days before leaf harvesting.

Imaging of agro-infiltrated leaves

FP visualization was realized at 5 dpa. Leaves were imaged with an AxioZoom V16 macroscope (Zeiss, Oberkochen, Germany) using 450- to 490-nm/500- to 550-nm excitation/emission wavelength filters for EG and of 538—562 nm/570—640 nm for TagRFP visualization. Images were processed using ImageJ (Schneider *et al.*, 2012) and GNU Image Manipulation Program (GIMP, www.gimp.org).

DAS-ELISA

Healthy, infected and agro-infiltrated leaves were ground at 1 : 5 w/v ratio in HEPES 100 mM pH 8, and saps were clarified for 5 min at 3000 *g*. GFLV or VLP detection was performed using a commercial DAS-ELISA kit (Bioreba, Reinach, Switzerland) according to the manufacturer's instructions. Briefly, plates were coated with polyclonal anti-GFLV antibodies diluted at a 1 : 1000 in coating buffer, incubated with clarified extracts before the addition of anti-GFLV monoclonal antibodies coupled to alkaline phosphatase at a 1 : 1000 dilution in conjugate buffer. Detection was realized using *para*-nitrophenylphosphate and absorbance at 405 nm measured with a Titertek Multiskan MCC/340 reader (Labsystems, Helsinki, Finland). Samples were considered positive when the absorbance values exceeded the control samples by at least a factor of three.

Negative staining, immunocapture and immunosorbent electron microscopy

Healthy, infected and agro-infiltrated leaves were ground in HEPES 100 mM pH 8, and saps were clarified by centrifugation at 3000 *g* for 5 min and processed for negative staining, immunocapture or ISEM. Negative staining was performed on 300 mesh nickel grids covered with carbon-coated Formvar (Electron Microscopy Science, Hatfield, PA) by incubation for 90 s with a 1% (w/v) ammonium molybdate solution. For immunocapture, grids were coated with polyclonal anti-GFLV antibodies (Bioreba) at a 1 : 100 dilution, incubated with plant saps for 2 h at 4 °C, washed in HEPES 25 mM pH 8 buffer and finally processed for negative staining. For ISEM, grids were coated with in-house monoclonal antibodies against GFLV at 0.05 mg/mL and incubated with VLPs for 1 h at room temperature. After blocking with 2% w/v BSA, 10% v/v normal goat serum, 0.05% Triton-X100 in 22.5 mM HEPES pH 8, grids were further incubated with either anti-GFLV (Bioreba) at a 1 : 100 dilution or anti-TR polyclonal antibodies (Evrogen) at 0.01 mg/mL for 1 h at room temperature. Immunogold labelling was performed using anti-rabbit antibodies conjugated to 10-nm colloidal gold particles at 1 : 50 dilution (British Biocell International). Washes with HEPES 25 mM pH 8 were performed between all steps. ISEM was performed in a similar manner except that anti-GFLV polyclonal antibodies (Bioreba) were used for capture and either a mix of anti-GFLV monoclonal antibodies (Gaire *et al.*, 1999) or anti-EGFP monoclonal antibodies (Roche Diagnostics GmbH, Mannheim, Germany) employed for detection. Finally, immunogold labelling was performed using anti-mouse antibodies conjugated with 10-nm colloidal gold particles (British Biocell International). Observations were realized using a Philips EM208 transmission electron microscope. Film-based photographs were acquired onto Kodak Electron Image Films SO-163 (Electron Microscopy Science) and developed. Micrographs were scanned and images were processed using GNU Image Manipulation Program (GIMP, www.gimp.org).

GFLV CP structure representation and analysis

CP subunit and capsid representations were made using the previously 3 Å resolved GFLV-F13 atomic structure (PDB ID: 4V5T, (Schellenberger *et al.*, 2011b) using the UCSF Chimera package (Pettersen *et al.*, 2004). The CP subunit ends accessibility data were obtained using VIPERdb (Carrillo-Tripp *et al.*, 2009).

Virus and VLP purification

GFLV was purified from *C. quinoa*-infected plants according to Schellenberger *et al.*, (2011a). VLPs were purified from agro-infiltrated *N. benthamiana* leaves following the same experimental procedure, except that the final discontinuous sucrose gradient was omitted. Briefly, a minimum of 350 g of leaves were ground in extraction buffer, the resulting extract was filtered, incubated with bentonite and finally clarified by centrifugation for 15 min at 1900 ***g***. VLPs were precipitated from clarified sap by adding PEG-20 000 and sodium chloride and further processed by centrifugation on a sucrose cushion followed by a sucrose density gradient fractionation. Two millilitres of fractions was collected from which aliquots at 1 : 500, 1 : 5000 and 1 : 10 000 dilutions were processed for a semiquantitative DAS-ELISA to identify VLP-enriched fractions that were further pooled before final ultracentrifugation at 290 000 ***g*** for 2 h. After resuspension in HEPES 25 mM pH 8, VLPs were quantified by DAS-ELISA (Vigne *et al.*, 2013) using purified GFLV as a standard.

SDS-PAGE, western blot and mass spectrometry

For SDS-PAGE analysis, 6 μg of GFLV particles equivalent from each purified sample was separated on an 8% acrylamide gel and stained with Coomassie blue using Instant Blue (Expedeon Inc., San Diego, CA). For mass spectrometry, SDS-PAGE bands of interest were excised and proteins were destained, reduced, alkylated, trypsin-digested overnight, chemotrypsin-digested and finally processed for nano-LC-MSMS analysis on a nanoU3000 (Dionex, Thermo Fisher Scientific)-ESI-MicroTOFQII (Bruker, Billerica, MA). Mass spectrometry data were analysed using Mascot (Matrix Science Ltd, London, UK) and Proteinscape (Bruker). For Western blot analyses, 0.05 μg of each sample was resolved on an 8% acrylamide gel and denatured proteins were electrotransferred onto Immobilon PVDF membranes. Membranes were incubated with rabbit polyclonal anti-GFLV antibodies at a 1 : 1000 dilution or with polyclonal anti-TR antibodies (Evrogen) at a 1 : 5000 dilution. Proteins were detected by chemiluminescence after incubation with goat anti-rabbit antibodies conjugated to horseradish peroxidase at a 1 : 12 500 dilution (Thermo Fisher Scientific) and with Lumi-Light solution (Roche). Images were taken with a G : Box imaging system (Syngene, Cambridge, UK), analysed with GeneTools (Syngene) and finally processed with GIMP (www.gimp.org).

Single-particle epifluorescence microscopy

Purified VLPs were diluted in HEPES 25 mM pH 8 to obtain individual spots upon imaging on an inverted epifluorescence microscope Axio Observer Z1 (Zeiss) equipped with an Orca Flash4.0 camera (Hamamatsu Photonics, Massy, France) and Spectra X light engine (Lumencor, Beaverton, OR). Excitation and emission wavelength filters were 455—495 nm and 505—555 nm for EG and of 532.5—557.5 nm and 570—640 nm for TR. Images were processed as described above.

Native agarose gel electrophoresis

Native gel electrophoresis of purified virions and VLPs was performed in 1% w/v agarose gels in 0.5X TAE buffer (20 mM TrisBase, 1.3 mM EDTA, 0.06% v/v acetic acid). For nucleic acid detection, 5 μg of virus particles or VLPs was diluted in loading buffer (10% v/v glycerol, HEPES 25 mM pH 8) supplemented with ethidium bromide (EtBr) at 0.1 μg/mL. After electrophoretic separation, the EtBr-prestained gel was first processed for nucleic acid content using the Gel Doc system (Bio-Rad, Hercules, CA) equipped with a 302-nm excitation source and a 520- to 640-nm band-pass emission filter before processing for Coomassie blue staining as mentioned previously.

For fluorescence imaging, 3 μg of purified VLPs was diluted in loading buffer and native gel electrophoresis was performed in the absence of EtBr. Imaging was performed with a G : Box imaging system (Syngene) equipped with a 450- to 485-nm excitation LED module and a 510- to 540-nm emission band-pass filter for EG visualization. TR visualization was realized upon 480- to 540-nm LED excitation and 590- to 660-nm band-pass emission filtering.

Acknowledgements

We thank Philippe Choquet for critical reading of the manuscript. The authors are grateful to the proteomic platform staff of the University of Strasbourg for mass spectrometry analyses and to the gardeners of the Experimental Unit at INRA Colmar. LB was supported by a contrat doctoral from the Ministère de l'Education Nationale, de l'Enseignement supérieur et de la Recherche. This work was supported by the Centre National de la Recherche Scientifique (CNRS), the Institut National de la Recherche Agronomique (INRA), the Université de Strasbourg and the Conseils Interprofessionels des vins d'Alsace, Champagne, Bourgogne and Bordeaux and the Agence Nationale de la Recherche (ANR) award ANR VinoBodies (ANR-14-CE19-0018).

References

Alonso, J., Górzny, M. and Bittner, A. (2013) The physics of tobacco mosaic virus and virus-based devices in biotechnology. *Trends Biotechnol.* **31**, 530–538.

Amari, K., Boutant, E., Hofmann, C., Schmitt-Keichinger, C., Fernandez-Calvino, L., Didier, P., Lerich, A. *et al.* (2010) A family of plasmodesmal proteins with receptor-like properties for plant viral movement proteins. *PLoS Pathog.* **6**, e1001119–e1001119.

Arcangeli, C., Circelli, P., Donini, M., Aljabali, A.A., Benvenuto, E., Lomonossoff, G.P. and Marusic, C. (2014) Structure-based design and experimental engineering of a plant virus nanoparticle for the presentation of immunogenic epitopes and as a drug carrier. *J. Biomol. Struct. Dyn.* **32**, 630–647.

Barbier, P., Demangeat, G., Perrin, M., Cobanov, P., Jacquet, C. and Walter, B. (1997) Grapevine genetically transformed with the coat protein gene of grapevine fanleaf virus: an analysis of transformants. In: *Proceedings of the 12th Meeting of the International Council for the Study of Virus and Virus-like Diseases of the Grapevine, Lisbon, Portugal*, p.131, http://icvg.org/general-assembly/.

Bertioli, D., Harris, R., Edwards, M., Cooper, J. and Hawes, W. (1991) Transgenic plants and insect cells expressing the coat protein of arabis mosaic virus produce empty virus-like particles. *J. Gen. Virol.* **72**, 1801–1809.

Bruckman, M.A., Hern, S., Jiang, K., Flask, C.A., Yu, X. and Steinmetz, N.F. (2013) Tobacco mosaic virus rods and spheres as supramolecular high-relaxivity MRI contrast agents. *J. Mater. Chem. B* **1**, 1482–1490.

Brumfield, S., Willits, D., Tang, L., Johnson, J.E., Douglas, T. and Young, M. (2004) Heterologous expression of the modified coat protein of Cowpea chlorotic mottle bromovirus results in the assembly of protein cages with altered architectures and function. *J. Gen. Virol.* **85**, 1049–1053.

Cardinale, D., Carette, N. and Michon, T. (2012) Virus scaffolds as enzyme nano-carriers. *Trends Biotechnol.* **30**, 369–376.

Carrillo-Tripp, M., Shepherd, C.M., Borelli, I.A., Venkataraman, S., Lander, G., Natarajan, P., Johnson, J.E. *et al.* (2009) VIPERdb2: an enhanced and web API enabled relational database for structural virology. *Nucleic. Acids. Res.* **37**, D436–D442.

Chandrasekar, V. and Johnson, J.E. (1998) The structure of tobacco ringspot virus: a link in the evolution of icosahedral capsids in the picornavirus superfamily. *Structure*, **6**, 157–171.

Chatterji, A., Ochoa, W., Shamieh, L., Salakian, S.P., Wong, S.M., Clinton, G., Ghosh, P. *et al.* (2004) Chemical conjugation of heterologous proteins on the surface of cowpea mosaic virus. *Bioconjug. Chem.* **15**, 807–813.

Chen, Q. and Lai, H. (2013) Plant-derived virus-like particles as vaccines. *Hum. Vaccin. Immunother.* **9**, 26–49.

Demangeat, G., Hemmer, O., Reinbolt, J., Mayo, M. and Fritsch, C. (1992) Virus-specific proteins in cells infected with tomato black ring nepovirus: evidence for proteolytic processing in vivo. *J. Gen. Virol.* **73**, 1609–1614.

Gaire, F., Schmitt, C., Stussi-Garaud, C., Pinck, L. and Ritzenthaler, C. (1999) Protein 2A of grapevine fanleaf nepovirus is implicated in RNA2 replication and colocalizes to the replication site. *Virology*, **264**, 25–36.

Glasgow, J. and Tullman-Ercek, D. (2014) Production and applications of engineered viral capsids. *Appl. Microbiol. Biotechnol.* **98**, 5847–5858.

Hassani-Mehraban, A., Creutzburg, S., van Heereveld, L. and Kormelink, R. (2015) Feasibility of cowpea chlorotic mottle virus-like particles as scaffold for epitope presentations. *BMC Biotechnol.* **15**, 80.

Hesketh, E.L., Meshcheriakova, Y., Dent, K.C., Saxena, P., Thompson, R.F., Cockburn, J.J., Lomonossoff, G.P. *et al.* (2015) Mechanisms of assembly and genome packaging in an RNA virus revealed by high-resolution cryo-EM. *Nat. Commun.* **6**, 10113. doi:10.1038/ncomms10113.

Ho, S.N., Hunt, H.D., Horton, R.M., Pullen, J.K. and Pease, L.R. (1989) Site-directed mutagenesis by overlap extension using the polymerase chain reaction. *Gene* **77**, 51–59.

Kratz, P.A., Böttcher, B. and Nassal, M. (1999) Native display of complete foreign protein domains on the surface of hepatitis B virus capsids. *Proc. Natl. Acad. Sci. U S A.* **96**, 1915–1920.

Lai-Kee-Him, J., Schellenberger, P., Dumas, C., Richard, E., Trapani, S., Komar, V., Demangeat, G. *et al.* (2013) The backbone model of the arabis mosaic virus reveals new insights into functional domains of nepovirus capsid. *J. Struct. Biol.* **182**, 1–9.

Lee, E.J., Lee, N.K. and Kim, I.-S. (2016) Bioengineered protein-based nanocage for drug delivery. *Adv. Drug Deliv. Rev.* doi: 10.1016/j.addr.2016.03.002.

Lewis, J.D., Destito, G., Zijlstra, A., Gonzalez, M.J., Quigley, J.P., Manchester, M. and Stuhlmann, H. (2006) Viral nanoparticles as tools for intravital vascular imaging. *Nat. Med.* **12**, 354–360.

Li, F. and Wang, Q. (2014) Fabrication of nanoarchitectures templated by virus-based nanoparticles: strategies and applications. *Small*, **10**, 230–245.

Liu, Z., Qiao, J., Niu, Z. and Wang, Q. (2012) Natural supramolecular building blocks: from virus coat proteins to viral nanoparticles. *Chem. Soc. Rev.* **41**, 6178–6194.

Lua, L.H., Connors, N.K., Sainsbury, F., Chuan, Y.P., Wibowo, N. and Middelberg, A.P. (2014) Bioengineering virus-like particles as vaccines. *Biotechnol. Bioeng.* **111**, 425–440.

Marmonier, A., Schellenberger, P., Esmenjaud, D., Schmitt-Keichinger, C., Ritzenthaler, C., Andret-Link, P., Lemaire, O. *et al.* (2010) The coat protein determines the specificity of virus transmission by *Xiphinema diversicaudatum*. *J. Plant. Pathol.* **92**, 275–279.

Merzlyak, E.M., Goedhart, J., Shcherbo, D., Bulina, M.E., Shcheglov, A.S., Fradkov, A.F., Gaintzeva, A., *et al.* (2007) Bright monomeric red fluorescent protein with an extended fluorescence lifetime. *Nat. Methods* **4**, 555–557.

Minten, I.J., Hendriks, L.J., Nolte, R.J. and Cornelissen, J.J. (2009) Controlled encapsulation of multiple proteins in virus capsids. *J. Am. Chem. Soc.* **131**, 17771–17773.

Montague, N.P., Thuenemann, E.C., Saxena, P., Saunders, K., Lenzi, P. and Lomonossoff, G.P. (2011) Recent advances of cowpea mosaic virus-based particle technology. *Hum. Vaccin.* **7**, 383–390.

Mueller, A., Eber, F.J., Azucena, C., Petershans, A., Bittner, A.M., Gliemann, H., Jeske, H. *et al.* (2011) Inducible site-selective bottom-up assembly of virus-derived nanotube arrays on RNA-equipped wafers. *ACS Nano* **5**, 4512–4520.

Pettersen, E.F., Goddard, T.D., Huang, C.C., Couch, G.S., Greenblatt, D.M., Meng, E.C. and Ferrin, T.E. (2004) UCSF Chimera—a visualization system for exploratory research and analysis. *J. Comput. Chem.* **25**, 1605–1612.

Peyret, H., Gehin, A., Thuenemann, E.C., Blond, D., El Turabi, A., Beales, L., Clarke, D. *et al.* (2015) Tandem fusion of hepatitis B core antigen allows assembly of virus-like particles in bacteria and plants with enhanced capacity to accommodate foreign proteins. *PLoS ONE*, **10**, e0120751.

Pokorski, J.K. and Steinmetz, N.F. (2010) The art of engineering viral nanoparticles. *Mol. Pharm.* **8**, 29–43.

Porta, C., Spall, V.E., Findlay, K.C., Gergerich, R.C., Farrance, C.E. and Lomonossoff, G.P. (2003) Cowpea mosaic virus-based chimaeras: effects of inserted peptides on the phenotype, host range, and transmissibility of the modified viruses. *Virology*, **310**, 50–63.

Quacquarelli, A., Gallitelli, D., Savino, V. and Martelli, G. (1976) Properties of grapevine fanleaf virus. *J. Gen. Virol.* **32**, 349–360.

Rhee, J.-K., Hovlid, M., Fiedler, J.D., Brown, S.D., Manzenrieder, F., Kitagishi, H., Nycholat, C. *et al.* (2011) Colorful virus-like particles: fluorescent protein packaging with the Qβ capsid. *Biomacromolecules*, **12**, 3977–3981.

Ritzenthaler, C. and Hofmann, C.(2007) Tubule-guided movement of plant viruses. In *Viral Transport in Plants*, Plant Cell Monographs, (Waigmann, E. and Heinlein, M., eds), Heidelberg: Springer. **7** 63–83.

Sainsbury, F., Thuenemann, E.C. and Lomonossoff, G.P. (2009) pEAQ: versatile expression vectors for easy and quick transient expression of heterologous proteins in plants. *Plant. Biotechnol. J.* **7**, 682–693.

Sainsbury, F., Cañizares, M.C. and Lomonossoff, G.P. (2010) Cowpea mosaic virus: the plant virus-based biotechnology workhorse. *Annu. Rev. Phytopathol.* **48**, 437–455.

Sainsbury, F., Saxena, P., Aljabali, A.A., Saunders, K., Evans, D.J. and Lomonossoff, G.P. (2014) Genetic engineering and characterization of cowpea mosaic virus empty virus-like particles. *Methods Mol. Biol.* **1108**, 139–153.

Sanfaçon, H., Wellink, J., Le Gall, O., Karasev, A., Van der Vlugt, R. and Wetzel, T. (2009) Secoviridae: a proposed family of plant viruses within the order Picornavirales that combines the families Sequiviridae and Comoviridae, the unassigned genera *Cheravirus* and *Sadwavirus*, and the proposed genus *Torradovirus*. *Arch. Virol.* **154**, 899–907.

Saunders, K. and Lomonossoff, G.P. (2013) Exploiting plant virus-derived components to achieve in planta expression and for templates for synthetic biology applications. *New Phytol.* **200**, 16–26.

Saunders, K., Sainsbury, F. and Lomonossoff, G.P. (2009) Efficient generation of cowpea mosaic virus empty virus-like particles by the proteolytic processing of precursors in insect cells and plants. *Virology*, **393**, 329–337.

Schellenberger, P., Andret-Link, P., Schmitt-Keichinger, C., Bergdoll, M., Marmonier, A., Vigne, E., Lemaire, O. *et al.* (2010) A stretch of 11 amino acids in the βB-βC loop of the coat protein of grapevine fanleaf virus is essential for transmission by the nematode *Xiphinema index*. *J. Virol.* **84**, 7924–7933.

Schellenberger, P., Demangeat, G., Lemaire, O., Ritzenthaler, C., Bergdoll, M., Oliéric, V., Sauter, C. *et al.* (2011a) Strategies for the crystallization of viruses: using phase diagrams and gels to produce 3D crystals of grapevine fanleaf virus. *J. Struct. Biol.* **174**, 344–351.

Schellenberger, P., Sauter, C., Lorber, B., Bron, P., Trapani, S., Bergdoll, M., Marmonier, A. *et al.* (2011b) Structural insights into viral determinants of nematode mediated Grapevine fanleaf virus transmission. *PLoS Pathog.* **7**, e1002034.

Schiller, J.T. and Lowy, D.R. (2012) Understanding and learning from the success of prophylactic human papillomavirus vaccines. *Nat. Rev. Microbiol.* **10**, 681–692.

Schneider, C.A., Rasband, W.S. and Eliceiri, K.W. (2012) NIH image to imageJ: 25 years of image analysis. *Nat. Methods* **9**, 671–675.

Seitsonen, J.J., Susi, P., Lemmetty, A. and Butcher, S.J. (2008) Structure of the mite-transmitted blackcurrant reversion nepovirus using electron cryo-microscopy. *Virology*, **378**, 162–168.

Servid, A., Jordan, P., O'Neil, A., Prevelige, P. and Douglas, T. (2013) Location of the bacteriophage P22 coat protein C-terminus provides opportunities for the design of capsid-based materials. *Biomacromolecules*, **14**, 2989–2995.

Shriver, L.P., Plummer, E.M., Thomas, D.M., Ho, S. and Manchester, M. (2013) Localization of gadolinium-loaded CPMV to sites of inflammation during central nervous system autoimmunity. *J. Mat. Chem. B.* **1**, 5256–5263.

Singh, S., Rothnagel, R., Prasad, B.V. and Buckley, B. (1995) Expression of tobacco ringspot virus capsid protein and satellite RNA in insect cells and three-dimensional structure of tobacco ringspot virus-like particles. *Virology*, **213**, 472–481.

Steinmetz, N.F., Ablack, A.L., Hickey, J.L., Ablack, J., Manocha, B., Mymryk, J.S., Luyt, L.G. *et al.* (2011) Intravital imaging of human prostate cancer using viral nanoparticles targeted to gastrin-releasing peptide receptors. *Small*, **7**, 1664–1672.

Suci, P.A., Varpness, Z., Gillitzer, E., Douglas, T. and Young, M. (2007) Targeting and photodynamic killing of a microbial pathogen using protein cage architectures functionalized with a photosensitizer. *Langmuir*, **23**, 12280–12286.

Takahashi, R.-U., Kanesashi, S.-N., Inoue, T., Enomoto, T., Kawano, M.-A., Tsukamoto, H., Takeshita, F. *et al.* (2008) Presentation of functional foreign peptides on the surface of SV40 virus-like particles. *J. Biotechnol.* **135**, 385–392.

Vigne, E., Gottula, J., Schmitt-Keichinger, C., Komar, V., Ackerer, L., Belval, L., Rakotomalala, L. *et al.* (2013) A strain-specific segment of the RNA-dependent RNA polymerase of Grapevine fanleaf virus determines symptoms in *Nicotiana* species. *J. Gen. Virol.* **94**, 2803–2813.

Zilian, E. and Maiss, E. (2011) An optimized mRFP-based bimolecular fluorescence complementation system for the detection of protein–protein interactions in planta. *J. Virol. Methods* **174**, 158–165.

2

Enhanced resistance in *Theobroma cacao* against oomycete and fungal pathogens by secretion of phosphatidylinositol-3-phosphate-binding proteins

Emily E. Helliwell[1,2,†], Julio Vega-Arreguín[3,†,§], Zi Shi[1,†,¶], Bryan Bailey[4], Shunyuan Xiao[5], Siela N. Maximova[1,‡], Brett M. Tyler[2,3,‡] and Mark J. Guiltinan[1,*,‡]

[1]*Department of Plant Science and Huck Institute of Life Sciences, The Pennsylvania State University, University Park, PA, USA*
[2]*Center for Genome Research and Biocomputing, and Department of Botany and Plant Pathology, Oregon State University, Corvallis, OR, USA*
[3]*Virginia Bioinformatics Institute and Department of Plant Pathology, Physiology and Weed Science, Virginia Polytechnic Institute and State University, Blacksburg, VA, USA*
[4]*United States Department of Agriculture, Agricultural Research Service, Beltsville, MD, USA*
[5]*Institute for Bioscience and Biotechnology Research & Department of Plant Science and Landscape Architecture, University of Maryland, College Park, MD, USA*

*Correspondence

§Current address: ENES Unidad León, Universidad Nacional Autónoma de México (UNAM), León, Gto. C.P.37684, Mexico.
¶Current address: Department of Crop and Soil Sciences, University of Georgia, Athens, GA, USA
†These authors contributed equally to this work.
‡These authors contributed equally to this work.

Summary

The internalization of some oomycete and fungal pathogen effectors into host plant cells has been reported to be blocked by proteins that bind to the effectors' cell entry receptor, phosphatidyl-inositol-3-phosphate (PI3P). This finding suggested a novel strategy for disease control by engineering plants to secrete PI3P-binding proteins. In this study, we tested this strategy using the chocolate tree *Theobroma cacao*. Transient expression and secretion of four different PI3P-binding proteins in detached leaves of *T. cacao* greatly reduced infection by two oomycete pathogens, *Phytophthora tropicalis* and *Phytophthora palmivora*, which cause black pod disease. Lesion size and pathogen growth were reduced by up to 85%. Resistance was not conferred by proteins lacking a secretory leader, by proteins with mutations in their PI3P-binding site, or by a secreted PI4P-binding protein. Stably transformed, transgenic *T. cacao* plants expressing two different PI3P-binding proteins showed substantially enhanced resistance to both *P. tropicalis* and *P. palmivora*, as well as to the fungal pathogen *Colletotrichum theobromicola*. These results demonstrate that secretion of PI3P-binding proteins is an effective way to increase disease resistance in *T. cacao*, and potentially in other plants, against a broad spectrum of pathogens.

Keywords: *Theobroma cacao*, disease resistance, phosphatidylinositol-3-phosphate-binding protein, effectors, oomycetes, fungi.

Introduction

Plant defence involves two overlapping tiers of responses (Jones and Dangl, 2006). The first is triggered when plants detect conserved microbial molecular signatures (microbe-associated molecular patterns—MAMPs), and is called PAMP-triggered immunity (PTI). PTI includes rapid production of reactive oxygen species, antimicrobial molecules such as phytoalexins, and pathogenesis-related (PR) proteins.

Successful pathogens have evolved effector proteins, which can inhibit host defence responses (Giraldo and Valent, 2013; Torto-Alalibo et al., 2010; Tyler and Rouxel, 2013). For example, in the oomycete *Phytophthora sojae*, 22 of 49 effectors screened could suppress PAMP-triggered responses (Wang et al., 2011). The second tier of plant defence involves detection of effector proteins by host resistance (R) proteins (usually but not always nucleotide-binding leucine-rich repeat proteins) and is termed effector-triggered immunity (ETI) (Jones and Dangl, 2006). ETI can be readily overcome if pathogen strains emerge that have lost expression of the effectors or carry variant effectors that are no longer recognized by the R

protein. Many effectors show high variability among pathogen species, so R-gene-mediated resistance is often ineffective against different species of a pathogen, even from the same genus (Giraldo and Valent, 2013; Tyler and Rouxel, 2013).

Given the uneven success of conventional R genes, other strategies targeting effectors, such as blocking their function, are of interest given the importance of effectors in establishment of disease. Effectors are typically delivered into the plant cell through either pathogen- or host-encoded machinery (Tyler et al., 2013). Effector delivery may occur from apoplastic hyphae, from specialized intracellular hyphae, or from haustoria (Tyler et al., 2013) which are specialized feeding structures developed from intracellular hyphae (Hahn and Mengden, 1997; Panstruga and Dodds, 2009).

Oomycete effectors carry a short conserved N-terminal motif, RXLR, followed by several acidic residues (dEER) (Jiang et al., 2008) that are required for entry of the effectors into host cells (Dou et al., 2008; Kale et al., 2010; Tyler et al., 2013; Whisson et al., 2007). Some fungal pathogen effectors may gain entry into plant cells through the same or similar processes (Kale et al., 2010; Plett et al.,

2011). The RxLR-dEER domain of oomycete effectors, and fungal effectors with RxLR-like motifs were found to bind to the lipid phosphatidylinositol-3-phosphate (PI3P) (Kale *et al.*, 2010; Plett *et al.*, 2011) which was demonstrated to be on the surface of plant and animal cells. PI3P is also necessary for endocytotic processes such as protein sorting and membrane trafficking (Corvera *et al.*, 1999; DeCamilli *et al.*, 1996; Kale *et al.*, 2010). By preventing effectors from binding PI3P using competing PI3P-binding proteins or inositol-1,3-diphosphate, it was demonstrated that binding of the effectors to PI3P was required for cell entry (Kale *et al.*, 2010; Plett *et al.*, 2011).

There are several classes of proteins that can recognize and bind to specific forms of phosphoinositides. Examples include pleckstrin homology (PH) domains, Phox homology (PX) domains, and Fab1, YOTB, Vac1 and EEA1 (FYVE) domains. Different PH domains can bind specifically to a diversity of phosphoinositides (Dowler *et al.*, 2000; Kutateladze, 2010; Lemmon, 2008). Phox homology (PX) domains usually bind to PI3P, and sometimes to PI4P (Lemmon, 2008). FYVE domains bind specifically to PI3P (Kutateladze, 2010). These PI-binding proteins play diverse roles in membrane trafficking, cell growth and signal transduction (Lemmon, 2008).

An example of a crop species in which no R-gene-mediated resistance has so far been found is *Theobroma cacao*, the chocolate tree. Diseases are a major cause of crop loss, reducing the cacao yield by an estimated 30% (about 810 000 tons) per year and causing much farmer hardship (Keane and Putter, 1992). One of the most damaging cacao diseases is black pod rot, which is caused predominantly by three *Phytophthora* species: *P. palmivora*, *P. megakarya* and *P. tropicalis* [cacao isolates of *P. tropicalis* were previously identified as *P. capsici* (Aragaki and Uchida, 2001), (Guest, 2007). These pathogens can attack all parts of the plant including leaves, but major losses result from infection of cacao pods resulting in necrosis, shrinkage and mummification. The genomes of these three pathogens encode numerous RxLR effectors (B. Tyler, B. Bailey, M. Guiltinan, unpublished data).

With the aim of developing a novel, broad-spectrum disease resistance technology for cacao and potentially for other crop plants, we targeted the cellular translocation of pathogen effector proteins mediated by PI3P. We engineered cacao plants to secrete PI3P-binding proteins with the aim of blocking effectors from binding PI3P and thus from entering plants to promote disease. We show here that both transient and stable expression of secreted PI3P-binding proteins in cacao substantially enhances resistance to *P. tropicalis* and *P. palmivora*, as well as to the fungal pathogen *Colletotrichum theobromicola*.

Results

Design and verification of PI3P-binding constructs

To test the efficacy of secreting PI3P-binding domains to confer disease resistance, a variety of domains representing different classes of PI3P-binding proteins were chosen (Table 1). The PI3P-binding proteins chosen were the PH domain proteins PEPP1 (human) and GmPH1 (soya bean) (Dowler *et al.*, 2000), the PX domain protein VAM7p (yeast) (Lee *et al.*, 2006) and the Fab1, YOTB, Vac1 and EEA1 (FYVE) domain protein Hrs (mouse) (Kutateladze, 2006). A tandem repeat of the Hrs FYVE domain (Hrs-2xFYVE) was used to increase its binding affinity (Gillooly *et al.*, 2000). As a control, we included the human PH domain protein FAPP1 which binds to PI4P (DiNitto and Lambright, 2006; Dowler *et al.*, 2000; Lemmon, 2008). To target the PI-P-binding domains to the apoplastic space upon expression, the domains were fused to the secretory leader from the soya bean PR1a protein (Cutt *et al.*, 1988; van Esse *et al.*, 2006; Honee *et al.*, 1998). Further, to allow visualization of protein expression and localization, the PI-P-binding domains were fused with enhanced green fluorescent protein (EGFP) (Figure 1a).

Transient expression of PI-P-binding domains

To test the efficacy of the proteins in cacao leaves, we developed an *Agrobacterium*-mediated transient gene expression system (agro-infiltration) capable of expressing substantial amounts of the proteins in a large percentage of leaf cells (Shi *et al.*, 2013). The transgenes were driven by a very strong modified E12-Ω CaMV35S promoter (Mitsuhara *et al.*, 1996). Two types of constructs were tested, those in which EGFP was directly fused to the test genes to create a fusion protein, and nonfusion constructs in which the two coding sequences were driven by separate promoters (Figure 1a). To quantify transcription of the transgenes after transient expression in the cacao leaves, quantitative reverse transcriptase–PCR (qRT-PCR) employing primer pairs separately spanning the PI-P-binding region and the EGFP region was used to measure transcript levels in leaves, 2 days after infiltration (Figure 2). Transcript levels following transient expression of each transgene were approximately the same in each case and similar to those of the endogenous controls *Theobroma cacao Acyl-Carrier Protein 1 (TcACP1)* and *TcTubulin1*. When the expression of the two components of the fused transcripts was measured individually using the specific primer sets (PI-P-binding domain and EGFP), as expected, there were no significant differences. Similar results were also observed when the PI-P-binding domain genes and EGFP were driven by separate promoters.

Cacao leaves transiently expressing apoplast-targeted PI3P-binding domains show increased resistance to *P. tropicalis*

To test the effect of PI3P-binding domain expression on pathogen resistance, the PI3P-binding proteins were expressed in cacao leaves by agro-infiltration, and the expression was verified using the EGFP reporter protein (Figure 3a). The reduced level of green fluorescence observed with the fusion proteins relative to the

Table 1 The names, species of origin, accession numbers, class type, binding specificity and size of all the domains used in this study

Domain name	Species	Accession number	Domain type	Binding specificity	Length (kb)
VAM7p-PX	*Saccharomyces cerevisiae*	P32912.1	Phox homology (PX)	PI3P	0.4
Hrs-2xFYVE	*Mus musculus*	D50050	Fab1, YOTB, Vac1, EEA1 (FYVE)	PI3P	0.5
PEPP1-PH	*Homo sapiens*	AAG01896.1	Pleckstrin Homology (PH)	PI3P	0.45
GmPh1-PH	*Glycine max*	NP_001235232.1	Pleckstrin Homology (PH)	PI3P	0.45
FAPP1-PH	*Homo sapiens*	AAG15199.1	Pleckstrin Homology (PH)	PI4P	0.3

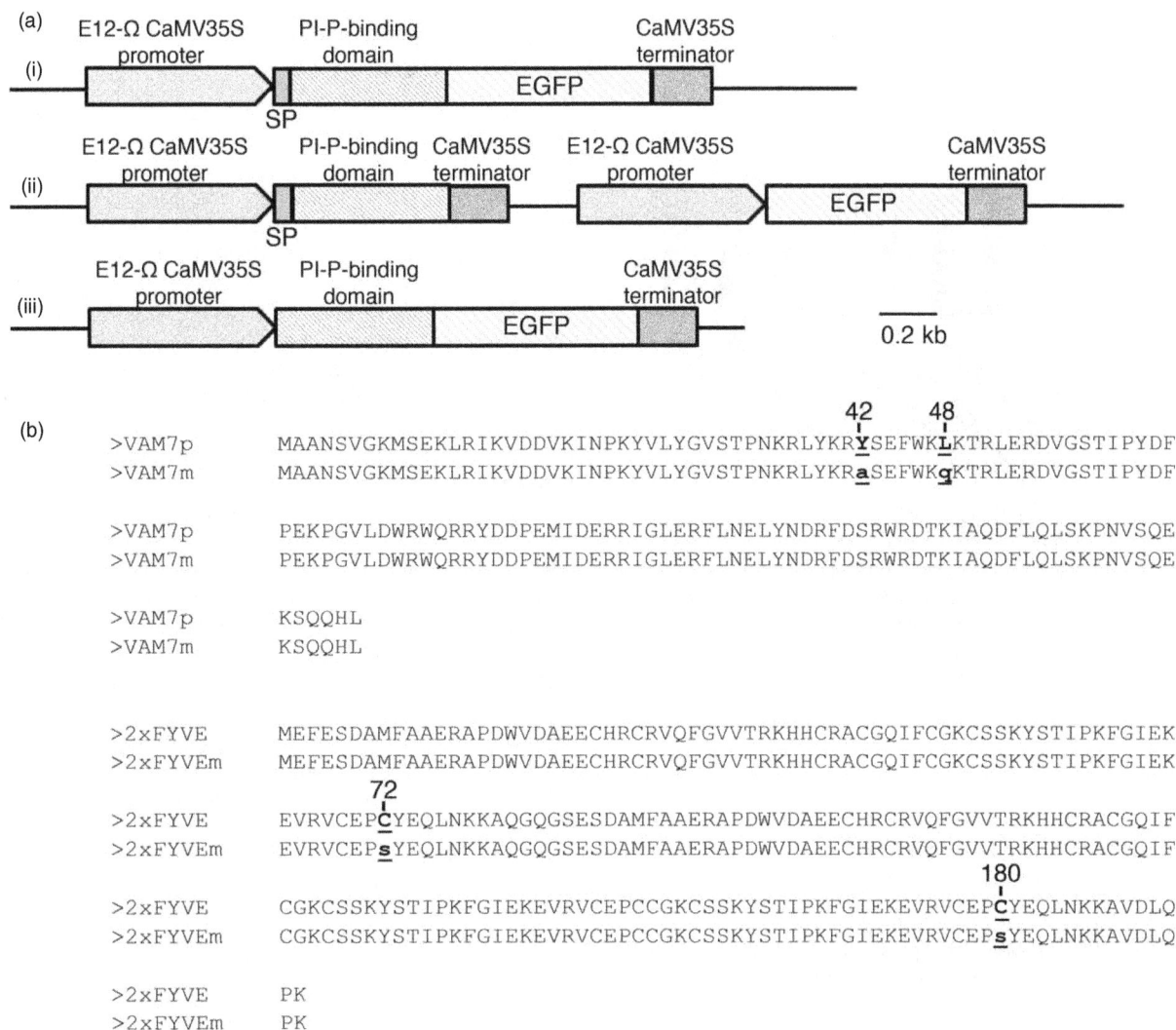

Figure 1 Design of constructs and mutant controls. (a) Structure of transgene cassettes used in this study. Constructs (i) and (ii) both include regions encoding a phosphoinositide (PI-P)-binding domain fused to a signal peptide (SP) under the control of a strong constitutive promoter (*35sE12*). In (i), there is an EGFP reporter gene fused to the SP+PI-P construct (VAM7p, Hrs, PEPP1, GmPH1, FAPP1). In (ii), the EGFP is under the control of a separate 35sE12 promoter (VAM7p, Hrs, PEPP1, VAM7 m). (iii) the PI-P-binding domain is fused to an EGFP reporter gene under the control of a 35sE12 promoter, but without the signal peptide (PEPP1). (b) Amino acid sequences of two PI3P-binding domains and their mutated versions: VAM7p and its mutant, VAM7 m (Cheever *et al.*, 2000); and the duplicated FYVE domain from Hrs and its mutant, Hrsm (Kutateladze, 2006; Raiborg *et al.*, 2001).

EGFP-only control is likely a result of export of the fusion protein into the apoplastic space where the lower pH in the apoplastic space would quench the GFP signal. Weak GFP fluorescence observed in the cytoplasm in these cases is likely due to incomplete secretion of the proteins and/or re-entry of the secreted proteins into the cells (Kale *et al.*, 2010). After 48 h, agar plugs containing *P. tropicalis* isolate 73–74 were placed on each leaf (Figure 3b) alongside a control consisting of sterile agar plugs. Pathogen infection was evaluated after 3 days by two methods, lesion area (Figure 3c) and the relative amount of pathogen genomic DNA (Figure 3d). Pathogen DNA levels were measured by quantitative PCR (qPCR) using DNA isolated from standard-sized leaf discs, using primers specific for pathogen and cacao *Actin* genes. The ratio of these two measurements was used as an indicator of the relative amount of pathogen DNA present in each lesion, which is a proxy for the pathogen biomass.

As compared to the control transformation lacking a PI3P-binding protein, cacao leaves expressing the PI3P-binding domains from VAM7p, Hrs, PEPP1 and GmPH1 showed 55%–85% reduction both in the areas of the lesions and in *P. tropicalis* colonization (as measured by genomic DNA levels) (Figure 3). Inoculation experiments with cacao leaves expressing SP::VAM7p, SP::Hrs and SP::PEPP1 unfused to an EGFP protein showed no difference to results compared to leaves expressing fusion proteins (Figure S1). The expression of the PI4P-binding domain from FAPP1 resulted in lesion sizes that were not significantly different than the control leaves. Lastly, transformation of constructs encoding two different mutated domains, VAM7 m and Hrsm, that could no longer bind PI3P (Cheever *et al.*, 2001; Kutateladze, 2006; Raiborg *et al.*, 2001) (Figure 1b) had no significant effect on lesion size, nor on pathogen biomass, confirming that the PI3P-binding activity of these two proteins was required to produce resistance.

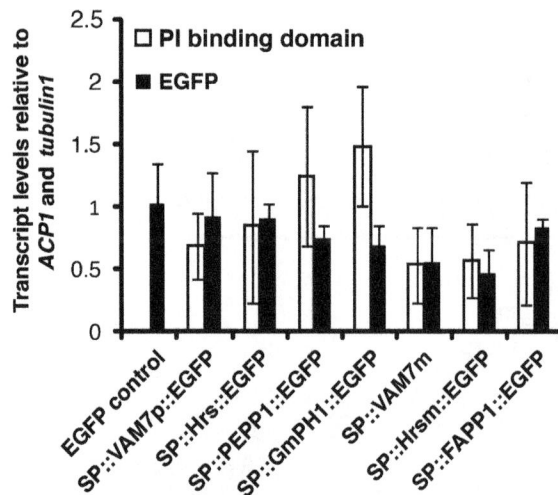

Figure 2 Measurement of the transcript levels from each construct following transient expression in detached cacao leaves using quantitative reverse transcriptase–PCR. Bars represent transcript levels as measured by primers for each respective PI-P-binding domain or for EGFP, standardized to the levels from internal control genes *TcACP1* and *TcTubulin1*. Measurements represent averages from three different leaves. Error bars represent standard errors. All levels show no statistically significant differences among each other.

Cacao leaves transiently expressing PI3P-binding domains show increased resistance to *P. palmivora*, a virulent agent of black pod rot

To test the efficacy of the constructs against a second cacao pathogen, leaves transiently expressing functional or mutated VAM7p domains or the EGFP-only control (Figure 4) were challenged with *P. palmivora*. Leaves expressing functional VAM7p showed significantly reduced lesion sizes and pathogen colonization as compared to the leaves expressing either mutated VAM7p domain or the EGFP-only control (Figure 4).

Apoplastic targeting of PI3P-binding domain proteins is required for resistance to *P. tropicalis*

The effector-blocking strategy assumes that the PI3P-binding domains must be targeted to the apoplastic space, to bind PI3P in the outer leaflet of the plasma membrane. To test this assumption, a PEPP1 PI3P-binding domain construct lacking a secretory leader was used (named PEPP1::EGFP in Figure 1a), which should result in protein accumulation solely in the cytoplasm. The expression of PEPP1::EGFP resulted in high levels of EGFP fluorescence in the cytoplasm approximately similar to the EGFP-only control, whereas, as expected, the SP::PEPP1::EGFP construct produced very little cytoplasmic EGFP (Figure 5a). Based on the lesion size and genomic qPCR bioassays, there was no significant difference in resistance conferred by the cytoplasmic-targeted PEPP1::EGFP compared with the control EGFP-only (Figure 5b). On the other hand, cacao leaves transformed with the apoplast-targeted SP::PEPP1::EGFP construct showed significantly smaller lesions and significantly less *P. tropicalis* colonization. These results demonstrate that to confer resistance to *P. tropicalis*, the PI3P-binding domains must be targeted to the apoplastic space.

Generation and verification of stable transgenic *T. cacao* plants expressing PI3P-binding domains from VAM7p, VAM7 mutant and Hrs

As a further test of the efficacy of the strategy, we generated stably transformed cacao plants expressing two of the PI3P-binding domains using *Agrobacterium*-mediated transformation (Maximova *et al.*, 2003). Stable transgenic plants carrying the SP::VAM7p construct (two independent lines), the SP::Hrs::EGFP construct (one line) and the SP::VAM7 m (non-PI3P-binding VAM7p mutant; one line) were produced and maintained in greenhouse conditions along with rooted cuttings of a control stable transgenic line carrying an EGFP-only construct.

The presence of the transgenes was confirmed using a PCR-based analysis (Maximova *et al.*, 2003) (Figure 6a). Three different sets of primers were designed to amplify a 100-bp sequence from the *TcActin* gene, a 500-bp sequence from the EGFP transgene and a 411-bp sequence (Bin19 backbone) located outside of the T-DNA region of the p126 transformation vector and not expected to be transferred to the transgenic plant genome. DNA was isolated from leaf tissues of five different genotypes: nontransformed cacao, EGFP-only transformants, SP::Hrs::EGFP transformants, SP::VAM7p transformants and SP::VAM7 m transformants. As expected, control p126A plasmid DNA produced amplified fragments with both Bin19 and EGFP primers. The nontransformed cacao samples produced amplified fragments only for cacao *actin*. Amplification of DNA from the transgenic cacao leaves using the three primer pairs resulted in products from *TcActin* and EGFP, but not from Bin19. The absence of Bin19 products from the stable transformants confirmed that there was no *Agrobacterium* present, nor any plasmid DNA contamination in the transgenic leaves. Therefore, it was inferred that genomic integration of the T-DNA had occurred in each of the transformants as expected, without inclusion of the flanking, non-T-DNA region of the Ti plasmids used.

The transcript levels of the constructs encoding the PI3P-binding domains and EGFP were measured using qRT-PCR (Figure 6b). The EGFP primers revealed transgene transcript levels about 90- to 180-fold higher than the endogenous *TcACP1* and *TcTubulin1* transcripts in the VAM7p-, Hrs- and VAM7 m-expressing lines and the control (EGFP-only). The *Hrs* primers revealed transgene transcript levels in the SP::Hrs::EGFP-expressing line about 300-fold higher than the *TcACP1* and *TcTubulin1* transcripts (the high signal from the *Hrs* primers relative to the EGFP primers likely results from the duplication of the FYVE domain). As expected, the *Hrs* primers did not amplify any sequences from the EGFP-only and SP::VAM7 lines. Likewise, the *VAM7* primers revealed transgene transcripts around 200-fold higher than *TcACP1* and *TcTubulin1* in the two SP::VAM7p lines and 100-fold higher in the SP::VAM7 m line, but failed to amplify any sequences from the EGFP-only and SP::Hrs::EGFP lines.

To confirm that the secreted proteins remained intact in the apoplastic space, a Western blot was performed to detect Hrs fused to EGFP. Total protein was extracted from three leaves each from Scavina6, from the EGFP-only transformant and from the SP::Hrs::EGFP transformant and blotted with a GFP-specific antibody. The blot revealed no GFP-positive proteins in Scavina6, proteins of expected size (~32 kDa) for EGFP in the EGFP-only samples, and two protein bands in the SP::Hrs::EGFP samples: larger bands of about 50–52 kDa, which is the expected size of a Hrs::EGFP fusion protein, and 32 kDa, showing that some cleavage occurs between the Hrs and the EGFP proteins (Figure S2).

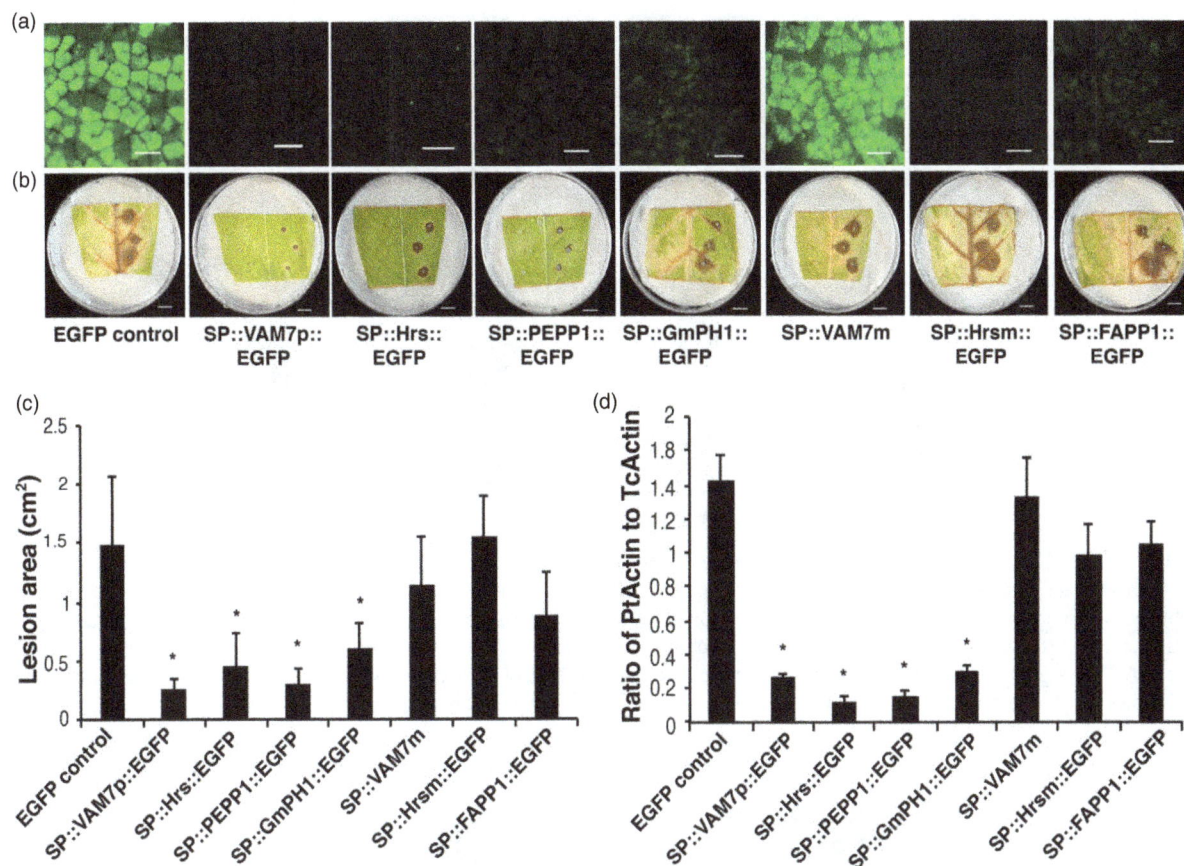

Figure 3 Transient expression of apoplast-targeted PI3P-binding domain proteins confers resistance to *Phytophthora tropicalis* in cacao. (a) Cacao leaf tissue transiently expressing EGFP 48 h after transformation by vacuum infiltration. Bars represent 250 μm. (b) Representative transiently transformed leaves, 3 days postinoculation with 3-mm^2 agar plugs containing *P. tropicalis* mycelia (right) and clean agar plugs (left). Bars represent 1 cm. (c) Average lesion areas from each genotype 3 days postinoculation, calculated by ImageJ. Each bar represents the mean ± standard error (SE) from three independent biological replicates, with 18 lesions per replicate. (d) Proliferation of *P. tropicalis* as determined by qPCR analysis of the genomic DNA ratio of *P. tropicalis* to *T. cacao* 3 days post-inoculation. Three lesions from each leaf piece were combined and extracted as a single genomic DNA sample. Each bar in the graph represents the mean ± SE of three independent biological replicates, with each replicate consisting of three inoculation sites from two pieces from different leaves. In (c) and (d), asterisks indicate significant differences ($P < 0.05$) to the EGFP-only control as determined by ANOVA.

At a gross morphological and developmental level, all transgenic plants appeared phenotypically normal. No obvious visible differences in growth rate or morphology were observed in any of the plants relative to nontransgenic plants grown in identical conditions (Figure S3a). Visualization EGFP in leaves from stable transgenic plants expressing EGFP (cytoplasmic-localized EGFP) or SP::Hrs::EGFP (apoplastic-localized EGFP) (Figure S3b) resembled the distribution of EGFP in transient expression experiments (Figures 3 and 5); cytoplasmically localized EGFP appeared much brighter than apoplastically localized EGFP due to quenching. Visualization of EGFP in SP::VAM7p and SP::VAM7 m stable transgenic plants revealed a bright cytoplasmic EGFP signal, as the EGFP protein is targeted to the cytoplasm and is not produced as a fusion to the signal peptide and PI-binding protein (Figure S3b). To verify that the expression of the PI-binding proteins caused no severe membrane dysfunction or cell death, leaf discs from untransformed Scavina6, SP::Hrs::EGFP and SP::VAM7p-1 lines were stained with propidium iodide and compared with stained leaf discs that were previously killed with CuSO$_4$. In healthy tissue, it is expected that propidium iodide is excluded from the inside of the cells, showing only staining of the membrane. In dead tissue, the cell membrane is permeable, allowing for staining of cellular

contents. As shown in Figure S3c, Scavina6 leaf tissue killed with CuSO$_4$ showed staining of the cellular contents, but not of the membrane, whereas staining of Scavina6 leaves and leaves from the SP::Hrs::EGFP and SP::VAM7p transformants showed staining of cell membranes, with the stain being excluded from the intracellular space.

Transgenic *T. cacao* plants expressing apoplast-targeted PI3P-binding domains show enhanced resistance to two *Phytophthora* species

Leaves from stably transformed cacao plants expressing VAM7p, VAM7 m, Hrs and EGFP-only constructs were inoculated with 3-mm agar plugs containing mycelia of *P. tropicalis* isolate 73–74 or *P. palmivora* isolate DUK23.1. Leaves from a *T. cacao* transformant expressing an antimicrobial peptide, D4E1, previously shown to exhibit enhanced resistance to *P. palmivora* (Mejía et al., 2012) were inoculated alongside as a positive control for resistance. Images of representative leaves 3 days after inoculation with the *Phytophthora* pathogens are shown in Figure 7a. Following inoculation with *P. tropicalis* 73–74, the *T. cacao* lines expressing Hrs showed a 69% reduction in lesion area, while the two independent VAM7p lines showed 74% and 79% reductions

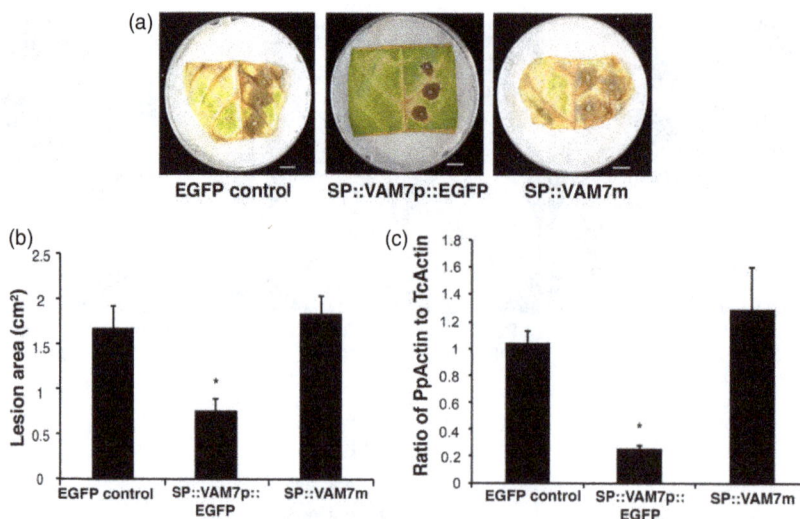

Figure 4 Transient expression of apoplast-targeted PI3P-binding domain VAM7p confers resistance to *Phytophthora palmivora*. (a) Representative transiently transformed leaves, 3 days postinoculation with agar plugs containing *P. palmivora* mycelium (right), and sterile agar plugs (left). Bars represent 1 cm. (b) Average lesion areas calculated with ImageJ, 3 days postinoculation by *P. palmivora*. Each bar represents the mean ± SE of two independent biological replicates, with 18 lesions each. (c) Proliferation of *P. palmivora* as determined by qPCR analysis of the ratio of *P. palmivora* and *T. cacao* genomic DNAs 3 days postinoculation. Each bar in the graph represents the mean ± SE of two independent replicates, with each replicate consisting three inoculation sites from two leaf pieces, each originating from different leaves. In (b) and (c), asterisks represent significant differences (*P* < 0.05) to the EGFP-only control as determined by ANOVA.

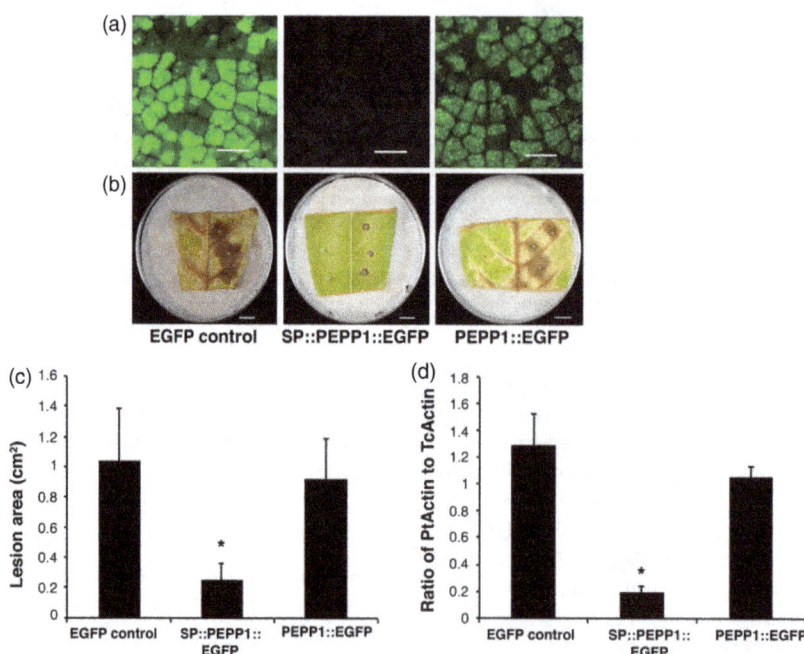

Figure 5 Resistance conferred by the PEPP1 PI3P-binding domain is dependent on targeting to the apoplast. (a) Cacao leaf tissue expressing EGFP, 48 h after transformation with each construct by vacuum infiltration. Bars represent 250 µm. (b) Representative transiently transformed leaves expressing the indicated construct, 3 days postinoculation with agar plugs either containing *P. tropicalis* mycelium (left) or sterile water agar (right). Bars represent 1 cm. (c) Lesion areas calculated by ImageJ, 3 days postinoculation by *P. tropicalis*. Each bar represents the mean ± SE from three independent biological replicates, with 18 lesions per replicate. (d) Proliferation of *P. tropicalis* as determined by qPCR analysis of the ratio of *P. tropicalis* to *T. cacao* genomic DNAs, 3 days postinoculation. Each bar in the graph represents the mean ± SE of three independent replicates, with each replicate consisting of three inoculation sites from two pieces from different leaves. In (c) and (d), asterisks represent significant differences (*P* < 0.05) to the EGFP-only vector control as determined by ANOVA.

in lesion area. This was significantly lower compared to EGFP-only-expressing leaves, and slightly better than the 60% reduction in the positive control line expressing the peptide D4E1. The

mutant VAM7 m line showed about a 50% increase in lesion size that was not statistically significant, which shows that the PI3P-binding activity of the protein is required for resistance (Figure 7a,

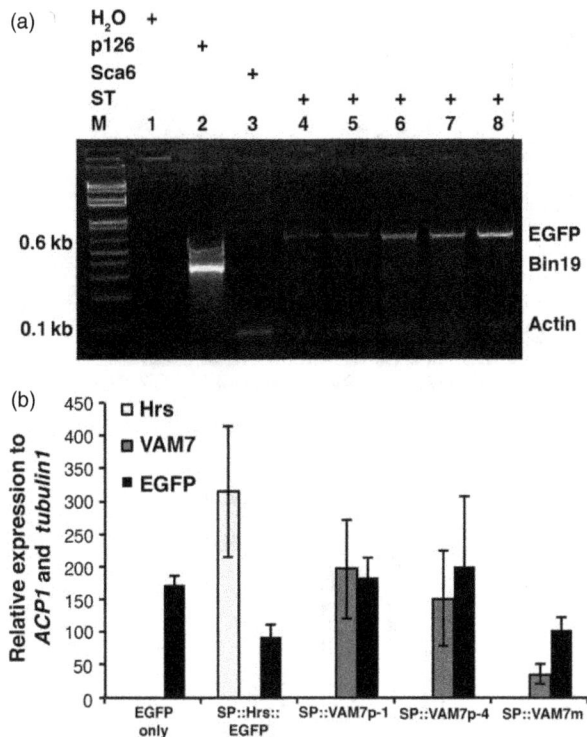

Figure 6 Verification of stable transformation of transgenic cacao lines with PI3P-binding protein constructs. (a) Test for transgene integration by genomic PCR analysis. Amplification of genomic DNA from nontransformed Scavina6 (Sca6, lane 3), stably transformed EGFP-only (lane 4), stably transformed SP::VAM7p::EGFP (line 1) (lane 5), stably transformed SP::VAM7p::EGFP (line 4) (lane 6), stably transformed SP:: Hrs::EGFP (lane 7) and stably transformed SP::VAM7 m (lane 8) using a mixture of primers to amplify a 500-bp fragment from *EGFP*, a 100-bp fragment from endogenous *actin* and a 400-bp fragment from the p126 plasmid non-T-DNA backbone (Bin19). Positive control for Bin19 was plasmid DNA from p126 (lane 2). (b) Measurement of transgene transcript levels in leaves of cacao stable transformants SP::Hrs::EGFP, SP::VAM7p-1, SP::VAM7p-4, SP::VAM7 m and EGFP-only by qRT-PCR. Bars represent the average expression of each respective PI3P-binding domain-containing fusion gene, or *EGFP* relative to internal control genes *TcACP1* and *TcTubulin1* in three different leaves. Error bars indicate standard errors.

b). The reductions in lesion sizes in the D4E1-, Hrs- and VAM7p-expressing leaves were accompanied by 70%–90% reductions in *P. tropicalis* genomic DNA content, whereas a nonsignificant 40% reduction in *P. tropicalis* genomic DNA content was observed in the mutant VAM7 m line (Figure 7c). None of the differences in lesion sizes or genomic DNA content among the four resistant transformants were statistically significant. Following inoculation by the more virulent pathogen isolate *P. palmivora* DUK23.1, *T. cacao* leaves expressing Hrs and the two independent VAM7p lines showed lesion reductions of 59%, 48% and 60% respectively, compared to the EGFP-only transformant. The mutant VAM7 m lines showed a 25% increase in *P. palmivora* lesion size, which was not significantly different from the EGFP-only control (Figure 7a,b). These results were supported by the qPCR data, which showed an 83%, 93% and 81% reduction in the content of *P. palmivora* genomic DNA in the Hrs- and two VAM7p-expressing lines, and a nonsignificant 9% reduction in the nonbinding VAM7 m line (Figure 7c).

Stable transgenic *T. cacao* plants expressing functional PI3P-binding domains show strong resistance to the fungal pathogen *C. theobromicola*

To test the resistance of the transformants to a fungal pathogen, leaves from the Hrs-, VAM7p-, VAM7 m- and EGFP-only-expressing lines were inoculated with 10-μL drops of a conidial suspension from two *C. theobromicola* isolates, 11–50 and 11–183. As for the *Phytophthora* resistance assays, leaves from the D4E1-expressing cacao line were used as a positive check for resistance (Mejía *et al.*, 2012). Images of representative leaves 4 days postinoculation by 11–50 and 11–183 are shown in Figure 8a. The three lines expressing the PI3P-binding domains from Hrs and VAM7p showed lesion areas after inoculation by *C. theobromicola* 11–50 that were reduced by 95%, 91% and 94%, respectively, compared to the EGFP-only control. The D4E1-expressing leaves showed a 75% reduction, which was significantly less resistance than that displayed by the PI3P-binding protein transformants. The mutant VAM7 m line showed about a 50% reduction in lesion size; however, this difference was not significant. When inoculated by a second *C. theobromicola* isolate, 11–183, the PI3P-binding protein transformants showed lesion areas reduced by 90%, 83% and 87%, respectively (Figure 8b), while the resistance conferred by the peptide was significantly weaker (55% reduction). The mutant VAM7 m line showed a 42% reduction in lesion area; however, this difference again was not significant. The strong resistance conferred by the PI3P-binding protein constructs was confirmed by qPCR quantification of *C. theobromicola* genomic DNA content: the genomic DNA content of 11–50 was reduced by 98% in the Hrs- and two VAM7p-expressing transgenic lines compared to the EGFP-only control line, while the expression of the peptide resulted in a 94% reduction of pathogen genomic DNA content. The genomic DNA content of 11–183 was reduced 97% in all three PI3P-binding lines and by 74% in the peptide-expressing line. The mutant VAM7 m line showed a 20% increase in both 11–50 and 11–183 genomic DNA (Figure 8c), showing that the slight decrease in lesion area was not associated with decreased pathogen colonization.

Discussion

Fungal and oomycete pathogens cause billions of dollars of economic loss each year throughout the world. For many crops, the lack of good resistance genes and difficult breeding systems means farmers must rely on chemical or cultural methods of disease management. Even when resistance genes are deployed in cultivars, the pathogen populations can often evolve rapidly to evade the resistance genes.

One critical step in the pathogen life cycle that is common across many bacterial, nematode, insect, fungal and oomycete species is the secretion and delivery of protein effectors into host cells, where they manipulate host physiology to promote the success of the pathogen (Torto-Alalibo *et al.*, 2010). The discovery that host cell entry by many oomycete effectors and several fungal effectors involves interactions with PI3P on the plasma membrane (Kale *et al.*, 2010; Plett *et al.*, 2011) suggested to us a possible target for blocking effector entry. We hypothesized that if sufficient amounts of PI3P-binding proteins could be expressed and targeted to the apoplastic space, the proteins could competitively inhibit the PI3P binding and internalization of pathogen effectors, resulting in enhanced disease resistance.

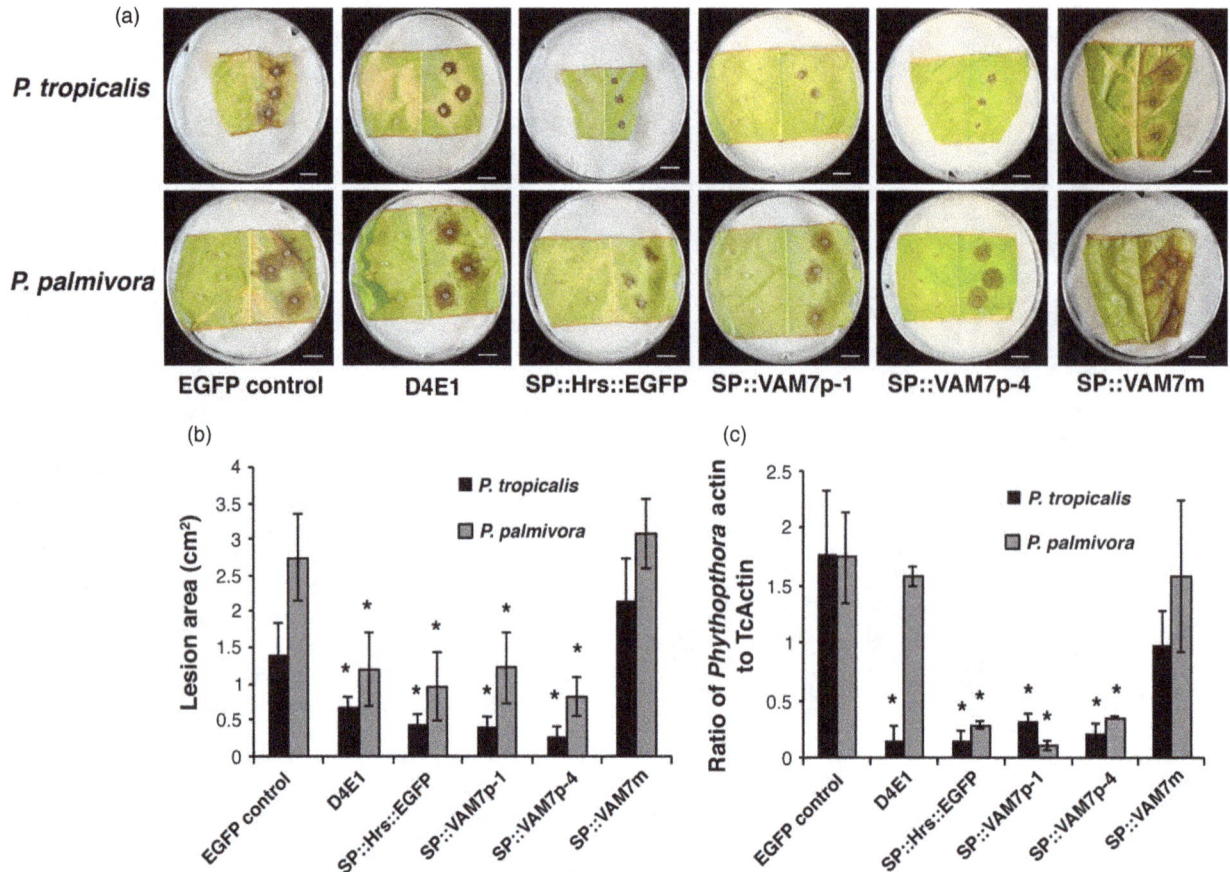

Figure 7 Expression of apoplast-targeted PI3P-binding domain proteins in cacao stable transformants confers resistance to *P. tropicalis* and *P. palmivora*. (a) Representative leaves from cacao stable transformants, 3 days postinoculation with *P. tropicalis* 73–74 (top row) or 3 days postinoculation with *P. palmivora* DUK23.1 (bottom row). Bars represent 1 cm. (b) Lesion area from each genotype calculated 3 days after inoculation with *P. tropicalis* or *P. palmivora*. Each bar represents the mean ± SE from three independent biological replicates, with 18 lesions per replicate. (c) qPCR quantification of pathogen proliferation (3 days after inoculation in each case). Each bar in the graph represents the mean ± SE of three independent replicates, with each replicate consisting of three inoculation sites from two pieces from different leaves. In (b) and (c), asterisks represent a significant difference ($P < 0.05$) to the EGFP-only control as determined by ANOVA.

Our results demonstrated that the expression of four completely different PI3P-binding proteins targeted to the apoplastic space did provide resistance against two oomycete pathogens and a fungal pathogen. Three of the four proteins were previously shown to block entry by oomycete and fungal effectors (Kale *et al.*, 2010). No resistance was conferred by a PI4P-binding protein, nor by two proteins with mutations that abolish PI3P binding. Furthermore, targeting of the proteins to the apoplast was required to confer resistance. Resistance was observed both in transiently transformed cacao leaves and in three stably transformed whole plants. Together, the results support our original hypothesis that apoplastic expression of PI3P-binding proteins could be capable of reducing infection.

Although secretion of PI3P-binding proteins results in resistance, the precise mechanisms by which they confer resistance remain to be investigated. Although the blocking of effector entry is the most obvious mechanism, it is also possible that the proteins trigger some kind of resistance or priming response. The purpose of external PI3P on plant cell membranes is currently unknown, and a role in defence signalling is not ruled out currently. Our data also do not rule out that the site of action of the PI3P-binding proteins is within the endomembrane system rather than in the apoplast. The very strong resistance against *C. theobromicola*

suggests either that this pathogen utilizes PI3P to deliver its effectors into host cells (which is currently unknown) or that the plants are expressing a broader mechanism of resistance.

So far, we have produced three stable cacao transformants. The resistance phenotypes of these plants are similar and are consistent with the transient expression results. We are producing additional stable transformants, including plants expressing other PI3P-binding proteins, to further confirm the efficacy of these transgenes, and to improve the levels of resistance. So far, the transgenic plants have not been tested for resistance against the three most destructive pathogens of cacao, namely the oomycete *P. megakarya* and the fungi *Moniliophthora perniciosa* (witches' broom) and *Moniliophthora roreri* (frosty pod), because this will require maturation of whole plants and fruit production.

In summary, our data suggest that this technology offers a promising level of resistance against diverse pathogens that may be applicable to a wide range of crop species.

Experimental procedures

Binary vector construction

All sequences of PI-P-binding domains were obtained from public databases and synthesized by GenScript Corporation, with codon

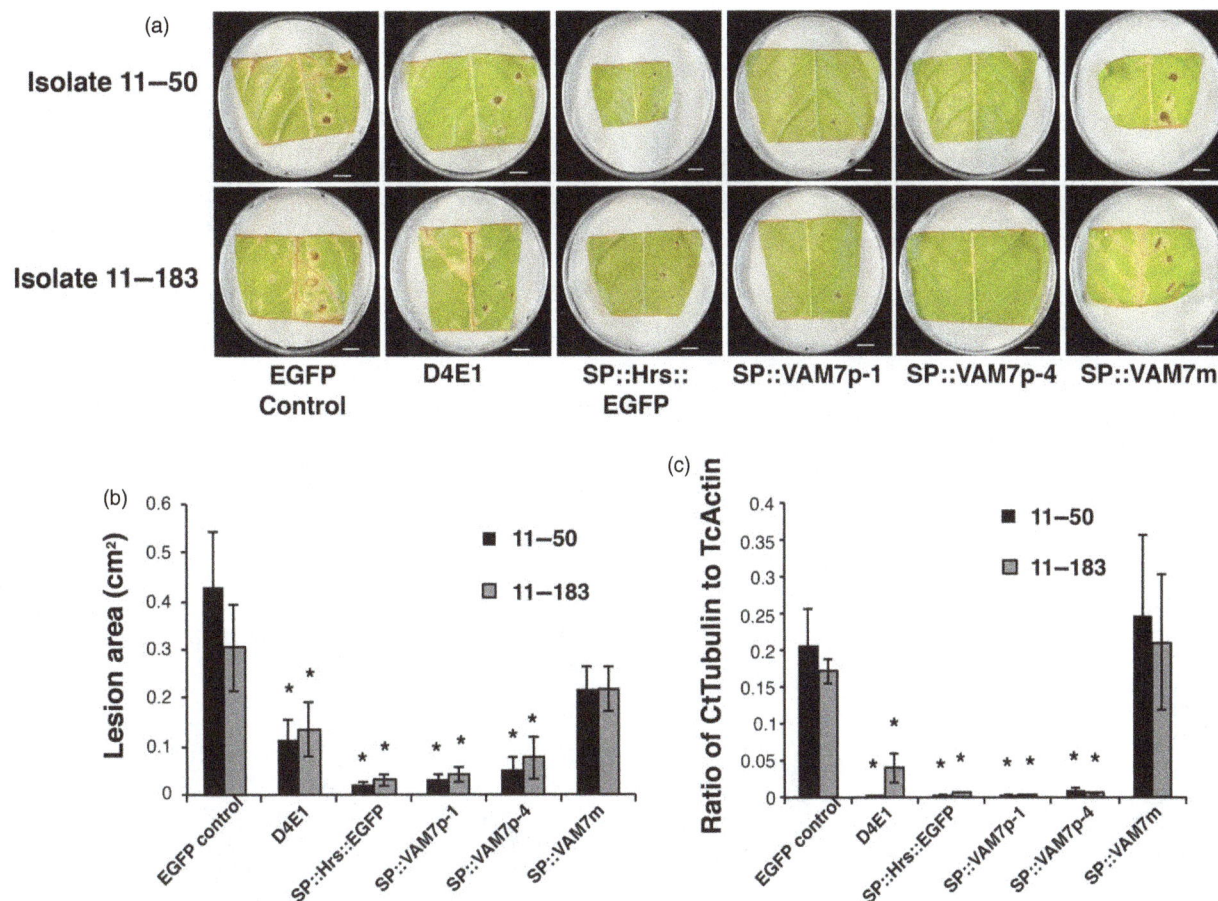

Figure 8 Expression of apoplast-targeted PI3P-binding domain proteins in cacao stable transformants confers resistance to two isolates of *Colletotrichum theobromicola*. (a) Representative leaves from cacao stable transformants 4 days after inoculation by *C. theobromicola* isolate 11–50 (top row) or *C. theobromicola* isolate 11–183 (bottom row). Bars represent 1 cm. (b) Lesion area of each *T. cacao* genotype measured 4 days after inoculation with *C. theobromicola* isolates 11–50 or 11–183. Bars represent the mean ± SE of two independent biological replicates, with 18 lesions each. (c) qPCR quantification of *C. theobromicola* proliferation 4 days after inoculation. Each bar in the graph represents the mean ± SE of two independent replicates, with each replicate consisting of three inoculation sites from two pieces from different leaves. In (b) and (c), asterisks represent a significant difference ($P < 0.05$) to the EGFP-only control as determined by ANOVA.

optimization for expression in *E. coli* and *N. tabacum*. These included human FAPP1-PH (AAG15199.1, residues 1–99), soya bean GmPH1 [NP_001235232.1, Glyma14g06560, residues 1–146; obtained by homology to AtPH1 (Dowler *et al.*, 2000)], human PEPP1-PH (AAG01896.1, residues 15–168), yeast VAM7p-PX (P32912.1, residues 1–134 with the substitution R73W) and mouse Hrs-2xFYVE [D50050, residues 147–223, as modified and duplicated by Gillooly *et al.* (2000)]. Site-specific mutagenesis of VAM7p-PX and Hrs-2xFYVE was performed using primer mutagenesis by PCR or by gene synthesis, respectively. Those mutations were previously described to abolish PI3P-binding by VAM7p-PX (Cheever *et al.*, 2001) or by Hrs (Kutateladze, 2006; Raiborg *et al.*, 2001). To create the EGFP-fused PI-P-binding domain vectors, binary vector pGH00-0126 (Maximova *et al.*, 2003) was made Gateway-compatible by adding the Gateway® cassette (Invitrogen, Waltham, MA) containing *attR* recombination sites flanking a *ccd*B gene and a chloramphenicol-resistance gene, generating vector 126gfp-gw. The PI-P-binding domains were cloned into 126gfp-gw, with (sp126gfp-gw) or without the signal peptide (SP) sequence from the *Glycine max* PR1a gene (NM_001251239, residues 1–27). Constructs where the PI-P-

binding domain was not fused to EGFP were created by adding SpeI and HpaI restriction sites onto the SP::PI-P-binding domain segment and subcloning into pGH00.0126 (GenBank: KF018690.1). All constructs were transformed into *Agrobacterium tumefaciens* strain AGL1 for both transient and stable transformations.

Transient and stable transformation of *T. cacao*

For transient transformation of detached cacao leaves, *A. tumefaciens* cells containing each transgene were grown as described in Maximova *et al.* (2003), induced with acetosyringone, and then vacuum-infiltrated into stage C leaves from cultivar Scavina6 as described in Shi *et al.* (2013). The Petri dishes were sealed and incubated at 25 °C for 2 days with light intensity of 145 m²/s and 14-h daylight. After 2 days, EGFP fluorescence was examined as described (Maximova *et al.*, 2003). Only leaves with green fluorescence coverage of 80% or more were further subjected to pathogen infection. Stable transformation of cacao was performed as described (Maximova *et al.*, 2003). Resulting stable transformants were grown for about 6 weeks in greenhouse conditions before leaves were sampled for further experiments.

Protein extraction and Western blot

Total protein was extracted from three leaves each of nontransformed Scavina6, stable transformed EGFP-only plants and stable transformed SP::Hrs::EGFP plants following the protocol of Pirovani et al. (2008). Western blotting was performed by electrophoresing 15 μg of total protein per sample on a 12% SDS-PAGE gel, followed by blotting onto a PVDF membrane (Millipore, Billerica, CA) using a wet transfer apparatus (BioRad, Hercules, CA). Post-transfer, the membrane was blocked with 2% bovine serum albumin in PBS-T (137 mM NaCl; 2.7 mM KCl; 10 mM Na_2HPO_4; 1.8 mM KH_2PO_4; 0.1% Tween-20), rinsed and incubated for 16 h with rabbit anti-GFP primary antibody (Immunology Consultants Lab, Inc., Portland, OR) at 4 °C. The membrane was rinsed in PBS-T, incubated for 1 h at room temperature with 1 : 10 000 dilution of HRP-linked anti-rabbit secondary antibody (GE Life Sciences, Buckinghamshire, UK) and then rinsed and exposed with West Pico Chemiluminescent Substrate (Thermo Scientific, Waltham, MA) before visualization with PhosphorImager (Storm 860; GE Healthcare, Buckinghamshire, UK).

Propidium iodide staining

Leaf discs (2.5 cm²) from stage C cacao leaves were vacuum-infiltrated using three 2-min applications of 100 μL each (sufficient to cover a leaf disc) of 30 μg/mL propidium iodide (PI), 0.02% Silwet in dH_2O. Mock-treated samples were prepared the same way, but were infiltrated with 100 μL dH_2O containing 0.02% Silwet without PI. To serve as a positive control for dead tissue, a set of leaf discs was infiltrated with 1 mM $CuSO_4$ 30 min prior to staining with propidium iodide. Leaf discs were mounted in water adaxial side up and imaged on a Zeiss AxioObserver (Carl Zeiss Microscopy GmbH, Jena, Germany) spinning disc confocal microscope, with a 561-nm excitation beam, 617/73 emission filter set and 20× objective. Z-stacks were generated that included the entire depth of the first epidermal cell layer, with 0.2 μm distance between slices. The same laser power, exposure time and detector gain were used for every slice and every sample.

Verification of transgene expression

To measure the levels of transcripts spanning the PI-P-binding domains and EGFP regions of the constructs, RNA was extracted from stage C leaves from stable transgenic lines or, in the case of transient expression experiments, from stage C Scavina6 leaves 5 days following infiltration with A. tumefaciens cells. In the case of transient expression experiments, the right-hand side of each leaf was inoculated with P. tropicalis, while the left side of each leaf, which was taken for RNA analysis, was mock-inoculated; this ensured that transcript levels were measured under the same conditions as the pathogen assays. RNA was extracted from the leaves using Plant RNA Reagent (Invitrogen) according to the manufacturer's protocol. cDNA synthesis was performed using the New England Biolabs (Ipswich, MA) cDNA Synthesis Kit. Transcript levels were measured by quantitative real-time PCR on a Step One Plus Real-Time PCR System (Applied Biosystems, Waltham, MA) with Takara SYBR Green reagent. PCR cycles were performed at 94 °C for 15 min, followed by 40 cycles of 94 °C for 30 s, 60 °C for 1 min and 72 °C for 1 min. Transcript levels from each transgene were measured relative to the transcript levels of TcACP1 (Tc01g039970) and TcTubulin1 (Tc06g000360) of each sample. Primer sequences are listed in Table S1.

To verify transgene insertion in the stable transformants, genomic DNA was extracted from cacao leaves using a modified CTAB method. Stage C leaves were frozen in liquid nitrogen and ground with a mortar and pestle with extraction buffer [10 mM Tris-HCl pH 8.0; 1.4 M NaCl, 10 mM Na_3EDTA pH 8.0, 2% polyvinylpyrrolidone (PVP), 2% cetyltrimethylammonium bromide (CTAB) and 0.2% β-mercaptoethanol] and then further homogenized using a TissueLyzer. Nucleic acids were separated using chloroform:isoamyl alcohol (24 : 1) and precipitated using isopropanol. Pellets were dissolved in TE buffer (10 mM Tris-HCl, 1 mM EDTA pH 8.0) and treated with RNase A (10 μg/mL) for 30 min at 37 °C, then re-extracted with phenol: chloroform: isoamyl alcohol (25 : 24 : 1) and precipitated using 2.5 M ammonium acetate and 70% ethanol. Resulting pellets were dissolved in sterile dH_2O, and the quantity and quality were checked using a Nanodrop spectrophotometer (Thermo Scientific). A total of 10 ng of each sample was amplified using all three sets of primers (T. cacao actin, Bin19 and EGFP, listed in Table S1) at 94 °C for 4 min, followed by 30 cycles of 94 °C for 30 s, 55 °C for 30 s and 72 °C for 1 min. Prior to termination, the samples were incubated at 72 °C for 7 min. Samples (20 μL) of each reaction were analysed by electrophoresis on a 2% agarose gel.

Pathogenicity assays

Oomycete pathogens P. tropicalis and P. palmivora were grown on 20% unclarified V8 medium (100 mL/L V8 juice, 3 g/L calcium carbonate and 15 g/L bacto agar) for 2 days at 27 °C, 12-h daylight. Stage C Scavina6 cacao leaves were inoculated, abaxial side up, on the right-hand side with three agar plugs containing actively growing Phytophthora mycelium from the margin of the colony, while the left-hand side was inoculated with sterile agar plugs as a negative control. Inoculated leaves were incubated at 27 °C and 12-h daylight cycle for 3 days before the evaluation of disease symptoms. The leaves were photographed with a Nikon D90 camera (Nikon, Tokyo, Japan) and lesion sizes were measured using ImageJ software tools (Imagej.nih.gov). Average lesion sizes were calculated from three replicates of 18 measurements each, and significance was determined by single-factor ANOVA.

Colletotrichum theobromicola isolates (11–50 and 11–183) were grown on potato dextrose agar (Sigma-Aldrich, St. Louis, MO) for 8 days at 27 °C and 12-h daylight. For inoculations, spores were suspended in sterile deionized water with 0.02% Tween-20, and the concentration was adjusted to 10^6 spores/mL with a hemocytometer. Leaf pieces, abaxial side up, were inoculated with three 10-μL drops on the right side, while three drops of deionized water were placed on the left side as control. Inoculated leaves were incubated at 27 °C and 12-h daylight for 4 days. Leaves were photographed and lesion sizes were measured as described above.

To assay pathogen DNA as a measure of virulence, the ratio of Phytophthora or Colletotrichum DNA to cacao DNA was determined by qPCR as follows. Tissue samples including the lesions (1.4 cm² for P. tropicalis and C. theobromicola, and 2.5 cm² for P. palmivora surrounding the inoculation site) were excised from the infected leaves and used for genomic DNA extraction. Tissue was ground using a TissueLyzer homogenizer followed by DNA purification with DNeasy Plant Mini Kit (Qiagen, Venlo, Netherlands). DNA qPCR was performed as described (Wang et al., 2011) using an ABI 7300 Real-Time PCR System (Applied Biosystems). The relative amount of Phytophthora or Colletotri-

chum genomic DNA in leaf discs was measured by amplification of the single-copy *PcActin* (BT031870.1), CtTubulin (KC512191.1) and *TcActin* (Tc01t010900) genes (Table S1), and the ratio of *Phytophthora* or *Colletotrichum* to cacao DNA was calculated as two to the power of the difference between Ct numbers.

Acknowledgements

We would like to thank Lena Sheaffer and Sharon Pishak for the technical assistance in maintenance of our cacao tissue culture and transformation pipelines, Germán Sandoya for assistance with statistical analyses, Daniel McClosky for assistance with staining and microscopy procedures, Dylan Storey for advice on distinguishing *P. tropicalis* from *P. capsici*, and Brent Kronmiller for bioinformatics assistance. We are also grateful to Andrew Fister and Yufan Zhang for valuable comments throughout this project. This work was supported in part by The Pennsylvania State University, College of Agricultural Sciences, The Huck Institutes of Life Sciences, the American Research Institute Penn State Endowed Program in the Molecular Biology of Cacao and grants from the National Science Foundation BREAD Program (IOS-0965353) to BT, MG, SM and SX, and to BT from the National Institute of Food and Agriculture, U.S. Department of Agriculture, under award number (2011-68004-30104).

References

Aragaki, M. and Uchida, J.Y. (2001) Morphological distinctions between *Phytophthora capsici* and *Phytophthora tropicalis* sp. nov. *Mycologia*, **193**, 137–145.

Cheever, M.L., Sato, T.K., de Beer, T., Kutateladze, T.G., Emr, S.D. and Overduin, M. (2001) Phox domain interaction with PtdIns(3)P targets the Vam7 t-SNARE to vacuole membranes. *Nat. Cell Biol.* **3**, 613–618.

Corvera, S., D'Arrigo, A. and Stenmark, H. (1999) Phosphoinositides in membrane traffic. *Curr. Opin. Cell Biol.* **11**, 460–465.

Cutt, J.R., Dixon, D.C., Carr, J.P. and Klessig, D.F. (1988) Isolation and nucleotide-sequence of cDNA clones for the pathogenesis-related proteins PR1a, PR1b and PR1c of *Nicotiana-tabacum* cv Xanthi Nc induced by TMV infection. *Nucleic Acids Res.* **16**, 9861.

DeCamilli, P., Emr, S.D., McPherson, P.S. and Novick, P. (1996) Phosphoinositides as regulators in membrane traffic. *Science*, **271**, 1533–1539.

DiNitto, J.P. and Lambright, D.G. (2006) Membrane and juxtamembrane targeting by PH and PTB domains. *Biochim. Biophys. Acta*, **1761**, 850–867.

Dou, D., Kale, S.D., Wang, X., Jiang, R.H., Bruce, N.A., Arredondo, F.D., Zhang, X. and Tyler, B.M. (2008) RXLR-mediated entry of *Phytophthora sojae* effector Avr1b into soybean cells does not require pathogen-encoded machinery. *Plant Cell*, **20**, 1930–1947.

Dowler, S., Currie, R.A., Campbell, D.G., Deak, M., Kular, G., Downes, C.P. and Alessi, D.R. (2000) Identification of pleckstrin-homology-domain-containing proteins with novel phosphoinositide-binding specificities. *Biochem. J.* **351**, 19–31.

van Esse, H.P., Thomma, B.P., van't Klooster, J.W. and de Wit, P.J. (2006) Affinity-tags are removed from *Cladosporium fulvum* effector proteins expressed in the tomato leaf apoplast. *J. Exp. Bot.* **57**, 599–608.

Gillooly, D.J., Morrow, I.C., Lindsay, M., Gould, R., Bryant, N.J., Gaullier, J.M., Parton, R.G. and Stenmark, H. (2000) Localization of phosphatidylinositol 3-phosphate in yeast and mammalian cells. *EMBO J.* **19**, 4577–4588.

Giraldo, M.C. and Valent, B. (2013) Filamentous plant pathogen effectors in action. *Nat. Rev. Microbiol.* **11**, 800–814.

Guest, D. (2007) Black pod: diverse pathogens with a global impact on cocoa yield. *Phytopathology*, **97**, 1650–1653.

Hahn, M, and Mengden, K. (1997) Characterization of in planta-induced rust genes isolated from a haustorium-specific cDNA library. *Mol Plant-Microbe Interact.* **10**, 427–437.

Honee, G., Buitink, J., Jabs, T., De Kloe, J., Sijbolts, F., Apotheker, M., Weide, R., Sijen, T., Stuiver, M. and De Wit, P.J.G.M. (1998) Induction of defense-related responses in Cf9 tomato cells by the AVR9 elicitor peptide of *Cladosporium fulvum* is developmentally regulated. *Plant Physiol.* **117**, 809–820.

Jiang, R.H.Y., Tripathy, S., Govers, F. and Tyler, B.M. (2008) RXLR effector reservoir in two Phytophthora species is dominated by a single rapidly evolving superfamily with more than 700 members. *Proc Natl Acad Sci USA*, **105**, 4874–4879.

Jones, J.D.G. and Dangl, J.L. (2006) The plant immune system. *Nature*, **444**, 323–329.

Kale, S.D., Gu, B., Capelluto, D.G., Dou, D., Feldman, E., Rumore, A., Arredondo, F.D., Hanlon, R., Fudal, I., Rouxel, T., Lawrence, C.B., Shan, W. and Tyler, B.M. (2010) External lipid PI3P mediates entry of eukaryotic pathogen effectors into plant and animal host cells. *Cell*, **142**, 284–295.

Keane, P.J. and Putter, C.A.J. (1992) *Cocoa Pest and Disease Management in Southeast Asia and Australasia.* Rome: Food and Agriculture Organization of the United Nations.

Kutateladze, T.G. (2006) Phosphatidylinositol 3-phosphate recognition and membrane docking by the FYVE domain. *Biochim. Biophys. Acta*, **1761**, 868–877.

Kutateladze, T.G. (2010) Translation of the phosphoinositide code by PI effectors. *Nat. Chem. Biol.* **6**, 507–513.

Lee, S.A., Kovacs, J., Stahelin, R.V., Cheever, M.L., Overduin, M., Setty, T.G., Burd, C.G., Cho, W. and Kutateladze, T.G. (2006) Molecular mechanism of membrane docking by the Vam7p PX domain. *J. Biol. Chem.* **281**, 37091–37101.

Lemmon, M.A. (2008) Membrane recognition by phospholipid-binding domains. *Nat. Rev. Mol. Cell Biol.* **9**, 99–111.

Maximova, S., Miller, C., Antunez de Mayolo, G., Pishak, S., Young, A. and Guiltinan, M.J. (2003) Stable transformation of *Theobroma cacao* L. and influence of matrix attachment regions on GFP expression. *Plant Cell Rep.* **21**, 872–883.

Mejia, L.C., Guiltinan, M.J., Shi, Z., Landherr, L. and Maximova, S.N. (2012) Expression of designed antimicrobial peptides in *Theobroma cacao* L. trees reduces leaf necrosis caused by *Phytophthora* spp. In: *Small Wonders: Peptides for Disease Control*, Vol. **1095** Washington, DC: American Chemical Society, pp. 379–395. 10.1021/bk-2012-1095.ch018.

Mitsuhara, I., Ugaki, M., Hirochika, H., Ohshima, M., Murakami, T., Gotoh, Y., Katayose, Y., Nakamura, S., Honkura, R., Nishimiya, S., Ueno, K., Mochizuki, A., Tanimoto, H., Tsugawa, H., Otsuki, Y. and Ohashi, Y. (1996) Efficient promoter cassettes for enhanced expression of foreign genes in dicotyledonous and monocotyledonous plants. *Plant Cell Physiol.* **37**, 49–59.

Panstruga, R. and Dodds, P.N. (2009) Terrific protein traffic: the mystery of effector protein delivery by filamentous plant pathogens. *Science*, **324**, 748–750.

Pirovani, C.P., Carvalho, H.A.S., Machado, R.C.R., Gomes, D.S., Alvim, F.C., Pomella, A.W.V., Gramacho, K.P., Cascardo, J.C., Amarante, G., Pereira, G. and Micheli, F. (2008) Protein extraction for proteome analysis from cacao leaves and meristems, organs infected by *Moniliophthora perniciosa*, the causal agent of witches' broom disease. *Electrophoresis*, **29**, 2391–2401.

Plett, J.M., Kemppainen, M., Kale, S.D., Kohler, A., Legue, V., Brun, A., Tyler, B.M., Pardo, A.G. and Martin, F. (2011) A secreted effector protein of *Laccaria bicolor* is required for symbiosis development. *Curr. Biol.* **21**, 1197–1203.

Raiborg, C., Bremnes, B., Mehlum, A., Gillooly, D.J., D'Arrigo, A., Stang, E. and Stenmark, H. (2001) FYVE and coiled-coil domains determine the specific localisation of Hrs to early endosomes. *J. Cell Sci.* **114**, 2255–2263.

Shi, Z., Zhang, Y., Maximova, S.N. and Guiltinan, M.J. (2013) TcNPR3 from *Theobroma cacao* functions as a repressor of the pathogen defense response. *BMC Plant Biol.* **13**, 204.

Torto-Alalibo, T., Collmer, C.W., Gwinn-Giglio, M., Lindeberg, M., Meng, S.W., Chibucos, M.C., Tseng, T.T., Lomax, J., Biehl, B., Ireland, A., Bird, D., Dean, R.A., Glasner, J.D., Perna, N., Setubal, J.C., Collmer, A. and Tyler,

B.M. (2010) Unifying themes in microbial associations with animal and plant hosts described using the gene ontology. *Microbiol. Mol. Biol. Rev.* **74**, 479–503.

Tyler, B.M., Kale, S.D., Wang, Q., Tao, K., Clark, H.R., Drews, K., Antignani, V., Rumore, A., Hayes, T., Plett, J.M., Fudal, I., Gu, B., Chen, Q., Affeldt, K.J., Berthier, E., Fischer, G.J., Dou, D., Shan, W., Keller, N.P., Martin, F., Rouxel, T. and Lawrence, C.B. (2013) Microbe-independent entry of oomycete RxLR effectors and fungal RxLR-like effectors into plant and animal cells is specific and reproducible. *Mol. Plant Microbe Interact.* **26**, 611–616.

Tyler, B.M. and Rouxel, T. (2013) Effectors of fungi and oomycetes: their virulence and avirulence functions and translocation from pathogen to host. In: *Molecular Plant Immunity*, (Sessa, G. ed.), John Wiley & Sons, Inc, pp. 123–167.

Wang, Q., Han, C., Ferreira, A.O., Yu, X., Ye, W., Tripathy, S., Kale, S.D., Gu, B., Sheng, Y., Sui, Y., Wang, X., Zhang, Z., Cheng, B., Dong, S., Shan, W., Zheng, X., Dou, D., Tyler, B.M. and Wang, Y. (2011) Transcriptional programming and functional interactions within the *Phytophthora sojae* RXLR effector repertoire. *Plant Cell*, **23**, 2064–2086.

Whisson, S.C., Boevink, P.C., Moleleki, L., Avrova, A.O., Morales, J.G., Gilroy, E.M., Armstrong, M.R., Grouffaud, S., van West, P., Chapman, S., Hein, I., Toth, I.K., Pritchard, L. and Birch, P.R.J. (2007) A translocation signal for delivery of oomycete effector proteins into host plant cells. *Nature*, **450**, 115–119.

3

Proteomic analysis of stress-related proteins and metabolic pathways in *Picea asperata* somatic embryos during partial desiccation

Danlong Jing[1,†], Jianwei Zhang[1,†], Yan Xia[1], Lisheng Kong[2], Fangqun OuYang[1], Shougong Zhang[1], Hanguo Zhang[3] and Junhui Wang[1,*]

[1]State Key Laboratory of Tree Genetics and Breeding, Key Laboratory of Tree Breeding and Cultivation of State Forestry Administration, Research Institute of Forestry, Chinese Academy of Forestry, Beijing, China
[2]Department of Biology, Centre for Forest Biology, University of Victoria, Victoria, BC, Canada
[3]State Key Laboratory of Tree Genetics and Breeding, Northeast Forestry University, Harbin, China

*Correspondence

email wangjh@caf.ac.cn
[†]Equal contributors to this work.

Summary

Partial desiccation treatment (PDT) stimulates germination and enhances the conversion of conifer somatic embryos. To better understand the mechanisms underlying the responses of somatic embryos to PDT, we used proteomic and physiological analyses to investigate these responses during PDT in *Picea asperata*. Comparative proteomic analysis revealed that, during PDT, stress-related proteins were mainly involved in osmosis, endogenous hormones, antioxidative proteins, molecular chaperones and defence-related proteins. Compared with those in cotyledonary embryos before PDT, these stress-related proteins remained at high levels on days 7 (D7) and 14 (D14) of PDT. The proteins that differentially accumulated in the somatic embryos on D7 were mapped to stress and/or stimuli. They may also be involved in the glyoxylate cycle and the chitin metabolic process. The most significant difference in the differentially accumulated proteins occurred in the metabolic pathways of photosynthesis on D14. Furthermore, in accordance with the changes in stress-related proteins, analyses of changes in water content, abscisic acid, indoleacetic acid and H_2O_2 levels in the embryos indicated that PDT is involved in water-deficit tolerance and affects endogenous hormones. Our results provide insight into the mechanisms responsible for the transition from morphologically mature to physiologically mature somatic embryos during the PDT process in *P. asperata*.

Keywords: *Picea asperata*, somatic embryo, partial desiccation treatment, proteomics, stress-related protein.

Introduction

The complete process of the vegetative propagation technique, somatic embryogenesis, in conifer includes embryonic callus initiation, proliferation, somatic embryo maturation and germination (Stasolla and Yeung, 2003). Among these phases, germination/conversion is regarded as the most important step to obtain plantlets; this determines the success of this technique. Morphologically, mature conifer somatic embryos cannot germinate or convert into viable plantlets unless the embryos undergo partial desiccation treatment (PDT) (Stasolla *et al.*, 2002). This treatment has been used effectively to improve the germination/conversion of somatic embryos in *Picea abies* (Bozhkov and Von Arnold, 1998; Find, 1997; Högberg *et al.*, 1998), *P. rubens* (Harry and Thorpe, 1991), *P. glauca* (Attree *et al.*, 1991; Kong and Yeung, 1995), *P. mariana* (Beardmore and Charest, 1995) and the *Pinus* species, *P. patula* (Jones and van Staden, 2001), *P. thunbergii*, *P. densiflora* and *P. armandii* var. *amamiana* (Maruyama and Hosoi, 2012), as well as *Abies nordmanniana* (Nørgaard, 1997; Salajova and Salaj, 2001; Vooková and Kormuťák, 2006).

Partial desiccation treatment that caused a gradual and limited loss of moisture content in conifer somatic embryos was first reported by Roberts (Roberts *et al.*, 1990). Some physiological and metabolic changes during PDT have been reported in conifer somatic embryos (Dronne *et al.*, 1997; Find, 1997; Kong and Yeung, 1995; Stasolla *et al.*, 2001). The somatic embryos of

P. mariana and *P. glauca* were dried at 97% or 88% relative humidity in the dark to reach a water content of 0.23 g H_2O/g d.wt before high rates of embryo germination/conversion were achieved (Bomal and Tremblay, 2000). The increased germination by PDT has been attributed to a substantial decrease in the endogenous levels of abscisic acid (ABA) (Find, 1997; Liao and Juan, 2015). Somatic embryos of white spruce produce less ethylene during the drying process (Kong and Yeung, 1994), while purine and pyrimidine metabolism is enhanced during PDT (Stasolla *et al.*, 2001). The fact that conifer somatic embryos can germinate and convert into viable plantlets only after PDT indicates that changes in gene expression may occur in these embryos during the process of PDT, resulting in the synthesis of sufficient stress-related proteins and germination-associated proteins. At present, the regulatory mechanisms underlying PDT remain unclear and this hypothesis needs further research.

Proteomics, which provides a global analysis of protein fluctuations, is a more effective technique than transcriptomics for inferring the role of stress-related proteins (Sano *et al.*, 2013). The isobaric tags for relative and absolute quantitation (iTRAQ) system is currently one of the most robust mass spectrometry techniques. This technology compares proteins on the basis of iTRAQ-tagged peptides, allowing identification and accurate quantification of proteins from multiple samples within dynamic ranges of protein abundance (Casado-Vela *et al.*, 2010; Chu *et al.*, 2015; Nogueira *et al.*, 2012). Comparative analysis of proteomics can provide further insight into the mechanisms regulating important processes during conifer somatic embryogenesis. The proteome analysis of Lippert et al. (2005) revealed that differentially accumulated proteins are involved in a variety of cellular processes in early somatic embryogenesis in *P. glauca*. During the somatic embryo development of *Larix principis-ruprechtii*, functional analysis of proteomics showed that the differentially accumulated proteins involved in primary metabolism, phosphorylation and oxidation reduction are up-regulated (Zhao et al., 2015). Although changes in the transcript levels of many genes have been reported using DNA microarrays during the maturation phase of somatic embryos in white spruce (Stasolla et al., 2003), global protein fluctuations in conifer somatic embryos during PDT have not yet been investigated. To identify stress-related proteins, germination-associated proteins, and metabolic pathways in conifer somatic embryos during PDT, proteomic analysis combined with measures of the physiological changes in the embryos is thus required.

Picea asperata Mast is a widely distributed native spruce in China. It has attracted increasing attention with regard to afforestation in barren regions due to its outstanding wood properties and adaptability (Fu et al., 1999; Luo et al., 2006). Our well-established somatic embryogenesis system in *P. asperata* provides plant materials of identical genotypes and highly synchronized embryos for accurate proteomics comparison during PDT. In this study, using embryogenic cell line 1931, we investigated the differences and changes in *P. asperata* somatic embryos at various stages (i.e. cotyledonary embryos before, during and after PDT) using iTRAQ-based proteomic and physiological analyses. With this most extensive proteomics analysis for *P. asperata* somatic embryos, we reveal important stress-related proteins and metabolic pathways that are associated with PDT in conifer somatic embryos.

Results

Effects of PDT on the morphology of *P. asperata* somatic embryos and their germination

To determine the effects of PDT on these embryos, the morphological characteristics of developing the embryos were analysed on days 0 (D0), 7 (D7), 14 (D14) of PDT and on germination day 1 after 7 days of PDT (G1). The mature somatic embryos were yellowish after separation from differential medium. The cotyledons were completely open and arranged circularly on the shoot apical pole (Figure 1a, e). During PDT, the embryos shrunk rapidly on the first day, and then the hypocotyls constantly thickened and elongated. On D7, the radicles became red, while the cotyledons and hypocotyls turned green (Figure 1b, f). After desiccation for 14 days, the cotyledons were dark green and closely attached to each other, while the hypocotyls elongated significantly and the radicles became dark red (Figure 1c, g). On G1, the hypocotyls elongated significantly and their sizes increased longitudinally. The radicles elongated markedly and accompanied by a colour change to white (Figure 1d, h).

A germination standard (Liao and Juan, 2015) was used to assess the germination performance after the embryos were placed on germination medium. From D0 to D14, the germination rate increased significantly corresponding to the duration of partial desiccation (Figure 2a). Compared with embryos without PDT, the germination rate increased significantly from 4.67% to 56.83% after 14 days of PDT. Without PDT, most of the hypocotyls became hyperhydric, and the radicles turned dark brown during germination (Figure 2b). After 7 days of PDT, embryo germination was stimulated with little hyperhydricity of the germinants and high germination rates. The plantlets converted from embryos after PDT showed hypocotyl extension and needle development at the shoot apical end (Figure 2c).

Figure 1 Effects of partial desiccation treatment (PDT) on the morphology of *P. asperata* somatic embryos. (a) and (e) Embryos before PDT were light yellowish on day 0 (D0) without colour change. (b) and (f) Embryos under PDT for 7 days (D7) had green cotyledons, green hypocotyls and red radicles. (c) and (g) Embryos under PDT for 14 days (D14) had dark green cotyledons, dark green hypocotyls and red radicles. (d) and (h) Embryos in germination medium on day 1 after PDT for 7 days (G1) showed growth of cotyledons and elongation of radicles and hypocotyls. All scale bar, 500 µm.

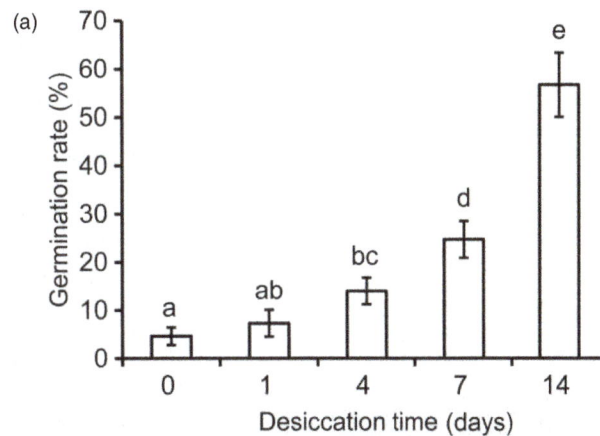

Figure 2 Effects of partial desiccation treatment (PDT) on the germination of *P. asperata* somatic embryos. (a) The germination rates of embryos after PDT for different times. Mean ± SD, *n* = 5. Significant differences are indicated by different letters (*P* < 0.05). (b) Photograph of embryos germinating without PDT. Most of the germinants were abnormal with clear hyperhydricity and poor radicle development. (c) Photograph of embryos germinating after 14 days of PDT. The germinants were strong with well-developed roots and shoots.

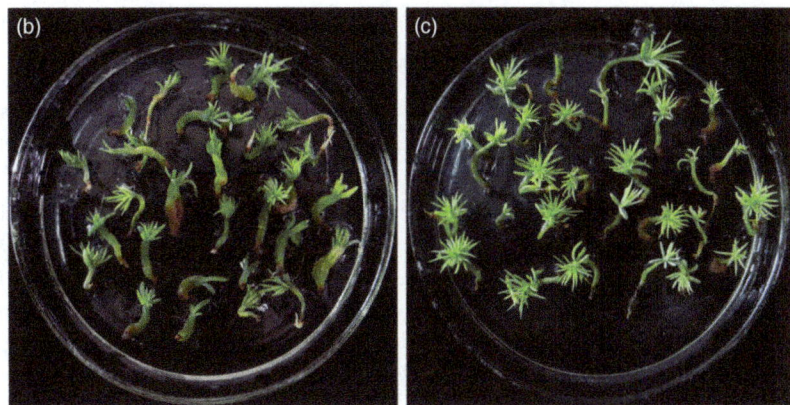

Identification of stress-related proteins and metabolic pathways in *P. asperata* somatic embryos

To determine the protein fluctuations during PDT, the total proteins in embryos on D0, D7, D14 and G1 were extracted and their profiles were explored using the iTRAQ technique. A total of 347 380 spectra were generated; 34 301 (9.87%) matched known peptides according to Mascot software; and 28 651 (83.53%) matched unique peptides. Ultimately, 10 216 peptides, 9297 (91.00%) unique peptides and 2773 proteins were identified. Meanwhile, the distributions of the lengths and numbers of peptides, mass and sequence coverage of the proteins, and the repeatability of replicates were assessed (Figures S1 and S2).

The annotated proteins were classified into three groups (cellular component, molecular function and biological process) on the basis of Gene Ontology (GO) enrichment analysis. The main cellular components were classified into cell (23.31%), cell part (23.31%), organelle (19.18%) and others (Figure S3). The molecular functions of proteins were mainly focused on catalytic activity (43.89%) and binding (41.25%) (Figure S4). The biological processes were mainly metabolic process (19.97%), cellular process (19.14%), response to stimulus (10.21%) and others (Figure S5).

Proteomic analysis of D7 *vs* D0, D14 *vs* D0, D14 *vs* D7 and G1 *vs* D7 was used to detect stress-related proteins in the embryos under PDT. A total of 636 proteins showed significant difference based on a 1.5-fold change at *P* < 0.05 (Figure S6). Comparisons of D7 *vs* D0 and D14 *vs* D0 showed a total of 337 differentially accumulated proteins, of which 66 (19.58%) were associated with stress tolerance (Figure 3a and Table S1). These stress-related proteins were mainly involved in osmosis, endogenous hormones, antioxidative proteins, molecular chaperones,

defence-related proteins, pyrimidine metabolism and embryogenesis-specific protein.

According to the physiological functions of stress-related differentially accumulated proteins, auxin-repressed 12.5 kDa protein, allene oxide synthase and abscisic stress-ripening protein were classified as endogenous hormones in the embryos under PDT. Three types of aquaporins (TIP2-1, TIP1-1 and PIP type) were associated with water transport by osmosis to prevent desiccation of the embryos during PDT. We identified 5 classes of defence-related proteins: pathogenesis-related protein, wound-induced protein, basic endochitinase, chitinase and osmotin. During PDT, 35 differentially accumulated proteins were enriched in antioxidative proteins, associated with the response to reactive oxygen species to diminish cytotoxic damage such as DNA damage, protein modification, lipid peroxidation and de-esterification (Table S1). Meanwhile, phosphoenolpyruvate carboxykinase, aldose 1-epimerase, NAD-dependent malic enzyme 62 kDa isoform, sucrose synthase 2 and xyloglucan endotransglucosylase/hydrolase protein A were identified as being involved in carbohydrate metabolism. The small heat-shock protein, class I heat-shock protein, splicing factor, heat-shock 70 kDa protein 10 and 18.1 kDa class I heat-shock protein were found to be molecular chaperones and involved in desiccation tolerance of the embryos. In addition, pyrimidine metabolism-related protein (deoxyuridine 5′-triphosphate nucleotidohydrolase) and embryogenesis-specific protein (provicilin, stem-specific protein, tubulin and 12S seed storage protein) were also identified.

Comparative proteomics of D14 *vs* D7 and G1 *vs* D7 were further analysed to evaluate the relevance of germination after PDT (G1) and a longer period of PDT (D14). From the sum of the numbers of differentially accumulated protein in both cases, we

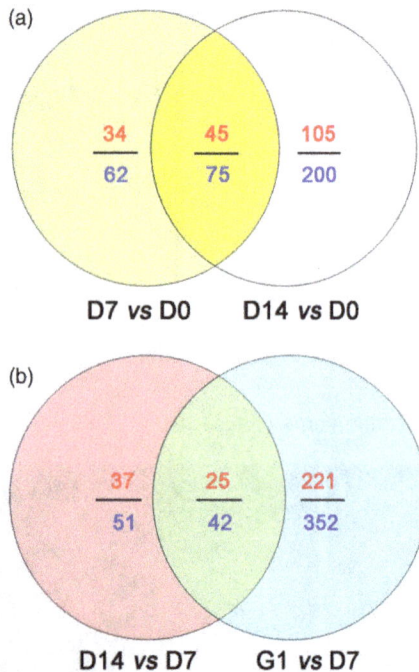

Figure 3 Venn diagram of differentially accumulated proteins in D7 *vs* D0 and D14 *vs* D0, D14 *vs* D7 and G1 *vs* D7. (a) Differentially accumulated proteins in D7 *vs* D0 and D14 *vs* D0. (b) Differentially accumulated proteins in D14 *vs* D7 and G1 *vs* D7. The numbers of differentially accumulated proteins are represented in blue. The numbers of up-regulated differentially accumulated proteins are indicated in red.

detected 445 such proteins, of which 52 (11.69%) were involved in photosynthesis (Figure 3b and Table S2). Among these photosynthesis-related proteins, 39 were identified in the D14 *vs* D7 comparison, while 28 were identified in the G1 *vs* D7 comparison. Meanwhile, the 15 photosynthesis-related proteins that were enriched in both D14 *vs* D7 and G1 *vs* D7 were mainly associated with light harvesting, chlorophyll synthesis and photosynthetic protection (Table S2).

To better visualize the differences in the metabolic pathways of the embryos during PDT, these differentially accumulated proteins were classified on the basis of GO enrichment analysis (Figure 4). The main biological functional categories for D7 *vs* D0 were response to stress, glyoxylate cycle, chitin metabolic process and response to stimulus. The biological processes for D14 *vs* D0 were mainly classified into photosynthesis categories, including photosynthetic electron transport chain, light reaction, chlorophyll biosynthetic process, tetrapyrrole biosynthetic process, porphyrin-containing compound biosynthetic process and photosynthetic electron transport in photosystem I (Figure 4a). The main biological functional categories for D14 *vs* D7 were photosynthesis, including light reaction, photosynthetic electron transport chain, electron transport chain, photosynthetic electron transport in photosystem I and photosystem II assembly. The biological processes for G1 *vs* D7 were mainly single-organism process, cell communication, DNA packaging, chromatin assembly, small-molecule metabolic process and translation (Figure 4b).

Changes in stress-related proteins in *P. asperata* somatic embryos during PDT

Fold changes in the protein profile of embryos during PDT were further analysed to estimate the stress-related proteins associated

with water-deficit tolerance. The fold changes of stress-related proteins accumulated in both D7 *vs* D0 and D14 *vs* D0 are shown in Table S3. Among these accumulated proteins, auxin-repressed 12.5 kDa protein and abscisic stress-ripening protein 2 were significantly up-regulated on D7 and D14. Compared with D0, aquaporin TIP2-1 was increased by 1.75-fold on D7 and 3.56-fold on D14. Four defence-related proteins, including wound-induced protein, osmotin, chitinase 4 and basic endochitinase C, were up-regulated 4.10-, 1.53-, 2.60- and 2.32-fold on D7, and 8.20-, 2.10-, 5.30- and 3.29-fold on D14 relative to D0, respectively. Twenty-eight accumulated proteins associated with antioxidation were significantly increased to reduce the effects of deleterious reactive oxygen species during PDT. For example, catalase and catalase isozyme 2 were increased 1.61- and 1.56-fold on D7, and 1.92- and 2.06-fold on D14, respectively. The change in catalase isozyme protein corresponded well with the catalase activity assays (Figure S7). Furthermore, we analysed changes in differentially accumulated proteins for D7 *vs* D0, D14 *vs* D0, D14 *vs* D7 and G1 *vs* D7 (Figure 5). Compared with D0, the stress-related proteins were up-regulated on D7 and D14, while the photosynthesis-related proteins were significantly up-regulated on D14.

Photosynthesis metabolism proteins and protein–protein interaction networks of differentially accumulated proteins

A total of 47 differentially accumulated proteins were significantly enriched in the photosynthesis pathway and were directly associated with photosynthesis (Table S2). Among these proteins, 36 were significantly up-regulated from D7 to D14, while 29 were up-regulated from D7 to G1. Notably, photosystem II, photosynthetic electron transport and F-type ATPase proteins were involved in photosynthetic metabolism under PDT (Figure S8). These patterns of differentially accumulated proteins showed that the embryos gradually carried out photosynthesis from D7.

A total of 1047 proteins were identified using *Arabidopsis* interaction data to evaluate the protein–protein interaction networks of the *P. asperata* somatic embryos under PDT. The protein interactions corresponded well to those from *Arabidopsis*. The 562 pairs of protein–protein interactions were used to build 498 vertices in the network (Figures S9–S12). Collectively, 13 interacting proteins were significantly differentially accumulated in the different PDT groups (Table 1). The interacting proteins that were up-regulated during PDT and the germination process were mainly associated with photosynthesis, glyoxylate and dicarboxylate metabolism, xenobiotics metabolism, protein processing in the endoplasmic reticulum and carbon fixation in photosynthetic organisms. However, proteins of the ribosome pathway were up-regulated during the PDT process, but down-regulated during the germination process, while proteins of the proteasome and protein-processing pathways were down-regulated during the PDT process, but up-regulated during the germination process.

Effects of PDT on water, ABA, indoleacetic acid (IAA) and H_2O_2 in *P. asperata* somatic embryos

To obtain an accurate understanding of the stress-related proteins and physiological changes in these embryos during PDT, we analysed their water, ABA, IAA and H_2O_2 content at D0, D1, D4, D7, D14 and G1. The water content declined ($P < 0.05$) drastically on the first day and then increased gradually (Figure 6a). On D1, the embryos were relatively dry as a result of

(a)

(b)

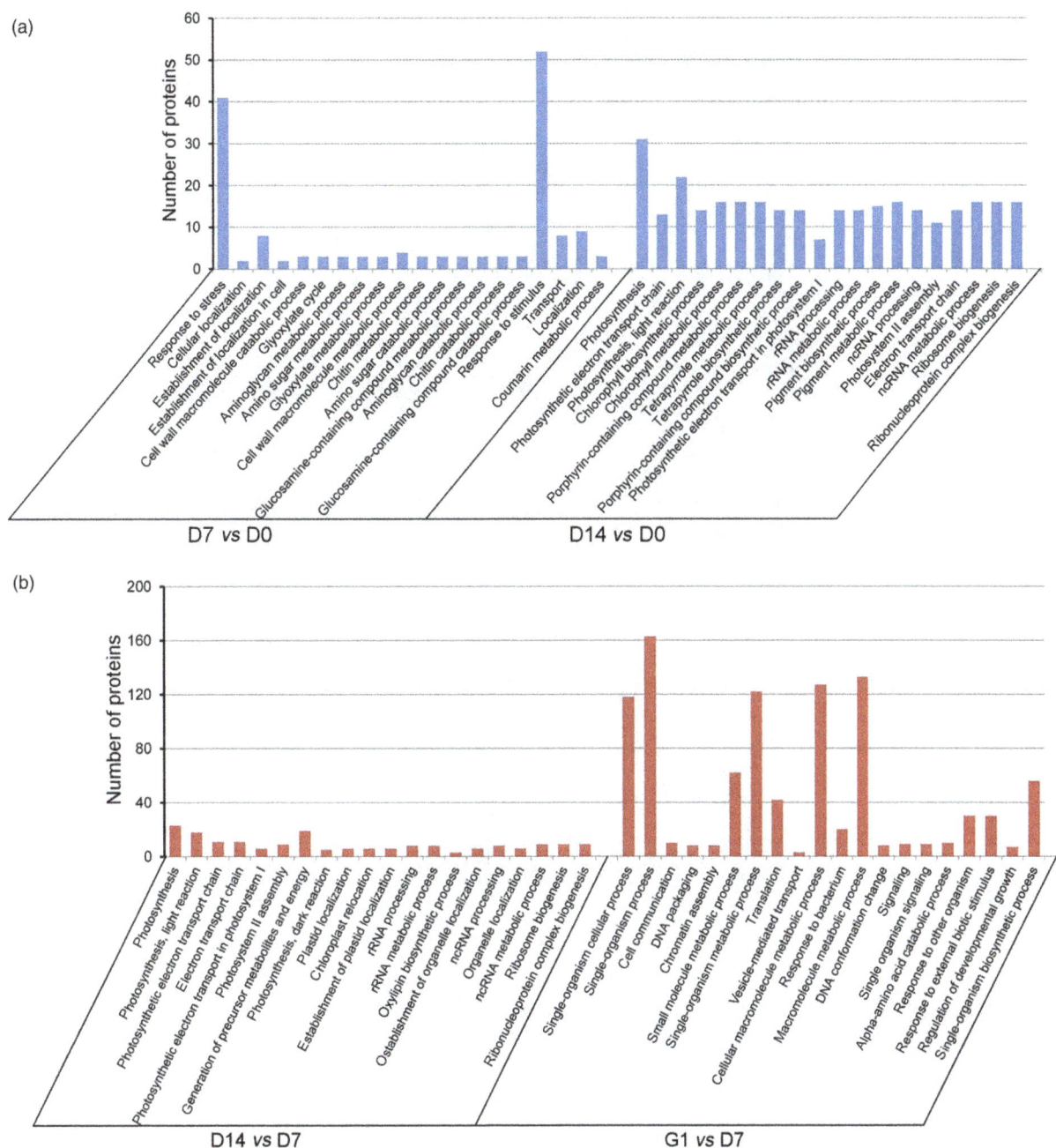

Figure 4 Gene Ontology (GO) enrichment analysis of differentially accumulated proteins. (a) Biological process classification for D7 *vs* D0 and D14 *vs* D0. (b) Biological process categories for D14 *vs* D7 and G1 *vs* D7.

rapid water loss to only 76.28% of the original fresh weight. However, the fresh weight recovered to 95.58% on D14. The embryos absorbed water rapidly from the culture medium when the embryos, after 7 days of PDT, were placed on germination medium for 1 day (i.e. G1), resulting in a fresh weight of 1.57-fold the initial value.

The dry weight of the embryos changed drastically with different durations of desiccation (Figure 6b). From D0 to D7, the dry weight declined due to transfer of the embryos from culture medium to an empty plate with filter paper. After D7, their dry weight began to increase until D14. Dry matter accumulated strongly on G1.

During PDT, the concentrations of ABA in the embryos increased slightly on D1, and decreased thereafter (Figure 7a),

reaching the lowest level on D14. The IAA level increased during PDT (Figure 7b). The H_2O_2 levels increased significantly during PDT, but then decreased quickly on G1 (Figure 7c).

Discussion

In this study, a large number of *P. asperata* somatic embryos of the same genetic background were used to investigate the effects of PDT on the proteome. The morphological, iTRAQ-based proteomic and physiological analyses highlighted the importance of PDT in stimulating embryo germination. This enabled us to identify the regulatory proteins and to clarify the molecular mechanisms underlying the PDT process. The results suggest that PDT acts by increasing the stress-related

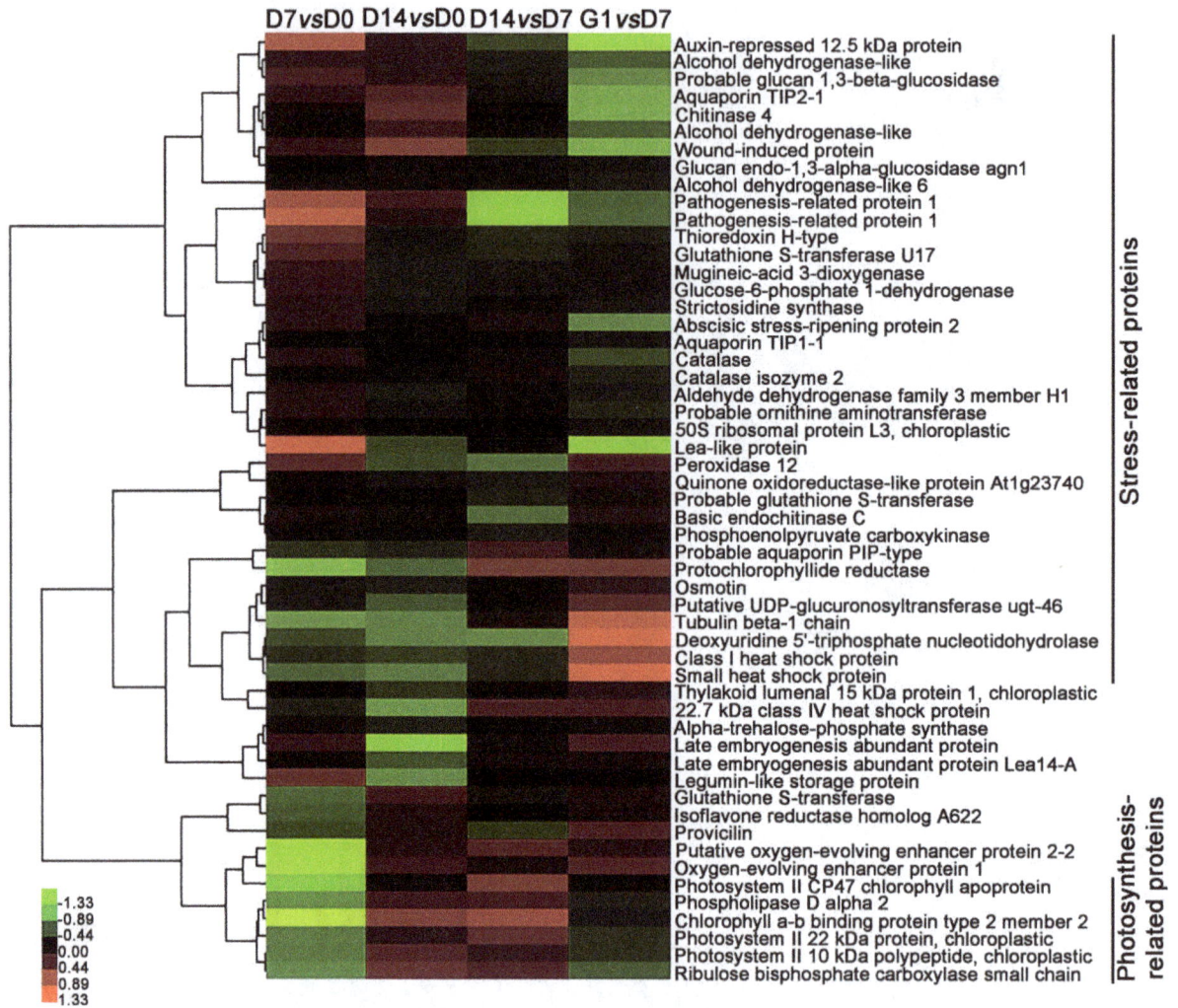

Figure 5 Heat map of normalized fold changes in accumulated proteins in D7 *vs* D0, D14 *vs* D0, D14 *vs* D7 and G1 *vs* D7.

Table 1 Common expression of protein–protein interaction among the different treatment

Accession ID	KEGG_Pathway	D7 *vs* D0	D14 *vs* D0	D14 *vs* D7	G1 *vs* D7
Pasi_224284112	Photosynthesis	Up	Up	Up	Up
Pasi_224285151	Photosynthesis	Up	Up	Up	
Pasi_116780862	Glyoxylate and dicarboxylate metabolism	Up	Up	Up	Up
Pasi_224286155	Proteasome	Down		Down	Up
Pasi_116793156	Protein processing in endoplasmic reticulum	Down		Down	Up
Pasi_224286155	Proteasome	Down		Down	Up
Pasi_116792087	Metabolism of xenobiotics by cytochrome P450	Up		Up	Up
Pasi_224286188	Protein processing in endoplasmic reticulum		Up	Up	Up
Pasi_148907259	Protein processing in endoplasmic reticulum		Up	Up	Up
Pasi_116786455	Carbon fixation in photosynthetic organisms			Up	Up
Pasi_116781903	RNA transport			Down	Up
Pasi_224285264	Ribosome	Up			Down
Pasi_294464103	RNA polymerase	Down		Down	

proteins in *P. asperata* somatic embryos that are deficient in water and nutrients. These proteins promote the transformation of these embryos from morphological maturity to physiological maturity and further induce photosynthesis in low light (Figure 8).

Proteins associated with PDT in *P. asperata* somatic embryos

This work confirms that somatic embryos of *P. asperata*, similar to other conifers (Hazubska-Przybył *et al.*, 2015; Percy *et al.*, 2001),

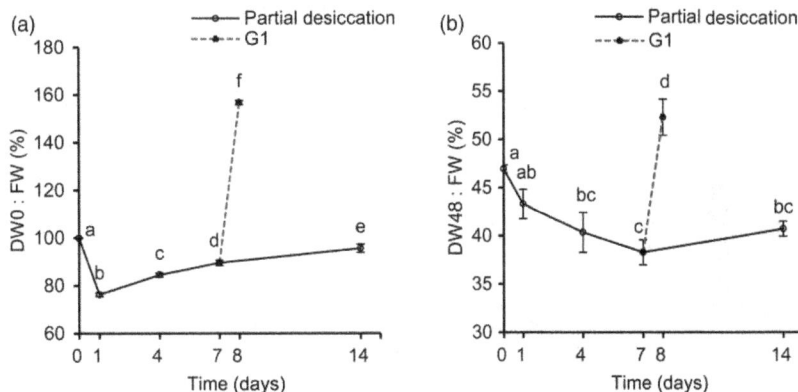

Figure 6 Changes of the water content in *P. asperata* somatic embryos during partial desiccation treatment (PDT). (a) Effect of the desiccation period on fresh weight. FW, fresh weight before PDT; DW0, fresh weight during PDT. (b) Effect of the desiccation period on dry weight. DW48, dry weight during PDT. Mean ± SD, $n = 4$. Significant differences are indicated by different letters ($P < 0.05$).

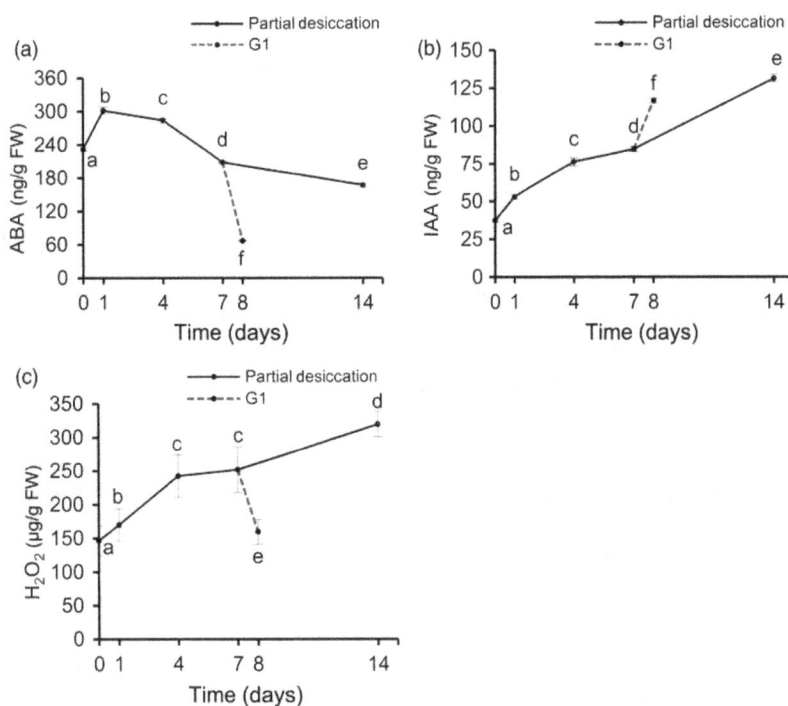

Figure 7 Changes in abscisic acid (ABA), indoleacetic acid (IAA) and H_2O_2 contents in *P. asperata* somatic embryos during PDT. (a) ABA content. (b) IAA content. (c) H_2O_2 content. Mean ± SD, $n = 4$. Significant differences are indicated by different letters ($P < 0.05$).

suffer from water-deficit stress under PDT. Comparative proteomics analysis of these embryos during PDT revealed that significantly differentially accumulated proteins are mainly involved in the biological regulation of water, plant hormones and the stress response. Furthermore, some important proteins, such as aquaporins, auxin-repressed proteins, catalase and heat-shock proteins, might play important roles during PDT, based on their physiological functions. Here, we found that the aquaporins (TIP2-1, TIP1-1 and PIP type) were significantly up-regulated during PDT; these proteins are important for water content and light induction. Similarly, previous studies have shown that aquaporin forms a 'tunnel' in the cell membrane to regulate the water transport under stress conditions (Boursiac et al., 2005), and can be induced by light (Loqué et al., 2005). As well as lowering the activation energy of water transport, aquaporin also enhances the permeability of the plasma membrane (Leitão et al., 2012).

Auxin-repressed proteins are involved in the response to salicylic acid signalling and are induced by IAA (Shi et al.,

2013). In our work, auxin-repressed protein was significantly up-regulated under PDT but down-regulated on G1, suggesting that the embryos had entered a stress state that was quickly released in the germination stage. A previous report also indicated that auxin-repressed protein is induced by abiotic stresses, and involved in growth arrest, possibly by inhibiting cell elongation (Lee et al., 2013). As a lytic enzyme of H_2O_2, catalase can purge active oxygen to reduce cell damage and its activity is markedly enhanced in the dried seed (Bailly, 2004; Berjak, 2006). In the somatic embryos of *P. asperata* during PDT, catalase remained at high levels on D7 and D14, coinciding with H_2O_2 content.

In our study, heat-shock proteins (Hsps) were up-regulated during the desiccation period; this might be related to rising H_2O_2 levels and embryonic development. However, Hsp22.7 belonged to the Hsp20 family and was down-regulated under PDT (Figure 5). Taken together, the four kinds of Hsps serving as molecular chaperones were up-regulated on G1, suggesting that they might help the protein refolding during germination.

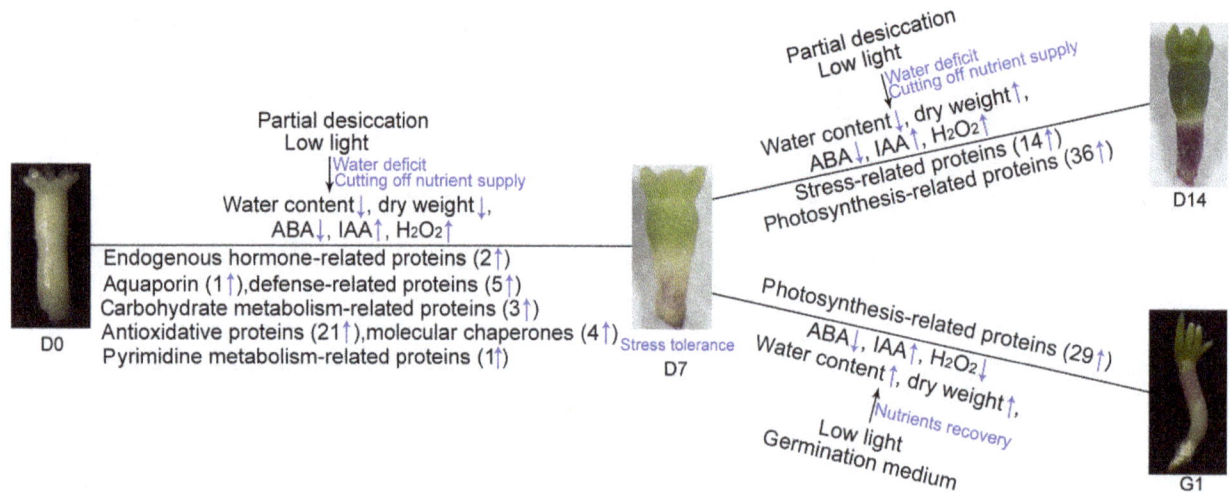

Figure 8 Schematic of changes in the regulatory proteins in response to partial desiccation treatment in *P. asperata* somatic embryos.

Previously, Hsps and small-molecule Hsps, which usually act as molecular chaperones, had been also reported to be involved in the stress response (Sun *et al.*, 2002) and the development of seed ripening (Wehmeyer *et al.*, 1996), as well as the repair and refolding of damaged proteins, thus protecting cells from damage by the stress and assisting seed or embryo maturation.

Proteins associated with embryo development and photosynthesis during PDT and germination

During PDT, chitinase was significantly up-regulated; this might be associated with physiological maturity of the somatic embryos in *P. asperata*. By contrast, in somatic embryos of *P. abies*, chitinase (class IV) has been reported to promote their transformation from embryonic cell mass (Wiweger *et al.*, 2003). It was also reported that chitinase was increased at the maturation stage of zygotic embryos in *Araucaria angustifolia* (Dos Santos *et al.*, 2006). Notably, glutamine synthase, which is involved in the development of somatic embryos, was up-regulated during the desiccation and germination in *P. asperata* somatic embryos. As reported for the late seed development of the Brazilian Pine, glutamine synthase accumulated in the early cotyledonary stage; it could facilitate the transformation of glutamate to glutamic acid and be acted as protein markers of embryonic maturity (Balbuena *et al.*, 2009).

The level of β-tubulin, a structural protein, is at its lowest in dry mature seeds. This protein has the highest expression during the dormancy release of Norway maple embryos and can be used as the indicator for dormancy breaking (Pawłowski *et al.*, 2004). In *P. asperata* somatic embryos, β-tubulin was significantly up-regulated during PDT and germination, indicating that these embryos might be switched to germination stage after 7 days of PDT. The same trend has been reported in the process of seed germination in *Arabidopsis* (Chibani *et al.*, 2006).

A major effect of PDT on the *P. asperata* somatic embryos was up-regulation of photosynthesis-related proteins. This could explain the increase in dry weight of the embryos after 7 days of PDT, demonstrated that organic matter was accumulating. Meanwhile, the photosynthesis-related proteins may be up-regulated to match the requirements of nutrient-deficit stress,

as well as indicating that the embryos have entered the germination stage.

Physiological responses of *P. asperata* somatic embryos to PDT

Many essential physiological processes for conifer somatic embryo development are affected by stress, exhibiting various defence mechanisms (Robinson *et al.*, 2009; dos Santos *et al.*, 2016; Zhao *et al.*, 2015). In accordance with the changes in stress-related proteins during the PDT process, the water content in the embryos decreased drastically due to water absorbance by the dry filter paper. Soon afterwards, as the filter paper constantly absorbed moisture from the air, the water content of the *P. asperata* somatic embryos increased. These sharp changes in water content were associated with stress-related proteins such as aquaporins, suggesting that these embryos suffered water-deficit stress under PDT. It is known that the cellular messenger H_2O_2 causes stress and induces catalase during somatic embryogenesis in *Larix leptolepis* (Zhang *et al.*, 2010). The rapidly increased H_2O_2 content in *P. asperata* somatic embryos during PDT indicates that these embryos were under oxidative stress and that this was associated with the changes in antioxidative proteins. The stress was released quickly during germination, because the H_2O_2 content of the embryos decreased significantly, with just 1 day of germination treatment (G1).

As an important osmotic regulator, ABA content changes when plant embryos are subjected to stress (Xiong and Zhu, 2003). At the beginning of PDT (D1), *P. asperata* somatic embryos suffered a strong desiccation stress, and the ABA content increased significantly. Then, the ABA content tended to decrease after D1, which might be due to rapid adaptation to the PDT condition with increasing humidity. However, our ABA results differed slightly from those of Find (1997), Liao and Juan (2015), probably because we carried out PDT under low light conditions. In somatic embryos of Norway spruce during PDT, the increasing germination frequency has been attributed to a substantial decrease in the ABA content, and one effect of PDT may be the breakdown of endogenous ABA (Find, 1997). On day 7 of PDT, the ABA content of *Picea morrisonicola* somatic embryos is substantially decreased (Liao and Juan, 2015). Dronne *et al.* (1997) also pointed out that PDT decreased the ABA content in hybrid larch somatic embryos. However, Kong (1994) found that PDT does not cause an increase in IAA concentration in white spruce somatic embryos. The increase in IAA concentration in our study may be due to the effect of light on PDT. Higher IAA concentrations could benefit embryo germination in addition to lower ABA concentrations after PDT.

Experimental procedures

Plant materials

Highly synchronized somatic embryos of *P. asperata* were obtained from embryogenic cell line 1931, which was initiated from an immature zygotic embryo of an elite mother tree in the National Spruce Germplasm Bank of China, in Gansu province. This cell line had been cultured in liquid medium for 1 year. This medium was consisted of half-strength salts of Litvay medium (LM) (Litvay *et al.*, 1985), 1% sucrose (Beijing, China), 0.1% casein enzymatic hydrolysate (Sigma, St. Louis, MO), 10 µM 2,4-dichlorophenoxyacetic acid (Sigma) and 5 µM 6-benzylaminopurine (Sigma), supplemented with 3.42 mM filter-sterilized L-glutamine (Sigma) at pH 5.8. Erlenmeyer flasks (250 mL) containing 80 mL liquid medium were used to culture the embryogenic tissue. The flasks of tissue were placed on a gyratory shaker (110 rpm) and cultured in darkness at 24 ± 1 °C. Suspension cultures were transferred to fresh medium every 12 days. Before being cultured on maturation medium for embryonic differentiation, embryogenic tissue was transferred to a piece of sterile filter paper (Whatman, Kent, UK) and the liquid medium was removed using a vacuum pump. Then, the filter paper with embryogenic tissue was transferred onto maturation medium. This medium contained half-strength salts of LM, 61 µM filter-sterilized (±) *cis, trans*-ABA (Gibco-BRL, Gaithersburg, MD), 5% polyethylene glycol 4000 (PEG4000, Merck, Darmstadt, Germany), 3% sucrose, 0.1% activated charcoal (Sigma) and 0.4% gellan gum (Sigma), supplemented with 3.42 mM filter-sterilized L-glutamine, 0.1% casein hydrolysate (Sigma), at pH 5.8. The cultures were kept in the dark at 24 ± 1 °C for 7 weeks. In total, more than 30 000 highly synchronized somatic embryos were obtained for later use.

Partial desiccation of *P. asperata* somatic embryos

For PDT, each batch of ~30 somatic embryos was transferred onto two layers of dry sterile filter paper (Whatman) in a small plastic Petri dish (35 × 12 mm) without a lid. Three of the dishes containing embryos were placed in a large Petri dish (90 × 15 mm) with 10 mL sterile deionized water. The large Petri dish was covered by a lid and sealed with parafilm, then incubated in a cultivation room at 24 ± 1 °C. Unlike PDT of conifer somatic embryos in the dark, we placed *P. asperata* somatic embryos under a 16-h photoperiod with a low light intensity of ~15 µmol/m²/s (LED fluorescent tubes) for 0 (D0), 1 (D1), 4 (D4), 7 (D7) and 14 days (D14). At the same time, some of the somatic embryos were partially desiccated for 7 days and then germinated at a low light intensity of ~15 µmol/m²/s (LED fluorescent tubes) for 1 day (G1).

After various treatments, embryos were collected, immediately frozen in liquid nitrogen and stored at −80 °C for water content, ABA, IAA and H_2O_2 analysis. Four independent biological replicates were acquired. Meanwhile, somatic embryos from D0, D7, D14 and G1 were assayed for iTRAQ analysis with two independent biological repeats for each treatment.

Plantlet regeneration capacity of *P. asperata* somatic embryos

For the germination stage, embryos after desiccation for 0, 1, 4, 7 and 14 days were transferred to germination medium. Based on our previous studies (unpublished data), the optimal germination medium contained ~30 mL of 1/4 LM with 0.6% gellan gum, 2% sucrose, 0.1% activated charcoal, 0.1% casein hydrolysate and 3.42 mM filter-sterilized L-glutamine. Thirty embryos were cultured in one Petri dish. Five independent biological replicates were used. In the first week, embryos were maintained in germination medium under a 16-h photoperiod at a low light intensity (~15 µmol/m²/s (LED fluorescent tubes)). Then, the light intensity was increased to 50 µmol/m²/s while the photoperiod was unchanged. The culture temperature was kept at 24 ± 1 °C during germination. Germinated embryos were counted after 3 weeks.

Protein extraction and digestion

Proteomic sequencing was performed by the Beijing Genome Institute (BGI). Total proteins were extracted according to the method of Qiao *et al.* (2012) with some modifications. Somatic embryos (200 mg) were ground to a fine powder in liquid nitrogen and suspended in 500 µL of lysis buffer (40 mM Tris–HCl, 2 M thiourea, 7 M urea, 4% CHAPS, pH 8.5) containing 1 mM phenylmethylsulfonyl fluoride (PMSF), 2 mM EDTA and 10 mM dithiothreitol (DTT), with supersonic extraction for 15 min. The homogenate was centrifuged at 30 000 *g* for 20 min at 4 °C. Then, the supernatant was mixed with a fivefold volume of chilled acetone containing 10% (w/v) trichloroacetic acid (TCA) and incubated at −20 °C overnight. The precipitate was vacuum-dried and again dissolved in 300 µL of lysis buffer (20 mM Tris–HCl, 2 M thiourea, 7 M urea, 4% Nonidet P-40, pH 8.5) again. After vortex mixing for 2 min, 10 mM DTT (final concentration) was added to the supernatant, then incubated at 56 °C for 1 h and alkylated with 55 mM iodoacetamide (IAM) at 45 °C in a darkroom for 1 h. The supernatant was mixed well with a fivefold volume of chilled acetone and incubated at −20 °C overnight. After centrifugation at 30 000 *g* at 4 °C for 20 min, the supernatant was discarded. The protein pellets were vacuum-dried and dissolved in 500 µL of 0.5 M triethylammonium bicarbonate (TEAB) and centrifuged at 30 000 *g* for 15 min at 4 °C. The protein content was assayed using the Bradford method (Bradford, 1976).

iTRAQ labelling and strong cation exchange (SCX) fractionation

Protein samples containing 100 µg of protein were digested using Trypsin Gold (Promega, Madison, WI) at a protein: trypsin ratio of 20 : 1 at 37 °C for 16 h to obtain peptides. After digestion, the peptides were dried by vacuum centrifugation and resuspended in 0.5 M TEAB. For the sample labelling, 8-plex iTRAQ reagents (Applied Biosystems, Foster City, CA) were used according to the manufacturer's protocol (Zieske, 2006). For each treatment, there were two biological replicates. Samples were labelled with the iTRAQ tags as follows: Sample D1 (113 and 114 tag), Sample D7 (115 and 116 tag), Sample D14 (117 and 118 tag) and Sample G1 (119 and 121 tag).

Strong cation exchange chromatography was performed on an LC-20AB HPLC Pump system (Shimadzu, Kyoto, Japan). The iTRAQ-labelled peptide mixtures were reconstituted in 4 mL buffer A (25% v/v acetonitrile, 25 mM NaH_2PO_4, pH 2.7) and loaded onto a 4.6 × 250 mm Ultremex SCX column containing 5-µm particles (Phenomenex, Torrance, CA). The peptides were eluted at 1 mL/min with elution buffer B (25% v/v acetonitrile, 25 mm NaH$_2$PO$_4$, 1 m KCl, pH 2.7). Elution was monitored by measuring the absorbance at 214 nm, and fractions were collected every 1 min. The eluted peptides were pooled into 20 fractions, desalted on a Strata X C18 column (Phenomenex) and vacuum-dried before LC-ESI-MS/MS analysis.

LC-ESI-MS/MS analysis

Each fraction was resolved in solvent A (5% acetonitrile, 0.1% formic acid) and centrifuged at 20 000 *g* for 10 min, and the average final concentration of peptide was ~0.5 µg/µL. The supernatant was separated using an LC-20AD Nano-HPLC (Shimadzu) with an autosampler onto a 2-cm C18 trap column. Then, the peptides were eluted onto a 10-cm analytical C18 column (inner diameter 75 µm) packed in-house. The samples were loaded at 8 µL/min for 4 min, and then the 35-min linear gradient was run at 300 nL/min starting from 2% to 35% solvent B (95% acetonitrile, 0.1% formic acid), followed by ramping up to 60% solvent B over 5 min, up to 80% in 2 min and maintained for 4 min, then finally restored to 5% in 1 min.

Data were acquired on a TripleTOF 5600 System (AB SCIEX, Concord, Ontario, Canada), using an ion spray voltage of 2.5 kV, nitrogen gas at 30 psi, nebulizer gas at 15 psi and an interface heater temperature of 150 °C. The MS was operated in high-resolution mode (>30 000 FWHM) for TOF MS scans. For information dependent data acquisition (IDA), survey scans were acquired in 250 ms, and as many as 30 product ion scans were collected if they exceeded a threshold of 120 counts/s and had a 2 + to 5 + charge state. The total cycle time was fixed at 3.3 s, and the Q2 transmission window was 100 Da for 100%. Four time bins were summed for each scan at a pulse frequency of 11 kHz, by monitoring of 40-GHz multichannel TDC detector using four-anode channel detection. A sweeping collision energy setting of 35 ± 5 eV, coupled with the iTRAQ adjust rolling collision energy, was applied to all precursor ions for collision-induced dissociation. Dynamic exclusion was set at 1/2 of the peak width (15 s), and then the precursor was refreshed off the exclusion list.

iTRAQ protein identification and quantification

Raw data files were converted into MGF files using Proteome Discoverer 1.2 (PD 1.2, Thermo, 5600 msconverter) and the files were searched. Proteins were identified using the Mascot search engine (Matrix Science, London, UK; version 2.3.02) against the TreeGenes nonredundant sequence database (http://dendrome.ucdavis.edu/treegenes/protein/prot_summary.php) containing 18 253 sequences. The search parameters were as follows: a mass tolerance of 0.1 Da (ppm) was permitted for intact peptide masses, and 0.05 Da for fragmented ions, with an allowance for one missed cleavage in the trypsin digests; Gln->pyro-Glu (N-term Q), oxidation (M) and iTRAQ 8-plex (Y) were the potential variable modifications, and carbamidomethyl (C), iTRAQ 8-plex (N-term) and iTRAQ 8-plex (K) were the fixed modifications. The charge states of the peptides were set to +2 and +3. Specifically, an automatic decoy database search was performed in Mascot by choosing the decoy checkbox in which a random sequence of the database was generated and tested for raw spectra as well as the real database. To reduce the probability of false peptide identification, only peptides with significance scores ≥20 at a 99% confidence interval by a Mascot probability analysis greater than 'identity' were counted as identified. Each confidently identified protein included at least one unique peptide. For protein quantization, we required that a protein contained at least two unique peptides. The quantitative protein ratios were weighted and normalized by the median ratio in Mascot. A 1.5-fold cut-off was set to determine quantitative changes of up-regulated and down-regulated proteins, with a *P*-value < 0.05.

Bioinformatics analysis of proteomic data

Functional analysis of identified proteins was conducted using GO annotation (http://www.geneontology.org/) and they were categorized according to their molecular function, biological process and cellular component. The identified proteins were further assigned to the Clusters of Orthologous Groups of proteins (COG) database (http://www.ncbi.nlm.nih.gov/COG/) and the Kyoto Encyclopedia of Genes and Genomes (KEGG) database (http://www.genome.jp/kegg/pathway.html). The flat files of *Picea abies*, *Picea glauca* and *Pinus taeda* proteins were each downloaded from UniProt as reference data sets. To use all identified proteins as a reference data set, the *P*-value that applied a hypergeometric distribution with FDR correction was calculated to obtain significant enrichment GO catalogues. The GO terms showing *P* < 0.05 were considered to be enriched.

All the identified proteins were taken as a reference for GO function entry enrichment significance analysis, which was used to determine whether the biological processes and functions of differentially accumulated proteins were significantly associated. Meanwhile, the corresponding relations between these proteins with KEGG Orthology (KO) IDs were selected from the protein annotation file. Then, these proteins were mapped to KEGG pathways by invoking the KEGG API according to the KO IDs. The KEGG terms showing *P* < 0.05 were considered to be enriched.

Network analysis

The protein–protein interactions (PPIs) of *A. thaliana* were downloaded from the IntAct database (http://www.ebi.ac.uk/intact/). Then, the protein sequences in these interaction pairs were also obtained from UniProt (www.uniprot.org). All identified proteins of *P. asperata* somatic embryos were mapped to the downloaded proteins of *A. thaliana* using the single-directional best hit method with an e-value of 1×10^{-10} and >30% similarity. By mapping the relations of PPIs in *A. thaliana*, the PPIs of the identified proteins in *P. asperata* somatic embryos were predicted. Furthermore, a predicted network of the identified proteins was constructed with Cytoscape (version: 2.8.3).

Determination of water, ABA, IAA and H$_2$O$_2$ content

The water content of somatic embryos was determined gravimetrically before and after partial desiccation. FW defined the weight before PDT, and DW0 defined the fresh weight during PDT. The embryos were dried in an oven at 60 °C for 48 h after desiccation. DW48 defined the dry weight during PDT. The changes of fresh weight during PDT were expressed as DW0/FW (%). The changes of embryonic dry weight during PDT were expressed as DW48/FW (%).

The contents of ABA and IAA were measured as described by Yang et al. (2001) and modified as follows. Samples of 100 mg were extracted with 10 mL 80% (v/v) cold methanol containing 1 mm butylated hydroxytoluene as an antioxidant. The extract was incubated at 4 °C overnight. After centrifugation at 10 000 *g* for 20 min at 4 °C, the supernatant was passed through Chromosep

C_{18} columns (C_{18} Sep-Park Cartridge; Waters Corp., Millford, MA), prewashed with 10 mL 100% (v/v) methanol, 5 mL 100% (v/v) ether and 5 mL 100% methanol, respectively. The hormone fractions were dried under N_2 and dissolved in 2 mL 0.01 M phosphate-buffered saline (PBS) containing 0.1% (v/v) Tween-20, 0.1% (w/v) gelatin (pH 7.4) for analysis by enzyme-linked immunosorbent assay (ELISA). The absorbance was recorded at 490 nm. The ABA and IAA contents were expressed as ng/g FW. H_2O_2 was assayed as described by Patterson et al. (1984) and was expressed as µg/g FW. Catalase activity was determined using the method described by Cui et al. (1999).

All data were analysed for significance using analysis of variance (ANOVA), and the differences were compared using Duncan's multiple range test. Percentage data were transformed by arcsine prior to analysis.

Acknowledgements

This work was supported by a grant from the China Twelfth Five-Year Plan for Science & Technology Support (2012BAD01B01). The authors thank Prof. Iain Charles Bruce (Peking University, China) for critical reading of the manuscript. The English in this document has also been checked by three professional editors, all native speakers of English. For a certificate, please see https://secure.es.acschemworx.acs.org/certificate/verify, and the certificate verification code is 8EF2-F8AC-B2E9-CE01-9ED7.

References

Attree, S., Moore, D., Sawhney, V. and Fowke, L. (1991) Enhanced maturation and desiccation tolerance of white spruce [Picea glauca (Moench) Voss] somatic embryos: effects of a non-plasmolysing water stress and abscisic acid. Ann. Bot. **68**, 519–525.

Bailly, C. (2004) Active oxygen species and antioxidants in seed biology. Seed Sci. Res. **14**, 93–107.

Balbuena, T.S., Silveira, V., Junqueira, M., Dias, L.L., Santa-Catarina, C., Shevchenko, A. and Floh, E.I. (2009) Changes in the 2-DE protein profile during zygotic embryogenesis in the Brazilian Pine (Araucaria angustifolia). J. Proteomics. **72**, 337–352.

Beardmore, T. and Charest, P.J. (1995) Black spruce somatic embryo germination and desiccation tolerance. I. Effects of abscisic acid, cold, and heat treatments on the germinability of mature black spruce somatic embryos. Can. J. For. Res. **25**, 1763–1772.

Berjak, P. (2006) Unifying perspectives of some mechanisms basic to desiccation tolerance across life forms. Seed Sci. Res. **16**, 1–15.

Bomal, C. and Tremblay, F.M. (2000) Dried cryopreserved somatic embryos of two Picea species provide suitable material for direct plantlet regeneration and germplasm storage. Ann. Bot. **86**, 177–183.

Boursiac, Y., Chen, S., Luu, D.T., Sorieul, M., van den Dries, N. and Maurel, C. (2005) Early effects of salinity on water transport in Arabidopsis roots. Molecular and cellular features of aquaporin expression. Plant Physiol. **139**, 790–805.

Bozhkov, P.V. and Von Arnold, S. (1998) Polyethylene glycol promotes maturation but inhibits further development of Picea abies somatic embryos. Physiol. Plant. **104**, 211–224.

Bradford, M.M. (1976) A rapid and sensitive method for the quantitation of microgram quantities of protein utilizing the principle of protein-dye binding. Anal. Biochem. **72**, 248–254.

Casado Vela, J., Martínez Esteso, M.J., Rodriguez, E., Borrás, E., Elortza, F. and Bru-Martínez, R. (2010) iTRAQ-based quantitative analysis of protein mixtures with large fold change and dynamic range. Proteomics, **10**, 343–347.

Chibani, K., Ali-Rachedi, S., Job, C., Job, D., Jullien, M. and Grappin, P. (2006) Proteomic analysis of seed dormancy in Arabidopsis. Plant Physiol. **142**, 1493–1510.

Chu, P., Yan, G.X., Yang, Q., Zhai, L.N., Zhang, C., Zhang, F.Q. and Guan, R.Z. (2015) iTRAQ-based quantitative proteomics analysis of Brassica napus leaves reveals pathways associated with chlorophyll deficiency. J. Proteomics. **113**, 244–259.

Cui, K., Xing, G., Liu, X., Xing, G. and Wang, Y. (1999) Effect of hydrogen peroxide on somatic embryogenesis of Lycium barbarum L. Plant Sci. **146**, 9–16.

Dos Santos, A.L.W., Wiethölter, N., El Gueddari, N.E. and Moerschbacher, B.M. (2006) Protein expression during seed development in Araucaria angustifolia: transient accumulation of class IV chitinases and arabinogalactan proteins. Physiol. Plant. **127**, 138–148.

Dronne, S., Label, P. and Lelu, M.A. (1997) Desiccation decreases abscisic acid content in hybrid larch (Larix × leptoeuropaea) somatic embryos. Physiol. Plant. **99**, 433–438.

Find, J.I. (1997) Changes in endogenous ABA levels in developing somatic embryos of Norway spruce (Picea abies (L.) Karst.) in relation to maturation medium, desiccation and germination. Plant Sci. **128**, 75–83.

Fu, L., Li, N. and Mill, R. (1999) Pinaceae. Flora China, **4**, 11–52.

Harry, I. and Thorpe, T. (1991) Somatic embryogenesis and plant regeneration from mature zygotic embryos of red spruce. Bot. Gaz. **152**, 446–452.

Hazubska-Przybyᵣ, T., Wawrzyniak, M., Obarska, A. and Bojarczuk, K. (2015) Effect of partial drying and desiccation on somatic seedling quality in Norway and Serbian spruce. Acta Physiol. Plant. **37**, 1–9.

Högberg, K., Ekberg, I., Norell, L. and Von Arnold, S. (1998) Integration of somatic embryogenesis in a tree breeding programme: a case study with Picea abies. Can. J. For. Res. **28**, 1536–1545.

Jones, N.B. and van Staden, J. (2001) Improved somatic embryo production from embryogenic tissue of Pinus patula. In Vitro Cell. Dev. Biol. Plant, **37**, 543–549.

Kong, L. (1994) Factors affecting white spruce somatic embryogenesis and embryo conversion. PhD dissertation, Calgary: University of Calgary.

Kong, L. and Yeung, E.C. (1994) Effects of ethylene and ethylene inhibitors on white spruce somatic embryo maturation. Plant Sci. **104**, 71–80.

Kong, L. and Yeung, E.C. (1995) Effects of silver nitrate and polyethylene glycol on white spruce (Picea glauca) somatic embryo development: enhancing cotyledonary embryo formation and endogenous ABA content. Physiol. Plant. **93**, 298–304.

Lee, J., Han, C.T. and Hur, Y. (2013) Molecular characterization of the Brassica rapa auxin-repressed, superfamily genes, BrARP1 and BrDRM1. Mol. Biol. Rep. **40**, 197–209.

Leitão, L., Prista, C., Moura, T.F., Loureiro-Dias, M.C. and Soveral, G. (2012) Grapevine aquaporins: gating of a tonoplast intrinsic protein (TIP2; 1) by cytosolic pH. PLoS ONE, **7**, e33219.

Liao, Y.K. and Juan, I.P. (2015) Improving the germination of somatic embryos of Picea morrisonicola Hayata: effects of cold storage and partial drying. J. For. Res. **20**, 114–124.

Lippert, D., Zhuang, J., Ralph, S., Ellis, D.E., Gilbert, M., Olafson, R., Ritland, K. et al. (2005) Proteome analysis of early somatic embryogenesis in Picea glauca. Proteomics, **5**, 461–473.

Litvay, J.D., Verma, D.C. and Johnson, M.A. (1985) Influence of a loblolly pine (Pinus taeda L.) Culture medium and its components on growth and somatic embryogenesis of the wild carrot (Daucus carota L.). Plant Cell Rep. **4**, 325–328.

Loqué, D., Ludewig, U., Yuan, L. and von Wirén, N. (2005) Tonoplast intrinsic proteins AtTIP2; 1 and AtTIP2; 3 facilitate NH3 transport into the vacuole. Plant Physiol. **137**, 671–680.

Luo, J., Sun, P., Wang, L. and Li, X. (2006) Growth trait variation of Picea asperata at seedling stage and provenance selection. J Southwest Forest. Coll. **4**, 003.

Maruyama, T.E. and Hosoi, Y. (2012) Post-maturation treatment improves and synchronizes somatic embryo germination of three species of Japanese pines. Plant Cell, Tissue Organ Cult. **110**, 45–52.

Nogueira, F.C., Palmisano, G., Schwämmle, V., Campos, F.A., Larsen, M.R., Domont, G.B. and Roepstorff, P. (2012) Performance of isobaric and isotopic labeling in quantitative plant proteomics. J. Proteome Res. **11**, 3046–3052.

Nørgaard, J.V. (1997) Somatic embryo maturation and plant regeneration in Abies nordmanniana Lk. Plant Sci. **124**, 211–221.

Patterson, B.D., MacRae, E.A. and Ferguson, I.B. (1984) Estimation of hydrogen peroxide in plant extracts using titanium (IV). *Anal. Biochem.* **139**, 487–492.

Pawłowski, T., Bergervoet, J., Bino, R. and Groot, S. (2004) Cell cycle activity and β-tubulin accumulation during dormancy breaking of *Acer platanoides* L. seeds. *Biol. Plant.* **48**, 211–218.

Percy, R.E., Livingston, N.J., Moran, J.A. and Von Aderkas, P. (2001) Desiccation, cryopreservation and water relations parameters of white spruce (*Picea glauca*) and interior spruce (*Picea glauca* × *engelmannii* complex) somatic embryos. *Tree Physiol.* **21**, 1303–1310.

Qiao, J., Wang, J., Chen, L., Tian, X., Huang, S., Ren, X. and Zhang, W. (2012) Quantitative iTRAQ LC-MS/MS proteomics reveals metabolic responses to biofuel ethanol in cyanobacterial *Synechocystis* sp. PCC 6803. *J. Proteome Res.* **11**, 5286–5300.

Roberts, D., Sutton, B. and Flinn, B. (1990) Synchronous and high frequency germination of interior spruce somatic embryos following partial drying at high relative humidity. *Can. J. Bot.* **68**, 1086–1090.

Robinson, A.R., Dauwe, R., Ukrainetz, N.K., Cullis, I.F., White, R. and Mansfield, S.D. (2009) Predicting the regenerative capacity of conifer somatic embryogenic cultures by metabolomics. *Plant Biotechnol. J.* **7**, 952–963.

Salajova, T. and Salaj, J. (2001) Somatic embryogenesis and plantlet regeneration from cotyledon explants isolated from emblings and seedlings of hybrid firs. *J. Plant Physiol.* **158**, 747–755.

Sano, N., Masaki, S., Tanabata, T., Yamada, T., Hirasawa, T. and Kanekatsu, M. (2013) Proteomic analysis of stress-related proteins in rice seeds during the desiccation phase of grain filling. *Plant Biotechnol. J.* **30**, 147–156.

dos Santos, A.L.W., Elbl, P., Navarro, B.V., de Oliveira, L.F., Salvato, F., Balbuena, T.S. and Floh, E.I. (2016) Quantitative proteomic analysis of *Araucaria angustifolia* (Bertol.) Kuntze cell lines with contrasting embryogenic potential. *J. Proteomics.* **130**, 180–189.

Shi, H.Y., Zhang, Y.X. and Chen, L. (2013) Two pear auxin-repressed protein genes, PpARP1 and PpARP2, are predominantly expressed in fruit and involved in response to salicylic acid signaling. *Plant Cell, Tissue Organ Cult.* **114**, 279–286.

Stasolla, C. and Yeung, E.C. (2003) Recent advances in conifer somatic embryogenesis: improving somatic embryo quality. *Plant Cell, Tissue Organ Cult.* **74**, 15–35.

Stasolla, C., Loukanina, N., Ashihara, H., Yeung, E.C. and Thorpe, T.A. (2001) Purine and pyrimidine metabolism during the partial drying treatment of white spruce (*Picea glauca*) somatic embryos. *Physiol. Plant.* **111**, 93–101.

Stasolla, C., Kong, L., Yeung, E.C. and Thorpe, T.A. (2002) Maturation of somatic embryos in conifers: morphogenesis, physiology, biochemistry, and molecular biology. *In Vitro Cell. Dev. Biol. Plant*, **38**, 93–105.

Stasolla, C., van Zyl, L., Egertsdotter, U., Craig, D., Liu, W. and Sederoff, R.R. (2003) The effects of polyethylene glycol on gene expression of developing white spruce somatic embryos. *Plant Physiol.* **131**, 49–60.

Sun, W., Van Montagu, M. and Verbruggen, N. (2002) Small heat shock proteins and stress tolerance in plants. *Biochim. Biophys. Acta Gene Struct. Expression*, **1577**, 1–9.

Vooková, B. and Kormuťák, A. (2006) Comparison of induction frequency, maturation capacity and germination of *Abies numidica* during secondary somatic embryogenesis. *Biol. Plant.* **50**, 785–788.

Wehmeyer, N., Hernandez, L.D., Finkelstein, R.R. and Vierling, E. (1996) Synthesis of small heat-shock proteins is part of the developmental program of late seed maturation. *Plant Physiol.* **112**, 747–757.

Wiweger, M., Farbos, I., Ingouff, M., Lagercrantz, U. and Von Arnold, S. (2003) Expression of Chia4-Pa chitinase genes during somatic and zygotic embryo development in Norway spruce (*Picea abies*): similarities and differences

between gymnosperm and angiosperm class IV chitinases. *J. Exp. Bot.* **54**, 2691–2699.

Xiong, L. and Zhu, J. (2003) Regulation of abscisic acid biosynthesis. *Plant Physiol.* **133**, 29–36.

Yang, Y., Xu, C., Wang, B. and Jia, J. (2001) Effects of plant growth regulators on secondary wall thickening of cotton fibres. *Plant Growth Regul.* **35**, 233–237.

Zhang, S., Han, S., Yang, W., Wei, H., Zhang, M. and Qi, L. (2010) Changes in H_2O_2 content and antioxidant enzyme gene expression during the somatic embryogenesis of *Larix leptolepis*. *Plant Cell, Tissue Organ Cult.* **100**, 21–29.

Zhao, J., Li, H., Fu, S., Chen, B., Sun, W., Zhang, J. and Zhang, J. (2015) An iTRAQ-based proteomics approach to clarify the molecular physiology of somatic embryo development in Prince Rupprecht's larch (*Larix principis-rupprechtii* Mayr). *PLoS ONE*, **10**, e0119987.

Zieske, L.R. (2006) A perspective on the use of iTRAQ reagent technology for protein complex and profiling studies. *J. Exp. Bot.* **57**, 1501–1508.

4

Fusarium oxysporum mediates systems metabolic reprogramming of chickpea roots as revealed by a combination of proteomics and metabolomics

Yashwant Kumar[1†], Limin Zhang[2†], Priyabrata Panigrahi[1], Bhushan B. Dholakia[1], Veena Dewangan[1], Sachin G. Chavan[1], Shrikant M. Kunjir[3], Xiangyu Wu[2], Ning Li[2], Pattuparambil R. Rajmohanan[3], Narendra Y. Kadoo[1], Ashok P. Giri[1], Huiru Tang[2,4]* and Vidya S. Gupta[1]*

[1]*Division of Biochemical Sciences, CSIR-National Chemical Laboratory, Pune, India*
[2]*Key Laboratory of Magnetic Resonance in Biological Systems, National Centre for Magnetic Resonance in Wuhan, Wuhan Institute of Physics and Mathematics, Chinese Academy of Sciences, Wuhan, China*
[3]*Central NMR Facility, CSIR-National Chemical Laboratory, Pune, India*
[4]*State Key Laboratory of Genetic Engineering, Metabolomics and Systems Biology Laboratory, School of Life Sciences, Fudan University, Shanghai, China*

*Correspondence
email vs.gupta@ncl.
res.in
email
huiru.tang@wipm.ac.cn
†The first two authors contributed equally to this work.

Keywords: Chickpea, *Fusarium oxysporum*, plant–pathogen interaction, proteomics, metabolomics, NMR.

Summary

Molecular changes elicited by plants in response to fungal attack and how this affects plant–pathogen interaction, including susceptibility or resistance, remain elusive. We studied the dynamics in root metabolism during compatible and incompatible interactions between chickpea and *Fusarium oxysporum* f. sp. *ciceri* (Foc), using quantitative label-free proteomics and NMR-based metabolomics. Results demonstrated differential expression of proteins and metabolites upon Foc inoculations in the resistant plants compared with the susceptible ones. Additionally, expression analysis of candidate genes supported the proteomic and metabolic variations in the chickpea roots upon Foc inoculation. In particular, we found that the resistant plants revealed significant increase in the carbon and nitrogen metabolism; generation of reactive oxygen species (ROS), lignification and phytoalexins. The levels of some of the pathogenesis-related proteins were significantly higher upon Foc inoculation in the resistant plant. Interestingly, results also exhibited the crucial role of altered Yang cycle, which contributed in different methylation reactions and unfolded protein response in the chickpea roots against Foc. Overall, the observed modulations in the metabolic flux as outcome of several orchestrated molecular events are determinant of plant's role in chickpea–Foc interactions.

Introduction

Chickpea (*Cicer arietinum* L.) is the second most widely grown legume in the world. As a top producer, India contributes about 90% of global chickpea production (http://faostat.fao.org/site/339/default.aspx). Chickpea is mainly used as a primary vegetarian source of human dietary protein and, thus, is of significance to food and nutritional security in the developing world. However, due to widespread occurrence of fungal pathogens, such as *Fusarium oxysporum* and *Ascochyta rabei*, the yield of chickpea has been constrained in spite of successive efforts of national and international breeding programmes. Annual yield losses due to wilt disease alone have been estimated to range from 10 to 90% (Anjaiah *et al.*, 2003; Jimenez-Diaz *et al.*, 1989). Wilt is caused by eight races of *Fusarium oxysporum* f. sp. *ciceri* (Foc) affecting all the major chickpea growing areas (Gurjar *et al.*, 2009). Foc infects roots and clogs the xylem, resulting in the obstruction of nutrient supply as wilting progresses, ultimately leading to plant death. This fungus can survive for many years in soil even without its host and, hence, poses a serious challenge for disease management (Haware *et al.*, 1996).

Many studies have been carried out to identify the molecular basis of Foc resistance or susceptibility in chickpea using various approaches including gene mapping, candidate gene identification, differential expression and biochemical analysis postfungal infection (Ashraf *et al.*, 2009; Giri *et al.*, 1998; Gowda *et al.*, 2009; Gupta *et al.*, 2010, 2013a; Gurjar *et al.*, 2012; Nimbalkar *et al.*, 2006). However, genomic scale dynamics of plant–fungus interaction is poorly understood. Further, due to variations in chromosomal rearrangements between nonmodel and model plants, differences in the plant–pathogen interaction with respect to signalling and immunity events are also expected and evinced (Gupta *et al.*, 2013a). Therefore, studies on nonmodel plants using unbiased modern high-throughput technologies are required to improve our knowledge of the plant–fungus interactions (Kushalappa and Gunnaiah, 2013; Mehta *et al.*, 2008). In recent years, a combination of metabolomics and gene-expression analysis has been employed to understand such interactions (Liu *et al.*, 2010) and the effects of gene manipulation on the systems metabolic changes of *Fusarium graminearum* (Chen *et al.*, 2011). A few studies using functional genomics (Cho *et al.*, 2012; Golkari *et al.*, 2007), proteomics (Lee *et al.*, 2009; Yang *et al.*, 2010) and metabolomics (Bollina *et al.*, 2011; Kumaraswamy *et al.*, 2011) approaches have also been reported on plant–pathogen systems of Fusarium head-blight and cereal crops.

As plant–fungus interactions are complex in nature and factors leading to resistance or susceptibility in plant remain largely obscure, the objective of current study was to assess overall modulations in the levels of proteins and metabolites in chickpea roots upon Foc inoculation. Time series profiling of proteome and metabolome of Foc-inoculated resistant and susceptible chickpea roots was performed using label-free quantitative proteomics and untargeted ^{1}H-NMR metabolomics. We observed highly orchestrated response with significant modulation in various metabolic processes. The results described here thus improve our fundamental knowledge of molecular dynamics associated with the chickpea–Foc interaction and potentially useful in designing strategies against wilt disease in chickpea.

Results

Protein identification and quantification in Foc-inoculated chickpea roots

Two days after inoculation (DAI) with Foc, the susceptible chickpea cultivar (JG62) showed yellowing phenotype followed by drooping of leaves that finally lead to complete wilting by 12 DAI. Whereas, Foc-inoculated resistant cultivar (Digvijay-DV) and the mock-inoculated wilt susceptible and resistant cultivars remained healthy throughout the experimental period. The JG62 plant could not sustain the fungal invasion beyond 12 DAI, while DV plants remained unaffected (Figure S1). Based on these phenotypes, 2 and 4 DAI were considered as early stage, while 8 and 12 DAI were considered as late stage. We conducted high-throughput label-free quantitative proteomics analysis with Foc- and mock-inoculated chickpea root tissues at various time points from 2 to 12 DAI. This analysis identified a total of 811 proteins (Table S1a) from which fungal proteins were excluded in further analysis (Lee et al., 2009). From these, 481 had statistically significant differential expression ($P < 0.05$ and fold change >1.2) across cultivars and over the course of infection (Tables S1b and c). The ratio of normalized intensity of proteins from the inoculated samples vis-a-vis respective controls revealed increased or decreased expression in the Foc-inoculated roots. The $\log_2$-transformed values of differentially expressed proteins were clustered using SplineCluster (Heard et al., 2006), which generated eight clusters using a prior precision of 1×10^{-4} (Figure 1).

Protein expression patterns in Foc-inoculated resistant and susceptible cultivars

The potential biological function for each cluster was deduced based on gene ontology enrichment analysis using BiNGO. Cluster 1 (C1) had 61 proteins enriched for isoflavonoids biosynthesis and response to oxidative stress (Figure 1), while there were 73 proteins enriched in lignin biosynthesis, s-adenosyl methionine biosynthetic process and glycolytic process in C2. Majority of proteins from C1 and C2 revealed higher expression at all the stages except 8 DAI in DV compared to JG62. The C3 cluster with 98 proteins was enriched for stress response, malate metabolism, oxidation–reduction process and glycolytic pathway. These proteins had increased expression at early stages which decreased at later stages in DV. Total 63 proteins from C4 cluster were enriched for the response to misfolded proteins, microtubule polymerization processes, osmotic stress response, proteosome core complex assembly and gluconeogenesis. The C5 cluster had 44 proteins enriched for protein degradation through ubiquitin, ATP biosynthesis, photorespiration and active proton (H$^+$) transport. Proteins from C4 and C5 clusters showed general trend of higher expression in all the stages except 12 DAI in DV. On the contrary, JG62 showed lower expression in majority of these proteins at later stages (Figure 1). Total 54 proteins had

high expression in C6 at all the stages except 12 DAI in DV compared to JG62 and did not show enrichment to any specific process through BiNGO. The C7 cluster had 52 proteins that were enriched in defence response and oxidative stress response. These proteins displayed overall contrasting expression patterns in DV and JG62 at early (2 DAI) and late (12 DAI) stages. The 36 proteins from C8 cluster were enriched in response to abiotic stress, fatty acid biosynthesis and nucleosome assembly processes.

Foc induced quantitative variation in proteins from important metabolic pathways

Upon Foc inoculation, we observed a complex response in chickpea from interconnected metabolic pathways including primary amino acid metabolism, glycolysis/gluconeogenesis, TCA cycle, phenylpropanoid pathway and increased lignification (Figure 2). Other cellular processes altered during Foc infection included unfolded protein response (UPR) and defence-related proteins. Enzymes such as, sucrose synthase, phosphoglucomutase, transaldolase, enolase, pyruvate dehydrogenase, citrate synthase, succinyl-coA ligase, fructose bis-phosphate aldolase, phosphogluco kinase, phosphoglyceratemutase, fumarate dehydratase and malate dehydrogenase were up-regulated up to 3.0-fold during early and late stages following Foc inoculation in DV. However, they showed up to 2.0-fold decrease in the roots of JG62. Other proteins from the same pathway such as fructokinase, succinate dehydrogenase and aconitase showed up to 2.56-fold higher level in DV, while JG62 exhibited reduction by 1.4-fold. Several proteins from the UPR pathway including Hsp70, luminal binding protein (BiP), calmodulin, component of SCF-for SKP1-Cullin-F box protein and protein disulfide isomerase (PDI) increased up to 2.6-fold in DV, while they showed 2-fold reduction in JG62 upon Foc inoculation. The majority of these proteins were high in the early stage and low in the late stage in DV, while JG62 showed the opposite expression trend. Further, proteins induced by abiotic stress such as profilin and aquaporin PIP-type 7a revealed 1.7 to 2.5-fold increase in DV upon Foc inoculation compared with JG62. However, nuclear transcription factor-Y (NF-Y) had >30-fold increase at both the stages in JG62 compared with DV.

Proteomic analyses further depicted that proteins involved in reactive oxygen species (ROS) generation such as ascorbate peroxidase, peroxiredoxin, dehydroascorbate reductase (DHAR), hydroxyacyl glutathione hydrolase, glutathione peroxidase, glutaredoxin, glutathione S-transferase (GST), quinone oxidoreductase and copper amine oxidase (CuAO) showed 3.0 to 6.0-fold increased expression in DV compared to JG62 exhibiting significant oxidative stress in chickpea after Foc inoculation. Additionally, enzymes involved in methionine metabolism, including methionine synthase, adenosine homocysteine hydrolase, adenosine kinase and AdoMet synthetase showed up to 1.6-fold higher levels in the Foc-inoculated DV at early and late stages, whereas JG62 displayed down-regulation by >1.3-fold at both the stages. In the present study, subsets of proteins participating in multiple branches of phenylpropanoid pathway such as lignin, flavonoid, isoflavonoid and phenolic biosynthesis were identified in Foc-inoculated chickpea cultivars. Caffeic acid O-methyltransferase (CAOMT) and caffeoyl-CoA O-methyltransferase (CCoAMT) showed up to 2.0 and 3.85-fold increase, respectively, in DV compared to JG62. Similarly, enzymes from isoflavonoid biosynthesis such as chalcone synthase (CHS), chalcone isomerase (CHI), isoflavone synthase (IFS) and isoflavone reductase (IFR) revealed 1.8 to 5.0-fold increase in DV than JG62.

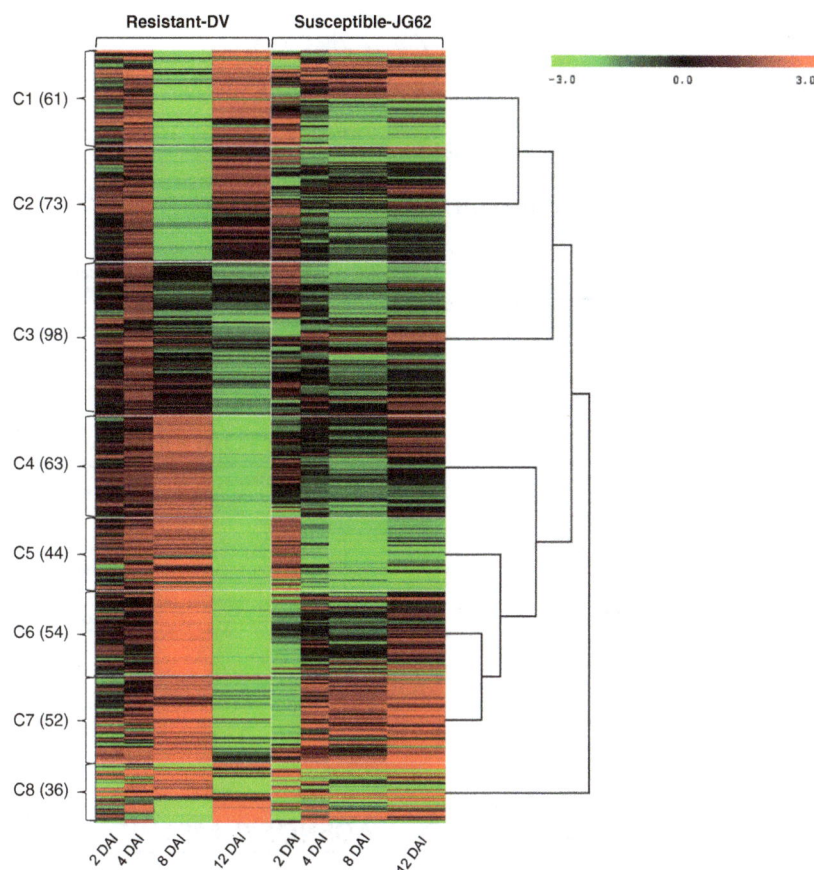

Figure 1 Clusters (C1 to C8) of 481 differential expressed proteins in chickpea root. For each protein, the ratio of Log$_2$-normalized expression of Foc inoculated with its respective control at various stages (2 to 12 DAI) and represented by a colour, according to the colour scale at the top. The number of proteins in a given cluster with similar expression trend is indicated in parentheses.

Additionally, pathogenesis-related (PR) proteins such as endo β-1,3-glucanase, major latex protein (MLP), major latex allergen hev b5 and Bet v1 showed up to 2.1-fold higher expression in DV compared with JG62. Similarly, β-gulcosidase, disease resistance response (DRR) protein-206 and DRR-49 had >5.0-fold higher accumulation in DV than in JG62. Proteolytic chitinases offer antifungal properties and confer resistance to fungal pathogens. Chitinase, selenium binding protein (SBP) and glycine-rich proteins revealed up to 1.8-fold increased expression in DV compared with JG62. Interestingly, some PR proteins such as PR protein STH-2 (2-fold), thaumatin-like protein PR-5b (>5-fold) and PR-4A (>10-fold) were very high at early stage in DV than JG62. However, these proteins exhibited reverse trend with 2.5 to 10-fold increase in JG62 at late stages of Foc infection indicating their response to heavy wounding in chickpea. Some of the proteins involved in the stress signalling process such as ABA-responsive protein, auxin-binding protein ABP19a and Ran-binding protein were increased by 2.1-fold in DV compared with JG62. Additionally, 14-3-3 and H$^+$-ATPase showed up to 2.0-fold higher levels in DV than in JG62, suggesting their role in defence response against Foc in chickpea roots.

Metabolic profiling in chickpea root

A typical annotated ^{1}H-NMR spectrum of chickpea root extract is depicted in Figure 3. The metabolite resonances were assigned according to the in-house databases and previous publication (Fan, 1996). These were further confirmed with a series of 2D NMR spectra including ^{1}H-^{1}H correlation spectroscopy (COSY), ^{1}H-^{1}H total correlation spectroscopy (TOCSY), J-resolved spectroscopy (JRES), ^{1}H-^{13}C heteronuclear single quantum coherence spectroscopy (HSQC) and ^{1}H-^{13}C heteronu-

clear multiple-bond correlation (HMBC) with both ^{1}H and ^{13}C chemical shifts and signal multiplicities were as shown in Table S2. A total of 52 dominant metabolites were identified including amino acids (Ala, Val, Ile, Leu, Gln, Glu, Asn, Trp, Lys and GABA), sugars (glucose, sucrose, fructose, trehalose and salicin), organic acids (pyruvate, lactate, acetate, citrate, succinate, formate, fumarate, malate and guanidoacetate), nucleosides (adenosine, uridine, inosine, 5CMP and hypoxanthine) and phytoalexins (genistein and luteolin). The identification of few of these metabolites was confirmed by spiking with their known standards (Table S2).

Differential metabolic alterations induced by Foc inoculation

Principal component analysis (PCA) revealed significant metabolic changes between Foc-inoculated resistant and susceptible cultivars at different time points. The averaged PCA scores were calculated for the first two PCs to construct the PCA trajectory that illustrated clear separation in metabolites from control and inoculated samples between two cultivars (Figure 4). Interestingly, the metabolic profiles obtained from the controls of both cultivars and the inoculated DV followed similar trajectory trends; however, the inoculated JG62 showed dramatic change in trajectory after 8 DAI (Figure 4). The metabolic changes resulted after Foc inoculations were further evaluated by constructing orthogonal projection to latent structures discriminant analysis (OPLS-DA) models. The quality of models was indicated by the values of R^2 and Q^2 and cross-validated with a CV-ANOVA approach ($P < 0.05$) (Figures 5a and b) and permutation tests (Figure S2). All the significantly changed metabolites are annotated in the OPLS-DA coefficient plots (Figures 5a and b) and summarized in Table S3.

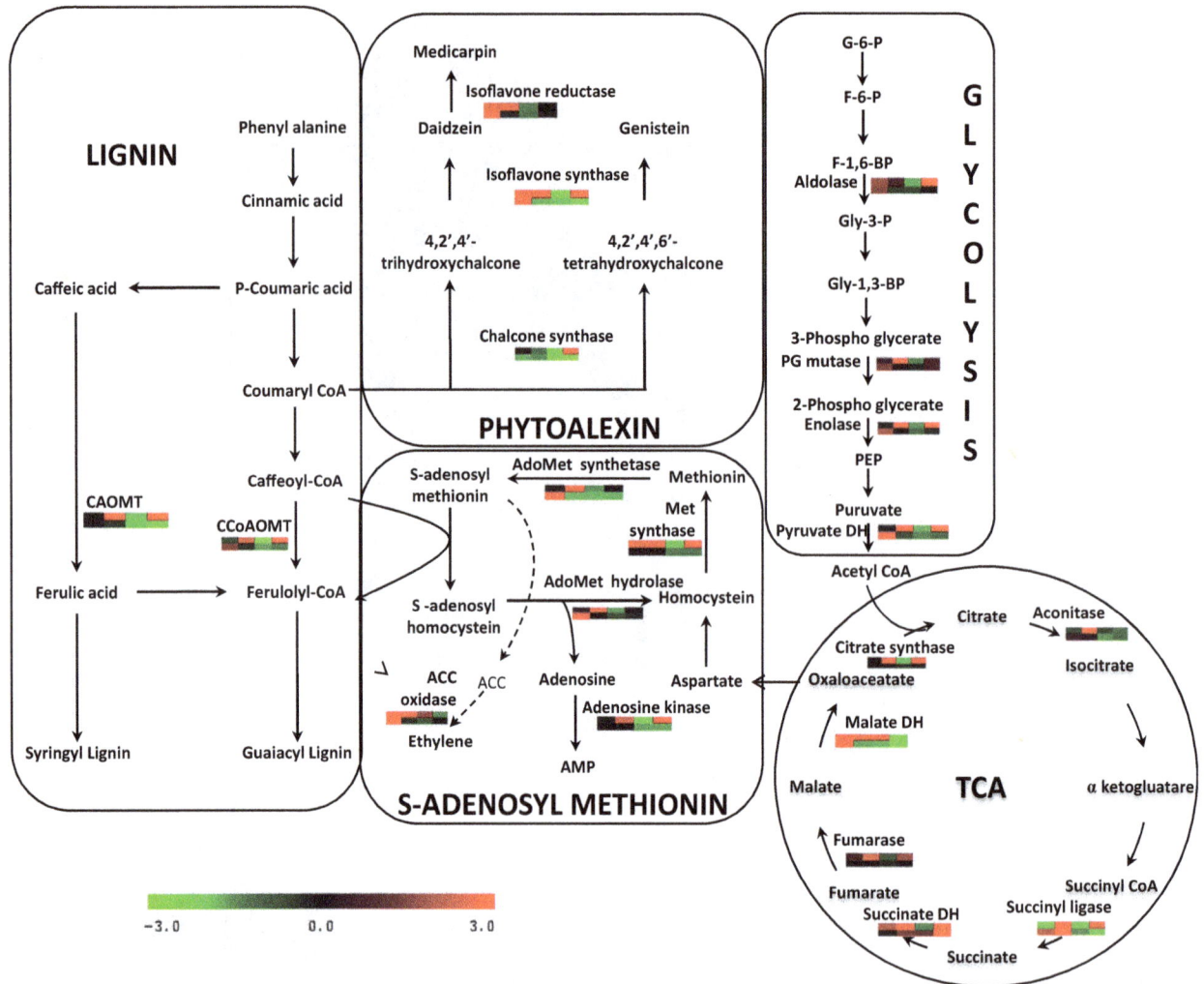

Figure 2 Interconnection between various metabolic processes in chickpea–Foc interaction. Each graph represents differential expression (fold change) pattern according to the colour scale; columns represent the four stages (2 to 12 DAI) after Foc inoculation. The first row represents the resistant cultivar, while the second row represents the susceptible cultivar.

Compared with the respective control, Foc inoculation induced marked reduction in the levels of amino acids including Thr, Ala, Lys and Asn at early stage in both the cultivars. However, their levels were significantly elevated in JG62 at late stage while DV remained unchanged (Figure 5). The decreased levels of Ile and Leu were observed at early stage and remained unchanged at late stage in JG62 after Foc inoculation. Interestingly, Val and Trp levels increased dramatically in the late stage of JG62 but remained unchanged in DV at both the stages (Figure 5). Similarly, two of the most important metabolites from nitrogen metabolism, such as Glu and Gln, were up-regulated only at early stage of Foc inoculation in DV. However, JG62 revealed significantly higher accumulation of both of these amino acids at early and late stages. In addition, Foc inoculation resulted in decreased level of glucose in DV at early stage; however, no significant change was observed in both the cultivars at late stage. The levels of sucrose and fructose were decreased at early stage in JG62, followed by significant reduction at the later stage. It is of particular interest that the levels of some nucleotides including uridine and orotate were significantly decreased at the later stage of JG62 compared with DV. However, the opposite

trend in the levels of adenosine and inosine was observed at the later stage of JG62. Compared with the controls, Foc inoculation induced significant reduction in the levels of malate and acetate in JG62 at the later stage. As for phytoalexins, genistein was increased in DV at the later stage but remained unchanged in JG62 and the level of luteolin decreased at the later stage of JG62 while DV was unaffected. A known antifungal compound, clotrimazole level increased in DV, but there was a marked reduction in JG62 at the later stage. In addition, Foc infection caused increase in quinone at early stage and phytosterol at the later stage of DV only.

Comparative expression of candidate genes in root

To obtain complementary information of transcriptional variations, we examined expression levels of key genes (Figure 6) involved in various metabolic pathways based on our proteomic and metabolomic findings. These included genes from nitrogen mobilization (glutamate dehydrogenase-GDH, glutamate synthase, glutamine synthase and asparagine synthase), stress response (NF-Y and SKP1-like protein 1A), methionine metabolism (methionine synthase and AdoMet synthetase), lignin and

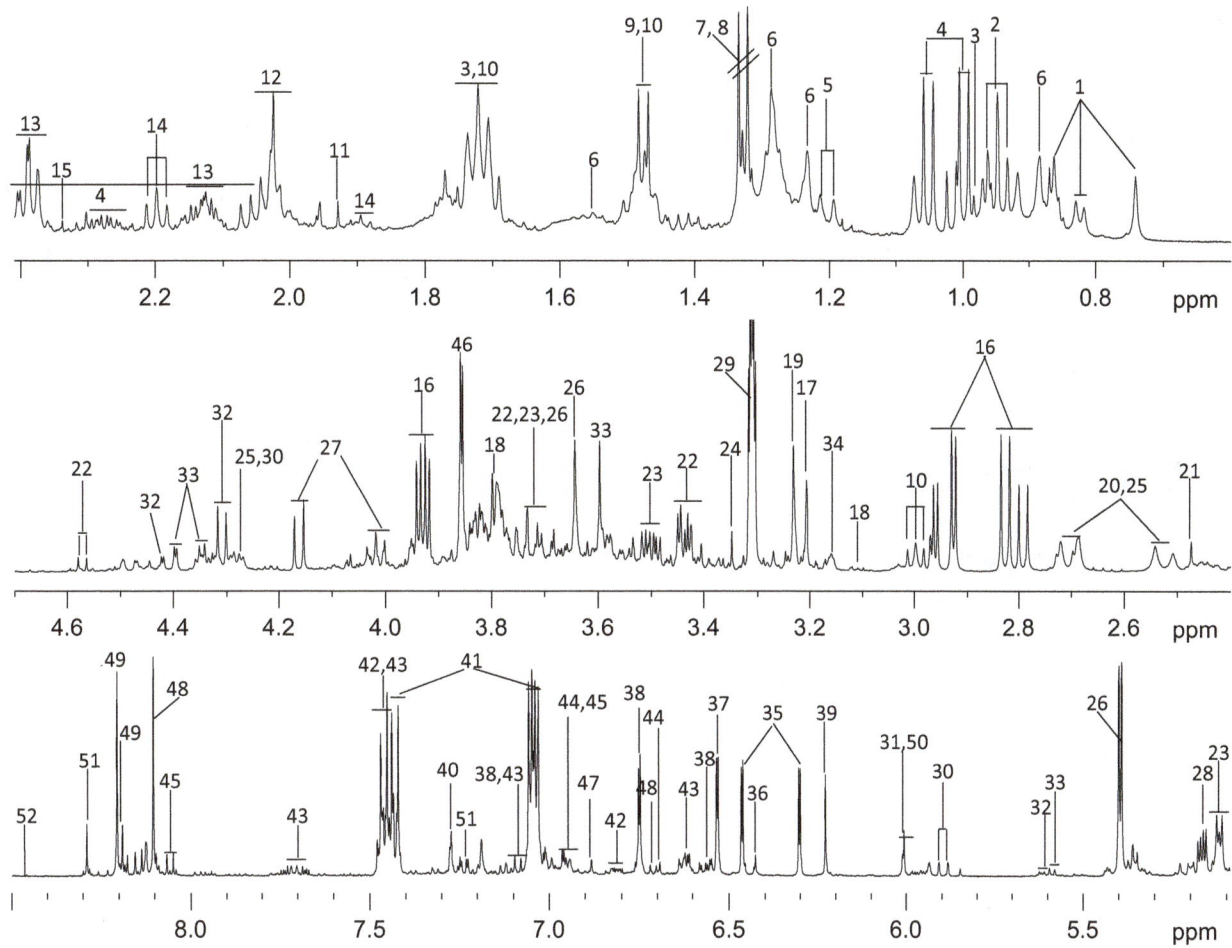

Figure 3 Typical NMR spectrum of chickpea root extract. Annotation with number and details of metabolites is provided in Table S2.

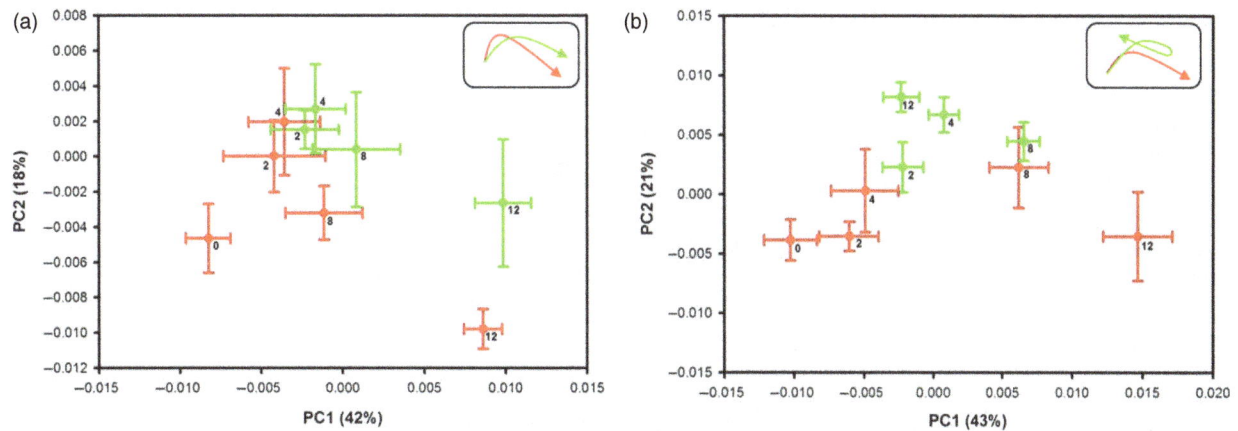

Figure 4 PCA trajectory plots. (a) resistant-DV and (b) susceptible-JG62 plants with their respective controls obtained from mean of PC1 and PC2 values at 2 to 12 DAI with error bars representing two standard deviations. Foc-inoculated samples are in green while respective controls in red. Top right corner box indicates overall pattern.

phytoalexin biosynthetic pathways (CCoAMT, CHS, CHI, iso-flavone 4′-O-methyltransferase, IFS and IFR). *GDH* expression markedly increased by >200 fold in the inoculated JG62 at the later stage. Similarly, significant up-regulation (up to 4-fold) of glutamine synthetase, asparagine synthetase and glutamate synthase was observed in the Foc-inoculated JG62 (Figure 6a–

d). However, candidate genes from methionine metabolism such as Adomet synthetase and methionine synthase revealed >3 and 10-fold enhanced expression, respectively, in DV (Figure 6e and f). Also, important genes from lignin and phytoalexin biosynthetic pathway displayed higher expression in DV compared with JG62 (Figures 6i–n). On the contrary, abiotic

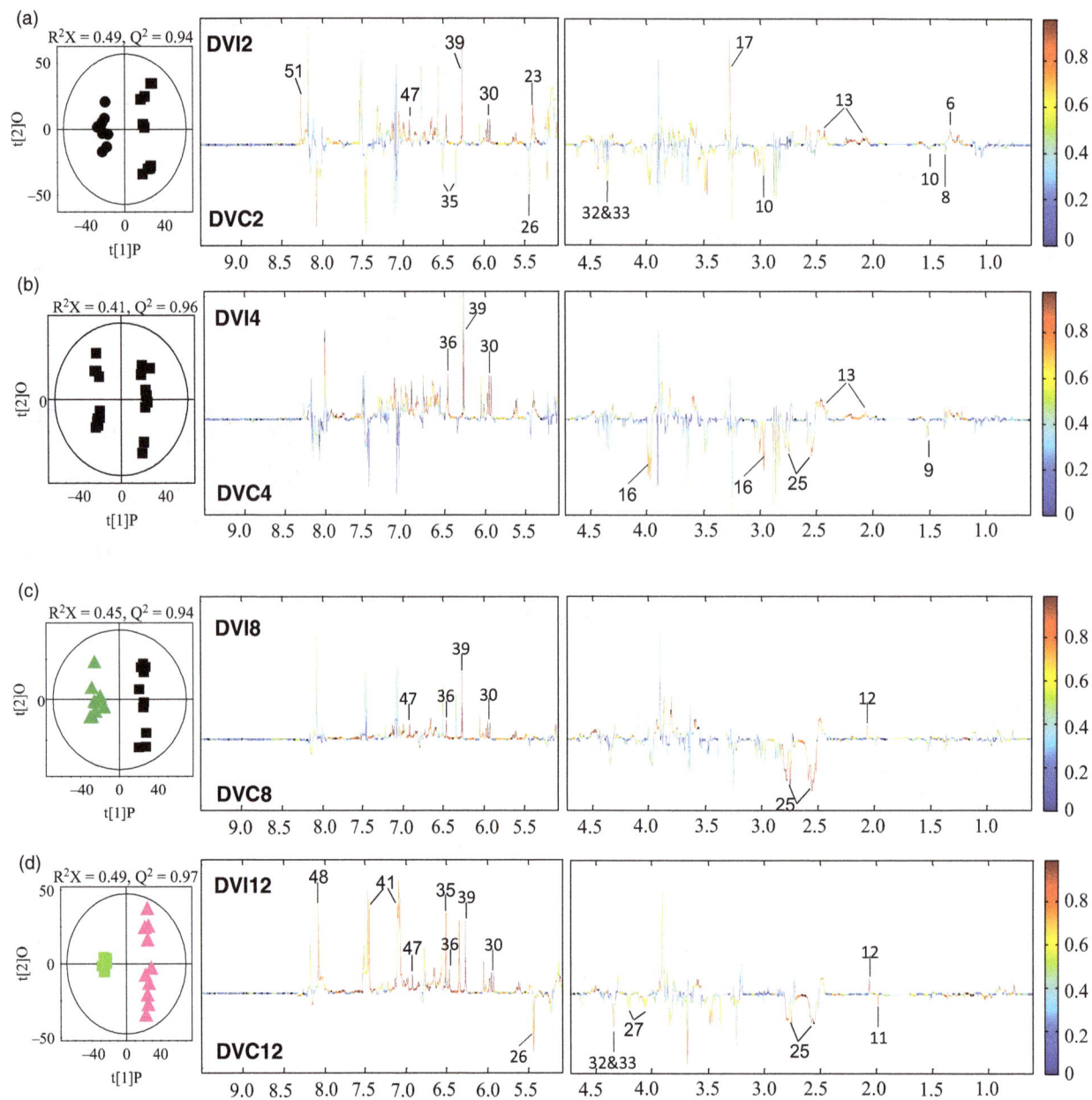

Figure 5 Pairwise comparison via OPLS-DA. OPLS-DA scores plots (left) and corresponding coefficient-coded loadings plots (right) obtained from metabolic profiles of Foc-inoculated (a–d) resistant-DV and (e–h) susceptible-JG62 cultivars and their respective controls at 2 to 12 DAI. The coloured scale in correlation coefficient (|r|: absolute values) plots shows the significance of metabolite variations discriminating between the Foc-inoculated and control plants.

stress induced *NF-Y* gene showed >8-fold higher expression in JG62 (Figure 6h), while *SKP*1 was expressed more in DV (Figure 6g).

Intense lignification is associated with Foc resistant phenotype

After Foc inoculation, differential lignin deposition in the root tissue of DV and JG62 was observed as the time progressed, wherein DV exhibited intense lignification compared with JG62 (Figure S3). The fungal pathogen invaded the susceptible cultivar and blocked the vascular tissue completely by 12 DAI, leading to wilting of JG62 plants.

Discussion

Foc induced remodelling in energy metabolism and nitrogen mobilization

Obligate biotrophs depend on host metabolism for nutrient uptake, which in turn determine their pathogenicity within the host. In the present study, proteins involved in glycolysis and TCA cycle were up-regulated in DV while down-regulated in JG62. We also observed steep alteration in primary metabolites (amino acids and sugars) specifically in JG62. Sugars affect disease susceptibility often favouring disease development while playing a critical role in innate defence pathways involving metabolic regulation (Bolouri

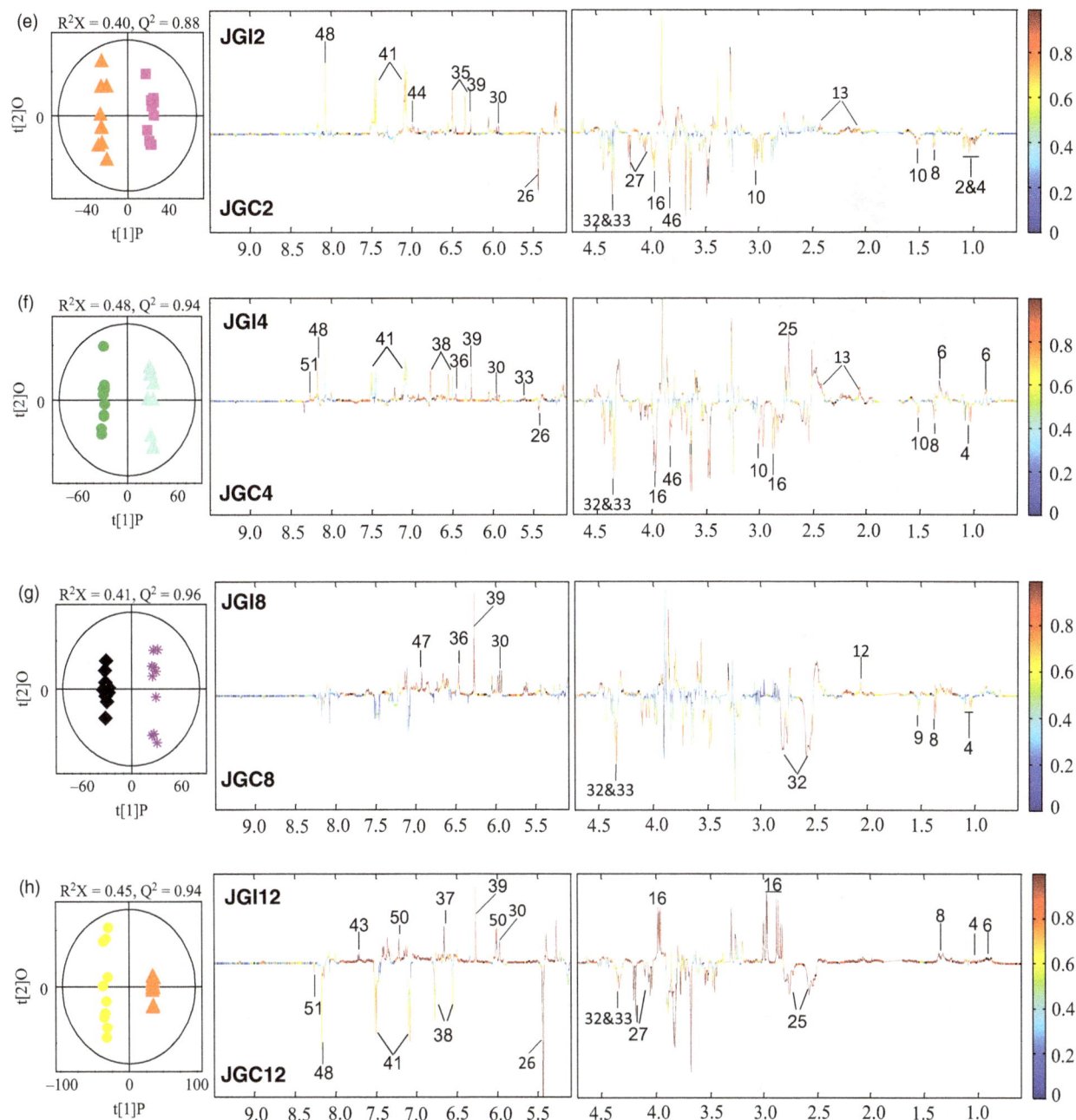

Figure 5 Continued.

Moghaddam and Van den Ende, 2012). In the present study, rapid decrease in sugars such as sucrose and fructose was observed upon Foc infection in both the cultivars; however, this was more predominant in the JG62. Similar rapid reduction in the levels of these sugars was also reported in sunflower upon infection with *Botrytis cinerea* (Dulermo *et al.*, 2009b). These results emphasized the regulatory role of sugars in the metabolic reprogramming and subsequently higher expression of proteins from glycolysis and TCA cycle endowed DV plants to successfully combat Foc invasion. Moreover, supportive evidence could be derived from earlier reports of *F. oxysporum* induced up-regulation of various ESTs from sugar metabolism in chickpea (Ashraf *et al.*, 2009; Gupta *et al.*, 2010, 2013a).

Nitrogen plays an essential role in the nutrient relationship between plants and pathogens. Modulation in amino acid concentration due to nitrogen mobilization after pathogen infection to plant has been reported (Dulermo *et al.*, 2009a; Tavernier *et al.*, 2007). Proteomics studies on wheat–*F. graminearum* exhibited increase expression in proteins from amino acid, carbon and nitrogen metabolism (Wang *et al.*, 2005; Zhou *et al.*, 2006). Such alteration with significantly low amino acid content was also observed in tomato and sunflower after infection of *B. cinerea* and *Sclerotinia sclerotiorum*, respectively (Berger *et al.*, 2004; Jobic *et al.*, 2007). Current study showed significant decrease in the concentration of various amino acids in the susceptible cultivar, which suggested that the fungus probably

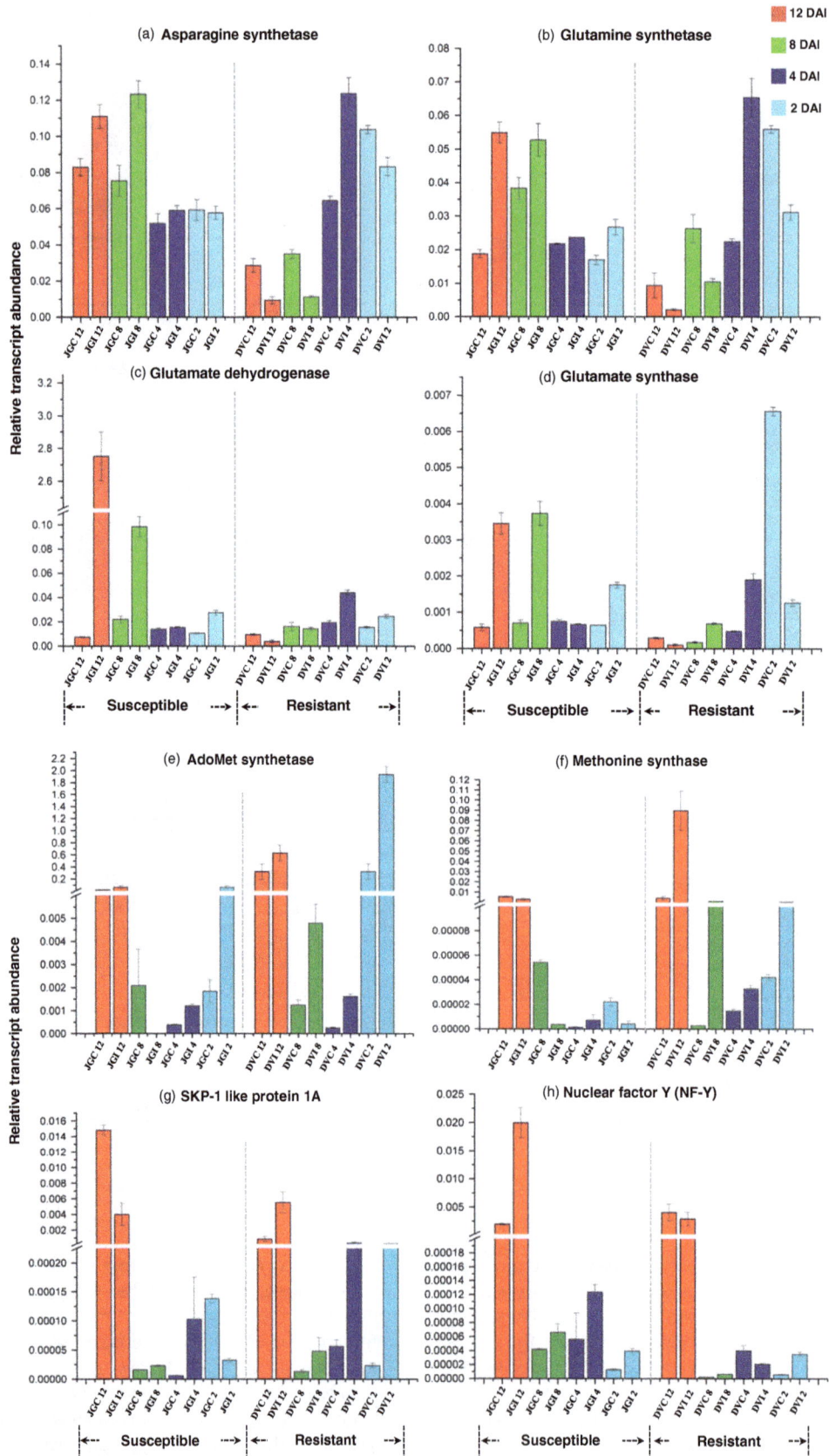

Figure 6 Quantitative real-time PCR of various candidate genes. (a–n) expression variation in each gene observed from chickpea root of Foc-inoculated resistant-DV and susceptible-JG62 cultivars as compared to their respective controls at 2 to 12 DAI.

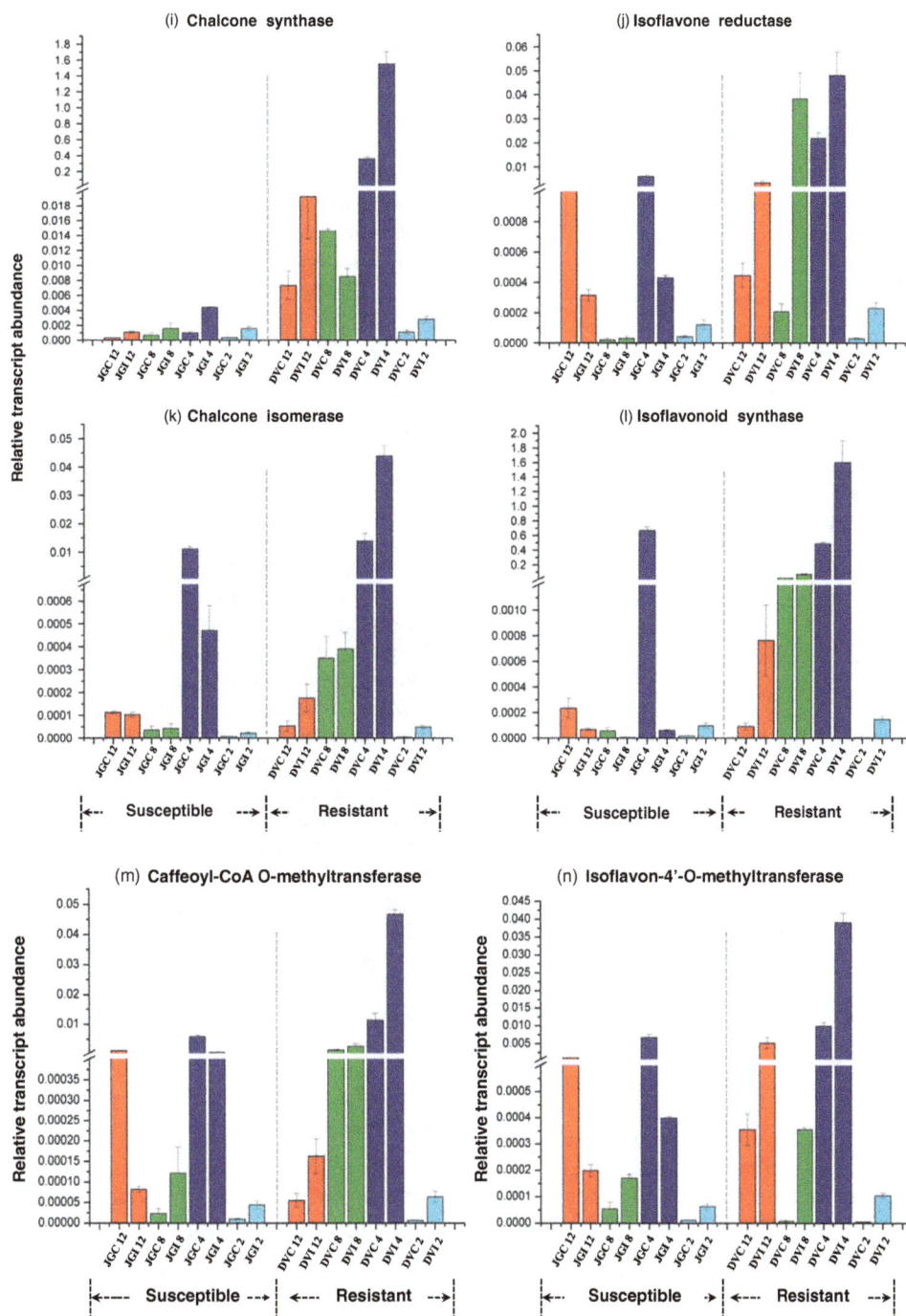

Figure 6 Continued.

utilized them for its establishment and proliferation inside the host. However, probably due to Foc sporulation, levels of amino acids were increased at the later stage in JG62. Hence, the role of nitrogen mobilization was further scrutinized in chickpea–Foc interaction by gene-expression analysis of four representative enzymes from nitrogen mobilization *viz.* glutamate synthase, glutamate dehydrogenase, glutamine synthetase and asparagine synthetase. Significant up-regulation of these genes in JG62 compared to DV correlated well with the metabolomics outcome. Consistently, proteomic analysis also revealed higher level of

sucrose synthase and glutamine synthase following Foc inoculation. Thus, proteomic, metabolomic and gene-expression results together indicated that the process of nitrogen mobilization could be critical in the establishment of Foc infection in the susceptible plants.

Stress responsive proteins in chickpea root

During stress, accumulation of unfolded proteins increases in endoplasmic reticulum (ER) and results in triggering the unfolded protein response (UPR) to remove the misfolded proteins by ubiquitin–proteasome pathway. Thus, UPR not only helps to avert the cytotoxic impact of misfolded proteins, but also assists to relieve stress and reinstate normal functions in ER (Ye et al., 2011). Up-regulation of critical proteins from UPR pathway such as Hsp70, BiP, calmodulin, SKP1 and PDI in DV could suggest their coordinate response. In *Arabidopsis thaliana* roots, swelling of ER and vacuolar collapse resulting in ER stress and cell death were observed during fungal colonization (Qiang et al., 2012). Similarly, PDI level was higher in wheat plants upon *Puccinia striiformis* inoculation (Maytalman et al., 2013). This study also reinforced that during defence response, there was constant requirement for proteins stabilization in the process of folding, assembly, vesicle trafficking and secretion. Collectively, these outcomes suggested the importance of an efficient UPR pathway utilization in the resistant chickpea against Foc.

During stress conditions, aquaporins such as PIP are involved in water transport in plant. Current investigation revealed significant increase in PIP-7a level at both the stages in DV, while drastic reduction was observed in JG62. This could have resulted in better water conductance in DV and helped the resistant plant to overcome fungal attack. On the contrary, diminished water transport with reduced aquaporin in xylem due to fungal spread in the susceptible plant could have eventually turned into wilting. Furthermore under water-limited conditions, nuclear factor-Y (NF-Y) family and ABA-responsive protein have shown to be up-regulated in *Arabidopsis* (Nelson et al., 2007). We observed high expression of these proteins at both the stages in JG62 compared with DV. Thus, present findings indicated constrain in water uptake in JG62 due to clogged xylem after fungal invasion that potentially increased the susceptibility to Foc.

Early recognition of Foc leads to ROS generation and lignosuberization

Generation of ROS is one of the earliest cellular responses to pathogen recognition and/or infection (Gupta et al., 2013b). The enzymes involved in ROS production such as peroxidase, DHAR, hydroxyacyl glutathione hydrolase, glutathione peroxidase, glutaredoxin, GST, quinone oxidoreductase and CuAO were significantly increased in DV compared with JG62 in the current investigation. It has been demonstrated that CuAO and peroxidases functionally correlate in lignosuberization process (Angelini et al., 1993; Scalet et al., 1991) and the inhibitors of CuAO result in decreased defence response (Rea et al., 1998, 2002). Additionally, Raju et al. (2008) found more lignification in the resistant chickpea cultivar compared with the susceptible one upon Foc infection. In the present study also, intense lignin deposition on the root cortex of Foc-inoculated resistant cultivar was observed (Figure S3). Taken together, ROS generation and higher expression of CuAO suggested that Foc triggered hydrogen peroxide generation and lignosuberization process leading to initiation of defence response in DV. Secondly, monolignol biosynthesis also plays critical role in the host defence mechanism through lignification, making cell wall more resistant to the pathogen penetration. Reduction in the monolignol biosynthesis through co-silencing of *CAOMT* and *CCoAMT* in wheat led to the higher penetration by *Blumeria graminis* (Bhuiyan et al., 2009). We also found up-regulation of these two lignin biosynthetic enzymes in DV than JG62 suggesting increased lignin deposition that might provide resistance against Foc.

Crucial role of methionine metabolism in Foc resistance

In the Yang cycle, methionine synthase converts homocysteine to methionine contributing in synthesis of Adomet by AdoMet synthetase. AdoMet can also lead to ethylene by 1-aminocyclopropane-1-carboxylic acid (ACC) synthase and ACC oxidase. However, down-regulation of ACC oxidase by 1.5-fold in both the cultivars indicated that preferential AdoMet pool was not channelized towards ethylene production. A recent proteomic study has shown that increased expression of AdoMet synthetase and lower level of ACC oxidase resulted in higher methionine recycling and lower ethylene biosynthesis in rice roots against *Herbaspirillum seropedicae* (Alberton et al., 2013). Silencing of AdoHcy hydrolase in transgenic tobacco plants confirmed its role in defence mechanism against pathogens (Masuta et al., 1995). Similarly, Kawalleck et al. (1992) identified mRNAs for AdoMet synthetase and AdoHcy hydrolase in parsley plant upon fungal infection. Altogether, current study indicated close association between pathogen defence and increased level of activated methyl groups during chickpea–Foc interplay.

Further, highly methylesterified pectin is required for normal plant cell wall, while it is de-esterified by pectin methyl esterase (PME) which leads to increased vulnerability of plant cell wall to the pathogen invasion (Lionetti et al., 2007). Our proteomics results showed increased expression of PME at both the stages in JG62 compared with DV. Earlier studies with either silencing of PME or overexpression of PME inhibitors in plants demonstrated negative role of PME in the pathogen resistance (An et al., 2008; Ma et al., 2013). Likewise, plant sterols are structurally related to cholesterol and control mechanical property of cell membrane (Hodzic et al., 2008) and also serve as substrate for several metabolic pathways (Itkin et al., 2013). Interestingly, our metabolomic data also revealed high phytosterol at later stages in DV. However, there was no significant change in JG62 except at 8 DAI. The stability of the plant cell wall is also maintained by actin binding cytoskeleton protein such as Profilin, which showed higher expression in DV compared with JG62 at late stage. Overall, altered methionine metabolism affected methyl esterification of pectin in Foc-infected susceptible roots, while stronger plant cell wall with normal pectin and cytoskeleton proteins could have helped the resistant chickpea cultivar with better defence against Foc.

Foc resistance is mediated by phenylpropanoid pathway

Plants respond to pathogen challenge by increased activation of the phenylpropanoid pathway leading to flavanoids, isoflavonoids and phenolics biosynthesis. They play multiple roles in plant–pathogen interaction including precursors for the defence-related phytoalexins and signal molecules in response to pathogen infection. As detailed in the results, enzymes involved in this pathway such as CHS, CHI, IFR and IFS were up-regulated in DV as compared to JG62. In the transgenic soybean roots, RNAi silencing of *CHS* gene showed decrease in total isoflavonoids as well as reduced resistance to fungal pathogens (Lozovaya et al., 2006; Subramanian et al., 2005). Additionally, Naoumkina et al. (2007) and Farag et al. (2008) showed that fungal extract induced higher levels of *CHI*, *IFS* and *IFR* leading to the production of phytoalexin in Medicago cell suspension culture to combat the pathogen. Consistently, the accumulation of genistein, luteolin and quinone identified by metabolome analysis in the present study correlated well with the proteomic data. Thus, the accumulation of isoflavonoid biosynthetic proteins and metabolites in DV suggested their potential involvement in Foc resistance.

Modulation of defence-related proteins and metabolites upon Foc inoculation

In the present study, we observed quantitative variation in many defence-related proteins such as endo β-1,3-glucanase, MLP, hev b5 and Bet v1, β-gulcosidase, DRR-206, DRR-49, chtinases, SBP, PR10, STH-2, PR4a, PR-5b, 14-3-3 and H$^+$-ATPase in Foc-inoculated chickpea cultivars. Many previous studies in other plant–pathogen interactions have also demonstrated the importance of such proteins in plant defence. Lytle et al. (2009) and Gurjar et al. (2012) have reported that Bet v1 protein is responsible for resistance towards pathogen. Similarly, β-1,3-glucanase is involved in plant defence response against pathogen (Shetty et al., 2009; Ward et al., 1991). Previously two independent studies have shown that β-glucosidase, DRR-206 and DRR-49 proteins contribute to lignification process (Burlat et al., 2001; Hosel and Barz, 1975). Increased chitinase activity was also reported earlier after Foc inoculation in the resistant chickpea cultivar (Giri et al., 1998). Similarly, overexpressed SBP in rice provided more resistance against rice blast fungus (Sawada et al., 2004). Another important defence-related protein, 14-3-3, is known to be associated with hypersensitive cell death in pathogen incompatible cultivars (Roberts, 2003). In our earlier study, up-regulation of 14-3-3 transcripts was detected in Foc-inoculated resistant chickpea roots (Nimbalkar et al., 2006). Thus, up-regulation of 14-3-3 and H$^+$-ATPase suggested their roles in activating hypersensitivity response in chickpea leading to Foc resistance. Apart from the above-mentioned proteins, our metabolite analysis revealed increase level of clotrimazole in Foc-inoculated DV at late stage compared to JG62. This is an interesting observation, clotrimazole being an anthropogenic antifungal compound and needs further experimentation. Thus, the pathogen attack triggered expression of defence-related genes, secondary metabolites with antimicrobial nature and PR proteins in the chickpea–Foc interactions.

In summary, our integrated approach of label-free quantitative proteomics, ^{1}H-NMR metabolomics and candidate gene-expression analysis has facilitated to understand the defence mechanism in chickpea against Foc. Most of the proteins/metabolites from various metabolic processes such as energy metabolism, isoflavonoid/flavonoid biosynthesis pathway and lignin biosynthesis that lead to defence were up-regulated in the resistant plant compared with the susceptible one. Further, elevated levels of some of the proteins from the UPR pathway during incompatible interaction indicated that proper protein folding and transport might be crucial for the plant's survival. Some of the proteins/metabolites such as isoflavone reductase and isoflavone synthase leading increased phytoalexins levels; CAOMT and CCoMT involved in lignosuberization and increased methionine synthase for efficient methylation process in the resistant plant could have helped against the pathogen. Overall, above findings improve our understanding on the metabolic reprogramming during the wilt disease progression.

Experimental procedures

Plant material

Seeds of wilt resistant (Digvijay, DV) and susceptible (JG62, JG) chickpea cultivars as well as pathogenic culture of Foc race 1 were obtained from Mahatma Phule Krishi Vidyapeeth (MPKV), Rahuri, India. All the steps of plant growth and pathogen inoculation were followed as previously described (Kumar et al., 2015) and divided into two groups, control or mock (sterile water) inoculated and Foc (~10^6 spores/mL) inoculated. The workflow for sample collection/data analysis is shown in Figure S4. As the pathogen is reported to colonize the xylem vessels 2 days postinoculation (Gupta et al., 2010), the root tissues were collected at 2, 4, 8 and 12 DAI. Whole roots of chickpea was used in equal amount from the resistant and susceptible cultivars at specific time points simultaneously for the proteomics and metabolomics analysis; because the fungal hyphae might not be localized but distributed in the whole root xylem tissue (Jimenez-Fernandez et al., 2013). In each group, a pool of 10 plants comprised one biological replicate. For proteomics analyses, tissues from three biological replicates were analysed, while ten biological replicates were analysed for metabolomics analysis. The harvested tissues were stored at −80 °C till further use.

Protein extraction and mass spectrometry analysis

Total proteome of root tissue was extracted as described by Isaacson et al. (2006). Protein pellets were solubilized in 50 mM ammonium bicarbonate buffer containing 0.1% Rapigest (Waters, Milford, MA). The dissolved proteins were reduced and alkylated by DTT and iodoacetamide, respectively, followed by overnight tryptic hydrolysis at 37 °C using Promega sequencing grade trypsin. The digested peptides were analysed with LC-MS^E workflow using nano-ACQUITY online coupled to a SYNAPT HDMS system (Waters). Nano-LC separation was performed with symmetry C18 trapping column (180 μm × 20 mm, 5 μm) and bridged-ethyl hybrid (BEH) C18 analytical column (75 μm × 250 mm, 1.7 μm). The binary solvent system comprised solvent A (0.1% formic acid in water), and solvent B (0.1% formic acid in acetonitrile). Each sample (500 ng) was initially applied to the trapping column and desalted by flushing with 1% solvent B for 1 min at a flow rate of 15 μL/min. Elution of the tryptic digested sample was performed at a flow rate of 300 nL/min by increasing the solvent B concentration from 3% to 40% over 90 min. Before data acquisition, the mass analyser was calibrated using Glu-fibrinopeptide B (Sigma-Aldrich, Steinheim, Germany) from m/z 50 to 1990. The Glu-fibrinopeptide B (GFP-B) was delivered at 500 fmole/μL to the mass spectrometer via a NanoLockSpray interface using the auxiliary pump of the nano-ACQUITY system at every 30 s interval for lock mass correction during data acquisition. Data-independent acquisition was performed (LC-MS^E) as described by Patel et al. (2009).

As accuracy and reproducibility in mass measurement are critical in data acquisition during large-scale proteomic experiments, PCA was used to assess the quality of the measurement in terms of replicate similarity. The replicates of each sample were clustered together reflecting inherent similarities between samples (Figure S5a). In addition, linear response and reproducibility of measurement of the quantitative proteomic data acquisition were tested by plotting two replicates (Figure S5b), whereas Figure S5c showed that data have been acquired below 3 ppm mass accuracy. Further, the percent coefficient of variance of retention time (% CV-RT) was calculated to assess the separation stability and coefficient of variance of 0.3 min, which also suggested stability in chromatographic separation (Figure S5d).

Analysis of quantitative proteomics data

The acquired LC-MS^E data were processed using the ProgenesisQI for Proteomics software (Waters). Protein identifications were obtained by searching the genomic databases of chickpea (http://www.icrisat.org/) and *Fusarium oxysporum* (http://www.broad-institute.org/). LC-MS^E data were searched with a fixed car-bamidomethyl modification for cysteine residues, along with a variable modification for oxidation of methionine, N-terminal acetylation, deamination of asparagine and glutamine and phosphorylation of serine, threonine and tyrosine. The ion accounting search algorithm within ProgenesisQI for Proteomics software was used which has been developed specifically for searching data-independent MS^E data sets and described by Li *et al.* (2009). The ion accounting search parameters were (a) precursor and product ion tolerance: automatic setting, (b) minimum number of product ion matches per peptide: 3, (c) minimum number of product ion matches per protein: 7, (d) minimum number of peptide matches per protein: 1 and (e) missed tryptic cleavage sites: 1. False-positive rate was set at 1%. Search results of the proteins and the individual MS/MS spectra with a confidence level at or >95% were accepted. Label-free quantitation of identified protein was done on the basis of spiked bovine serum albumin (BSA) protein.

Clustering of identified proteins

Data were normalized by spiked BSA (50 fmoles), and relative accumulation differences were determined for proteins having differential expression. Sum of three replicates of inoculated samples was divided by that of the respective controls. This established a ratio of accumulation of a protein in plants upon Foc inoculation compared with mock-inoculated control plants. The log$_2$-transformed ratio (susceptible/control and resistant/control) pairs were clustered by the application of SplineCluster (Heard *et al.*, 2006), a Bayesian model-based hierarchical clustering algorithm for time series data.

Gene ontology enrichment analysis

Protein functional annotation was determined using Blast2GO (Conesa *et al.*, 2005) and for each cluster, GO enrichment analysis was carried out using BiNGO 2.3 plugin tool in Cytoscape version 2.8 (Maere *et al.*, 2005). Over-represented GO biological process categories were identified using a hypergeometric test with a significance threshold of 0.05 after Benjamini and Hochberg false discovery rate correction (Benjamini and Hochberg, 1995) using the annotated chickpea genome as the reference set.

Metabolite extraction and NMR measurement

Plant root tissue was ground well in liquid nitrogen by using bead beater (Retsch GmbH, Retsch-Allee, Germany) and lyophilized. The powdered root tissues (~50 mg) were extracted with 0.75 mL of CD$_3$OD and 0.75 mL of 10 mM KH$_2$PO$_4$ buffer (pH 6.0) containing sodium3-trimethlysilyl [2,2,3,3-D$_4$] propionate (TSP) as described previously (Kim *et al.*, 2010). After ultrasonication for 20 min and centrifugation at 12 000 g for 10 min at room temperature (~25 °C), 0.5 mL of supernatant was collected for NMR detection. ^{1}H NMR spectra of root extracts were acquired at 25 °C on a Bruker AV II 500 spectrometer (Bruker Biospin, Rheinstetten, Germany) operating at 500.13 MHz for ^{1}H. A standard water-suppressed one-dimensional NMR spectrum was recorded using *noesypr1d* pulse sequence (RD-90°-t_1-90°-tm-90°-acquisition) with the recycle delay of 6 s and the mixing time (t_m), of 50 ms. Typically, 90° pulse was set to about 15 μs and 256 transients were collected into 48K data points for each spectrum with a spectral width of 16 ppm. All spectra were referenced to

chemical shift of TSP (δ = 0.00). For the metabolite assignment purpose, a range of two-dimensional NMR spectra were recorded for selected samples including COSY, TOCSY, HSQC and HMBC. In COSY and TOCSY experiments, respective 64 and 32 transients were collected into 2K data points for each of 256 increments with the spectral width of 2426 Hz for both dimensions. Magnitude mode was used with gradient selection for the COSY experiments, whereas the *mlevgpphw5* pulse program was employed as the spin-lock scheme in the phase sensitive mode, with the mixing time of 60 ms, for TOCSY. Both HSQC and HMBC spectra were acquired using the gradient-selected sequences. In HSQC experiment, 80 transients were collected into 1K data points for each of 140 increments. In HMBC experiment, 160 transients were collected into 2K data points for each of 256 increments. The spectral widths were 2426 Hz for ^{1}H and 9809 Hz for ^{13}C in HSQC and HMBC experiments. Confirmation of resonance assignments were performed for few metabolites by spiking the samples with known standards (Table S2).

NMR spectra processing and multivariate data analysis

All the ^{1}H NMR spectra were manually corrected for phase and baseline distortions using TOPSPIN (v2.1; Bruker Biospin) and calibrated for chemical shift drifting by in-house-developed script for MATLAB (The Mathworks, Natick, MA). The spectral region δ 0.5–9.5 was divided into bins with width of 0.002 ppm (1.0 Hz) using AMIX software (v3.8.3; Bruker Biospin GmbH, Germany). The region δ 4.727–5.089 ppm was discarded to remove the effects of imperfect water presaturation. The areas of the remaining bins were normalized to total sum of intensity for each spectrum to compensate for the overall concentration differences prior to statistical data analysis. Multivariate data analyses were carried out with SIMCA-P+ v 12.0 software package (Umetrics, Umeå, Sweden). PCA was performed on the mean-centred NMR data to inspect overall data distributions and possible outliers. Using the NMR data as the X-matrix and group information as Y-matrix, OPLS-DA was carried out with unit variance scaling (Trygg, 2002; Xiao *et al.*, 2008). The OPLS-DA models were 7-fold cross-validated and the quality of the model was described by the parameters R^{2}X, representing the total explained metabolic variables and Q^2, indicating the model predictability. The models were further evaluated with a CV-ANOVA approach ($P < 0.05$) and permutation tests. To facilitate interpretation of the results, back-transformation (Cloarec *et al.*, 2005) of the loadings generated from the OPLS-DA was performed prior to generating the loadings plots, which were colour-coded with the Pearson linear correlation coefficients of variables (or metabolites) using an in-house-developed script for MATLAB (The Mathworks) (Wang *et al.*, 2007). The colour-coded correlation coefficient indicates the significance of the metabolite contribution to the class separation, with hot colours (e.g. red) being more significance of the metabolite contributions to the group classification than cold ones (e.g. blue). In this study, a correlation coefficient cut-off value of 0.602 (i.e., $N = 10$, $|r| > 0.602$) was used for the statistical significance based on the discrimination significance at the level of $P < 0.05$, which was determined according to the discriminating significance of the Pearson's product–moment correlation coefficient (Cloarec *et al.*, 2005).

Quantitative real-time PCR analysis

Total RNA was extracted from 100 mg root tissue by using TRI Reagent (Sigma-Aldrich). First strand cDNA synthesis was performed using the High Capacity cDNA Reverse Transcription Kits (Applied Biosystems, Foster City, CA) with 3 µg of DNase1 treated total RNA using oligo (dT) primer following manufacturer's protocol. Gene-specific primers were designed using Primer Express (v2.0) software (Applied Biosystems) and listed in Table S4. Real-time PCR was carried out as earlier described (Barvkar et al., 2012) using FastStart universal SYBR green master mix (Roche, Mannheim, Germany) with 7900HT Fast real-time PCR system (Applied Biosystems). The Initiation factor 4α ($IF4\alpha$) gene was used as internal standard or reference gene (Garg et al., 2010).

Lignin staining

Lignin accumulation within root tissue after Foc inoculation was detected using the phloroglucinol/hydrochloric acid stain as described by MauchMani and Slusarenko (1996). The control and Foc-inoculated chickpea roots were subjected to transverse sections with a scalpel and immersed in 1 mL of 1% phloroglucinol in 6N HCl for 5 min, and the lignin staining was visualized under light microscope.

Acknowledgements

Y.K. acknowledges Senior Research Fellowship from the Council of Scientific and Industrial Research (CSIR), India. The authors acknowledge Dr. Moskau Detlef (Bruker BioSpin AG, Switzerland) for the help rendered during data acquisition; Dr. Nicholas Heard (Imperial College London, UK) for help in SplineCluster analysis and Waters, UK for providing 'Progenesis QI for proteomics' software and support. Financial support to H.R.T. by National Natural Science Foundation, China (21175149 and 91439102) and to CSIR-NCL (BSC 0117) is gratefully acknowledged. Authors have no conflict of interest.

References

Alberton, D., Muller-Santos, M., Brusamarello-Santos, L.C.C., Valdameri, G., Cordeiro, F.A., Yates, M.G., Pedrosa, F.D. et al. (2013) Comparative proteomics analysis of the rice roots colonized by Herbaspirillum seropedicae strain SmR1 reveals induction of the methionine recycling in the plant host. J. Proteome Res. 12, 4757–4768.

An, S.H., Sohn, K.H., Choi, H.W., Hwang, I.S., Lee, S.C. and Hwang, B.K. (2008) Pepper pectin methylesterase inhibitor protein CaPMEI1 is required for antifungal activity, basal disease resistance and abiotic stress tolerance. Planta, 228, 61–78.

Angelini, R., Bragaloni, M., Federico, R., Infantino, A. and Portapuglia, A. (1993) Involvement of polyamines, diamine oxidase and peroxidase in resistance of chickpea to Ascochyta rabiei. J. Plant Physiol. 142, 704–709.

Anjaiah, V., Cornelis, P. and Koedam, N. (2003) Effect of genotype and root colonization in biological control of Fusarium wilts in pigeonpea and chickpea by Pseudomonas aeruginosa PNA1. Can. J. Microbiol. 49, 85–91.

Ashraf, N., Ghai, D., Barman, P., Basu, S., Gangisetty, N., Mandal, M.K., Chakraborty, N. et al. (2009) Comparative analyses of genotype dependent expressed sequence tags and stress-responsive transcriptome of chickpea wilt illustrate predicted and unexpected genes and novel regulators of plant immunity. BMC Genom. 10, 415.

Barvkar, V.T., Pardeshi, V.C., Kale, S.M., Kadoo, N.Y., Giri, A.P. and Gupta, V.S. (2012) Proteome profiling of flax (Linum usitatissimum) seed: characterization of functional metabolic pathways operating during seed development. J. Proteome Res. 11, 6264–6276.

Benjamini, Y. and Hochberg, Y. (1995) Controlling the false discovery rate - a practical and powerful approach to multiple testing. J. R. Stat. Soc. Series B Stat. Methodol 57, 289–300.

Berger, S., Papadopoulos, M., Schreiber, U., Kaiser, W. and Roitsch, T. (2004) Complex regulation of gene expression, photosynthesis and sugar levels by pathogen infection in tomato. Physiol. Plant. 122, 419–428.

Bhuiyan, N.H., Selvaraj, G., Wei, Y.D. and King, J. (2009) Gene expression profiling and silencing reveal that monolignol biosynthesis plays a critical role in penetration defense in wheat against powdery mildew invasion. J. Exp. Bot. 60, 509–521.

Bollina, V., Kushalappa, A.C., Choo, T.M., Dion, Y. and Rioux, S. (2011) Identification of metabolites related to mechanisms of resistance in barley against Fusarium graminearum, based on mass spectrometry. Plant Mol. Biol. 77, 355–370.

Bolouri Moghaddam, M.R. and Van den Ende, W. (2012) Sugars and plant innate immunity. J. Exp. Bot. 63, 3989–3998.

Burlat, V., Kwon, M., Davin, L.B. and Lewis, N.G. (2001) Dirigent proteins and dirigent sites in lignifying tissues. Phytochemistry, 57, 883–897.

Chen, F.F., Zhang, J.T., Song, X.S., Yang, J., Li, H.P., Tang, H.R. and Liao, Y.C. (2011) Combined metabonomic and quantitative real-time PCR analyses reveal systems metabolic changes of Fusarium graminearum induced by Tri5 gene deletion. J. Proteome Res. 10, 2273–2285.

Cho, S.H., Lee, J., Jung, K.H., Lee, Y.W., Park, J.C. and Paek, N.C. (2012) Genome-wide analysis of genes induced by Fusarium graminearum infection in resistant and susceptible wheat cultivars. J. Plant Biol. 55, 64–72.

Cloarec, O., Dumas, M.E., Trygg, J., Craig, A., Barton, R.H., Lindon, J.C., Nicholson, J.K. et al. (2005) Evaluation of the orthogonal projection on latent structure model limitations caused by chemical shift variability and improved visualization of biomarker changes in H^1 NMR spectroscopic metabonomic studies. Anal. Chem. 77, 517–526.

Conesa, A., Gotz, S., Garcia-Gomez, J.M., Terol, J., Talon, M. and Robles, M. (2005) Blast2GO: a universal tool for annotation, visualization and analysis in functional genomics research. Bioinformatics, 21, 3674–3676.

Dulermo, T., Bligny, R., Gout, E. and Cotton, P. (2009a) Amino acid changes during sunflower infection by the necrotrophic fungus B. cinerea. Plant Signal. Behav. 4, 859–861.

Dulermo, T., Rascle, C., Chinnici, G., Gout, E., Bligny, R. and Cotton, P. (2009b) Dynamic carbon transfer during pathogenesis of sunflower by the necrotrophic fungus Botrytis cinerea: from plant hexoses to mannitol. New Phytol. 183, 1149–1162.

Fan, W.M.T. (1996) Metabolite profiling by one- and two-dimensional NMR analysis of complex mixtures. Prog. Nucl. Magn. Reson. Spectrosc. 28, 161–219.

Farag, M.A., Huhman, D.V., Dixon, R.A. and Sumner, L.W. (2008) Metabolomics reveals novel pathways and differential mechanistic and elicitor-specific responses in phenylpropanoid and isoflavonoid biosynthesis in Medicago truncatula cell cultures. Plant Physiol. 146, 387–402.

Garg, R., Sahoo, A., Tyagi, A.K. and Jain, M. (2010) Validation of internal control genes for quantitative gene expression studies in chickpea (Cicer arietinum L.). Biochem. Bioph. Res. Commun. 396, 283–288.

Giri, A.P., Harsulkar, A.M., Patankar, A.G., Gupta, V.S., Sainani, M.N., Deshpande, V.V. and Ranjekar, P.K. (1998) Association of induction of protease and chitinase in chickpea roots with resistance to Fusarium oxysporum f.sp. ciceri. Plant. Pathol. 47, 693–699.

Golkari, S., Gilbert, J., Prashar, S. and Procunier, J.D. (2007) Microarray analysis of Fusarium graminearum-induced wheat genes: identification of organ-specific and differentially expressed genes. Plant Biotechnol. J. 5, 38–49.

Gowda, S.J.M., Radhika, P., Kadoo, N.Y., Mhase, L.B. and Gupta, V.S. (2009) Molecular mapping of wilt resistance genes in chickpea. Mol. Breeding 24, 177–183.

Gupta, S., Chakraborti, D., Sengupta, A., Basu, D. and Das, S. (2010) Primary metabolism of chickpea is the initial target of wound inducing early sensed Fusarium oxysporum f. sp ciceri race 1. PLoS ONE, 5, e9030.

Gupta, S., Bhar, A. and Das, S. (2013a) Understanding the molecular defense responses of host during chickpea-Fusarium interplay: where do we stand? Funct. Plant Biol. 40, 1285–1297.

Gupta, S., Bhar, A., Chatterjee, M. and Das, S. (2013b) Fusarium oxysporum f. sp ciceri race 1 induced redox state alterations are coupled to downstream defense signaling in root tissues of chickpea (Cicer arietinum L.). PLoS ONE, 8, e73163.

Gurjar, G., Barve, M., Giri, A. and Gupta, V. (2009) Identification of Indian pathogenic races of *Fusarium oxysporum* f. sp *ciceris* with gene specific, ITS and random markers. *Mycologia*, **101**, 484–495.

Gurjar, G., Giri, A.P. and Gupta, V.S. (2012) Gene expression profiling during wilting in chickpea caused by *Fusarium oxysporum* f. sp. *ciceri*. *Am. J. Plant Sci.* **3**, 190–201.

Haware, M.P., Nene, Y.L. and Natarajan, M. (1996) Survival of *Fusarium oxysporum* f. sp. *ciceri* in the soil in the absence of chickpea. *Phytopathol. Mediterr.* **35**, 9–12.

Heard, N.A., Holmes, C.C. and Stephens, D.A. (2006) A quantitative study of gene regulation involved in the immune response of anopheline mosquitoes: an application of Bayesian hierarchical clustering of curves. *J. Am. Stat. Assoc.* **101**, 18–29.

Hodzic, A., Rappolt, M., Amenitsch, H., Laggner, P. and Pabst, G. (2008) Differential modulation of membrane structure and fluctuations by plant sterols and cholesterol. *Biophys. J* . **94**, 3935–3944.

Hosel, W. and Barz, W. (1975) Beta-glucosidases from *Cicer arietinum* L - purification and properties of isoflavone-7-o-glucoside specific beta-glucosidases. *Eur. J. Biochem.* **57**, 607–616.

Isaacson, T., Damasceno, C.M.B., Saravanan, R.S., He, Y., Catala, C., Saladie, M. and Rose, J.K.C. (2006) Sample extraction techniques for enhanced proteomic analysis of plant tissues. *Nat. Protoc.* **1**, 769–774.

Itkin, M., Heinig, U., Tzfadia, O., Bhide, A.J., Shinde, B., Cardenas, P.D., Bocobza, S.E. *et al.* (2013) Biosynthesis of antinutritional alkaloids in solanaceous crops is mediated by clustered genes. *Science*, **341**, 175–179.

Jimenez-Diaz, R.M., Trapero-Casas, A. and de Cabrera la Colina, J. (1989) Races of *Fusarium oxysporum* f. sp. *ciceri* infecting chickpea in southern Spain. In *Vascular Wilt Diseases of Plants* (Tjamos, E.C. and Beckman, C.H., eds), NATO ASI Series, H28, pp. 515–520. Berlin: Springer Verlag.

Jimenez-Fernandez, D., Landa, B.B., Kang, S., Jimenez-Diaz, R.M. and Navas-Cortes, J.A. (2013) Quantitative and microscopic assessment of compatible and incompatible interactions between chickpea cultivars and *Fusarium oxysporum* f. sp. *ciceri* races. *PLoS ONE*, **8**, e61360.

Jobic, C., Boisson, A.M., Gout, E., Rascle, C., Fevre, M., Cotton, P. and Bligny, R. (2007) Metabolic processes and carbon nutrient exchanges between host and pathogen sustain the disease development during sunflower infection by *Sclerotinia sclerotiorum*. *Planta*, **226**, 251–265.

Kawalleck, P., Plesch, G., Hahlbrock, K. and Somssich, I.E. (1992) Induction by fungal elicitor of s-adenosyl-l-methionine synthetase and s-adenosyl-l-homocysteine hydrolase messenger-RNAs in cultured-cells and leaves of *Petroselinum crispum*. *Proc. Natl Acad. Sci. USA* **89**, 4713–4717.

Kim, H.K., Choi, Y.H. and Verpoorte, R. (2010) NMR-based metabolomic analysis of plants. *Nat. Protoc.* **5**, 536–549.

Kumar, Y., Dholakia, B.B., Panigrahi, P., Kadoo, N.Y., Giri, A.P. and Gupta, V.S. (2015) Metabolic profiling of chickpea-Fusarium interaction identifies differential modulation of disease resistance pathways. *Phytochemistry*, **116**, 120–129.

Kumaraswamy, K.G., Kushalappa, A.C., Choo, T.M., Dion, Y. and Rioux, S. (2011) Mass spectrometry based metabolomics to identify potential biomarkers for resistance in barley against fusarium head blight (*Fusarium graminearum*). *J. Chem. Ecol.* **37**, 846–856.

Kushalappa, A.C. and Gunnaiah, R. (2013) Metabolo-proteomics to discover plant biotic stress resistance genes. *Trends Plant Sci.* **18**, 522–531.

Lee, J., Feng, J., Campbell, K.B., Scheffler, B.E., Garrett, W.M., Thibivilliers, S., Stacey, G. *et al.* (2009) Quantitative proteomic analysis of bean plants infected by a virulent and avirulent obligate rust fungus. *Mol. Cell Proteomics* **8**, 19–31.

Li, G.Z., Vissers, J.P.C., Silva, J.C., Golick, D., Gorenstein, M.V. and Geromanos, S.J. (2009) Database searching and accounting of multiplexed precursor and product ion spectra from the data independent analysis of simple and complex peptide mixtures. *Proteomics*, **9**, 1696–1719.

Lionetti, V., Raiola, A., Camardella, L., Giovane, A., Obel, N., Pauly, M., Favaron, F. *et al.* (2007) Overexpression of pectin methylesterase inhibitors in *Arabidopsis* restricts fungal infection by *Botrytis cinerea*. *Plant Physiol.* **143**, 1871–1880.

Liu, C.X., Hao, F.H., Hu, J., Zhang, W.L., Wan, L.L., Zhu, L.L., Tang, H.R. *et al.* (2010) Revealing different systems responses to brown planthopper

infestation for pest susceptible and resistant rice plants with the combined metabonomic and gene-expression analysis. *J. Proteome Res.* **9**, 6774–6785.

Lozovaya, V., Ulanov, A., Lygin, A., Duncan, D. and Widholm, J. (2006) Biochemical features of maize tissues with different capacities to regenerate plants. *Planta*, **224**, 1385–1399.

Lytle, B.L., Song, J., de la Cruz, N.B., Peterson, F.C., Johnson, K.A., Bingman, C.A., Phillips, G.N. Jr *et al.* (2009) Structures of two *Arabidopsis thaliana* major latex proteins represent novel helix-grip folds. *Proteins*, **76**, 237–243.

Ma, L., Jiang, S., Lin, G., Cai, J., Ye, X., Chen, H., Li, M. *et al.* (2013) Wound-induced pectin methylesterases enhance banana (Musa spp. AAA) susceptibility to *Fusarium oxysporum* f. sp. *cubense*. *J. Exp. Bot.* **64**, 2219–2229.

Maere, S., Heymans, K. and Kuiper, M. (2005) BiNGO: a Cytoscape plugin to assess overrepresentation of gene ontology categories in biological networks. *Bioinformatics*, **21**, 3448–3449.

Masuta, C., Tanaka, H., Uehara, K., Kuwata, S., Koiwai, A. and Noma, M. (1995) Broad resistance to plant viruses in transgenic plants conferred by antisense inhibition of a host gene essential in S-adenosylmethionine-dependent transmethylation reactions. *Proc. Natl Acad. Sci. USA* **92**, 6117–6121.

MauchMani, B. and Slusarenko, A.J. (1996) Production of salicylic acid precursors is a major function of phenylalanine ammonia-lyase in the resistance of arabidopsis to *Peronospora parasitica*. *Plant Cell*, **8**, 203–212.

Maytalman, D., Mert, Z., Baykal, A.T., Inan, C., Gunel, A. and Hasancebi, S. (2013) Proteomic analysis of early responsive resistance proteins of wheat (*Triticum aestivum*) to yellow rust (*Puccinia striiformis* f. sp *tritici*) using ProteomeLab PF2D. *Plant Omics*, **6**, 24–35.

Mehta, A., Brasileiro, A.C.M., Souza, D.S.L., Romano, E., Campos, M.A., Grossi-De-Sa, M.F., Silva, M.S. *et al.* (2008) Plant-pathogen interactions: what is proteomics telling us? *FEBS J.* **275**, 3731–3746.

Naoumkina, M., Farag, M.A., Sumner, L.W., Tang, Y.H., Liu, C.J. and Dixon, R.A. (2007) Different mechanisms for phytoalexin induction by pathogen and wound signals in *Medicago truncatula*. *Proc. Natl Acad. Sci. USA* **104**, 17909–17915.

Nelson, D.E., Repetti, P.P., Adams, T.R., Creelman, R.A., Wu, J., Warner, D.C., Anstrom, D.C. *et al.* (2007) Plant nuclear factor Y (NF-Y) B subunits confer drought tolerance and lead to improved corn yields on water-limited acres. *Proc. Natl Acad. Sci. USA* **104**, 16450–16455.

Nimbalkar, S.B., Harsulkar, A.M., Giri, A.P., Sainani, M.N., Franceschi, V. and Gupta, V.S. (2006) Differentially expressed gene transcripts in roots of resistant and susceptible chickpea plant (*Cicer arietinum* L.) upon *Fusarium oxysporum* infection. *Physiol. Mol. Plant Pathol.* **68**, 176–188.

Patel, V.J., Thalassinos, K., Slade, S.E., Connolly, J.B., Crombie, A., Murrell, J.C. and Scrivens, J.H. (2009) A comparison of labeling and label-free mass spectrometry-based proteomics approaches. *J. Proteome Res.* **8**, 3752–3759.

Qiang, X.Y., Zechmann, B., Reitz, M.U., Kogel, K.H. and Schafer, P. (2012) The mutualistic fungus *Piriformospora indica* colonizes arabidopsis roots by inducing an endoplasmic reticulum stress-triggered caspase-dependent cell death. *Plant Cell*, **24**, 794–809.

Raju, S., Jayalakshmi, S.K. and Sreeramulu, K. (2008) Comparative studies on induction of defense related enzymes in two different cultivars of chickpea (*Cicer arietinum* L) genotypes by salicylic acid, spermine and *Fusarium oxysporum* f. sp. *ciceri*. *Aust. J. Crop Sci.* **2**, 121–140.

Rea, G., Laurenzi, M., Tranquilli, E., D'Ovidio, R., Federico, R. and Angelini, R. (1998) Developmentally and wound-regulated expression of the gene encoding a cell wall copper amine oxidase in chickpea seedlings. *FEBS Lett.* **437**, 177–182.

Rea, G., Metoui, O., Infantino, A., Federico, R. and Angelini, R. (2002) Copper amine oxidase expression in defense responses to wounding and *Ascochyta rabiei* invasion. *Plant Physiol.* **128**, 865–875.

Roberts, M.R. (2003) 14-3-3 Proteins find new partners in plant cell signalling. *Trends Plant Sci.* **8**, 218–223.

Sawada, K., Hasegawa, M., Tokuda, L., Kameyama, J., Kodama, O., Kohchi, T., Yoshida, K. *et al.* (2004) Enhanced resistance to blast fungus and bacterial blight in transgenic rice constitutively expressing OsSBP, a rice homologue of mammalian selenium-binding proteins. *Biosci. Biotechnol. Biochem.* **68**, 873–880.

Scalet, M., Federico, R. and Angelini, R. (1991) Time courses of diamine oxidase and peroxidase-activities, and polyamine changes after mechanical injury of chickpea seedlings. *J. Plant Physiol.* **137**, 571–575.

Shetty, N.P., Jensen, J.D., Knudsen, A., Finnie, C., Geshi, N., Blennow, A., Collinge, D.B. *et al.* (2009) Effects of beta-1,3-glucan from *Septoria tritici* on structural defence responses in wheat. *J. Exp. Bot.* **60**, 4287–4300.

Subramanian, S., Graham, M.Y., Yu, O. and Graham, T.L. (2005) RNA interference of soybean isoflavone synthase genes leads to silencing in tissues distal to the transformation site and to enhanced susceptibility to *Phytophthora sojae. Plant Physiol.* **137**, 1345–1353.

Tavernier, V., Cadiou, S., Pageau, K., Lauge, R., Reisdorf-Cren, M., Langin, T. and Masclaux-Daubresse, C. (2007) The plant nitrogen mobilization promoted by *Colletotrichum lindemuthianum* in Phaseolus leaves depends on fungus pathogenicity. *J. Exp. Bot.* **58**, 3351–3360.

Trygg, J. (2002) O2-PLS for qualitative and quantitative analysis in multivariate calibration. *J. Chemom.* **16**, 283–293.

Wang, Y., Yang, L.M., Xu, H.B., Li, Q.F., Ma, Z.Q. and Chu, C.G. (2005) Differential proteomic analysis of proteins in wheat spikes induced by *Fusarium graminearum. Proteomics*, **5**, 4496–4503.

Wang, Y.L., Lawler, D., Larson, B., Ramadan, Z., Kochhar, S., Holmes, E. and Nicholson, J.K. (2007) Metabonomic investigations of aging and caloric restriction in a life-long dog study. *J. Proteome Res.* **6**, 1846–1854.

Ward, E.R., Payne, G.B., Moyer, M.B., Williams, S.C., Dincher, S.S., Sharkey, K.C., Beck, J.J. *et al.* (1991) Differential regulation of beta-1,3-glucanase messenger-RNAs in response to pathogen infection. *Plant Physiol.* **96**, 390–397.

Xiao, C.N., Dai, H., Liu, H.B., Wang, Y.L. and Tang, H.R. (2008) Revealing the metabonomic variation of rosemary extracts using H^1 NMR spectroscopy and multivariate data analysis. *J. Agric. Food Chem.* **56**, 10142–10153.

Yang, F., Jensen, J.D., Svensson, B., Jorgensen, H.J.L., Collinge, D.B. and Finnie, C. (2010) Analysis of early events in the interaction between *Fusarium graminearum* and the susceptible barley (*Hordeum vulgare*) cultivar Scarlett. *Proteomics*, **10**, 3748–3755.

Ye, C.M., Dickman, M.B., Whitham, S.A., Payton, M. and Verchot, J. (2011) The unfolded protein response is triggered by a plant viral movement protein. *Plant Physiol.* **156**, 741–755.

Zhou, W.C., Eudes, F. and Laroche, A. (2006) Identification of differentially regulated proteins in response to a compatible interaction between the pathogen *Fusarium graminearum* and its host, *Triticum aestivum. Proteomics*, **6**, 4599–4609.

5

Genome-wide transcriptomic and proteomic analyses of bollworm-infested developing cotton bolls revealed the genes and pathways involved in the insect pest defence mechanism

Saravanan Kumar[1,**], Mogilicherla Kanakachari[2,†,**], Dhandapani Gurusamy[2], Krishan Kumar[1], Prabhakaran Narayanasamy[2,‡], Padmalatha Kethireddy Venkata[2,§], Amolkumar Solanke[2], Savita Gamanagatti[3], Vamadevaiah Hiremath[3], Ishwarappa S. Katageri[3], Sadhu Leelavathi[1], Polumetla Ananda Kumar[2,¶] and Vanga Siva Reddy[1,*]

[1]International Centre for Genetic Engineering and Biotechnology, New Delhi, India
[2]National Research Centre on Plant Biotechnology, Indian Agricultural Research Institute (IARI), New Delhi, India
[3]University of Agricultural Sciences, Dharwad, India

*Correspondence
email vsreddy@icgeb.res.in

[†]Present address: College of Agriculture, Department of Entomology, S225 Agriculture Science Centre-N, University of Kentucky, Lexington, KY 40546-0091, USA.
[‡]Present address: Division of Plant Pathology, Indian Agricultural Research Institute (IARI), New Delhi 110012, India
[§]Present address: Department of Crop Physiology, University of Agricultural Sciences (UAS), G.K.V.K, Bangalore 560065, Karnataka, India
[¶]Present address: Division of Biotechnology, Indian Institute of Rice Research (IIRR), Hyderabad 500030, India
[**]These authors contributed equally to this work.

Keywords: *Gossypium hirsutum*, *Helicoverpa armigera*, biotic stress response, transcriptome, proteome, defence mechanism.

Summary

Cotton bollworm, *Helicoverpa armigera*, is a major insect pest that feeds on cotton bolls causing extensive damage leading to crop and productivity loss. In spite of such a major impact, cotton plant response to bollworm infection is yet to be witnessed. In this context, we have studied the genome-wide response of cotton bolls infested with bollworm using transcriptomic and proteomic approaches. Further, we have validated this data using semi-quantitative real-time PCR. Comparative analyses have revealed that 39% of the transcriptome and 35% of the proteome were differentially regulated during bollworm infestation. Around 36% of significantly regulated transcripts and 45% of differentially expressed proteins were found to be involved in signalling followed by redox regulation. Further analysis showed that defence-related stress hormones and their lipid precursors, transcription factors, signalling molecules, etc. were stimulated, whereas the growth-related counterparts were suppressed during bollworm infestation. Around 26% of the significantly up-regulated proteins were defence molecules, while >50% of the significantly down-regulated were related to photosynthesis and growth. Interestingly, the biosynthesis genes for synergistically regulated jasmonate, ethylene and suppressors of the antagonistic factor salicylate were found to be up-regulated, suggesting a choice among stress-responsive phytohormone regulation. Manual curation of the enzymes and TFs highlighted the components of retrograde signalling pathways. Our data suggest that a selective regulatory mechanism directs the reallocation of metabolic resources favouring defence over growth under bollworm infestation and these insights could be exploited to develop bollworm-resistant cotton varieties.

Introduction

Plants and insects have coexisted for about 350 million years leading to the evolution of both positive and negative interactions (Gatehouse, 2002). Positive interactions include insect-mediated pollination, seed dispersion, etc. that offer mutual benefit to the insect and the host, while negative interactions include insect predation (Gatehouse, 2002) that often causes detrimental effects to the host. In view of the long standing relationship, it is quite obvious that plants have evolved a diverse set of stress-specific constitutive and/or inducible defence mechanisms to resist and coexist with the insect pests (Gatehouse, 2002). The response pattern in both the mechanisms might either be activated locally at the infected site, systemically in the uninfected regions or by both the aforementioned through signalling

molecules (Gatehouse, 2002). Signal perception and activation results in a vast cascade of events at cellular and molecular levels ultimately contributing to the defence mechanism. Specialized defence mechanisms that protect plants from insects include physical barriers such as cell wall and cuticle (Kempema et al., 2007), cellular processes including lignifications, cross-linking of cell wall components, release of volatile and nonvolatile metabolites (Kempema et al., 2007) and molecular processes like activation of defence-related genes and pathways (Ramirez et al., 2009; Ryan, 1990). In addition, hormones, transcription factors (TFs) and redox regulators play major role in stress response and defence signalling. Hormones are secondary signals that amplify primary elicitor signals during biotic stress (Yang et al., 1997). Salicylic acid (SA), ethylene (ET), jasmonic acid (JA) and systemin are the major phytohormones that are often quoted as stress-specific signalling molecules (Arimura et al., 2005; Loake and Grant, 2007; Sun et al., 2011; Yang et al., 1997). Moran and Thompson (2001) suggested that there is a complex crosstalk among hormonal pathways that control the plant responses to wounds, insect pest and pathogen attacks. In addition to hormones, pathogen elicitors also activate TFs that interacts with the pathogen-responsive cis elements present in the promoters of the defence-related genes. Even a single pathogen elicitor is capable of activating multiple TFs that can interact with the cis elements present within the same or different promoter regions ultimately leading to stimulation of vast set of defence-related genes and gene products (Yang et al., 1997). Oxidative burst is one of the major processes that occur during biotic stress condition (Lamb and Dixon, 1997). Maintenance of the redox balance within the plant cell plays a crucial role in modulating redox sensitive genes and proteins including many TFs (Torres, 2010). Release of reactive oxygen species (ROS) such as O_2^- and H_2O_2 induces many defence-related events including cell wall reinforcement through lignifications and cross-linking of glycoproteins in the extracellular matrix, activation of defence-related genes, molecules, etc. (Jabs et al., 1997). The above-mentioned cascade of events demand and consume considerable amount of energy. During stress conditions, plant systems manage their biological energy and resources by either partitioning or favouring the molecular machineries towards defence and/or growth. As a result, most plants do survive insect predation; however, it is often accompanied by reduced growth and yield penalty depending on the site of insect predation.

Cotton (Gossypium spp.) is the leading contributor of natural fibre and is an important source of textile commodity, oil and protein meal (Han et al., 2004; Mei et al., 2004). Cotton bolls are crucial tissues that harbour lint (textile fibre) which is of huge economic value. The effect of insect pest infestation followed by secondary infection could lead up to 80% loss in cotton fibre production (Oerke, 2006). Around 1326 species of insects have been reported worldwide as cotton pests, and among them, bollworm (Helicoverpa armigera) is the major pest that directly feed and destroy the developing fibre tissue within the cotton bolls (Dua et al., 2006; Matthews, 1994). Biotic stress induced through insect pest attack regulates cellular events majorly driven by expression changes of genes and their associated pathways. Earlier efforts to understand plant diseases caused by insect, pathogen infestations have identified certain genes and pathways involved in the biotic stress tolerance in cotton (Artico et al., 2014; Dubey et al., 2013; Gao et al., 2013). Systems level analysis at the transcript and protein levels more often reveals the near-complete status of an organism subjected to stress or disease conditions (Komatsu et al., 2009; Srivastava et al., 2013). Such analyses are yet to be employed to understand molecular and cellular mechanisms operational during cotton plant and bollworm interactions. Further understanding of these interactions using high-throughput approaches might reveal stress-induced responses and endogenous resistance mechanisms operational in the host. In this context, we have made an attempt to understand the mechanisms adapted by cotton plant during bollworm attack using both transcriptomic and proteomic tools. As bolls are the target site for fibre synthesis as well as bollworm feeding, we have performed comparative analyses of developing cotton bolls subjected to bollworm infestation. Our comprehensive genome-wide analyses have revealed several new and interesting insights about cotton plant and bollworm interactions. Knowledge gained through this

study could be further exploited to develop bollworm-resistant cotton varieties.

Results

Differentially regulated transcripts and proteins in bollworm-infested cotton bolls

Cotton plants were grown in the field conditions following common agronomic practices. Only bolls that were infected by cotton bollworm (H. armigera) insect larvae were used for further analysis (Figure 1a,b). To study the effect of insect stress in developing boll tissue, microarray-based transcriptome profiling and two-dimensional gel electrophoresis (2D PAGE) followed by MALDI TOF/TOF-based proteome analyses were carried out at different developmental stages (Figure 1b,c). Labelled mRNA was hybridized to Affymetrix cotton GeneChip Genome array. Identified transcripts with a false discovery rate (FDR) adjusted P value ≤0.01 and fold change ≥3 were considered as differentially expressed transcripts (DETs) (Figure 1d; Table S1). In total, 8694 transcripts comprising 39% of the total transcripts present on the cotton GeneChip showed differential expression under bollworm infestation. Transcripts were annotated using the Arabidopsis TAIR protein database version 10 through BLASTX with E value cut-off ≤e-10. Identification of transcripts related to TFs, phytohormones and signal transduction were attained using Arabidopsis transcription factor and Arabidopsis hormone databases, respectively, as mentioned in Materials and methods. Classification of DETs under different functional categories was attained using MIPS functional catalogue. Gene expression patterns in response to bollworm infestation were classified using hierarchical clustering (Figure 1e). Differentially expressed transcripts showing consistent up- and down-regulation among boll developmental stages are tabulated (Figure 2a,b; Tables S4 and S5).

Two-dimensional gel electrophoresis (2D PAGE)-based proteome analysis of bollworm-infested (biotic stress-induced BS) and noninfested (control, CN) cotton bolls showed an average of 393 reproducibly detected protein spots across developmental stages (0, 2 and 5 dpa) (Figures 3a–c and S1a–c). At least two independent replicate gels were generated per sample (Control, CN vs bollworm infested, BS). Only those spots that were reproducibly detected were further considered for comparative analysis. Quantification of the protein spots was attained using the per cent volume criterion that corresponds to the expression level of the detected spot regions. Protein spots showing fold change of ±1.5 with a P value <0.5 as mentioned in Materials and methods were considered as differentially expressed spots. Comparative analysis revealed that around 35% of the detected spots (137 spots) were differentially expressed (±1.5-fold), and among them, 98 spots were identified using MALDI TOF/TOF (Table S14). Further Gene Ontology (GO)-based annotation and functional classification of the DEPs under various categories were attained using BLAST2GO platform version 2.7 (Figure 2e,f).

Comparative analysis of the transcriptome and proteome data sets highlighted 37 overlapping unique accessions that were quantified at both the transcript and protein levels (Table S15). Among them, only 10 genes (accessions) were found to have similar expression pattern in at least one of the developmental stages that were analysed in this study. Such poor corelation among the transcriptome and proteome data sets could either be attributed to the post-transcriptional regulation of the identified genes or be the experimental

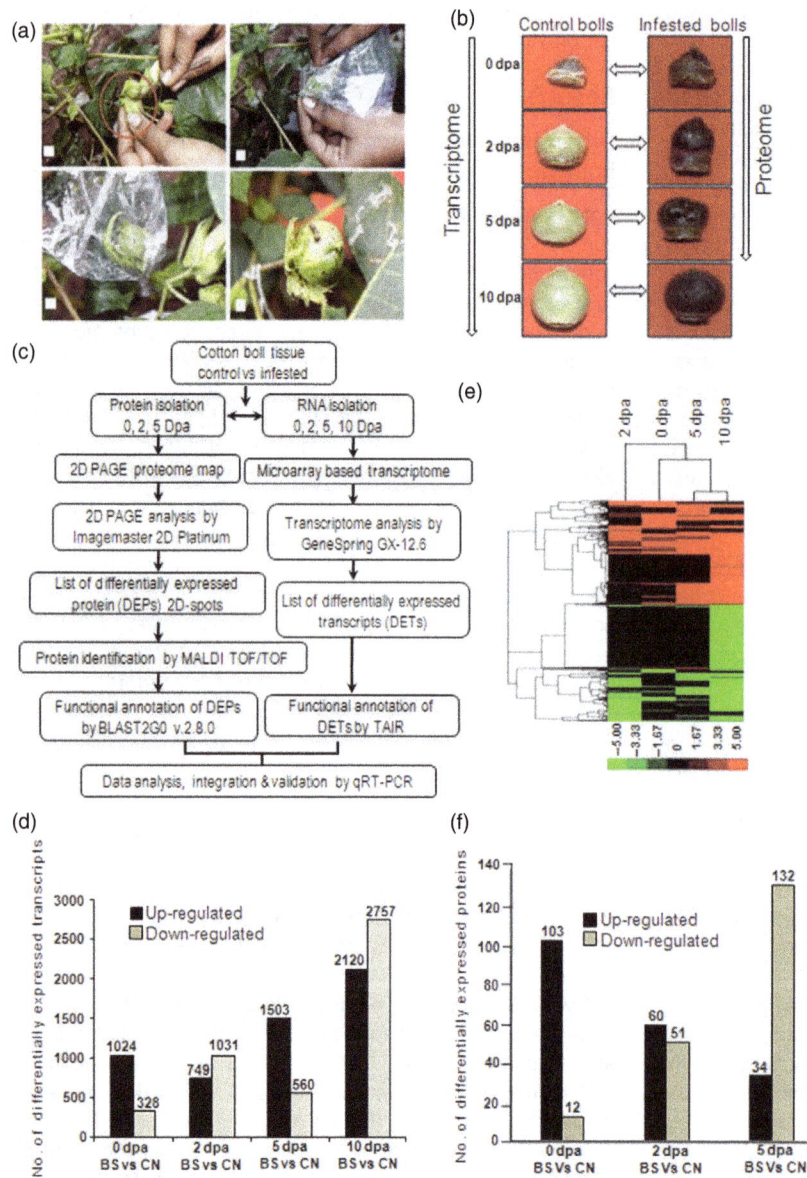

Figure 1 Bollworm infested biotic stress induction in cotton bolls, G. hirsutum L. cv. Bikaneri Narma. (a) Method of biotic stress induction in cotton bolls under field conditions. (b) Boll developmental stages of control (CN) and bollworm infected tissues (BS) used in the current study (0, 2, 5 and 10 dpa/days post anthesis). (c) Schematic overview of proteome and transcriptome data generation and analyses workflow. (d) Number of differentially expressed transcripts (DETs) during boll development stages under BS as compared to their respective stages of CN. (e) Cluster analysis showing the differentially expressed transcripts related to biotic stress. (f) Number of differentially expressed proteins (DEPs) during boll development stages under BS as compared to their respective stages of CN.

limitations associated with the transcript and protein turnover measurements.

Bollworm infestation induces early and consistent response in developing cotton bolls

Analysis of the transcriptome and proteome profile of the bolls that were subjected for only 8 h of bollworm infestation (0 dpa) showed 1352 (15.55%) DETs and 115 (36.37%) DEPs (Figure 1d, f). Majority of those transcripts (75.73%) and proteins (89.5%) were exclusively up-regulated in biotic stress-induced bolls (BS) at 0 dpa (Figure 1d,f and 4). Cluster analysis of the DETs and DEPs revealed that the gene expression pattern gradually changed upon boll development (Figure 5a,b). Briefly, 42.07% of 2-dpa, 72.85% of 5-dpa, 43.46% of 10-dpa transcripts were up-regulated, while 54.05% of 2 dpa and only 25.75% of 5 dpa proteins were up-regulated in BS-induced bolls (Figure 1d,f). Analysis of the transcriptome data sets revealed an increase in the number of up-regulated transcripts at 5 and 10 dpa (Figure 1d). However, such pattern was not observed at protein levels as the proteome profile showed a steady decline towards development

(0–5 dpa, Figure 1f). In support of such drastic decline in transcript and protein populations, infested bolls showed compromised growth in terms of size accompanied by infection-related symptoms such as browning and rotting (Figure 1b). The discrepancies among transcriptome and proteome pattern on the one hand and growth of infested boll on the other suggest that the plant system's continuous effort to encounter pest attack at transcript level is somehow not been translated to the protein level. Nevertheless, the above-mentioned discrepancies can also be attributed to the limitations associated with transcript and protein quantitation procedures.

Defence signalling involves major reprogramming of metabolic and biosynthetic pathways

Gene Ontology-based functional annotation revealed that 62% of the DEPs were enzymes involved in regulating metabolic processes such as carbohydrate (10%), amino acid (19%) and lipid (18%) metabolisms (Figure 2e). Cellular component-based classification showed that DEPs were distributed among plastid (20%), mitochondria (14%), nucleus (15%), extracellular (19%)

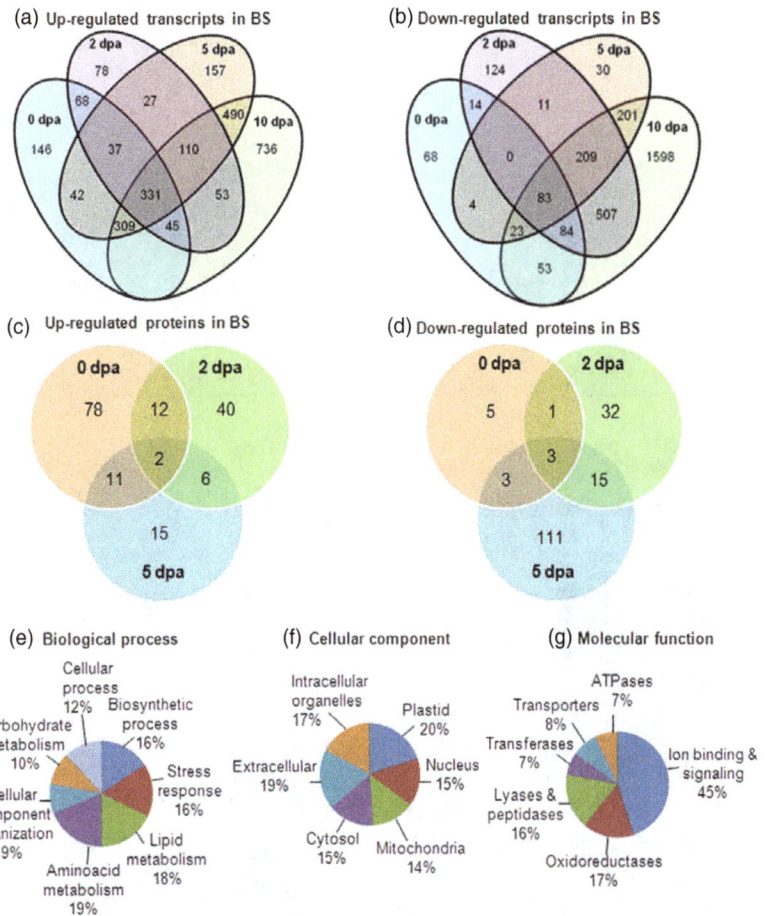

Figure 2 Diagrammatic view of the up- and down-regulated transcripts (a, b) and proteins (c, d) among the boll developmental stages under bollworm infestation in comparison with their respective controls. Gene Ontology-based annotation and classification of the differentially expressed proteins into the (e) biological process, (f) cellular component and (g) molecular function categories.

and cytosolic (15%) localizations (Figure 2f). Molecular function-based classification showed that majority of the DEPs were involved in signalling (45%) and redox regulation (17%) (Figure 2g). Significantly enriched pathways corresponding to up- and down-regulated transcripts are presented in Table 1. Hierarchical clustering clearly showed that there were certain population of transcripts and proteins that were down-regulated upon development and certain others that were consistently up-regulated across the developmental stages (Figure 5a,b). To explain the reprogramming pattern, we have focussed on the key metabolic pathways that showed major differences in terms of their expression exclusively during biotic stress conditions. Differentially expressed transcripts and DEPs related to trehalose, raffinose, malate, starch and cell wall metabolism majorly accounted for the carbohydrate metabolism (Table S9). Briefly, up-regulated transcripts related to trehalose phosphate synthase (TPS) and trehalose phosphatase (TPP) that are involved in trehalose biosynthesis (Tables 2 and S9) were identified in this study (Wingler, 2002). Two unique isoforms of galactinol synthases (GolS) involved in galactinol synthesis were found to be consistently up-regulated in our data set (Tables 2, S4, and S9). Further, our study also showed the up-regulated transcripts of malate synthase (MS) and down-regulated malate dehydrogenase (MDH) transcripts and proteins (Tables S9 and S14). Malate synthase is involved in the synthesis of malate, whereas MDH catalyses the oxidation of malate to pyruvate and CO_2. The above-mentioned pattern in turn correlates with stress-responsive accumulation of trehalose, raffinose and malate in the infested bolls. In plants, synthesis and degradation of nonstructural

carbohydrates (NSCs) such as starch plays a major role in the regulation of carbon source availability during growth and unfavourable conditions (Sulpice et al., 2009). Data curation revealed the down-regulated transcripts involved in starch biosynthesis such as starch synthase, ADP-glucose pyrophosphorylase (AGPase) and up-regulated enzymes of starch degradation pathway such as starch excess 1 and starch binding domain-containing glycoside hydrolase (Yano et al., 2005) (Table 2, S1). Further analysis showed the down-regulated group of carbohydrate active enzymes (CAZymes) that catalyse cell wall metabolism-related processes such as loosening, elongation, grafting and maturation (Table S7). Analysis of nitrogen, amino acid and protein metabolism genes revealed that glutamine metabolism was up-regulated and ubiquitin cascade-related genes were differentially regulated. Pathway curation revealed that glutamine synthase isoform 1 (GS1), glutamate dehydrogenase (GDH) and nitrate reductase (NiR) involved in nitrogen metabolism were found to be up-regulated (Tables S4 and S14). In case of ubiquitin cascade, we observed that ubiquitin ligase (E3) was up-regulated, whereas ubiquitin conjugating enzyme (E2) was down-regulated in BS condition at transcript and protein levels (Tables S4, S11, and S14). This in turn suggests the onset of rate limiting pattern or controlled proteolysis through ubiquitin-mediated protein degradation during stress. In addition to the metabolic enzymes, we also observed up-regulated members of both sugar and amino acid transporters throughout the developmental stages suggesting active transportation processes (Table S12). Further, our study also revealed that fatty acid and lipid metabolism-related genes were differentially regulated (Table S10). Pathway mapping and

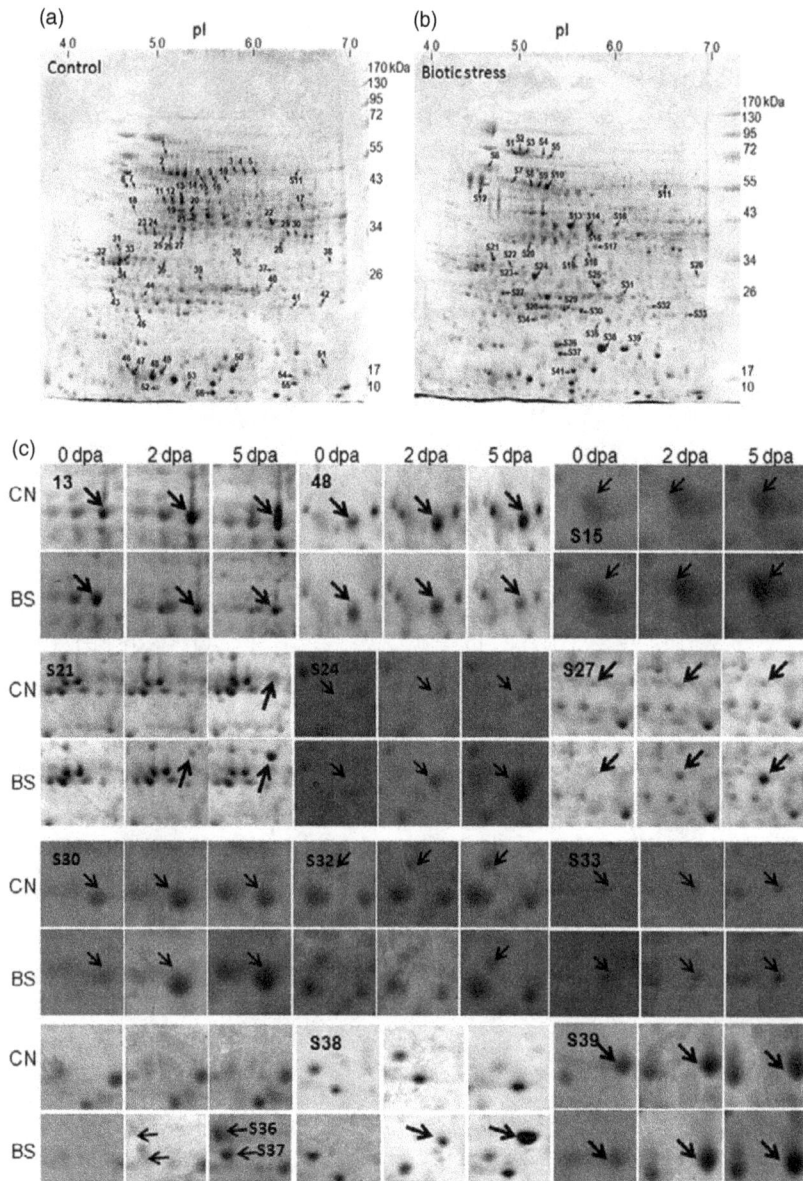

Figure 3 Representative Coomassie stained 2D PAGE proteome profile of Control, CN (a) and Boll worm infested, BS (b) cotton bolls. Annotated 2D spots corresponding to the differentially expressed proteins identified using MALDI TOF/TOF. Detailed list of identified proteins are tabulated in Table S14. (c) 2D Spot profile of representative proteins showing differential expression under bollworm infestation in comparison with their respective control bolls during boll developmental stages (0, 2, 5 dpa). 2D PAGE proteome profile of infested and control bolls during developmental stages are presented in Figure S1.

curation suggested stimulation as well as repression of different lipid precursor pathways. Briefly, majority of the up-regulated genes were involved in glycolipid and phospholipid metabolism including enzymes related to α-linolenic acid metabolism that ultimately lead to JA biosynthesis (Tables S10 and S14). On the other hand, we observed that down-regulated genes such as 3-oxo-5-alpha-steroid 4-dehydrogenase (DET 2) and 24-sterol C-methyltransferase (SMT2-2) were involved in sterol biosynthesis. Sterols are membrane lipids that serve as precursor molecules for brassinosteroid (BR) biosynthesis (Table S10).

Repression of chloroplast, mitochondrial metabolism and cellular growth is evident during stress

In this study, the redundant expression pattern observed in majority of the metabolic pathways included both up- and down-regulated genes. However, in case of photosynthesis, most of the DETs and proteins were found to be majorly down-regulated throughout the developmental stages. Briefly, genes related to photosystem I, II, ATP synthase, light harvesting complex,

chlorophyll binding proteins, etc. were consistently down-regulated under BS condition (Tables S8 and S14). Also genes encoding enzymes involved in carbon fixation including ribulose-1, 5-bisphosphate carboxylase/oxygenase (RuBisCO) and members of tricarboxylic acid cycle were found to be down-regulated (Table S8). However, we observed transcripts related to pentatricopeptide repeat containing protein (GUN1) were up-regulated (Tables S2 and S11; Figure 6). Further, transcripts related to mitochondria such as ATP synthase (mitochondrial), phosphate transporter, mitochondria-associated membrane gly-coprotein (MAM33), etc. were found to be down-regulated (Tables S9 and S12), while transcripts related to mitochondrial alternative oxidase (AOX) involved in alternative respiratory pathway were found to be consistently up-regulated (Tables 2 and S1) (Vanlerberghe and McIntosh, 1997). In addition, cellular growth-related genes such as cytoskeleton proteins, annexins, profilins and expansins were found to be down-regulated upon development (Tables S7 and S14). Analysis of cell cycle and DNA replication-related genes revealed that members of cyclin-

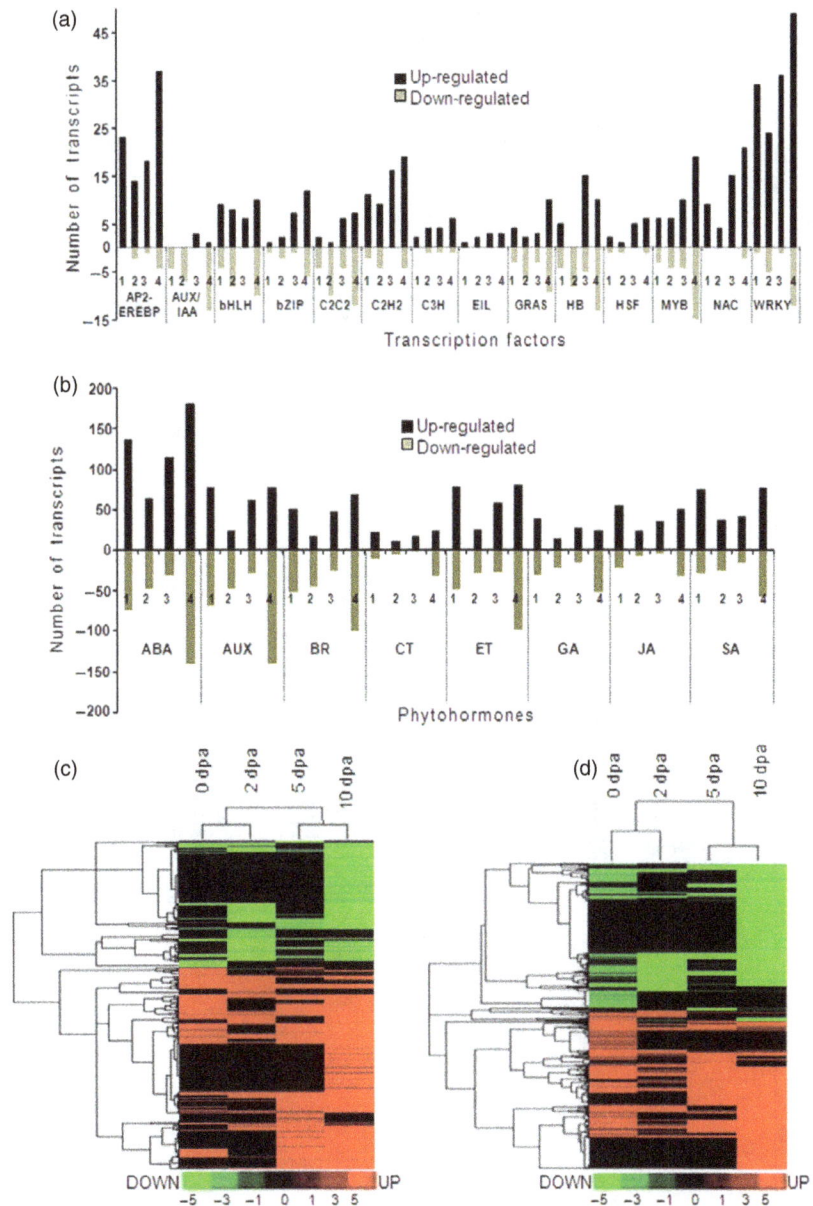

Figure 4 Differentially expressed transcription factors (a) and phytohormone (b) under bollworm infestation as compared to their respective control. Numbers 1–4 represents different stages; (1)—0 dpa, (2)—2 dpa, (3)—5 dpa and (4)—10 dpa. Cluster analysis performed using log_2-transformed fold change values showing the differentially expressed transcripts related to transcription factors (c) and phytohormones (d) at boll developmental stages. Putative transcription factors and phytohormones at each stage are presented in Tables S2 and S3.

dependent protein kinase family, cell division-related proteins, histones, DNA replication factors, etc. were down-regulated in BS-induced bolls (Table S7).

Calcium and redox signalling pathways are stimulated to regulate host response

Calcium (Ca^{2+}) is often quoted as the second important messenger that activates signalling cascades in response to various stimuli including biotic stress (Sanders et al., 2002). Genes encoding Ca^{2+}/calmodulin-binding proteins, calcineurin B-like proteins, Ca^{2+} binding EF-hand family proteins, Ca^{2+}-dependent ATPases were found to be up-regulated under BS conditions (Tables 1 and S14). Protein kinase cascades account for the major contributors of induced immunity in plants. In this study, we observed the up-regulation of two major stress-activated protein kinases such as calcium-dependent protein kinases and mitogen-activated protein kinases. Also, other kinases such as SNF1(sucrose nonfermenting)-related protein kinase, calcineurin B-like interacting protein kinase-6 (CIPK6),

ankyrin protein kinases were found to be significantly up-regulated in BS condition. The above-mentioned Ca^{2+} signalling molecules can be broadly classified into nonenzymatic sensor proteins and enzymatic proteins. These proteins together constitute a network that not only maintains the intracellular calcium levels but are also involved in other signalling events including protein activation. Among the Ca^{2+}-dependent enzymes, majority were observed to be kinases involved in phosphorylation reaction that determines the activity of substrate protein molecules. So we further analysed the data sets for kinase substrates and the pathways regulated by these kinases during stress. Interestingly, we observed CDPK-activated protein substrates such as NADPH oxidases and oxidoreductases that were up-regulated during later stages of development (10dpa) (Dubiella et al., 2013). Manual curation of data sets revealed that members of peroxidases, superoxide dismutases (SODs), glutathione S-transferases (GSTs), etc. were found to be differentially regulated at transcript and protein levels (Tables S13 and S14). These enzymes are actively involved in the ROS

Table 1 Significantly regulated metabolic pathways under biotic stress

Bin no.	Probe set ID	Cotton gene accessions	Arabidopsis ortholog ID	Gene function	Boll developmental stages (log 2-transformed fold change values)			
					0 dpa	2 dpa	5 dpa	10 dpa
Cell wall								
10.1.2	ghiaffx.19354.1.s1_s_at	DW488498.1	AT4G10960	UGE5 (UDP-D-glucose/UDP-D-galactose 4-epimerase 5)	35.22	3.95	5.46	10.6
10.1.30.3	ghiaffx.59913.1.a1_s_at	DW501774.1	AT3G01640	GHMP kinase family protein	6.05	3.93	3.41	10.81
10.2	ghi.3263.1.a1_at	DT467895		CSLE6—Cellulose synthase-like family E	67.36	14.35	3.28	23.75
10.5.1.1	ghi.2198.1.s1_s_at	DT550268	AT5G55730	FLA1 (Fasciclin-like arabinogalactan 1)	-46.39	-4.13	-15.9	-3.85
10.6.3	gra.2038.1.a1_s_at	CO124831	AT1G49320	BURP domain-containing protein	-13.63	-3.18	-5.12	-3.59
10.6.3	ghi.7933.1.s1_at	AF410458.1	AT3G07970	QRT2 (QUARTET 2); polygalacturonase	8.16	33.95	3.78	198.2
10.6.3	ghi.4648.2.a1_s_at	AY464057.1	AT1G49320	BURP domain-containing protein	-11.98	-3.17	-4.78	-3.57
10.7	ghi.10493.1.s1_s_at	DT466412	AT5G57560	TCH4 (Touch 4)	12.36	7.98	4.52	7.42
10.7	ghi.6188.1.a1_at	CO493635	AT2G39700	ATEXPA4 (Arabidopsis thaliana Expansin A4)	-76.94	-3.49	-4.22	-3.12
10.7	graaffx.27319.1.s1_s_at	CO089724	AT2G36870	Xyloglucan: xyloglucosyl transferase	-20.46	-7.63	-9.77	-3.68
10.8.1	ghi.3458.1.a1_at	DT465599	AT1G02810	Pectinesterase family protein	15.41	12.31	14.09	11.89
Secondary metabolism								
16.1.5	ghi.3337.1.a1_at	DT467161		DT467161; DB_XREF = GH_CHX19P12.r	20.89	12.27	10.09	20.18
16.2	ghi.5889.2.s1_s_at	Ghi.5889	AT3G26040	Transferase family protein	-32.25	-4.95	-9.9	-7.21
16.2	graaffx.15820.1.s1_s_at	CO123702	AT5G23940	EMB3009 (embryo defective 3009)	-15.72	-3.68	-5.99	-7.07
16.7	ghi.9734.1.s1_at	AW186914	AT5G55340	Long-chain alcohol O-fatty-acyltransferase family protein	-20.5	-4.1	-4.88	-11.25
16.1	ghiaffx.2668.1.a1_at	DT463855	AT5G05390	LAC12 (laccase 12)	117.15	28.42	6.33	23.95
16.1	ghi.3427.1.a1_s_at	DT465993	AT5G09360	LAC14 (laccase 14)	13.57	21.96	8.45	7.89
16.1	ghi.9266.1.a1_at	DT468530	AT4G39830	L-ascorbate oxidase	17.25	4.82	5.36	5.68
Hormone signalling								
17.2.3	gra.275.1.s1_s_at	CO090219	AT1G60710	ATB2; oxidoreductase	25.46	5.4	4.28	13.3
17.1.3	ghi.2169.1.a1_at	DT468931	AT5G13200	GRAM domain-containing protein/ABA-responsive protein-related	130.07	11.99	10.05	19.8
17.5.1	graaffx.20159.1.a1_s_at	CO110779	AT1G05010	ACO4, 1-aminocyclopropane-1-carboxylate oxidase	29.6	4.69	18.42	13.57
17.5.1	ghiaffx.4691.1.a1_at	DW517948.1	AT1G78440	ATGA2OX1 (gibberellin 2-oxidase 1)	14.1	26.98	14.1	18.17
17.5.1	ghi.6898.1.s1_at	CA992714	AT5G24530	DMR6 (Downy mildew-resistant 6)	20.92	3.44	3.17	6
17.5.1.1	ghi.5451.1.s1_at	DQ122174.1	AT4G11280	ACS6 (1-Aminocyclopropane-1-carboxylic acid (ACC) synthase 6)	17.73	29.71	5.56	44.56
17.5.2	ghiaffx.39399.1.s1_at	DW499761.1	AT5G47220	ERF2 (ethylene-responsive element binding factor 2)	18.53	9.83	6.81	12.43
17.8.1	ghiaffx.43038.1.s1_at	DW497938.1	AT1G19640	JMT (Jasmonic acid carboxyl methyltransferase)	10.54	36.36	18.85	14.13
17.7.1.2	ghi.1739.4.s1_x_at	DT464580	AT1G17420	LOX3; lipoxygenase	22.56	7.96	14.44	7.55
17.7.1.2	gra.38.1.s1_s_at	CO123081	AT3G45140	LOX2 (lipoxygenase 2)	11.39	5.32	15.23	5.91
17.7.2	ghi.10327.1.s1_s_at	CO123081	AT1G19180	JAZ1 (Jasmonate zim-domain protein 1)	65.2	4.87	5.13	14.28
Respiratory function								
20.1.1	ghi.9151.2.s1_at	DT049218	AT5G51060	RHD2 (Root hair defective 2)	11.00	10.03	8.79	4.52
Defence genes								
20.1.7	ghiaffx.8053.1.a1_at	DW518810.1	AT1G64160	Disease resistance-responsive family protein/Dirigent family protein	10.77	77.58	5.85	7.53
20.1.7.6.1	ghiaffx.61299.1.s1_at	DW508705.1	AT1G17860	Trypsin and protease inhibitor family protein/Kunitz family protein	45.7	24.03	14.33	153.28
Heat-shock proteins								

Table 1 Continued

Bin no.	Probe set ID	Cotton gene accessions	Arabidopsis ortholog ID	Gene function	Boll developmental stages (log 2-transformed fold change values)			
					0 dpa	2 dpa	5 dpa	10 dpa
20.2.1	graffx.28942.1.s1_s_at	CO085044	AT4G36990	HSF4 (Heat-shock factor 4)	13.22	3.58	4.07	5.71
20.2.1	gfiaffx.11624.1.s1_s_at	DW505128.1	AT4G27670	HSP21 (Heat-shock protein 21)	3.18	3.84	3.79	10.39
20.2.1	gfi.10640.1.s1_s_at	DN780720	AT3G12580	HSP70 (Heat-shock protein 70)	5.7	3.17	6.44	14.67
20.2.1	gfi.9308.1.s1_at	DT527234	AT5G59720	HSP18.2 (Heat-shock protein 18.2)	4.47	5.41	7.22	15.12
Redox regulators								
21.3	gfi.8085.1.s1_at	AF329368.1	AT2G16060	AHB1 (Arabidopsis haemoglobin 1)	26.19	12.11	20.59	4.35
21.3	gcraffx.24390.1.s1_s_at	BF273948	AT3G10520	AHB2 (Arabidopsis haemoglobin 2)	−8.71	−10.68	−4.49	−3.05
21.4	gfiaffx.62078.1.s1_at	DW517172.1	AT1G28480	GRX480; Electron carrier/Protein disulphide oxidoreductase	4.03	4.43	3	7.31
26.12	gfi.7950.1.s1_at	AY366083.1	AT5G06720	Peroxidase	61.08	400.1	3.34	16.95
26.9	gfiaffx.2752.1.s1_at	CA993163	AT2G29420	ATGSTU7 (Arabidopsis thaliana glutathione s-transferase tau 7)	91.46	8.58	12.83	9.66
Transcription factors								
27.3.3	gfi.10443.1.s1_at	DT049130	AT1G19210	AP2 domain-containing transcription factor	8.38	14.35	19.22	3.42
27.3.3	gtaaffx.196.1.a1_s_at	AY572462.1	AT3G16770	ATEBP (Ethylene-responsive element binding protein)	45.61	58.16	10.37	21.92
27.3.3	gfiaffx.25508.1.s1_at	DW496030.1	AT5G50080	ERF110, DNA binding/Transcription factor	183.63	28.72	49.8	15.07
27.3.35	gfiaffx.4349.1.a1_s_at	DW509077.1	AT3G62420	ATBZIP53 (Basic region/Leucine zipper motif 53)	6.93	4.37	3.72	6.81
27.3.35	gfi.5267.1.a1_at	DT046626	AT4G34590	GBF6 (G-Box binding factor 6)	−144	−7.71	−23	−5.66
27.3.25	gfiaffx.8694.1.s1_a_at	DW499451.1	AT4G05100	AtMYB74 (Myb domain protein 74)	233.09	28.85	14.35	77.79
27.3.25	gfi.10620.1.s1_at	DT465545	AT4G37260	MYB73 (Myb domain protein 73)	12.84	10.67	3.75	8.2
27.3.32	gra.1530.2.s1_s_at	CO125258	AT2G03340	WRKY3; Transcription factor	12.89	3.59	3.8	9.18
27.3.32	gfi.9240.1.s1_s_at	DT468893	AT2G38470	WRKY33; Transcription factor	14.81	3.33	5.33	5.47
27.3.32	gfi.9192.1.s1_at	DT468825	AT1G80840	WRKY40; Transcription factor	8.88	11.55	4.95	4.86
27.3.32	gfiaffx.30199.1.s1_at	DW506814.1	AT2G47260	WRKY23; Transcription factor	43.52	30.64	3.93	22.82
27.3.32	gfiaffx.6177.1.s1_at	DW505740.1	AT5G13080	WRKY75; Transcription factor	3.91	4.33	3.28	4.62
27.3.32	gfi.9182.3.s1_at	DT461494	AT1G62300	WRKY6; Transcription factor	59.11	8.2	5.34	14.9
Protein degradation								
29.5.1	gra.1032.1.s1_s_at	CO123218	AT5G67360	ARA12; Serine-type endopeptidase	−7.45	−3.52	−6.69	−3.79
29.5.1	gfiaffx.53444.1.a1_s_at	DW229324.1	AT2G05920	Subtilase family protein	−47.35	−3.35	−17.5	−3.71
29.5.4	gfi.7891.1.s1_s_at	DT462224	AT1G03220	Extracellular dermal glycoprotein	322.82	34.58	9.5	54.08
29.5.4	gfi.10117.2.a1_s_at	DT462947	AT1G44130	Nucellin protein	7.56	7.06	4.6	3.93
29.5.7	gfi.6869.1.a1_at	CA992801	AT1G70170	MMP (Matrix metalloproteinase)	37.2	3.78	3.15	13.9
29.5.7	gfi.6849.1.a1_at	CA992877	AT5G15250	FTSH6 (FTSH protease 6)	42.13	4.35	4.19	31.63
29.5.9	gfiaffx.12228.1.s1_at	DW512183.1	AT2G46620	AAA type ATPase family protein	18.34	3.48	3.78	12.66
29.5.11.4.2	gfi.6875.2.a1_at	DT466966	AT3G02840	Immediate-early fungal elicitor family protein	43.34	7.19	21.82	14.16
29.5.11.4.2	gfiaffx.13045.1.s1_at	DW226240.1	AT1G29340	PUB17 (Plant U-box 17)	16.02	3.63	5.23	7.17
29.5.11.4.2	gfi.6875.1.s1_at	CA992786	AT1G66160	U-box domain-containing protein	26.45	3.69	9.54	9.73
29.5.11.4.2	gfiaffx.15780.1.a1_s_at	DW224570.1	AT4G03510	RMA1; protein binding/Ubiquitin–protein ligase/Zinc ion binding	13.96	3.93	6.46	9.12
29.5.11.4.2	gfi.2766.2.s1_s_at	DT550860	AT3G05200	ATL6; protein binding/Zinc ion binding	6	5.23	3.45	4.26
29.5.11.4.2	gfi.4550.1.a1_at	DT051102	AT5G42200	Zinc finger (C3HC4-type RING finger) family protein	15.82	7.93	3.08	21.12

Table 1 Continued

Bin no.	Probe set ID	Cotton gene accessions	Arabidopsis ortholog ID	Gene function	Boll developmental stages (log 2-transformed fold change values)				
					0 dpa	2 dpa	5 dpa	10 dpa	
29.5.11.4.2	ghi.965.2.s1_at	DT467116	AT2G37150	Zinc finger (C3HC4-type RING finger) family protein	4.48	3.46	3.25	5.48	
29.5.11.4.3.2	ghi.8366.1.s1_s_at	DT462696	AT1G30200	F-box family protein	37.35	15.69	15.9	16.16	
Signalling									
30.1.1	ghi.10316.1.s1_s_at	DT051688	AT2G35150	EXL1 (Exordium-like 1)	−83.47	−3.57	−8.79	−6.26	
30.1.1	gra.1243.1.s1_s_at	CO123292	AT2G17230	EXL5 (Exordium-like 5)	−66.14	−9.46	−11.3	−3.47	
30.1.1	ghiaffx.45244.1.s1_x_at	DT051989	AT5G51550	EXL3 (Exordium-like 3)	−29.46	−5.73	−15.9	−5.58	
30.2.3	ghi.8869.1.s1_s_at	DT457129	AT2G26730	Leucine-rich repeat transmembrane protein kinase	−18.69	−3.85	−13.9	−6.78	
30.2.11	ghi.5385.1.s1_s_at	DT047980	AT5G66330	Leucine-rich repeat family protein	−24.57	−7.2	−26.1	−3.48	
30.2.17	ghiaffx.15325.1.s1_at	DT466359	AT2G38090	MYB family transcription factor	4.07	12.7	11.78	4.74	
30.2.17	graaffx.26883.1.a1_s_at	CO090919	AT1G56130	Leucine-rich repeat family protein/Protein kinase family protein	14.95	9.38	4.26	4.78	
30.2.20	ghiaffx.24594.1.s1_at	DT466937	AT1G18390	ATP binding/protein tyrosine kinase	7.52	7.5	4.9	4.09	
30.2.23	ghi.9233.1.a1_at	DT465825	AT1G11050	Protein kinase family protein	43.61	9.59	6.17	8.26	
30.3	ghiaffx.25338.1.a1_at	DW515811.1	AT5G37780	CAM1 (Calmodulin 1)	15.32	8.21	10.95	4.95	
30.3	ghiaffx.12687.1.a1_at	DW489633.1	AT2G41860	CPK14; ATP binding/Protein tyrosine kinase	4.24	4.07	7.55	3.22	
30.3	ghiaffx.40480.2.s1_at	DW497946.1	AT5G24270	SOS3 (Salt overly sensitive 3)	8.35	4.03	4.02	6.62	
30.3	ghi.4855.1.a1_at	CA993640	AT5G49480	ATCP1 (Ca²⁺-binding protein 1)	13.48	3.85	3.21	7.31	
30.3	ghi.1114.1.s1_s_at	DN817396	AT5G39670	Calcium-binding EF-hand family protein	49	14.56	4.36	9.21	
30.3	ghi.3763.1.a1_s_at	DT461952	AT3G63380	Calcium-transporting ATPase, plasma membrane-type	7.63	8.69	3.3	4.96	
30.3	ghiaffx.42473.1.s1_at	DW508297.1	AT5G13460	IQD11 (IQ domain 11); Calmodulin-binding	−17.58	−3.82	−12.7	−7.35	
30.3	ghi.2692.1.s1_at	DT465311	AT2G15760	Calmodulin-binding protein	13.32	6.84	3.88	4.16	
30.3	ghiaffx.6766.1.s1_at	DW504268.1	AT5G54490	PBP1 (Pinoid-binding protein 1)	13.38	34.38	19.25	11.08	
30.6	ghi.6088.2.s1_at	DT467539	AT3G45640	ATMPK3 (Arabidopsis thaliana mitogen-activated protein kinase 3)	54.54	9.53	3.27	21.63	
30.8	ghiaffx.23436.1.s1_x_at	DR455118	AT5G67070	RALFL34 (Ralf-like 34)	−18.79	−3.06	−6.19	−7.91	

Table 2 Expression pattern and biological role of crucial genes identified in the bollworm-infested cotton bolls through transcriptome and proteome approaches

S no.	Accession	Transcript ID/Protein ID	Expression pattern (fold change values)				Biological significance
			0 dpa	2 dpa	5 dpa	10 dpa	
Carbohydrate metabolism—Stabilizers & osmoprotectants							
01	AY628139.1	Trehalose 6-phosphate synthase (TPS)	4.79	–	10.26	9.43	Trehalose biosynthesis.
02	DT462221		3.57	4.32	10.51	13.89	Stabilization of macromolecular structures
03	DT465672		18.25	–	58.87	82.78	
04	DT464172	Trehalose-6-phosphate phosphatase (TPP)	–	–	3.12	25.58	
05	DW238688.1	Galactinol synthase 1 (GolS)	23.33	4.21	8.38	5.78	Raffinose biosynthesis.
06	DW227458.1	Galactinol synthase 2 (GolS)	3.98	29.09	7.78	84.01	Raffinose–osmoprotectants
07	X52305.1	Malate synthase	5.77	–	134.42	355.63	Malate biosynthesis
08	DW232560.1	Malate dehydrogenase	-3.54	-3.22	–	25.44	
	gi|11133373*	Malate dehydrogenase	-7.356	-1.868	-5.188	–	
09	DT455780	Starch synthase III	-4.55			-5.64	Starch synthesis.
10	DN760794	Putative starch synthase				-14.69	Starch-carbon source during unfavourable conditions
11	CO121872	Starch Excess 1 (SEX-1)			-3.27		
12	DW501538.1	Starch binding Glycoside hydrolase			4.24	6.70	
Amino acid metabolism—Stress-specific protein isoforms, nitrogen mobilization							
13	gi|21196462* CA993194	Glutamine synthase	–	1.76	1.61	-5.280	Biotic stress-specific protein isoform GS1 (Figure 3c, spot 15)
14	gi|121334*	Glutamine synthetase PR-1 (GS1)	1.53	1.54	1.72	–	
15	gi|9969*	Glutamate-ammonia ligase	2.1002	1.3649	1.2754	–	
16	DT462524	Glutamate dehydrogenase (GDH)	3.72	3.22	9.32	8.51	
17	CO085887	Nitrate reductase (NiR)	6.91	10.53	18.72	70.65	
Lipid metabolism—Phytohormone biosynthesis							
18	DT047194	Allene Oxide synthase (AOS)	3.57	5.11	4.13	–	α-Linolenic acid metabolism. Jasmonic acid biosynthesis
20	gi|40642247*	Allene oxide cyclase (AOC)	8.10	2.59	1.36	–	
21	DQ116446.1	3-oxo-5-alpha-steroid	–	–	-3.83	-9.09	Brassinosteroid biosynthesis
22	AJ513325	4-dehydrogenase (DET2)	–	–	-5.14	-5.23	
23	DT564348	Sterol methyl transferase (SMT-2)	-6.84	-14.19	-3.02	-14.51	
24	CO122079	Zeaxanthin epoxidase	-3.21	–	–	–	ABA biosynthesis
25	DQ122174.1	Aminocyclopropane-1-carboxylate synthase (ACC)	29.71	5.56	44.57	17.73	Ethylene biosynthesis
Retrograde signalling components and salicylic acid suppressors							
26	DT462103	WRKY 40	124.03	61.51	1.29	29.29	Re Retrograde signalling (Mitochondria)
27	DT466107	Alternative Oxidase (AOX)	33.53	3.66	19.82	94.72	
28	CA993875	Pentatricopeptide repeat containing protein (GUN1)	14.25	63.25	21.79	59.99	Retrograde signalling (Chloroplast)
29	DT468893	WRKY 33	3.33	5.33	5.48	14.81	
30	DW506256.1	Ethylene insensitive 3 (EIN 3)	3.15	4.24	7.72	13.69	Suppressors of Salicylic Acid
31	DW241764.1	Enhanced Disease Resistance 1(EDR1)	–	–	–	3.49	
32	CO085044	Heat-shock trancription factor (HSf-1)	3.59	4.07	5.71	13.23	Regulator of GolS

Table 2 Continued

S no.	Accession	Transcript ID/Protein ID	Expression pattern (fold change values)					Biological significance
			0 dpa	2 dpa	5 dpa	10 dpa	
Defence molecules							
33	DT554033	Chitinase	40.76	21.29	8.05	49.51	Defence, (Figure 3c, spot 27)
34	gi|1729760*	Chitinase	–	1.73	2.28	–	
35	DN780414	Osmotin	75.69	–	32.98	382.36	
36	gi|595836886*	Osmotin	4.47	–1.1568	1.58	–	
37	CF932178	Pathogenesis-related protein 4 (PR 4)	11.19	3.87	5.12	29.86	Figure 3c, spots 36, 37
38	gi|10505374*	Pathogenesis-related protein 10 (PR 10)	Protein spot detected only under biotic stress condition				

Fold change values with –ve sign indicate down-regulation.

–: No fold change value was determined.

*Protein IDs and expression values obtained from proteome analysis.

metabolism leading to detoxification reactions and activation of downstream signalling cascades.

Transcription factor analysis revealed components of stress, retrograde signalling and suppressors of SA

Analysis of the transcriptome data revealed 1048 differentially expressed TFs accounting for 12% of the DETs under BS conditions (Table S2). Further, around 8.1% (705) of the up-regulated and 3.94% (343) of the down-regulated transcripts correspond to transcription factor families (TFs). Stress-responsive TFs belonging to WRKY, AP2-EREBP, NAC, bHLH, MYB, C2H2 and ethylene insensitive three families were found to be up-regulated (Figure 6a). Further, a number of development-related TFs belonging to AUX/IAA, C2C2, GRAS and HB families were down-regulated during boll developmental stages (Figure 4a). Among the TFs, WRKY family accounted for about 13% of differentially expressed TFs in all of the analysed developmental stages (0–10 dpa). Following WRKY, AP2-EREBP, NAC and MYB TFs were found to be relatively more in the transcriptome data set. Among the stress-related TFs, AP2-EREBP plays a central role in the abscisic acid (ABA)-dependent stress signalling pathways. Literature survey and manual curation of the TFs composition and regulation pattern revealed that the TFs such as WRKY 40, NAC along with AOXs as mentioned elsewhere constitutively account for the components of mitochondrial retrograde signalling pathways (Sophia et al., 2013; Van Aken et al., 2013). Further analysis revealed the up-regulated members of WRKY33 and EIN3 that act as suppressors of SA.

Bollworm attack induces synthesis of synergistically regulated phytohormones

Phytohormone-related DETs constituted for about 64.5% of 0 dpa and <25% of 2, 5 and 10 dpa. Classification and annotation of phytohormone-related transcripts showed that around 24% of them were related to ABA followed by auxin (AUX/IAA), ethylene (ET), BR, SA, gibberellic acid (GA), JA and cytokinin. Manual curation revealed that transcripts corresponding to zeaxanthin epoxidase that catalyses the first step of ABA biosynthesis were found to be down-regulated (Table S3) (Marin et al., 1996). Transcripts corresponding to tryptophan amino-transferase (TAA1), aldehyde dehydrogenase of indole 3-pyruvic acid pathway (IPA) and cytochrome P450 enzyme-CYP79B2 of the indole-3-acetaldoxime pathway (IAOX) were found to be down-regulated. In addition, flavin monooxygenase (YUC), tryptophan decarboxylase of the indole-3-acetamide (IAM) pathway, transcripts related to nitrilases were found to be up-regulated (Table S3). The above-mentioned pathways (IPA, IAOX and IAM) ultimately lead to auxin synthesis in plants. The aforementioned expression pattern suggests that auxin biosynthesis is partially inhibited during bollworm attack. Further, transcripts related to 1-aminocyclopropane-1-carboxylic acid (ACC) synthase and ACC oxidase involved in ethylene biosynthesis were found to be consistently up-regulated (Table S4). Transcripts related to sterol biosynthesis were found to be down-regulated (Tables S3 and S10). Sterols serve as precursor for BR synthesis. In addition, transcripts related to 3-oxo-5-alpha-steroid 4-dehydrogenase and BR biosynthetic protein DWARF1 involved in BR biosynthesis also were found to be down-regulated. Isochorismatase hydrolase (ISH) was the only SA biosynthetic pathway-related enzyme that was found to be up-regulated in our data set (Table S1). However, pathway annotation revealed that ISH catalyses the conversion of isochorismic acid to 2, 3-

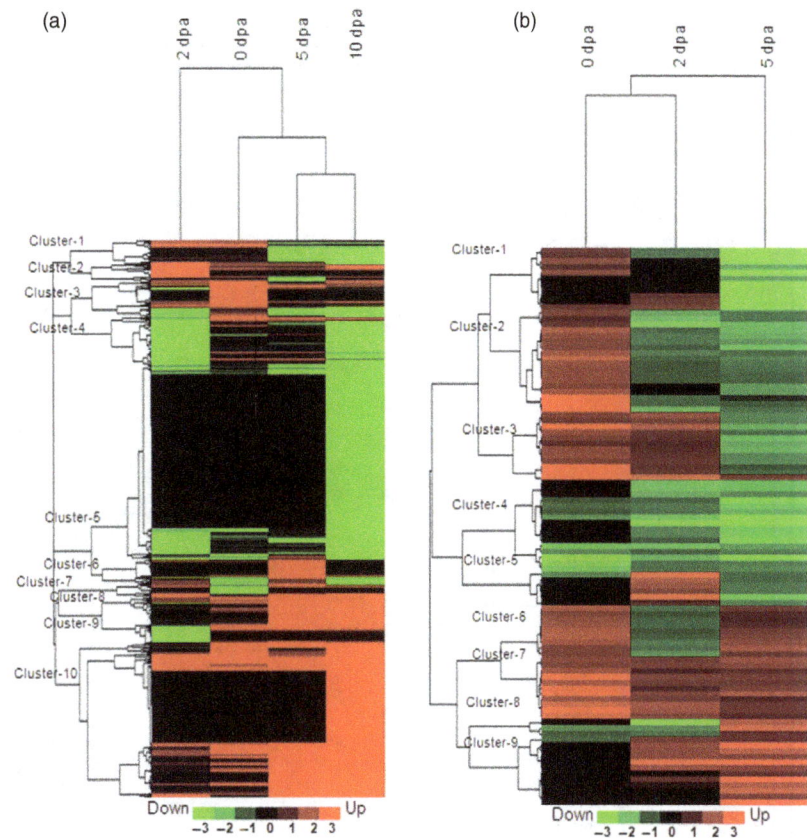

Figure 5 Heat map view of the cluster analysis depicting the expression pattern of differentially expressed transcripts (DETs) (a) and differentially expressed proteins (DEPs) (b). Hierarchical cluster analyses of DETs (fold change ±3) and DEPs (fold change ±1.5) under biotic stress as compared to their respective control samples during fibre development stages (0, 2, 5 and 10 dpa). List of Affymetrix cotton probe set IDs, and fold change for transcripts present in each cluster are presented in Table S16. List of Spot IDs, protein accessions and fold change for proteins are presented in Table S17. The hierarchical clustering was performed using complete linkage method with Euclidean distance based on fold change data compared to control samples using Cluster 3.0.

dihydroxybenzoic acid (DHB) and this reaction does not necessarily lead to SA biosynthesis in plants (Figure 6). Interestingly, most of the crucial enzymes involved in JA biosynthesis including phospholipase A, lipoxygenase, allene oxide synthase (AOS), allene oxide cyclase (AOC) and acyl-coA oxidase were found to be up-regulated throughout the developmental stages in BS-induced bolls at transcript and protein levels (Tables S10 and S14). Expression pattern of the hormonal biosynthesis-related genes suggests that biotic stress has induced JA and ET that are reported to act in a cooperative fashion (Figure 6). Likewise, ABA being the major contributor for DETs showed down-regulated pattern leading to its suppression during biotic stress. Down-regulation of BR and partial stimulation of auxin provides clues about the rate limiting pattern for allowing growth during stress.

Bona fide defence molecules and pathways are stimulated in response to biotic stress

The ultimate output of metabolic reprogramming, transcription factor regulation, ROS, Ca^{2+} signalling, hormonal biosynthesis, etc. leads to stimulation and synthesis of defence molecules and processes. In our data set, defence-related proteins such as the members of pathogenesis-related protein family (PR4, PR10, osmotin and thaumatin), chitinase, beta-glucanase and proteinase inhibitors were found to be up-regulated at transcript and protein levels (Figure 3c; Tables 2, S4, and S14). In addition, transcripts related to polyamine biosynthesis pathway such as SAM decarboxylase, spermidine synthase, arginine decarboxylase also were found to be up-regulated. Polyamines are cited as phytohormone like molecules that accumulate in response to stress and they also play major role in defence (Gill and Tuteja, 2010; Hussain et al., 2011; Waie and Rajam, 2003). In addition,

ROS regulating enzymes such as peroxidases (cytosolic and extracellular), SODs, NADPH oxidase and oxidative stress-specific GSTs were found to be up-regulated (Table S13).

Comparative analysis highlights concordant and discordant members of transcript-protein pairs

Comparative analysis revealed 37 unique accessions that were commonly identified in both the transcriptome and proteome approaches (Table S15). Gene Ontology-based annotation and classification of these genes revealed that majority of them were involved in metabolic, cellular and biosynthetic processes including amino acid, nucleotide and osmolyte metabolism, stress and defence response and phytohormone biosynthesis (Figure S3a–c). Manual curation of the data sets showed two distinct groups of transcript and protein pairs such as the genes with concordant (similar) expression patterns and the genes with discordant (dissimilar) expression patterns at transcript and protein levels (Table S15). Concordant members included genes such as chaperonin, actins, phosphoglycerate kinase, gibberellin oxidase, MDH and SODs that were found to be down-regulated, while defence and stress response-specific genes such as chitinase, PR protein, protease inhibitor and carbonic anhydrase were found to be up-regulated across developmental stages. Discordant members included ATP synthases, RuBisCO, glutamine synthase, proteasome subunits, etc. that showed poor corelation in their transcript and protein expression patterns.

Validation of microarray and proteome data by qRT-PCR analysis

To validate the data, semi-quantitative real-time PCR (qRT-PCR) analysis was performed on 42 selected differentially expressed genes (32 up-regulated and 10 down-regulated) during boll

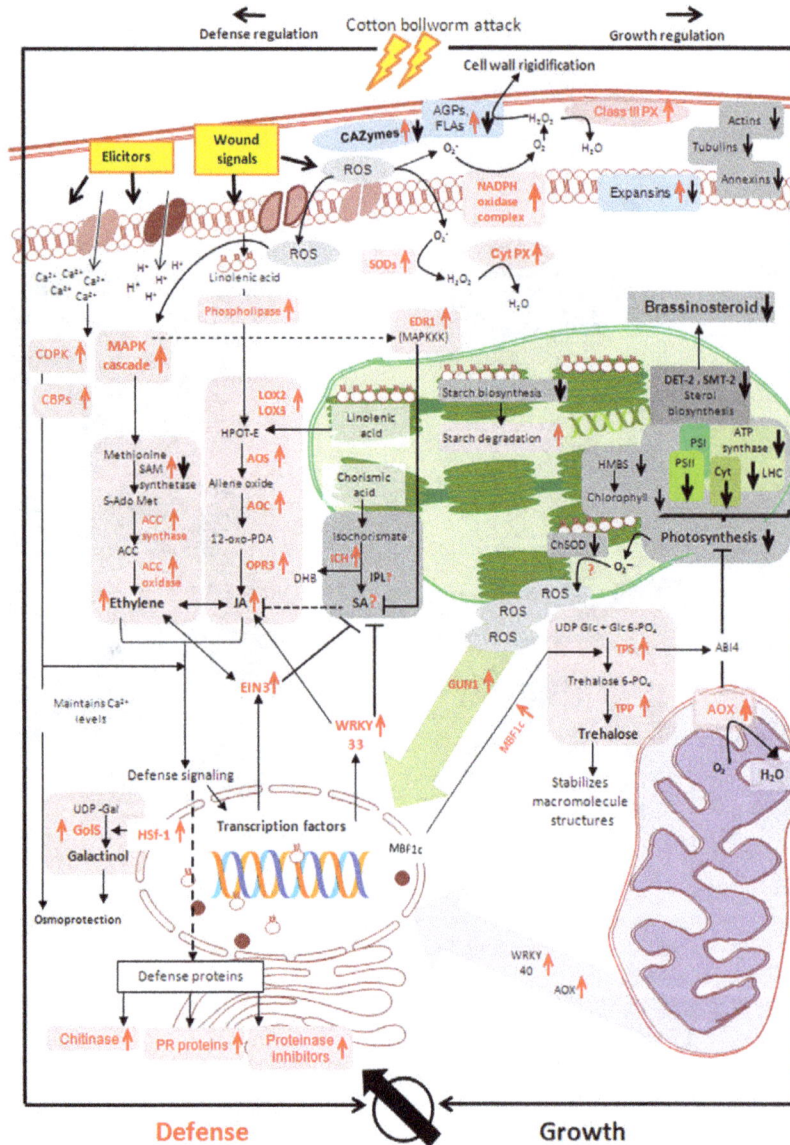

Figure 6 Putative model depicting the regulation of molecular events in cotton bolls subjected to bollworm infestation. Genes related to the redox regulation—peroxidase (PX), super oxide dismutase (SODs), metabolic process—trehalose phosphate synthase (TPS), trehalose phosphate phosphatase (TPP), galactinol synthase (GolS), signalling cascades—calcium-dependent protein kinase (CDPK), calcium-binding proteins (CBPs), mitogen-activated protein kinase (MAPK), enhanced disease resistance 1 (EDR1), phytohormone synthesis—lipoxygenase (LOX2, LOX3), allene oxide synthase (AOS), allene oxide cyclase (AOC), 12-oxo-phytodienoic acid reductase (OPR3), S-adenosylmethionine synthetase (SAM), 1-aminocyclopropane-1-carboxylate (ACC) synthase/oxidase, isochorismatase hydralase (ICH), isochorismate pyruvate lyase (IPL), 3-oxo-5-alpha-steroid 4-dehydrogenase (DET 2), 24-sterol C-methyltransferase (SMT2-2), transcription factors—ethylene insensitive 3(EIN3), heat-shock transcription factor (HSf-1), multiprotein bridging factor 1c (MBF1c), retrograde signalling—alternative oxidase (AOX), pentatricopeptide repeat containing protein (GUN1), defence—pathogenesis-related (PR) protein, photosynthesis—photosystem (PS I, PS II), cytochrome complex (Cyt), light harvesting complex (LHC), hydroxy methyl bilane synthase (HMBS) and growth—carbohydrate active enzymes (CAZymes) are annotated along with their expression pattern. Upward pointing arrow indicates up-regulation and downward pointing arrow indicates down-regulation of respective genes. The overall pattern suggests the selective regulation of signalling cascades favouring defence over growth in bollworm-infested cotton bolls.

developmental stages under cotton bollworm infestation (Figure S2). The results showed that the expression patterns of transcripts and proteins observed through microarray and proteome analyses were in parallel with those obtained by qRT-PCR (Figure S2).

Discussion

In the current study, transcriptomic approach has resulted in the identification of relatively more number of differentially expressed genes as compared to the proteomic approach. Nevertheless, the expression pattern of crucial phytohormone biosynthesis genes including S-adenosylmethionine synthase and AOC, cytoskeleton proteins such as annexin isoforms and actins, cytosolic ascorbate peroxidase involved in redox regulation, signalling and stress-specific response genes such as calcium-binding protein and GS1 were exclusively identified at the protein levels. Transcriptomic approach revealed a vast set of TFs which are otherwise difficult to be identified at protein levels. The concordant pattern observed among the transcript-protein pairs such as chitinase,

PR proteins, carbonic anhydrase and protease inhibitors adds significance and direct evidence for active defence signalling during bollworm infestation in developing cotton bolls. Also, discordance observed among transcript-protein pairs might reflect true biological discordance that could be attributed to post-transcriptional regulations, protein/transcript stability, miRNAs, etc. and this needs to be investigated further. On the whole, our data suggest that employing two complementary approaches have increased the overall coverage of the differentially expressed genes that in turn has aided in filling crucial gaps in the above-mentioned processes.

Host–pest interactions induce synthesis of additional metabolic resources ensuring survival

Transcriptomic and proteomic data obtained in this study revealed major changes in the carbohydrate metabolisms such as the significantly up-regulated genes encoding TPS, TPP and GolS involved in trehalose and raffinose biosynthesis (Table 2)

and down-regulated genes involved in cell wall metabolisms (Table S7; Figure 6). Trehalose is a nonreducing disaccharide that plays a major role in the stabilization of proteins and molecular structures during stress (Garg et al., 2002). Likewise, raffinose belongs to a family of oligosaccharides that accumulates during stress and acts as osmoprotectants (Unda et al., 2012). Both trehalose and raffinose family of saccharides are referred to as compatible solutes responding to stress conditions in plants (Zhou et al., 2014). In addition to that, carbohydrate active and associated proteins like CAZymes, AGPs and FLAs were found to be differentially regulated in our data set (Table S7). AGPs are a heterogeneous class of abundant proteoglycans localized in both cell wall and cytosolic regions (Kumar et al., 2013). Reactive oxygen species molecules (H_2O_2) released during stress conditions, cross-link the AGPs to the cell wall leading to rigidity and thereby render protection against pest and pathogen invasion. The rigidity caused in the cell wall matrix also acts as a negative regulator of cell growth (Cleland and Karlsnes, 1967; Gille et al., 2009; Sadava and Chrispeels, 1973).

Analysis of the nitrogen and amino acid metabolism highlighted additional clues on BS regulation. Briefly, GS1 and glutamate-ammonia ligase were found to be consistently up-regulated throughout the developmental stages (Table S14). In plants, glutamine (Gln) serves as the primary source for inorganic nitrogen, N (NO_3^- and NH_4^+) that gets subsequently utilized for biosynthesis of major amino acids like Glu, Asp and Asn. Among the genes involved in nitrogen/amino acid metabolism, glutamine synthetase (GS), glutamate synthase, GDH and NiR play primary roles in the assimilation of NH_4^+. Two isoforms of GS are reported in plants among which GS1 is induced and the other isoform, GS2 is suppressed during pathogen attack (Pageau et al., 2006). Interestingly, our study revealed the consistent up-regulation of GS1, GDH and NiR under BS condition. Among the above mentioned, GS2 and NiR are involved in primary nitrogen assimilation, whereas GS1 and GDH are involved in organic nitrogen remobilization (Pageau et al., 2006). In addition, GS1 and GDH are also cited as senescence-related markers in plants (Pageau et al., 2006). Expression pattern observed in the current study in turn suggests that nitrogen remobilization and senescence leading to stress regulation is active, whereas signals related to primary nitrogen assimilation leading to growth are not evidenced.

Lipid metabolism serves as one of the major contributor for energy, membrane biogenesis, signalling molecules, etc. Our study revealed a bias in the stimulation of certain lipid metabolic pathways. Briefly, linolenic acid metabolic enzymes were up-regulated, whereas sterol biosynthesis genes were down-regulated. Linolenic acid and sterols are lipid molecules colocalized within the plastid and thylakoid membranes (Schwertner and Biale, 1973). Our data showed that the factors related to linolenic acid metabolism leading to biosynthesis of stress-responsive JA were induced, while the genes related to sterol biosynthesis that leads to growth-related Br synthesis were down-regulated. Interestingly, cellular component-based annotation of the above-mentioned pathways and processes revealed that growth and defence-related molecules are colocalized within the same compartment such as chloroplast. However, under biotic stress, only defence-related factors are positively regulated leaving behind the growth-related factors (Figure 6). Such a switch over in the regulation of lipid metabolism at subcellular level further ensures resistance during bollworm attack.

Diverse pathway regulation and association delineates bollworm infestation-specific signalling pattern

In addition to above-mentioned metabolic pathways, signalling molecules such as calcium, redox regulators, phytohormones, TFs and protein kinases that play independent role through diverse pathways were found to be differentially regulated in our data set (Figures S3 and S4). Briefly, up-regulated members of Calcium (Ca^{2+}) binding proteins and Ca^{2+} transporting ATPases are involved in maintaining cytosolic calcium levels during stress conditions. Redox regulators and oxidative stress regulators such as SODs, GSTs, peroxidases, NADPH oxidases, respiratory burst oxidase homologs (Rbohs), DHARs, etc. were found to be temporally regulated in our data set (Figures 6 and S4; Table S13). Among them, NADPH oxidases, Rbohs and extracellular peroxidases are the major regulators of the primary apoplastic oxidative burst during insect attack (Torres, 2010). These enzymes catalyse reactions leading to ROS release which further stimulates downstream enzymes like SODs and GSTs localized at other cellular components. Superoxide dismutases and GSTs catalyse detoxification reactions, while PCBRs are involved in antioxidant synthesis ultimately protecting cells from oxidative stress (Niculaes et al., 2014). Temporal expression of the ROS scavengers indicates a compromised pattern executed by cotton bolls in response to bollworm attack. Phytohormones are secondary signals that often regulate development and stress conditions in plants. Our study showed the positive regulation of stress-related hormones such JA and ethylene accompanied by repression or down-regulation of growth-related Auxin, BR (Figure 5). Interestingly, we did not find genes related to SA biosynthesis which is also a biotic stress-specific hormone; however, positively regulated suppressors for SA synthesis such as EIN3 and WRKY 33 have been evidenced in the current study (Table 2; Figure 6). Such observations in turn suggest that bollworm attack favours synergistic JA–ethylene synthesis over the antagonistic SA. Further, our study also revealed the up-regulation of nuclear encoded plastid and mitochondrial components including TFs and enzymes. Among them, WRKY TF family is often linked with biotic stress response and pathogen-associated molecular pattern (PAMP) (Eulgem and Somssich, 2007; Rushton et al., 1996). In addition, WRKY TFs also regulate the expression of nuclear encoded mitochondrial proteins (Van Aken et al., 2013). For example; up-regulated members of WRKY40 identified in the current study are known to regulate the expression of AOX enzyme during stress conditions (Ivanova et al., 2014). These factors along with ROS molecules together constitute the plastid and mitochondrial signalling pathway components that regulates nuclear gene expression related to cell cycle and growth. Such pattern further reveals the retrograde trend operational during biotic stress conditions (Figure 6).

Bollworm infestation induces major reallocation of metabolic resources favouring defence over growth

In response to insect pest attack, the host plant initiates several layers of defence including PAMP-triggered immunity (PTIs), effector-triggered immunity (Dangl and Jones, 2001). Following stress perception, a series of signal transduction events that include metabolic pathway regulation, defence molecule synthesis, etc. are stimulated to encounter the external attack (Figure S3). The above-mentioned coordinated events suggest an energy intensive mechanism that needs to be executed in order to exert the defence response. Knowledge gained through the current study highlights that the source for such additional energy could be attained by suppressing growth locally (boll tissue growth). In short, we observed a hierarchy of factors and processes that specifically suppresses growth-related events and stimulates defence-related processes. Expression trend of carbohydrate, amino acid and lipid metabolism also reveals the synthesis of stress response-related molecules such as trehalose, raffinose, linolenic acid and suppression of growth-related factors such as cell wall elongation enzymes, and sterols. Suppression of fundamental processes such as photosynthesis and cell cycle not only retards further growth but also regulates the amount of ROS released by them and also preserves considerable amount of energy that could be channelized for defence signalling. Likewise, defence response such as lignifications, cross-linking of proteoglycans to cell wall matrix offers cell wall rigidification on the one hand and negatively regulates cellular expansion on the other. All these factors together suggest a major reallocation of metabolic resources favouring defence over growth through selective regulation of specific pathways and processes during bollworm attack.

Conclusion

The present study delineates boll-specific endogenous defence mechanisms adapted by cotton plants under bollworm attack. The vast number of coordinated events including stimulations and repressions of major biological processes ultimately suggest major reallocation of metabolic resources that favours defence over growth in developing cotton bolls. Taking such insights into account, strategies targeting stimulation of multiple phytohormones and better sustainment of defence as well as growth signals could aid in developing resistant varieties against insect pest.

Materials and methods

Plant material and biotic stress treatment

Cotton (*Gossypium hirsutum* cv. Bikaneri Narma) plants were grown at Agricultural Research Station, Dharwad farm, Dharwad, during 2012–2013 Kharif seasons following recommended agronomic practices. Two separate plots of the same genotype were maintained with a space of 90 cm between rows and 20 cm between plants. The plots were covered with nylon nets to protect from any external pest incidence. Plot designated as control (no infection from any class of insects including bollworms) and plot designated as infested (infested with *H. armigera*) were protected in early stage (45 days after sowing) from incidence of sucking pests by spraying recommended insecticides. During peak flowering stage (65–85 days after sowing), 2nd–3rd instar larvae of *H. armigera*, raised on bendi (*Abelmoschus esculentus* L. Syn. *Hibiscus esculentus*) fruits in the Entomology laboratory maintained at 25 °C in 65%–70% relative humidity on a 14/10-h light/dark cycle, were released on buds of cotton on the day of pollination. The buds with larva were covered using paper bag with proper aeration to prevent larvae movement from the bud. Such a set-up ensured maximum damage of bolls by larva. The cotton bolls used as control(s) were also covered with paper bags with pores in order to prevent

predation by insect pests and to ensure similar microenvironment as that of biotic stress-induced bolls. Samples were collected after 8 h of infection and labelled as 0 dpa, likewise samples collected after 2 and 5 days of insect infestation were labelled as 2 and 5 dpa, respectively. After 5 days of infestation, the insect was removed; the bolls were collected after 10 days of further growth and were labelled as 10 dpa. Harvested cotton boll samples were frozen immediately in liquid nitrogen and stored at −70 °C until further use.

Total RNA isolation and Microarray experiments

Infected bolls (complete) from 0 and 2 dpa and only infested portion of 5 and 10 dpa boll samples along with their respective controls were used for RNA extraction. In order to minimize plant to plant and mode of infection variations, boll samples were collected and pooled from five independent plants and considered as one biological replicate. Total RNA isolation, analysis and quality check were performed as previously described. Affymetrix Cotton GeneChip Genome array (Affymetrix, Santa Clara, California) having 23 977 probe sets representing 21 854 cotton transcripts was used for transcriptome analysis (http://www.affymetrix.com/catalog/131430/AFFY/Cotton-Genome-Array#1_1). Three biological replicates were maintained to test the reproducibility and quality of the chip hybridization. Microarray hybridization, staining and washing procedures were carried out as described in the Affymetrix protocols with minor modifications (Padmalatha et al., 2012). The arrays were scanned with a GeneChip scanner 3000.

GeneChip data processing and analysis

After scanning of each array, DAT, CEL, CHP, XML and JPEG image files were generated using GeneChip Operating Software platform. The CEL files having estimated probe intensity values were analysed with GeneSpring GX-12.6 software (Agilent Technologies, Santa Clara, California) to get DETs. The robust multiarray average algorithm was used for the back ground correction; quantile normalization and median polished probe set summarization to generate single expression value for each probe set. Normalized expression values were $\log_2$-transformed, and differential expression analysis was performed using unpaired *t*-test. The *P* values were corrected by applying the FDR correction (Benjamini and Hochberg, 2000). Differentially expressed transcripts with FDR corrected *P* value ≤0.01 and fold change ≥3 were included for further data analysis. The hierarchical clustering was performed using complete linkage method with Euclidean distance based on log fold change data compared to control samples using Cluster 3.0 (Eisen et al., 1998) to display the expression pattern and tree diagram of DETs. The DETs were annotated using NetAffx annotation data for Cotton GeneChip (http://www.affymetrix.com, release 26).

Functional annotation of probe sets and pathways

To obtain functional annotation of transcripts, the consensus sequences of probe sets present in the Cotton GeneChip were mapped to the Arabidopsis TAIR protein database version 10 (http://www.arabidopsis.org) by BLASTX with *E* value cut-off ≤e-10. To identify the putative TFs and transcripts related to phytohormone biosynthesis and signal transduction pathways, the consensus sequences of all probe sets presented in cotton GeneChip were searched against the Arabidopsis transcription factor database (http://plntfdb.bio.uni-potsdam.de, version 3.0) and Arabidopsis hormone database (http://ahd.cbi.pku.edu.cn, version 2.0), respectively, by BLASTX with *E* value cut-off ≤e-10. Differentially expressed transcripts were grouped into functional categories based on MIPS functional catalogue (http://mips.gsf.de/projects/funcat). Further, MapMan software version 3.5.0 (http://gabi.rzpd.de/projects/MapMan/) was used to visualize the expression of differentially regulated cotton transcripts onto metabolic pathways (Usadel *et al.*, 2005). The microarray data are deposited in the Gene Expression Omnibus (GEO) database (http://www.ncbi.nlm.nih.gov/geo) at the NCBI under the series accession numbers GSE55511.

Two-dimensional gel electrophoresis (2D SDS-PAGE)-based proteome analysis

The total protein from cotton bolls was isolated using phenol extraction method. The protein pellets were dissolved in 2D SDS-PAGE rehydration buffer (7 M urea, 2 M thiourea, 2% CHAPS, 0.5% ampholytes, 40 mM DTT), and an aliquot of protein sample from two independent replicates was subjected to two-dimensional gel electrophoresis (2D SDS–PAGE) as described previously (Kumar *et al.*, 2013), briefly for the first-dimensional separation, the sample was loaded onto a 13-cm immobilized pH gradient (IPG) linear (pI 4–7) strips (GE Healthcare Life Sciences, U.S.A), and isoelectric focusing was performed according to manufacturer's instructions. Strips were then equilibrated, and second-dimensional separation was carried out on 12% SDS–polyacrylamide gel (13 cm, 1.5 mm). Gels were stained with Coomassie blue staining to visualize the protein spots and were stored in 1% acetic acid at 4 °C until further use. Gels were scanned using GE Image scanner III (GE Healthcare Life Sciences, U.S.A) through Labscan software version 6.0.1 and analysed using Imagemaster 2D Platinum software version 6.0.1 (GE Healthcare Life Sciences, U.S.A). Protein spot detection parameters were set as: Smooth: 3, Minimum area: 11 and Saliency: 200. Detected protein spots were manually re-evaluated to remove artefacts such as dust particles and streaks. Reproducibly detected protein spots were quantified using the per cent volume criterion. The relative volume corresponding to the detected spot region was considered to represent the expression level. Protein spots that showed normalized expression values of ±0.6-, 1.5-fold (biotic stress/control) were considered for statistical evaluation. To define the significant difference, *P* value <0.05 was set through Student's *t*-test and one-way ANOVA. Protein expression values within the above-mentioned thresholds were considered as differentially expressed.

Protein identification, annotation and classification from 2D gel spots

Differentially expressed protein spots were subjected to in-gel tryptic digestion followed by MALDIT TOF-based identification procedure as previously described (Kumar *et al.*, 2013). Peptide MS/MS spectrum processing was achieved through Flexanalysis software version 3 and database search using Biotools software version 3.2. The database search parameters were set as described: fragment masses were searched in three independent databases: they were (i) NCBInr database (06/03/2010) containing 10 551 781 sequences (total) including 290 173 sequences from green plants (Viridiplantae), (ii) *Gossypium raimondii* protein database containing 40 976 sequences downloaded from CottonGen website (ftp://ftp.bioinfo.wsu.edu/species/Gossypium_raimondii/CGP-BGI_G.raimondii_Dgenome/genes/), (iii) *Gossypium arboreum* protein database containing 40 134 sequences downloaded from Cotton Genome Project website (ftp://cotton:cotton321$@public.genomics.org.cn/Ca_all_Ver-

sion2.GENE.pep.gz) through mascot search engine, taxonomy was set as Viridiplantae, enzyme was set as trypsin, fixed modifications included carbamidomethylation of cysteine, variable modifications included oxidation of methionine, protein mass was unrestricted, missed cleavage was set to 1, MS tolerance of ±100 ppm and MS/MS tolerance of ±/−0.75 da. Only peptides with an individual ion score of >40 (*P* < 0.05) were considered for protein identification. Identified proteins were sequences that were exported into BLAST2GO platform version 2.7 (www.blast2go.com/b2ghome) to attain GO-based annotation, classification and pathway mapping (Conesa *et al.*, 2005).

Transcriptome and proteome data set integration

Unique protein sequences corresponding to the differentially expressed proteins were subjected to GEO Nucleotide Translated BLAST: tblastn analysis to obtain accessions corresponding to *Gossypium hirsutum* (taxid: 3635) transcripts. The search parameters for tblastn were set as follows: database—GEO; organism: *G. hirsutum* (taxid: 3635). Transcript accessions with E value cut-off ≤e-10 and ≥70% sequence identity were considered as matched sequence.

The quantitative real-time PCR (qRT-PCR) analysis

The qRT-PCR analysis was performed on selected differentially expressed genes to validate the microarray and proteome expression data. RNA isolation followed by cDNA synthesis, qRT-PCR analysis and fold change calculations were performed as previously described (Padmalatha *et al.*, 2012). The list of primers used in the current study is presented in Table S6. The *GhPP2A1* gene (accession no: DT545658) from *G. hirsutum* was used as reference gene to normalize the expression values (Artico *et al.*, 2010).

Acknowledgements

This work was supported by funds from the Indian Council of Agricultural Research (ICAR) under the National Agricultural Innovation Project (NAIP/C4/C10103), Component-4 and funds from the Department of Biotechnology (DBT), Government of India, and International Centre for Genetic Engineering and Biotechnology, New Delhi. We thank Prof N. K. Singh, National Research Centre on Plant Biotechnology, for providing the Microarray facility used in the study. KK acknowledges Senior Research Fellowship from Department of Biotechnology, India.

References

Arimura, G., Kost, C. and Boland, W. (2005) Herbivore-induced, indirect plant defences. *Biochim. Biophys. Acta*, **1734**, 91–111.

Artico, S., Nardeli, S.M., Brilhante, O., Gross-de-Sa, M.F. and Alves-Ferreira, M. (2010) Identification and evaluation of new reference genes in *Gossypium hirsutum* for accurate normalization of real-time quantitative RT-PCR data. *BMC Plant Biol.* **10**, 49.

Artico, S., Ribeiro-Alves, M., Oliveira-Neto, O.B., de Macedo, L.L.P., Silveira, S., Grossi-de-Sa, M.F., Martinelli, A.P. *et al.* (2014) Transcriptome analysis of *Gossypium hirsutum* flower buds infested by cotton boll weevil (*Anthonomus grandis*) larvae. *BMC Genom.* **15**, 854.

Benjamini, Y. and Hochberg, Y. (2000) On the adaptive control of the false discovery rate in multiple testing with independent statistics. *J. Educ. Behav. Stat.* **25**, 60–83.

Cleland, R.E. and Karlsnes, A. (1967) A possible role for hydroxyproline containing proteins in the cessation of cell elongation. *Plant Physiol.* **42**, 669–671.

Conesa, A., Götz, S., García-Gómez, J.M., Terol, J., Talón, M. and Robles, M. (2005) Blast2GO: a universal tool for annotation, visualization and analysis in functional genomics research. *Bioinformatics*, **21**, 3674–3676.

Dangl, J.L. and Jones, J.D. (2001) Plant pathogens and integrated defence responses to infection. *Nature*, **411**, 826–833.

Dua, I.S., Kumar, V. and Bhavneet, D.E.A. (2006) Genetically modified cotton and its biosafety concerns: a review. In *Current Concepts in Botany* (Mukerji, K.G., Manoharachary, C., I.K. International Publishing House, New Delhi, India.), ed. pp. 447–459.

Dubey, N.K., Goel, R., Ranjan, A., Idris, A., Singh, S.K., Bag, S.K., Chandrashekar, K. *et al.* (2013) Comparative transcriptome analysis of *Gossypium hirsutum* L. in response to sap sucking insects: aphid and whitefly. *BMC Genom.* **14**, 241.

Dubiella, U., Seybold, H., Durian, G., Komander, E., Lassig, R., Witte, C.P., Schulze, W.X. *et al.* (2013) Calcium-dependent protein kinase/NADPH oxidase activation circuit is required for rapid defense signal propagation. *Proc. Natl Acad. Sci. USA*, **110**, 8744–8749.

Eisen, M.B., Spellman, P.T., Brown, P.O. and Botstein, D. (1998) Cluster analysis and display of genome-wide expression patterns. *Proc. Natl Acad. Sci. USA*, **95**, 14863–14868.

Eulgem, T. and Somssich, I.E. (2007) Networks of WRKY transcription factors in defense signaling. *Curr. Opin. Plant Biol.* **10**, 366–371.

Gao, W., Long, L., Zhu, L.F., Xu, L., Gao, W.H., Sun, L.Q., Liu, L.L. *et al.* (2013) Proteomic and virus-induced gene silencing (VIGS) analyses reveal that gossypol, brassinosteroids, and jasmonic acid contribute to the resistance of cotton to *Verticillium dahliae*. *Mol. Cell Proteomics*, **12**, 3690–3703.

Garg, A.K., Kim, J.-K., Owens, T.G., Ranwala, A.P., Choi, Y.D., Kochian, L.V. and Wu, R.J. (2002) Trehalose accumulation in rice plants confers high tolerance levels to different abiotic stresses. *Proc. Natl Acad. Sci. USA*, **99**, 15898–15903.

Gatehouse, J.A. (2002) Plant resistance towards insect herbivores: a dynamic interaction. *New Phytol.* **156**, 145–169.

Gill, S.S. and Tuteja, N. (2010) Polyamines and abiotic stress tolerance in plants. *Plant Signal Behav.* **5**, 26–33.

Gille, S., Haensel, U., Ziemann, M. and Pauly, M. (2009) Identification of plant cell wall mutants by means of a forward chemical genetic approach using hydrolases. *Proc. Natl Acad. Sci. USA*, **106**, 14699–14704.

Han, Z.G., Guo, W.Z., Song, X.L. and Zhang, T.Z. (2004) Genetic mapping of EST-derived microsatellites from the diploid *Gossypium arboreum* in allotetraploid cotton. *Mol. Genet. Genomics*, **272**, 308–327.

Hussain, S.S., Ali, M., Ahmad, M. and Siddique, K.H. (2011) Polyamines: natural and engineered abiotic and biotic stress tolerance in plants. *Biotechnol. Adv.* **29**, 300–311.

Ivanova, A., Law, S.R., Narsai, R., Duncan, O., Lee, J.-H., Zhang, B., Van Aken, O. *et al.* (2014) A functional antagonistic relationship between auxin and mitochondrial retrograde signaling regulates alternative oxidase 1a expression in Arabidopsis. *Plant Physiol.* **165**, 1233–1254.

Jabs, T., Tschope, M., Colling, C., Hahlbrock, K. and Scheel, D. (1997) Elicitor-stimulated ion fluxes and O2—from the oxidative burst are essential components in triggering defense gene activation and phytoalexin synthesis in parsley. *Proc. Natl Acad. Sci. USA*, **94**, 4800–4805.

Kempema, L.A., Cui, X., Holzer, F.M. and Walling, L.L. (2007) Arabidopsis transcriptome changes in response to phloem-feeding silver leaf whitefly nymphs. Similarities and distinctions in responses to aphids. *Plant Physiol.* **143**, 849–865.

Komatsu, S., Yamamoto, R., Nanjo, Y., Mikami, Y., Yunokawa, H. and Sakata, K. (2009) A Comprehensive analysis of the soybean genes and proteins expressed under flooding stress using transcriptome and proteome techniques. *J. Proteome Res.* **8**, 4766–4778.

Kumar, S., Kumar, K., Pandey, P., Rajamani, V., Padmalatha, K.V., Dhandapani, G., Kanakachari, M. *et al.* (2013) Glycoproteome of elongating cotton fiber cells. *Mol. Cell Proteomics*, **12**, 3677–3689.

Lamb, C. and Dixon, R.A. (1997) The oxidative burst in plant disease resistance. *Annu. Rev. Plant Physiol. Plant Mol. Biol.* **48**, 251–275.

Loake, G. and Grant, M. (2007) Salicylic acid in plant defence-the players and protagonists. *Curr. Opin. Plant Biol.* **10**, 466–472.

Marin, E., Nussaume, L., Quesada, A., Gonneau, M., Sotta, B., Hugueney, P., Frey, A. *et al.* (1996) Molecular identification of zeaxanthin epoxidase of *Nicotiana plumbaginifolia*, a gene involved in abscisic acid biosynthesis and corresponding to the ABA locus of *Arabidopsis thaliana*. *EMBO J.* **15**, 2331–2342.

Matthews, G.A. (1994) Cultural control. In *Insect Pests of Cotton* (Matthews, G.A. and Tunstall, J.P., eds), pp. 455–461. Wallingford, UK: CABI Publishing.

Mei, M., Syed, N.H., Gao, W., Thaxton, P.M., Smith, C.W., Stelly, D.M. and Chen, Z.J. (2004) Genetic mapping and QTL analysis of fiber-related traits in cotton (*Gossypium*). *Theor. Appl. Genet.* **108**, 280–291.

Moran, P.J. and Thompson, G.A. (2001) Molecular responses to aphid feeding in Arabidopsis in relation to plant defense pathways. *Plant Physiol.* **125**, 1074–1085.

Niculaes, C., Morreel, K., Kim, H., Lu, F., McKee, L.S., Ivens, B., Haustraete, J. *et al.* (2014) Phenylcoumaran benzylic ether reductase prevents accumulation of compounds formed under oxidative conditions in poplar xylem. *Plant Cell*, **9**, 3775–3791.

Oerke, E.C. (2006) Crop losses to pests. *J. Agric. Sci.* **144**, 31–43.

Padmalatha, K.V., Dhandapani, G., Kanakachari, M., Kumar, S., Dass, A., Patil, D.P., Rajamani, V. *et al.* (2012) Genome-wide transcriptomic analysis of cotton under drought stress reveal significant down-regulation of genes and pathways involved in fiber elongation and up-regulation of defense responsive genes. *Plant Mol. Biol.* **78**, 223–246.

Pageau, K., Reisdorf-Cren, M., Morot-Gaudry, J.F. and Masclaux-Daubresse, C. (2006) The two senescence-related markers, GS1 (cytosolic glutamine synthetase) and GDH (glutamate dehydrogenase), involved in nitrogen mobilization, are differentially regulated during pathogen attack and by stress hormones and reactive oxygen species in *Nicotiana tabacum* L. leaves. *J. Exp. Bot.* **57**, 547–557.

Ramirez, C.C., Guerra, F.P., Zuniga, R.E. and Cordero, C. (2009) Differential expression of candidate defense genes of poplars in response to aphid feeding. *J. Econ. Entomol.* **102**, 1070–1074.

Rushton, P.J., Torres, J.T., Parniske, M., Wernert, P., Hahlbrock, K. and Somssich, I.E. (1996) Interaction of elicitor-induced DNA-binding proteins with elicitor response elements in the promoters of parsley PR1 genes. *EMBO J.* **15**, 5690–5700.

Ryan, C.A. (1990) Protease inhibitors in plants: genes for improving defenses against insects and pathogens. *Annu. Rev. Phytopathol.* **28**, 425–449.

Sadava, D. and Chrispeels, M.J. (1973) Hydroxyproline-rich cell wall protein (extensin): role in the cessation of elongation in excised pea epicotyls. *Dev. Biol.* **30**, 49–55.

Sanders, D., Pelloux, J., Brownlee, C. and Harper, J.F. (2002) Calcium at the crossroads of signaling. *Plant Cell*, **14**(Suppl), S401–S417.

Schwertner, H.A. and Biale, J.B. (1973) Lipid composition of plant mitochondria and of chloroplasts. *J. Lipid Res.* **14**, 235–242.

Sophia, Ng., Ivanova, A., Duncan, O., Law, S.R., Aken, O.V., Clercq, I.D., Wang, Y. *et al.* (2013) A membrane-bound NAC transcription factor, ANAC017, mediates mitochondrial retrograde signaling in Arabidopsis. *Plant Cell*, **25**, 3450–3471.

Srivastava, V., Obudulu, O., Bygdel, J., Löfstedt, T., Rydén, P., Nilsson, R., Ahnlund, M. *et al.* (2013) OnPLS integration of transcriptomic, proteomic and metabolomic data shows multi-level oxidative stress responses in the cambium of transgenic hipl-superoxide dismutase Populus plants. *BMC Genom.* **14**, 893.

Sulpice, R., Py, E.T., Ishihara, H., Trenkamp, S., Steinfath, M., Witucka-Wall, H., Gibon, Y. *et al.* (2009) Starch as a major integrator in the regulation of plant growth. *Proc. Natl Acad. Sci. USA*, **106**, 10348–10353.

Sun, J.Q., Jiang, H.L. and Li, C.Y. (2011) Systemin/jasmonate-mediated systemic defense signaling in tomato. *Mol. Plant*, **4**, 607–615.

Torres, M.A. (2010) ROS in biotic interactions. *Physiol. Plant.* **138**, 414–429.

Unda, F., Canam, T., Preston, L. and Mansfield, S.D. (2012) Isolation and characterization of galactinol synthases from hybrid poplar. *J. Exp. Bot.* **63**, 2059–2069.

Usadel, B., Nagel, A., Thimm, O., Redestig, H., Blaesing, O.E., Palacios-Rojas, N., Selbig, J., *et al.* (2005) Extension of the visualisation tool MapMan to allow statistical analysis of arrays, display of co-responding genes and comparison with known responses. *Plant Physiol.* **138**, 1195–1204.

Van Aken, O., Zhang, B., Law, S., Narsai, R. and Whelan, J. (2013) AtWRKY40 and AtWRKY63 modulate the expression of stress-responsive nuclear genes encoding mitochondrial and chloroplast proteins. *Plant Physiol.* **162**, 254–271.

Vanlerberghe, G.C. and McIntosh, L. (1997) ALTERNATIVE OXIDASE: from gene to function. *Annu. Rev. Plant Physiol. Plant Mol. Biol.* **48**, 703–734.

Waie, B. and Rajam, M.V. (2003) Effect of increased polyamine biosynthesis on stress response in transgenic tobacco by introduction of human S-adenosylmethionine gene. *Plant Sci.* **164**, 727–734.

Wingler, A. (2002) The function of trehalose biosynthesis in plants. *Phytochemistry* **60**, 437–440.

Yang, Y., Shah, J. and Klessig, D.F. (1997) Signal perception and transduction in plant defense responses. *Genes Dev.* **11**, 1621–1639.

Yano, R., Nakamura, M., Yoneyama, T. and Nishida, I. (2005) Starch-related alpha-glucan/water dikinase is involved in the cold-induced development of freezing tolerance in Arabidopsis. *Plant Physiol.* **138**, 837–846.

Zhou, M., Suna, G., Suna, Z., Tanga, Y. and Wu, Y. (2014) Cotton proteomics for deciphering the mechanism of environment stress response and fiber development. *J. Proteomics*, **105**, 74–84.

6

Phaseolin expression in tobacco chloroplast reveals an autoregulatory mechanism in heterologous protein translation

Francesca De Marchis, Michele Bellucci and Andrea Pompa*

Research Division of Perugia, Institute of Biosciences and Bioresources, National Research Council, Perugia, Italy

*Correspondence

email andrea.pompa@ibbr.cnr.it

Keywords: control by epistasy of synthesis, chloroplast transformation, negative feedback, phaseolin, protein folding, protein translation.

Summary

Plastid DNA engineering is a well-established research area of plant biotechnology, and plastid transgenes often give high expression levels. However, it is still almost impossible to predict the accumulation rate of heterologous protein in transplastomic plants, and there are many cases of unsuccessful transgene expression. Chloroplasts regulate their proteome at the post-transcriptional level, mainly through translation control. One of the mechanisms to modulate the translation has been described in plant chloroplasts for the chloroplast-encoded subunits of multiprotein complexes, and the autoregulation of the translation initiation of these subunits depends on the availability of their assembly partners [control by epistasy of synthesis (CES)]. In *Chlamydomonas reinhardtii*, autoregulation of endogenous proteins recruited in the assembly of functional complexes has also been reported. In this study, we revealed a self-regulation mechanism triggered by the accumulation of a soluble recombinant protein, phaseolin, in the stroma of chloroplast-transformed tobacco plants. Immunoblotting experiments showed that phaseolin could avoid this self-regulation mechanism when targeted to the thylakoids in transplastomic plants. To inhibit the thylakoid-targeted phaseolin translation as well, this protein was expressed in the presence of a nuclear version of the phaseolin gene with a transit peptide. Pulse–chase and polysome analysis revealed that phaseolin mRNA translation on plastid ribosomes was repressed due to the accumulation in the stroma of the same soluble polypeptide imported from the cytosol. We suggest that translation autoregulation in chloroplast is not limited to heteromeric protein subunits but also involves at least some of the foreign soluble recombinant proteins, leading to the inhibition of plastome-encoded transgene expression in chloroplast.

Introduction

Transgene integration into the plastid genome (plastome) is a very promising tool for producing recombinant proteins in plants, and many successful examples have been described (Bock and Warzecha, 2010; Daniell, 2006). In transplastomic plants, foreign proteins can be accumulated in the chloroplast soluble fraction, formed by stroma and thylakoid lumen, or in the chloroplast membranous fraction, composed of thylakoid membranes and envelope. For example, the chloroplast-expressed *Arabidopsis* inner envelope membrane protein Tic40 has been inserted in the tobacco chloroplast inner envelope (Singh *et al.*, 2008), while other proteins are targeted to thylakoid membranes (Ahmad *et al.*, 2012; Shanmugabalaji *et al.*, 2013). Sorting foreign polypeptides to the tobacco thylakoid lumen improve their accumulation, like the bacterial alkaline phosphatase (Bally *et al.*, 2008) or a camelid antibody fragment (Lentz *et al.*, 2012), but the most abundant proteins expressed in transformed chloroplasts are all accumulated in the stroma. Examples of this recombinant protein hyperexpression include an insecticidal toxin expressed at 46% of the plant's total soluble proteins (TSPs) (De Cosa *et al.*, 2001), or a proteinaceous antibiotic expressed at >70% of the plant's TSPs (Oey *et al.*, 2009). However, up to now, it is almost impossible to predict the accumulation rate of heterologous protein in transplastomic plants (Bock, 2014).

Indeed, in several cases, the expression level of recombinant proteins in the transplastomic plants appears to be very poor (Bellucci *et al.*, 2005; Birch-Machin *et al.*, 2004; Wirth *et al.*, 2006). It is well known that over evolutionary time most of the chloroplast genes have been either eliminated or transferred to the nucleus of the host cell, so today a highly integrated modulation between the nucleus and the plastome is needed in cell development (Timmis *et al.*, 2004; Woodson and Chory, 2008). This coordinate expression in chloroplasts takes place mainly through translational regulation, which is a major feature of plastome gene expression, and the expression of heterologous genes inserted into the plastome is also mainly regulated by post-transcriptional mechanisms (De Marchis *et al.*, 2012; Manuell *et al.*, 2007; Tiller and Bock, 2014). However, accumulation of plastome-encoded foreign protein requires several key steps including rate of transcription, translation and protein stability (Scotti *et al.*, 2013). Many studies conducted with chimeric gene fusions have identified combinations of promoters, 5'-UTRs and 3'-UTRs, which can be used to achieve a high level of recombinant protein expression in chloroplasts, regulating transcript stability and translatability (Tangphatsornruang *et al.*, 2011; Yang *et al.*, 2013). In addition, the N-terminal sequence of recombinant proteins expressed in the chloroplasts is a key factor for both mRNA stability/translatability (Kuroda and Maliga, 2001) and protein stability (Ye *et al.*, 2001). Unfortunately, there are no precise rules for the best performing sequences, and empiric attempts have to be made. For example, Elghabi et al. (2011) have fused N-terminal segments of highly expressed proteins in plastids to the transgene coding region, stabilizing the cyanovirin-N mRNA. In other cases, significant accumulation of foreign proteins has been achieved fusing an 11.6-kDa N-terminal cholera toxin B subunit (CTB) to therapeutic proteins (Kwon et al., 2013). Proteins fused to CTB have various accumulation levels, from 1% to 70% of total leaf protein (Kwon et al., 2013; Ruhlman et al., 2010), and the sole removal of a protease cleavage site between CTB and coagulation factor IX has enhanced fusion protein accumulation by 20-fold (Verma et al., 2010). In spite of the similarity to its prokaryotic ancestors of the gene expression machinery, regulation of translation in the chloroplast results to be more complex than in bacteria and it is ensured, for example by many nucleus-encoded RNA-binding translational factors (Barkan and Small, 2014; Stern et al., 2010). Furthermore, the subunits of multiprotein complexes, formed by both nuclear-encoded and plastome-encoded proteins, reveal a mechanism in which the translation rate of plastome-encoded proteins is self-regulated by the availability of their assembly partners. This process, called control by epistasy of synthesis (CES), has been described in tobacco and maize, as well as in senescent rice leaves, for Rubisco large subunit (LS) synthesis, which depends on the presence of its assembly partner Rubisco small subunit (Suzuki and Makino, 2013; Wostrikoff and Stern, 2007; Wostrikoff et al., 2012). Also in Arabidopsis, a CES process may regulate synthesis of the PSII protein CP47 (Levey et al., 2014). It is not yet clear whether autoregulatory mechanisms in plants represent a general feature of chloroplast gene expression like in Chlamydomonas reinhardtii (Zoschke et al., 2013), where CES regulates the synthesis of PSII, PSI, cytochrome b_6f and H^+-ATP synthase proteins (Boulouis et al., 2011; Drapier et al., 2007; Minai et al., 2006; Wostrikoff et al., 2004). However, it should be considered that chloroplast gene expression in C. reinhardtii is different from that in plants in many aspects. For example, with a multisubunit plastid-encoded RNA polymerase present in both C. reinhardtii and plants, no nucleus-encoded plastid RNA polymerase seems to exist in C. reinhardtii whereas it has been additionally present in plant plastids (Shiina et al., 2005). Moreover, foreign protein levels in this alga chloroplast are generally an order of magnitude lower than in plants (Michelet et al., 2010). Translational autoregulation processes similar to CES have been reported in prokaryotic and other eukaryotic systems (Fontanesi et al., 2010). In the bacterium Borrelia burgdorferi, for example, the Bpur polypeptide is able to interact with the 5' region of its own mRNA, thereby inhibiting translation (Jutras et al., 2013;).

Protein stability is often the key aspect that determines foreign protein accumulation in transplastomic plants (Elghabi et al., 2011), but other factors can be the cause of poor or undetectably low expression levels of heterologous polypeptides in these plants. Our study aimed to understand whether, in transplastomic plant chloroplasts, translational autoregulation mechanisms can be described even for at least some of the foreign soluble stromal proteins not involved in the formation of large heteromeric protein complexes. We previously tried to overexpress in transplastomic plants a recombinant protein, termed zeolin (De Marchis et al., 2011b), which could be fused to proteins of biotechnological interest to enhance their accumulation (de Virgilio et al., 2008). Zeolin is a chimeric polypeptide composed of the bean seed protein phaseolin with or without its own signal peptide for ER lumen targeting (Frigerio et al., 1998; Vitale et al.,

1995), fused to a maize γ-zein domain. Unfortunately, the synthesis of zeolin, when expressed without its signal peptide in the chloroplast stroma, was strongly inhibited, whereas the same protein with the N-terminal signal peptide was transported to the thylakoid compartment and accumulated there to a much higher extent (De Marchis et al., 2011b). This last result confirmed the previous reports, which demonstrated that signal peptides could target to thylakoids recombinant proteins expressed by transformed chloroplasts (Bally et al., 2008; Hennig et al., 2007). We observed that zeolin was subjected to partial fragmentation at the junction between the phaseolin and the zein portions in tobacco transplastomic plants (Bellucci et al., 2007). Therefore, to investigate the reason for zeolin synthesis inhibition in the chloroplast stroma, we generated several tobacco plants transformed in the nucleus, in the plastome or in both genomes of supertransformed plants, with a phaseolin gene, with or without its signal peptide, which codes for the N-terminal zeolin portion. We discovered that in supertransformed plants phaseolin mRNA translation was inhibited due to the accumulation in the stroma of the same soluble protein imported from the cytoplasm and that phaseolin could avoid this autoregulation mechanism when targeted to the thylakoids in transplastomic plants. These results show the existence of negative feedback acting at the level of heterologous protein accumulation in the plant chloroplast. Indeed, phaseolin does not take part in the formation of heteromeric protein complexes into the chloroplast; hence, negative feedback seems not to be limited to the CES process, which concerns endogenous proteins involved in the formation of heteromeric complexes.

Results

Expression of different phaseolin genes in tobacco chloroplasts and analysis of phaseolin polypeptides folding

To prove the existence of a mechanism that controls the accumulation of plastome-encoded foreign proteins by negative feedback, we generated transplastomic tobacco plants expressing plastome-inserted transgenes' coding for phaseolin both with (P) and without its N-terminal signal peptide (ΔP). The accumulation of these two proteins in tobacco transformants was verified in Western blots with antiphaseolin antiserum (Figure 1a). The signal intensity of the 46-kDa bands corresponding to the two phaseolin forms was quantified indicating that P accumulated almost a hundred times more than ΔP in the chloroplast. However, the low ΔP accumulation was not attributable to a defect in the transcription of its gene as demonstrated by northern blotting analysis on the same two transplastomic plants in Figure 1a. Following hybridization with an antiphaseolin probe, three phaseolin-encoding transcripts were detected (monocistronic phaseolin, dicistronic aadA/phaseolin and polycistronic 16S/trnI/aadA/phaseolin), as previously described (Bellucci et al., 2007). Moreover, there were comparable recombinant mRNA levels between plants transformed with P or ΔP constructs (Figure 1b). The failure of ΔP accumulation inside tobacco chloroplast could be due to an intense proteolytic activity towards this protein; therefore, we generated nuclear-transformed tobacco plants expressing a phaseolin variant (tpΔP) with a C-terminal FLAG tail, in which the signal peptide had been replaced by the tobacco Rubisco small subunit (SS) transit peptide, which directed the protein into the chloroplast. The correct plastidial localization of tpΔP was confirmed through immunofluorescence experiments (Fig-

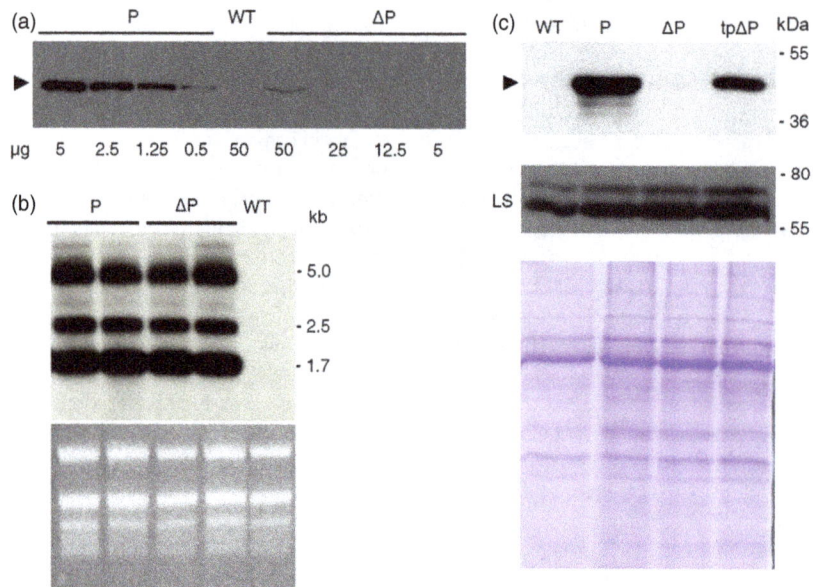

Figure 1 Transcription and protein accumulation of different phaseolin genes in tobacco chloroplasts. (a) Total proteins extracted from leaves of a WT plant, or transplastomic plants expressing P or ΔP, were analysed by SDS-PAGE and Western blotting using antiphaseolin antiserum. The arrowhead marks the position of mature phaseolin, and μg stands for total protein extract. (b) Northern blot analysis was performed on total RNA extracted from P, ΔP and WT plants and hybridized with the *phaseolin* gene. Numbers at right indicate molecular mass markers in kb. rRNA stained by ethidium bromide is shown as a loading control. (c) Eight micrograms of total proteins extracted from leaves of a WT plant, transplastomic plants expressing P or ΔP, or phaseolin with a transit peptide (tpΔP) was separated by SDS-PAGE and immunoblotted with antiphaseolin or anti-Rubisco large subunit (LS) antiserum. The arrowhead marks the position of mature phaseolin. Protein stained by Coomassie is shown as a loading control. Numbers at right indicate molecular mass markers in kDa.

ure S1). tpΔP was synthesized on cytoplasmic ribosomes and then imported into the chloroplast, where the cleavage of the transit peptide originated a polypeptide virtually identical to the ΔP expressed by the transplastomic tobacco plant and synthesized on plastid ribosomes. We postulated that if the chloroplast proteolytic degradation specifically limited ΔP accumulation instead of P accumulation the amount of tpΔP protein detected inside the chloroplast should have been much lower than the amount of P accumulated in the same organelle. Conversely, as the amount of tpΔP was roughly 1/3–1/5 of that of P (Figure 1c, upper panel), there should have been other reasons for the large reduction of ΔP accumulation in chloroplast. A possible cause of the low accumulation of ΔP protein could be ascribed to a general decrease in chloroplast proteins' synthesis. To exclude this possibility, the same samples were analysed for both the content of the endogenous plastid protein Rubisco large subunit (LS) (Figure 1c, middle panel), and for total protein content (Figure 1c, lower panel). We can conclude from these results that there was not a decrease in proteins synthesis. The difference in the accumulation of the two phaseolin forms might find the reason in ΔP polypeptide instability caused by an alteration in protein folding. To verify this hypothesis, the three-dimensional conformation of P and ΔP was examined. When properly folded, phaseolin assembled in the endoplasmic reticulum (ER) into homotrimers which, being resistant to *in vitro* trypsin digestion, produced 20- to 30-kDa peptide fragments that did not undergo further degradation (Deshpande and Nielsen, 1987; Pompa *et al.*, 2010). In Figure 2a, total leaf proteins extracted from tobacco plants expressing the two chloroplast phaseolin forms, or a nuclear-encoded phaseolin (Pnu) translocated into the ER lumen, were subjected to trypsin digestion followed by an immunoblot analysis with antiphase-

olin antiserum. All the plants produced 18- to 30-kDa peptide fragments suggesting that phaseolin polypeptides were properly folded. In particular, while the Pnu control digestion showed several fragments of 18–22 kDa, chloroplast P and ΔP showed a quite similar resistant fragment profile after trypsin treatment, with a 25-kDa fragment and a 18-kDa doublet (Figure 2a). Indeed, the intensity difference of the 18-kDa doublet between P and ΔP did not mean a difference in protein conformation because in other trypsin digestions the intensity of this 18-kDa doublet was almost identical. Furthermore, the proper achievement of the homotrimeric three-dimensional structure was also confirmed by sucrose sedimentation velocity gradient experiments (Figure 2b). These show that ΔP polypeptides had the same peak of migration of the Pnu trimers (phaseolin molecular weight 46 kDa × 3 = 138 kDa), whereas P migrated as trimers and oligomers constituted by the association of two or more trimers. These results demonstrated that ΔP low expression did not originate by its unfolded status.

Accumulation of plastome-encoded phaseolins depends on their intraplastidial localization

We also investigated the effect of intraplastidial localization on P and ΔP accumulation. Purified chloroplasts were fractionated into the thylakoid and stroma compartments and subjected to immunoblot experiments with antiphaseolin antiserum. P was almost completely recovered in the thylakoid fraction, whereas ΔP was present in both thylakoid and stromal fractions, albeit to a greater extent in the first one (Figure 3a). To verify whether P and ΔP were inserted in the thylakoid fraction or only associated with it, thylakoids were subjected to sequential washes with different saline concentration buffers. While ΔP, already detached from thylakoids after the first wash with an

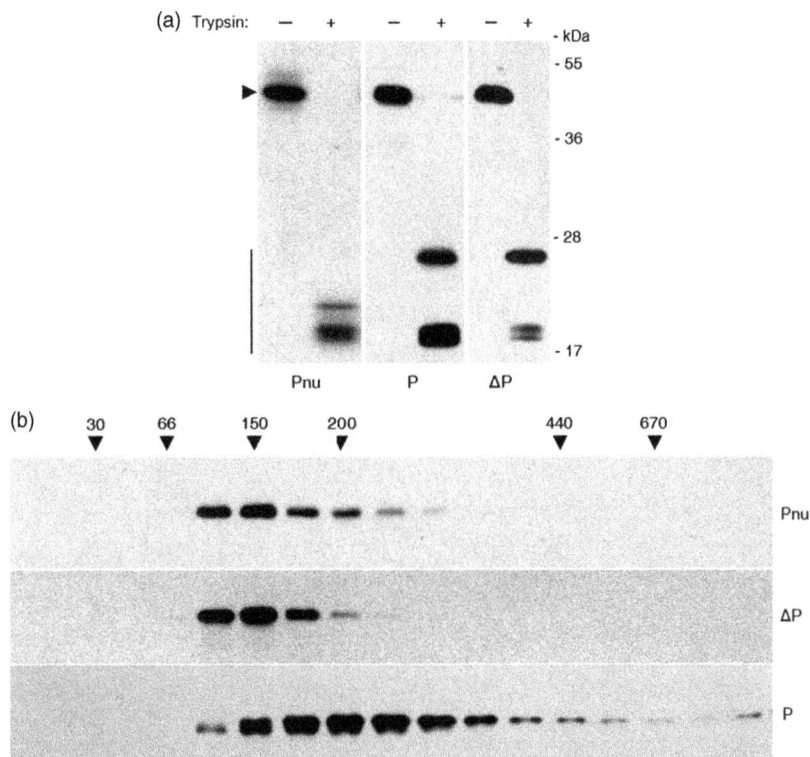

Figure 2 Three-dimensional conformation of P and ΔP phaseolins in the chloroplast. (a) Total leaf proteins extracted from nuclear-transformed plants expressing the entire phaseolin (Pnu, protein aliquots of 8 μg), as a positive control for trypsin assay, and from transplastomic plants expressing P (protein aliquots of 8 μg) or ΔP (protein aliquots of 40 μg) were subjected to trypsin digestion *in vitro* for 30 min or incubated without enzyme. Protein samples were then analysed by SDS-PAGE and Western blotting using antiphaseolin antiserum. The arrowhead marks the position of intact phaseolin, while vertical bar marks the trypsin-resistant phaseolin fragments. Numbers at right indicate molecular mass markers in kDa. (b) The same proteins extracted from (a) were fractionated by centrifugation on velocity sucrose gradient. A leaf homogenate aliquot of 0.12 mL was loaded on the sucrose gradient, whereas 0.60 mL of leaf homogenate was loaded for the ΔP sample due to the low transgene expression of this transplastomic line. Different fractions were collected, and each fraction was analysed by SDS-PAGE and Western blotting using antiphaseolin antiserum. Numbers on top indicate molecular mass, in kDa, of sedimentation markers. Protein blots in (a) and (b) were exposed to different times (a few seconds for the Pnu and P samples and a few minutes for the ΔP sample).

isotonic solution (WB), was completely removed by the addition of a highly concentrated saline buffer (NaCl), P remained strictly associated with thylakoid membranes without being affected by these treatments (Figure 3b). This demonstrated that ΔP was only weakly associated with the thylakoids, whereas P was likely inserted into the thylakoid lipid bilayer or translocated into the thylakoid lumen, with its signal peptide removed after the insertion/translocation (Figure 3c). Thus, the localization of P in the thylakoids and its high accumulation level in comparison with ΔP (Figure 1a) suggest that ΔP localization to the stromal side of thylakoids could lead to its low accumulation. To understand whether the presence in the stroma per se was a sufficient condition for ΔP-reduced expression, we verified whether the chloroplast-imported mutant of phaseolin, tpΔP, after the transit peptide removal, was recovered in the stroma as expected. Total proteins from leaves of transgenic tpΔP and transplastomic P plants were separated in two fractions containing the soluble- and the membrane-associated polypeptides. Proteins recovered in the soluble fraction (Figure 3d, Sol) were extracted with a buffer containing a high NaCl concentration which, as shown in Figure 3b, was not able to solubilize the integral membrane

proteins. Intrinsic thylakoid polypeptides were then extracted from the pelleted membranous fraction with a buffer containing Triton X-100 as detergent (Figure 3d, Pel). Both the stromal control Rubisco LS and the thylakoid control CP47 were localized in the soluble or membranous fractions, respectively. The results showed that tpΔP was a soluble stromal protein; on the contrary P, as expected, was strictly associated with the membranes or translocated into the thylakoid lumen (Figure 3d). As tpΔP synthesized in the cytoplasm possessed the same stromal localization of ΔP but an accumulation level almost comparable to that of P, the logical reason for the big difference in P and ΔP accumulation must be found in the mechanism regulating their mRNA translation or protein synthesis in the stroma.

Phaseolin synthesis in chloroplast is repressed at the translational level by an autoregulation mechanism

The regulation of the expression of the plastome transgenes' coding for P and ΔP was investigated at the translational level by pulse–chase experiments with radioactive amino acids. With this technique, it was possible to monitor the synthesis of a protein in the unit of time along with its half-life. Protoplasts isolated from P

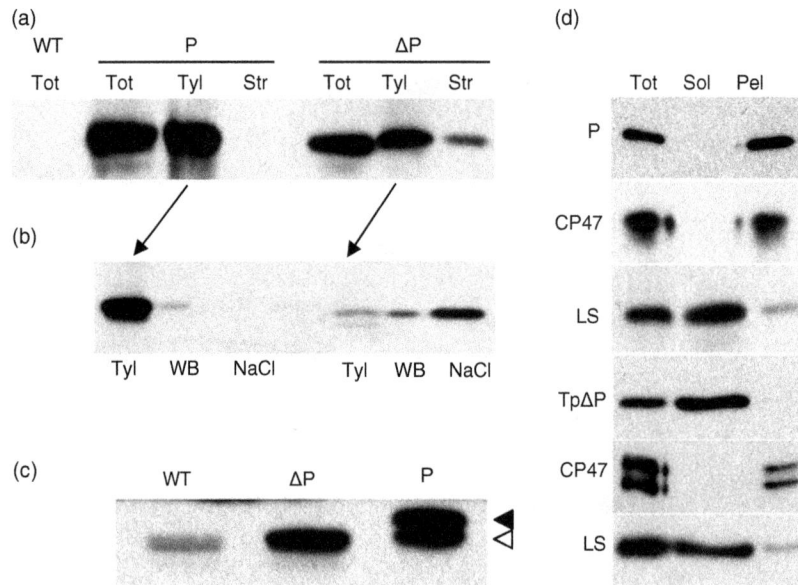

Figure 3 Subplastidial localization of P, ΔP and tpΔP. (a) Sucrose-purified chloroplasts derived from transplastomic plants expressing P and ΔP proteins were fractionated into stroma (Str) and thylakoid fractions (Tyl). (b) The thylakoid fractions of (a) were subjected to two sequential washes with an isotonic buffer (WB) and then with a saline buffer containing 2 M NaCl. To increase the ΔP signal intensity, 20 μg of protein was loaded in the stromal sample (while 2 μg of protein was loaded for the P sample), and 5 μg of chlorophyll was loaded in the Tot and Tyl samples (while 0.5 μg of protein was loaded for the P sample). (c) Cleavage of the phaseolin signal peptide in chloroplast. Proteins extracted from young leaves of a WT plant and transplastomic plants expressing ΔP and P were analysed by SDS-PAGE and Western blotting with antiphaseolin antiserum. The gel was run for a time longer than that in Figure 1a to separate the entire phaseolin from phaseolin without the signal peptide. Black arrowhead indicates the entire phaseolin polypeptide, and empty arrowhead indicates phaseolin without signal peptide. (d) Proteins, extracted from lysed leaves and homogenated with a saline buffer supplemented with 2 M NaCl, were separated in two fractions: the soluble (including stromal) proteins and the integral membrane (including thylakoids) proteins. Samples were analysed by immunoblot using antibodies against phaseolin, CP47 and LS. The CP47 antibody detected a doublet signal in the tpΔP plant sample that in the P plant sample is not resolved due to short run of the SDS-PAGE. While 2 μg of protein was loaded in the stromal samples, 0.5 μg of chlorophyll was loaded in the Tot and Tyl samples. All protein blots were exposed few seconds to decrease the background signals.

and ΔP transplastomic tobacco plants were pulse-labelled for 1 h and chased for the indicated periods of time. Protoplasts were homogenated, immunoprecipitated with the antiphaseolin antiserum and analysed by SDS-PAGE and fluorography. The P protein was detected at the pulse and displayed a half-life of about 4 h. Conversely, it was impossible to detect ΔP, likely because this protein was under the detection limit of the experiment (Figure 4a). To rule out that a rapid degradation of newly synthesized ΔP protein makes it impossible to be detected, the same experiment was carried out by decreasing the pulse to 15 min, obtaining the same result. Moreover, no phaseolin polypeptide was detected in the supernatant fractions recovered after protoplast immunoprecipitation and subjected to a second round of immunoprecipitation with the same antibody (data not shown). The translation activity of these two genes was further investigated using polysome analysis (Figure 4b). Extracts from P and ΔP transplastomic leaves were fractionated in sucrose density gradients, with and without EDTA, and analysed by northern blotting experiments with a phaseolin probe. EDTA treatment released associated ribosomes from mRNAs, and comparison between EDTA-containing and EDTA-free gradient samples allowed us to determine the monosome- versus polysome-containing fractions. A difference in polysome loading was detected between P and ΔP transplastomic plants, suggesting that the corresponding genes' translation had been not carried out with comparable efficiency. P mRNA was prevalently associated with actively translating polysomes (Figure 4b, sample P,

lanes 6–8), while ΔP mRNA was poorly translated because it was largely detected in the top fractions of the gradient (Figure 4b, sample ΔP, lanes 3–5). These results revealed a significant reduction in ΔP mRNA translation activity.

To understand the mechanism regulating ΔP mRNA translation, we hypothesized that in the chloroplast stroma ΔP protein would repress its own translation with an autoregulation mechanism. In this hypothetic case, the P protein could escape this self-regulation because its signal peptide mediated the rapid translocation into the thylakoid compartment, subtracting P polypeptides from the stroma. Thus, we increased the abundance of a phaseolin protein in the stroma of transformed chloroplasts to trigger P autoregulation. A transplastomic P plant was supertransformed with the nuclear tpΔP construct, and the resulting plants were named supertransformants (SupT). The SupT plants were analysed for the presence of both nuclear and plastidial phaseolin genes by PCR, and for the resistance to both kanamycin and spectinomycin, which were used for the two different transformation events' selection (Figure S2). While the plastome-encoded P could be visualized only with the antiphaseolin antiserum, the nuclear-encoded tpΔP protein was C-terminal Flag-tagged, thus detectable with both anti-Flag and antiphaseolin antibodies. To assess the relative amount of the two phaseolin polypeptides in SupT plants, total leaf proteins from tobacco tpΔP, P and SupT transformants were separated, as described in Figure 3d, into soluble and membranous fractions and subjected to SDS-PAGE followed by Western blot analysis with anti-Flag or antiphaseolin

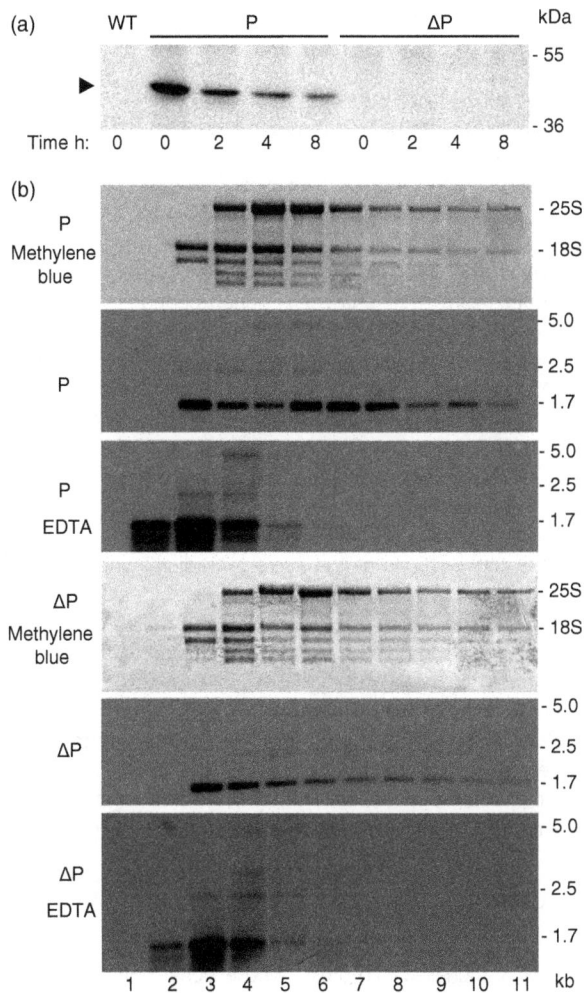

Figure 4 Modulation of P and ΔP synthesis in chloroplasts. (a) Protoplasts from transplastomic P and ΔP tobacco plants were pulse-labelled for 1 h and chased for the indicated periods of time. Homogenated cells were immunoprecipitated with antiphaseolin antiserum and analysed by SDS–PAGE and fluorography. Black arrowhead indicates phaseolin polypeptide. Numbers at right indicate molecular mass markers in kDa. (b) Polysome analysis was performed from leaves of P and ΔP plants following sedimentation through a 15%–55% sucrose gradient, with or without EDTA. Methylene blue staining is used to visualize the ribosomal RNA fractionation profile for the samples without EDTA, while the methylene blue staining for the samples with EDTA is not shown. An equal proportion of RNA isolated from each fraction was analysed by northern blot with the phaseolin probe. Numbers at right indicate molecular mass markers in kb.

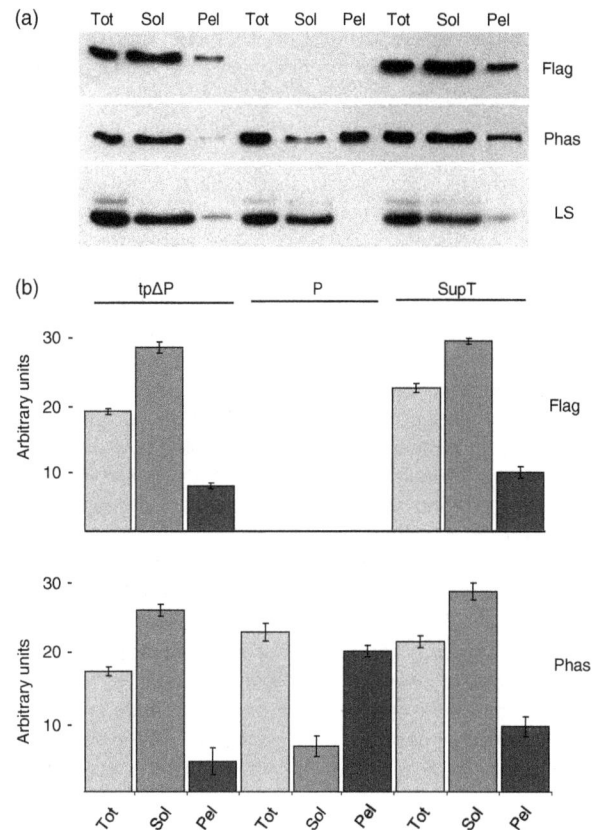

Figure 5 Comparison of relative phaseolin amounts in soluble and membrane fractions in P, tpΔP and supertransformants (SupT) tobacco plants. (a) Total proteins (Tot) from tpΔP, P and SupT tobacco plants were separated with a saline buffer (2 M NaCl) in soluble (Sol) and integral membrane proteins (Pel) as in Figure 3d, then analysed by immunoblot using antibodies against FLAG, phaseolin and LS. (b) Quantitative analysis of phaseolin proteins detected by immunoblot in (a). Error bars represent the standard deviation calculated for each chloroplast fraction derived from three independent measurements per fraction using anti-Flag or antiphaseolin antibodies.

antisera. The anti-Flag antibody, in both tpΔP and SupT plants, detected almost all the tpΔP protein in the soluble fraction except for a small amount detected in the membranous compartment, which could also be due to the contamination by the stroma as judged by the use of an anti-LS antibody (Figure 5a, upper and lower panel). As expected, no signal was shown with anti-Flag antibody in the P transplastomic plant. The use of the antiphaseolin antiserum in the same plant samples allowed us to reveal both P and tpΔP polypeptides. In this case, as in Figure 3d, in P plants the protein was mostly recovered in the pellet containing the thylakoid membranes (Figure 5a, middle panel). Assuming that SupT plants should express the same amount of plastidial phaseolin of the original transplastomic plant, we expected to

recover in the SupT membranous fraction (Pel) at least the identical amount of phaseolin polypeptide detected in same fraction of P plant. On the contrary, the amount of phaseolin measured on the SupT membranes was roughly less than half of that measured in P membranes (Figure 5b), suggesting that in SupT the contribution given by P protein on the whole phaseolin content is reduced with respect to the nuclear-encoded tpΔP. Consequently, the total amount of phaseolin detected in the SupT samples was almost entirely located into the soluble fraction, and it was likely attributable to the sole contribution of tpΔP. These data suggested that P protein was no longer present in the SupT fraction containing the thylakoid membranes.

Two hypotheses were formulated to explain the fate of the plastome-encoded P protein in the SupT plants: the protein could have been relocated in the stroma, or its synthesis could have been strongly inhibited. To verify these last assumptions, we performed both sucrose sedimentation velocity gradient experiments and pulse–chase assays. The sucrose gradient confirmed that in SupT plant the large part of phaseolin was likely constituted by trimeric tpΔP polypeptides, whereas P trimers and oligomers were a minor fraction of the total phaseolin polypeptides (Figure S4). The pulse–chase experiments were performed in the presence of the cytosolic protein synthesis inhibitor cycloheximide (CHX) (Figure 6). This allowed us to unveil whether the production of P synthesized in the stroma had really been suppressed in SupT plants. Protoplasts isolated from the three transformed plants tpΔP, P and SupT were pulse-labelled for 1 h, in the presence or absence of CHX, and then immuno-precipitated with antiphaseolin antibody and analysed by SDS-PAGE and fluorography (Figure 6a). To verify that all the radioactive-labelled phaseolin polypeptides were immunoprecip-itated after the pulse–chase, the supernatant fractions derived from the previously described immunoprecipitation were sub-jected to a second round of immunoprecipitation with the same antibody and then analysed in the same way, but they did not show any presence of phaseolin polypeptides (Figure S3). In tpΔP plants, in the presence of CHX, phaseolin is not recovered after immunoprecipitation because this drug inhibited tpΔP synthesis in the cytoplasm (Figure 6a, lanes 1–2). CHX should not have inhibited translation on plastid ribosomes, but an approximate reduction of 50% of the phaseolin synthesized by transplastomic P plants was observed, possibly due to drop in the amount of some nuclear-encoded factors targeting the *psbA* 5′UTR that are required for D1 expression (Figure 6a, lanes 3–4). The total amount of phaseolin recovered after immunoprecipitation from SupT protoplasts was bigger (Figure 6a, lane 5) than the phaseolin present in tpΔP and P protoplasts (Figure 6a, lanes 1 and 3), but it disappeared in the presence of CHX (Fig 6a, lane 6). As phaseolin loading control, an immunoblot analysis with an

antiphaseolin antibody was performed on total proteins extracted from the same number of pulse-labelled protoplasts used for the immunoprecipitation (Figure 6b, c). In this way, all the phaseolins immunoprecipitated from SupT protoplasts was synthesized in the cytosol. This strongly suggested that, in SupT transformants, the presence of tpΔP in the stroma repressed the P synthesis. Therefore, the regulation of the plastidial P expression occurred at the translational level, as already seen for ΔP in the transplastomic plants (Figure 4b). To verify this hypothesis, an analysis on polysomes extracted from P, ΔP and SupT tobacco leaves was performed with the phaseolin signal peptide sequence as radioactive probe. This probe was able to discriminate between the two phaseolin mRNAs of the SupT plants, because the P mRNA would be hybridized, whereas tpΔP mRNA would not due to the lack of the signal peptide sequence in the corresponding gene. In the same way, the P mRNA of the P plants, used as positive control, would be hybridized to this probe, but no signal would result from the ΔP mRNA in ΔP plants, used as negative control. There was a shift in polysome association of the P mRNA with the top fractions of the sucrose gradient when the SupT plant (Figure 7, sample SupT, lanes 4–5) was compared with the P plant. Thus, the translation efficiency of the P mRNA was reduced in SupT plants in comparison with the P transformants, where the same transcript was mostly polysomal associated (Figure 7, sample P, lanes 6–8 and Figure 4b). As expected, no signal was visualized in ΔP plants. These data indicate the existence of an autoregulation mechanism managing the plastome-encoded phaseolin expression, whose mRNA translation is controlled by the amount of soluble phaseolin in the stroma (Figure 8).

Discussion

In this work, we have demonstrated that the translation of a recombinant phaseolin protein, whose gene is inserted in the tobacco plastome, is down-regulated by the presence in the stroma of soluble phaseolin polypeptides. Many chloroplast-encoded proteins control their own production in *C. reinhardtii* and plants when the availability of their assembly partners is reduced (CES process), but this autoregulation has always been reported for polypeptides involved in the formation of hetero-meric protein complexes. The stoichiometric imbalance of these complexes is the determinant for triggering the CES process, which affects only a limited number of proteins. Here we show that a negative feedback mechanism is able to regulate the translation in the stroma of a heterologous protein, which is not a part of any chloroplast endogenous heteromeric complexes. Therefore, autoregulation of translation in chloroplast can be also extended at least to the expression of some of the foreign proteins in transplastomic plants, and it is likely a more basic biological phenomenon than the previous belief. In particular, accumulation of mature phaseolin in the stroma of transplastomic plants is very low not due to polypeptide instability caused by protease degradation or alteration in protein folding (Figure 2), but because the translation of its mRNA is strongly reduced (Figure 4b). This inhibition occurs specifically on plastid ribosomes because when the same phaseolin protein, expressed from nuclear-transformed tobacco plants, is synthesized on cytoplas-mic ribosomes and imported into the chloroplast, its accumula-tion significantly increased. Similarly, the P protein, which is the chloroplast-encoded full-length phaseolin, including its signal peptide, can reach accumulation levels comparable to that of the nuclear-encoded mutant (Figure 1c). This has been possible as

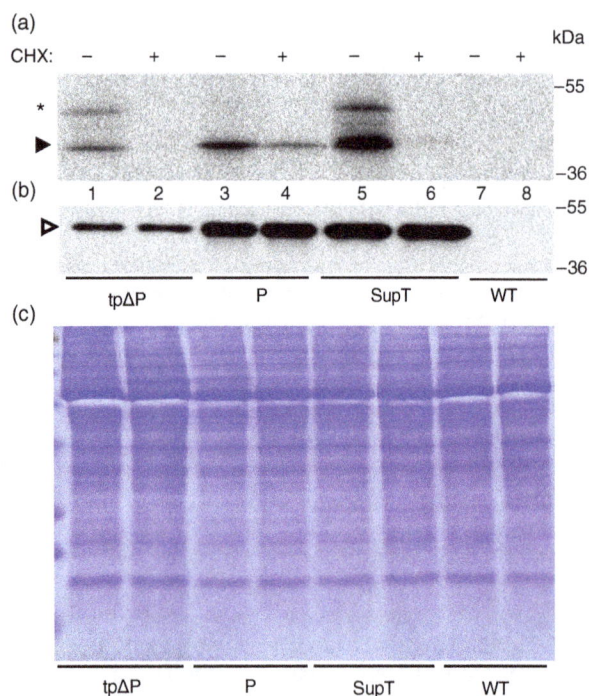

Figure 6 Repression of P synthesis in supertransformants (SupT) tobacco plants. Protoplasts from leaves of a WT plant, or from transformed tobacco plants expressing tpΔP, P and SupT proteins, were pulse-labelled for 1 h in the presence or absence of cycloheximide (CHX). An equal number of homogenated protoplasts were immunoprecipitated (a) or detected by immunoblot (b), with antiphaseolin antiserum. (c) Protein stained by Coomassie is shown as a loading control of (b). Black arrowhead indicates radiolabelled phaseolin polypeptide, empty arrowhead indicates phaseolin and asterisk refers to the entire tpΔP protein with the transit peptide not yet cleaved.

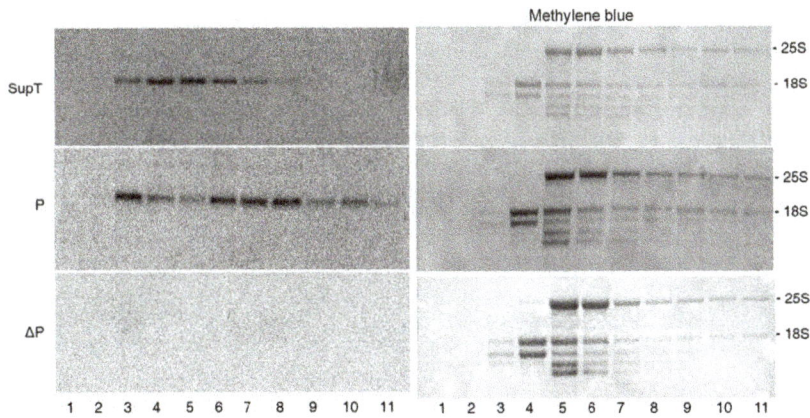

Figure 7 Polysome analysis in supertransformants (SupT) plants. Left panels: total leaf RNA from ΔP, P and SupT plants was fractionated through a 15%–55% sucrose gradient. The RNA present in the different fractions was extracted and analysed by northern blot with the phaseolin signal peptide sequence as a probe. Right panels: methylene blue staining is used to visualize the ribosomal RNA fractionation profile.

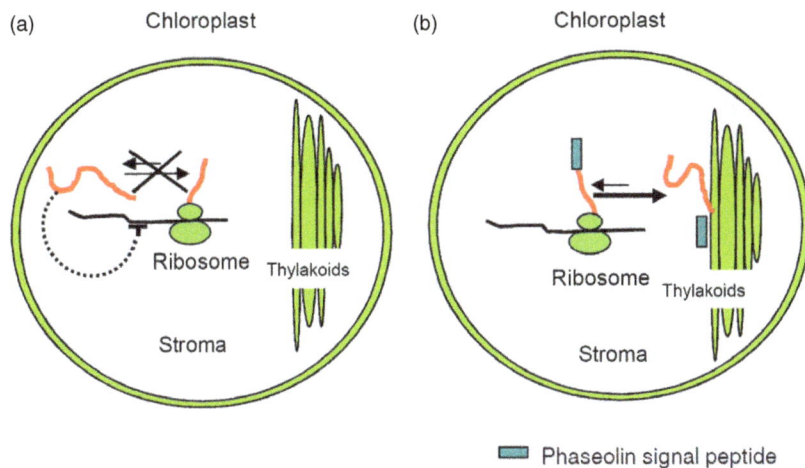

Figure 8 Schematic representation of the autoregulatory translation mechanism in tobacco chloroplast described in this study for phaseolin transgenes inserted into the plastome. (a) Negative regulatory feedback loop is revealed through repression of translation triggered by the presence of stromal phaseolin (ΔP transplastomic plants). (b) When they are synthesized together with their signal peptide, phaseolin polypeptides are targeted to the thylakoid membranes decreasing the amount of recombinant protein localized in the stroma, thus avoiding the activation of the autoregulatory translation mechanism.

the phaseolin with the signal peptide is synthesized in the stroma but relocalized into the thylakoids after cleavage of the transit peptide (Figure 3), likely performed by the lumenal peptidase on the lumenal face of the thylakoid membrane. Considering that phaseolin does not have the physicochemical properties of an intrinsic membrane protein, we suggest that it has likely been translocated into the thylakoid lumen. Therefore, no matter which molecular determinant inhibits phaseolin translation in the chloroplast, this mechanism relies on the amount of soluble phaseolin in the stroma as a mode of activation. These results offer the interpretation for another report, where we observed that the accumulation of the zeolin fusion protein with phaseolin at the N-terminal was determined by its intraplastidial localization (De Marchis *et al.*, 2011b). Reasonably, the scarcity of other transplastomic proteins could be attributed to autoregulation mechanisms, for example the human papillomavirus E7 antigen (Morgenfeld *et al.*, 2014). To prove necessity of an excess over a threshold level of stromal phaseolin to induce its autoregulation, we supertransformed a P transplastomic plant with the nuclear phaseolin construct (expressing the chloroplast-targeted phaseolin). While the P mRNA was actively translated in transplastomic plants, in SupT P transcript resulted to be less efficiently translated, indicating that in these plants the phaseolin accumulation in the chloroplast was almost exclusively due to the phaseolin polypeptides imported into the stroma from the cytosol (Figures 6 and 7). Moreover, this also means that the chloroplast translational regulation apparatus cannot distinguish between the phaseolin synthesized by the plastid ribosomes and the cytoplas-

mic phaseolin. This last phaseolin is able to trigger the self-regulation mechanism that ultimately represses the synthesis of the P phaseolin protein. We ignore how this mechanism works and it may resemble the CES process in *C. reinhardtii*, where down-regulation of translation seems to be mediated by unidentified ternary translational activators capable of competitive binding to both the unassembled CES subunits and the 5′ untranslated region (5′UTR) of their corresponding mRNAs. In the transplastomic plants generated in this study, phaseolin Open reading frame (ORF) is fused to the tobacco *psbA* 5′UTR. The product of the chloroplast *psbA* gene is the PSII subunit D1, and in *C. reinhardtii* its decreased synthesis in the absence of protein assembly is due to inhibition of translation mediated by the 5′UTR of its mRNA (Minai *et al.*, 2006). Although D1, which has an amino acid sequence completely different from that of phaseolin, is not reported as a CES subunit in tobacco, it is still possible that a tobacco protein binding to both phaseolin and *psbA* 5′ UTR-phaseolin transcript regulates heterologous phaseolin translation in the stroma. However, the *psbA* 5′UTR resulting in overexpression of several foreign proteins should also be considered (Verma and Daniell, 2007). We think that the here-suggested autoregulation machinery for phaseolin expression in the chloroplast could derive from a defence mechanism of bacterial origin. Horizontal gene transfer plays an important role in the evolution of bacteria; for example, it is responsible for antibiotic resistance transfer (Koonin *et al.*, 2001). Horizontally transferred genes can either confer a selective advantage or result dangerous; therefore, bacteria have developed defence mechanisms like the bacterial nucleoid-associated protein H-NS that transcriptionally represses horizontally acquired genes in *Salmonella* (Ali *et al.*, 2013; Navarre *et al.*, 2006). Thus, it is possible that the chloroplast has developed a biological system starting from an ancient defence pathway based on autorepressed translation of proteins that are soluble in the stroma (Figure 8). The CES feedback process may have evolved from this general regulation mechanism to coordinate the assembly of large heteromeric complexes. Recently, the existence of multiple negative regulatory feedback loops has been revealed in chloroplast, which compensates the decreased translation level or plastid mRNAs transcription in *C. reinhardtii* (Ramundo *et al.*, 2013). In conclusion, we propose an autoregulation mechanism regulating heterologous protein accumulation in the stroma of transplastomic plants that is working for phaseolin and likely for other soluble heterologous proteins with low expression levels. This mechanism should be protein-specific considering that many foreign proteins have been hyperexpressed in the stroma of transplastomic plants thanks to their accumulation as insoluble aggregates (Kwon *et al.*, 2013) and crystals (De Cosa *et al.*, 2001), or due to the plant physiological adaptations (Bally *et al.*, 2009).

Experimental procedures

Growth conditions

Nicotiana tabacum (cv. Petit Havana) was grown at 24 °C with a 16-h light/8-h dark period under 60 µE/m/s^2. Transplastomic plants were propagated on MSO medium supplemented with 500 mg/L spectinomycin. The nuclear transformants Pnu, tpΔP and SupT were maintained on MSO containing 50 mg/L kanamycin. T0 seeds were obtained from all the transgenics grown in the greenhouse. T1 plants were obtained after the germination of T0 seeds on agar-solidified MS medium plus 500 mg/L spectinomycin (transplastomic plants) or 50 mg/L kanamycin (Pnu, tpΔP plants) or 500 mg/L spectinomycin/50 mg/L kanamycin (SupT).

Gene constructs and plant transformation

The ORF of phaseolin was amplified from plasmid pDHA.T343F (Pedrazzini *et al.*, 1997), digested with NdeI/NotI and cloned into pCR2.1-5′UTR (Watson *et al.*, 2004), to obtain pCR2.1-5′UTR-P and pCR2.1-5′UTR-ΔP (where the 72-bp phaseolin signal sequence was deleted) intermediate plasmids, in which the ORFs were under the *psbA* promoter/5′UTR control. The *psbA*/5′UTR-P and *psbA*/5′UTR-ΔP cassettes were obtained by EcoRV/NotI digestion of pCR2.1-5′UTR-P and pCR2.1-5′UTR-ΔP, respectively, and cloned into pLD-CTV (Dhingra *et al.*, 2004), generating pLD-CTV-P and pLD-CTV-ΔP. Homoplasmic transplastomic plants were obtained as described (De Marchis *et al.*, 2011b).

To prepare phaseolin with a transit peptide (tpΔP), the ORF coding for ΔP was PCR-amplified from pCR2.1-5′UTR-ΔP using primers SphIΔP/EcoRIΔP-FLAG (this primer adds to the ΔP C-terminus a Flag epitope), digested with SphI/EcoRI and cloned into pJIT117 (Guerineau *et al.*, 1988), containing the transit peptide of the tobacco Rubisco small subunit (SS). The tpΔP-FLAG fragment, obtained by EcoRI/HindIII digestion of pJIT117.tpΔP-FLAG, was blunted by treatment with the Klenow fragment of DNA polymerase I and inserted into BamHI/blunted-linearized pDHA vector (Hellens *et al.*, 2000), under the control of the 35S promoter, obtaining pDHA.tpΔP-FLAG. The DNA fragment excised by EcoRI digestion of pDHA.tpΔP-FLAG, including the 35S promoter, the tpΔP-FLAG sequence and the 35S terminator,

was cloned into the *EcoRI* site of the pGreenII binary vector (Tabe *et al.*, 1995), generating pGreenII.tpΔP-FLAG, which was introduced into the GV3101 strain of *Agrobacterium tumefaciens*. WT or pLD-CTV-P transplastomic plants were transformed as described (De Marchis *et al.*, 2011a). The oligonucleotides used in this study were described in Table S1. Transgenic tobacco plants expressing phaseolin (Pnu) were obtained by seeds from Alessandro Vitale's laboratory.

Protein analysis

Total proteins were extracted from 0.3 g of leaves grounded in liquid nitrogen, homogenized in 0.8 mL of extraction buffer and analysed as reported previously (Bellucci *et al.*, 2007), except for the antiphaseolin and anti-Rubisco antiserum which were diluted 1 : 10 000 or 1 : 7500, respectively. When subjected to trypsin assay, protein aliquots of 8 µg (Pnu and P plants) or 40 µg (ΔP) were digested for 30 min at 37 °C with 10 µL of trypsin (Roche Diagnostics GmbH, Mannheim, Germany) from a 0.5 µg/µL solution in HCl 1 mm, or with 10 µL of 1 mm HCl as control. The samples were transferred on ice to stop the digestion and then treated as described above.

Chloroplasts were isolated from 25 g batches of leaves and fractionated as described (Salvi *et al.*, 2008), with minor modifications. Intact chloroplast was lysed and loaded on the top of a discontinuous sucrose gradient. The tubes were centrifuged at 70 000 *g* for 1 h and 4 °C to separate the soluble stromal proteins from the pelleted thylakoids. For further thylakoid purification, the pellet was washed with 10 volumes of washing buffer and centrifuged at 110 000 *g* for 1 h and 4 °C and then a minimum volume of washing buffer was added to the thylakoid pellet. To verify the association between thylakoids and proteins in the P and ΔP plants, the thylakoid pellet was washed with the washing buffer and centrifuged at 14 000 *g* for 10 min and 4 °C. The supernatant was recovered, the thylakoid pellet was washed again with the washing buffer plus 2 m NaCl and the saline supernatant was obtained by centrifugation. Chloroplast fractions were analysed by immunoblot assay using antiphaseolin antiserum (1 : 10 000).

Alternatively, to isolate chloroplast fractions, another method was established by grinding in liquid nitrogen 300 mg of leaf tissue and adding 1.2 mL of chloroplast lysis buffer [10 mm MOPS-NaOH, pH 7.8, 4 mm MgCl2, 1× protease inhibitor mix COMPLETE (Roche Diagnostics GmbH)]. The sample was divided in two aliquots of 0.6 mL. To obtain the total sample, the first aliquot was supplemented with 1% Triton X-100, vortexed and centrifuged at 12 000 *g* for 10 min and 4 °C, and the supernatant was recovered. The second aliquot was freeze–thawed in liquid nitrogen and vortexed three times, supplemented with 2 m NaCl and 0.5 m DTT, vortexed again and centrifuged at 12 000 *g* for 10 min and 4 °C to obtain the soluble fraction. The pellet was resuspended in lysis buffer plus 1% Triton X-100 and centrifuged at 12 000 *g* for 10 min and 4 °C to obtain the membrane fraction. Chloroplast fractions were analysed by immunoblot assay using antiphaseolin (1 : 10 000), anti-Flag (1 : 1000; Sigma-Aldrich, St. Louis, MO), anti-Rubisco (1 : 7500; Jackson ImmunoResearch Inc., West Grove, PA) or anti-CP47 (1 : 5000) antiserum. Protein band intensities (arbitrary units) were measured with the public-domain ImageJ software (US National Institute of Health, http://rsb.info.nih.gov/ij) on three independent immunoblots, using anti-Flag antibody, of soluble and membrane fractions in P, tpΔP and SupT tobacco plants. Other three independent measurements were obtained using antiphaseolin antibody.

Velocity centrifugation on sucrose gradients was performed from young leaves of tobacco plants as described (Pompa et al., 2010), except that to increase the ΔP signal intensity on the Western blot, 0.60 mL of ΔP leaf homogenate was loaded on the sucrose gradient, while 0.12 mL aliquots were used for the other plant samples.

Gene expression analysis

Total RNA was extracted with the GenEluteTM Plant Genomic DNA Kit (Sigma-Aldrich), and 2.5 μg was analysed by northern blot as described (Bellucci et al., 2007). The phaseolin ORF was used as probe.

Polysomes were analysed from an extract prepared by grinding 300 mg of leaf tissue in 1 mL of polysome extraction buffer as described (Barkan, 1993). The ORF of the phaseolin gene or its PCR-amplified 72-bp signal peptide sequence was used as probes.

Protoplast preparation, pulse–chase labelling, immunoprecipitation and immunofluorescence

Protoplasts were analysed as described by Pedrazzini et al. (1997), with minor modifications. Briefly, protoplasts were prepared from young tobacco leaves, subjected to pulse–chase labelling and, after overnight recovery, subjected to pulse–chase labelling with Pro-Mix (a mixture of [35S]Met and [35S]Cys; GE Healthcare Little Chalfont, Buckinghamshire, United Kingdom). Homogenization of the protoplasts was performed by adding to frozen samples protoplast homogenation buffer (150 mM Tris–Cl, pH 7.5, 150 mM NaCl, 1.5 mM EDTA, 1.5% Triton X-100 and Complete protease inhibitor cocktail [Roche]). Proteins were immunoselected using rabbit polyclonal antisera against phaseolin. The immunoprecipitates were analysed by SDS-PAGE. After electrophoresis, gels were treated with AmplifyTM fluorography reagent (GE Healthcare), dried and exposed for fluorography. When treated with CHX, protoplasts were supplemented before the pulse with 10 μg/mL of the antibiotic for 30 min. Aliquots of untreated protoplasts from the same protoplast preparation were used to extract and analyse total proteins as described in the paragraph 'Protein analysis', except for the extraction buffer to which 1% Triton X-100 was added.

After overnight recovery, protoplasts were fluorescence-labelled according to Frigerio et al. (2001) with minor modifications. Protoplasts were treated with antiphaseolin (1 : 10 000) antiserum, and the primary antibody was detected using fluorescein isothiocyanate-conjugated (FITC) anti-rabbit secondary antibody at 1 : 200 dilution (Jackson Immunoresearch). Cells were visualized with a Zeiss PALM Microbeam Axio-observer.Z1 fluorescence microscope equipped with a 63× oil immersion objective. Images were collected with an AxioCam MRm 60N-C 1"1, ox camera (Zeiss, Oberkochen, Germany) and visualized with Axiovision software.

Acknowledgements

We thank Alessandro Vitale for kindly providing antibodies against phaseolin protein and Stefano Cristiani for technical assistance.

References

Ali, S.S., Whitney, J.C., Stevenson, J., Robinson, H., Howell, P.L. and Navarre, W.W. (2013) Structural insights into the regulation of foreign genes in Salmonella by the Hha/H-NS complex. J. Biol. Chem. **10**, 13356–13369.

Ahmad, N., Michoux, F. and Nixon, P.J. (2012) Investigating the production of foreign membrane proteins in tobacco chloroplasts: expression of an algal plastid terminal oxidase. PLoS ONE, **7**, e41722.

Bally, J., Paget, E., Droux, M., Job, C., Job, D. and Dubald, M. (2008) Both the stroma and thylakoid lumen of tobacco chloroplasts are competent for the formation of disulphide bonds in recombinant proteins. Plant Biotechnol. J. **6**, 46–61.

Bally, J., Nadai, M., Vitel, M., Rolland, A., Dumain, R. and Dubald, M. (2009) Plant physiological adaptations to the massive foreign protein synthesis occurring in recombinant chloroplasts. Plant Physiol. **150**, 1474–1481.

Barkan, A. (1993) Nuclear mutants of maize with defects in chloroplast polysome assembly have altered chloroplast RNA metabolism. Plant Cell, **5**, 389–402.

Barkan, A. and Small, I. (2014) Pentatricopeptide repeat proteins in plants. Annu. Rev. Plant Biol. **65**, 415–442.

Bellucci, M., De Marchis, F., Mannucci, R., Bock, R. and Arcioni, S. (2005) Cytoplasm and chloroplasts are not suitable subcellular locations for b-zein accumulation in transgenic plants. J. Exp. Bot. **56**, 1205–1212.

Bellucci, M., De Marchis, F., Nicoletti, I. and Arcioni, S. (2007) Zeolin is a recombinant storage protein with different solubility and stability properties according to its localization in the endoplasmic reticulum or in the chloroplast. J. Biotechnol. **131**, 97–105.

Birch-Machin, I., Newell, C.A., Hibberd, J.M. and Gray, J.C. (2004) Accumulation of rotavirus VP6 protein in chloroplasts of transplastomic tobacco is limited by protein stability. Plant Biotechnol. J. **2**, 261–270.

Bock, R. and Warzecha, H. (2010) Solar-powered factories for new vaccines and antibiotics. Trends Biotechnol. **28**, 246–252.

Bock, R. (2014) Genetic engineering of the chloroplast: novel tools and new applications. Curr. Opin. Biotechnol. **26**, 7–13.

Boulouis, A., Raynaud, C., Bujaldon, S., Aznar, A., Wollman, F.A. and Choquet, Y. (2011) The nucleus-encoded trans-acting factor MCA1 plays a critical role in the regulation of cytochrome f synthesis in Chlamydomonas chloroplasts. Plant Cell, **23**, 333–349.

Daniell, H. (2006) Production of biopharmaceuticals and vaccines in plants via the chloroplast genome. Biotechnol. J. **1**, 1071–1079.

De Cosa, B., Moar, W., Lee, S.B., Miller, M. and Daniell, H. (2001) Overexpression of the Bt cry2Aa2 operon in chloroplasts leads to formation of insecticidal crystals. Nat. Biotechnol. **19**, 71–74.

De Marchis, F., Balducci, C., Pompa, A., Riise Stensland, H.M., Guaragno, M., Pagiotti, R., Menghini, A.R., Persichetti, E., Beccari, T. and Bellucci, M. (2011a) Human α-mannosidase produced in transgenic tobacco plants is processed in human α-mannosidosis cell lines. Plant Biotechnol. J. **9**, 1061–1073.

De Marchis, F., Pompa, A., Mannucci, R., Morosinotto, T. and Bellucci, M. (2011b) A plant secretory signal peptide targets plastome-encoded recombinant proteins to the thylakoid membrane. Plant Mol. Biol. **76**, 427–441.

De Marchis, F., Pompa, A. and Bellucci, M. (2012) Plastid proteostasis and heterologous protein accumulation in transplastomic plants. Plant Physiol. **160**, 571–581.

Deshpande, S.S. and Nielsen, S.S. (1987) In vitro enzymatic hydrolysis of phaseolin, the major storage protein of Phaseolus vulgaris L. J. Food Sci. **52**, 1326–1329.

Dhingra, A., Portis, A.R. Jr and Daniell, H. (2004) Enhanced translation of a chloroplast-expressed RbcS gene restores small subunit levels and photosynthesis in nuclear RbcS antisense plants. Proc. Natl Acad. Sci. USA, **20**, 6315–6320.

Drapier, D., Rimbault, B., Vallon, O., Wollman, F.A. and Choquet, Y. (2007) Intertwined translational regulations set uneven stoichiometry of chloroplast ATP synthase subunits. EMBO J. **8**, 3581–3591.

Elghabi, Z., Karcher, D., Zhou, F., Ruf, S. and Bock, R. (2011) Optimization of the expression of the HIV fusion inhibitor cyanovirin-N from the tobacco plastid genome. Plant Biotechnol. J. **9**, 599–608.

Fontanesi, F., Soto, I.C., Horn, D. and Barrientos, A. (2010) Mss51 and Ssc1 facilitate translational regulation of cytochrome c oxidase biogenesis. *Mol. Cell. Biol.* **30**, 245–259.

Frigerio, L., de Virgilio, M., Prada, A., Faoro, F. and Vitale, A. (1998) Sorting of phaseolin to the vacuole is saturable and requires a short C-terminal peptide. *Plant Cell*, **10**, 1031–1042.

Frigerio, L., Pastres, A., Prada, A. and Vitale, A. (2001) Influence of KDEL on the fate of trimeric or assembly defective phaseolin: selective use of an alternative route to vacuoles. *Plant Cell*, **13**, 1109–1126.

Guerineau, F., Woolston, S., Brooks, L. and Mullineaux, P. (1988) An expression cassette for targeting foreign proteins into chloroplasts. *Nucleic Acids Res.* **9**, 11380.

Hellens, R.P., Edwards, E.A., Leyland, N.R., Bean, S. and Mullineaux, P.M. (2000) pGreen: a versatile and flexible binary Ti vector for Agrobacterium-mediated plant transformation. *Plant Mol. Biol.* **42**, 819–832.

Hennig, A., Bonfig, K., Roitsch, T. and Warzecha, H. (2007) Expression of the recombinant bacterial outer surface protein A in tobacco chloroplasts lead to thylakoid localization and loss of photosynthesis. *FEBS J.* **274**, 5749–5758.

Jutras, B.L., Jones, G.S., Verma, A., Brown, N.A., Antonicello, A.D., Chenail, A.M. and Stevenson, B. (2013) Posttranscriptional self-regulation by the Lyme disease bacterium's BpuR DNA/RNA-binding protein. *J. Bacteriol.* **195**, 4915–4923.

Koonin, E.V., Makarova, K.S. and Aravind, L. (2001) Horizontal gene transfer in prokaryotes: quantification and classification. *Annu. Rev. Microbiol.* **55**, 709–742.

Kuroda, H. and Maliga, P. (2001) Complementarity of the 16S rRNA penultimate stem with sequences downstream of the AUG destabilizes the plastid mRNAs. *Nucleic Acids Res.* **29**, 970–975.

Kwon, K.C., Verma, D., Singh, N.D., Herzog, R. and Daniell, H. (2013) Oral delivery of human biopharmaceuticals, autoantigens and vaccine antigens bioencapsulated in plant cells. *Adv. Drug Deliv. Rev.* **65**, 782–799.

Lentz, E.M., Garaicoechea, L., Alfano, E.F., Parreño, V., Wigdorovitz, A. and Bravo-Almonacid, F.F. (2012) Translational fusion and redirection to thylakoid lumen as strategies to improve the accumulation of a camelid antibody fragment in transplastomic tobacco. *Planta*, **236**, 703–714.

Levey, T., Westhoff, P. and Meierhoff, K. (2014) Expression of a nuclear-encoded psbH gene complements the plastidic RNA processing defect in the PSII mutant hcf107 in *Arabidopsis thaliana*. *Plant J.* **80**, 292–304.

Manuell, A.L., Quispe, J. and Mayfield, S.P. (2007) Structure of the chloroplast ribosome: novel domains for translation regulation. *PLoS Biol.* **5**, e209.

Minai, L., Wostrikoff, K., Wollman, F.A. and Choquet, Y. (2006) Chloroplast biogenesis of photosystem II cores involves a series of assembly-controlled steps that regulate translation. *Plant Cell*, **18**, 159–175.

Michelet, L., Lefebvre-Legendre, L., Burr, S.E., Rochaix, J. and Goldschmidt-Clermont, M. (2010) Enhanced chloroplast transgene expression in a nuclear mutant of Chlamydomonas. *Plant Biotechnol. J.* **9**, 565–574.

Morgenfeld, M., Lentz, E., Segretin, M.E., Alfano, E.F. and Bravo-Almonacid, F. (2014) Translational fusion and redirection to thylakoid lumen as strategies to enhance accumulation of human papillomavirus e7 antigen in tobacco chloroplasts. *Mol. Biotechnol.* **56**, 1021–1031.

Navarre, W.W., Porwollik, S., Wang, Y., McClelland, M., Rosen, H., Libby, S.J. and Fang, F.C. (2006) Selective silencing of foreign DNA with low GC content by the H-NS protein in Salmonella. *Science*, **14**, 236–238.

Oey, M., Lohse, M., Kreikemeyer, B. and Bock, R. (2009) Exhaustion of the chloroplast protein synthesis capacity by massive expression of a highly stable protein antibiotic. *Plant J.* **57**, 436–445.

Pedrazzini, E., Giovinazzo, G., Bielli, A., de Virgilio, M., Frigerio, L., Pesca, M., Faoro, F., Bollini, R., Ceriotti, A. and Vitale, A. (1997) Protein quality control along the route to the plant vacuole. *Plant Cell*, **9**, 1869–1880.

Pompa, A., De Marchis, F., Vitale, A., Arcioni, S. and Bellucci, M. (2010) An engineered C-terminal disulfide bond can partially replace the phaseolin vacuolar sorting signal. *Plant J.* **61**, 782–791.

Ramundo, S., Rahire, M., Schaad, O. and Rochaix, J.D. (2013) Repression of essential chloroplast genes reveals new signaling pathways and regulatory feedback loops in chlamydomonas. *Plant Cell*, **25**, 167–186.

Ruhlman, T., Verma, D., Samson, N. and Daniell, H. (2010) The role of heterologous chloroplast sequence elements in transgene integration and expression. *Plant Physiol.* **152**, 2088–2104.

Salvi, D., Rolland, N., Joyard, J. and Ferro, M. (2008) Purification and proteomic analysis of chloroplasts and their sub-organellar compartments. *Methods Mol. Biol.* **432**, 19–36.

Scotti, N., Bellucci, M. and Cardi, T. (2013) The chloroplasts as platform for recombinant proteins production. In *Translation in Mitochondria and Other Organelle* (Duchêne, A.-M., ed.), pp. 225–262. Berlin Heidelberg: Springer-Verlag.

Shanmugabalaji, V., Besagni, C., Piller, L.E., Douet, V., Ruf, S., Bock, R. and Kessler, F. (2013) Dual targeting of a mature plastoglobulin/fibrillin fusion protein to chloroplast plastoglobules and thylakoids in transplastomic tobacco plants. *Plant Mol. Biol.* **81**, 13–25.

Shiina, T., Tsunoyama, Y., Nakahira, Y. and Khan, M.S. (2005) Plastid RNA polymerases, promoters, and transcription regulators in higher plants. *Int. Rev. Cytol.* **244**, 1–68.

Singh, N.D., Li, M., Lee, S.B., Schnell, D. and Daniell, H. (2008) Arabidopsis Tic40 expression in tobacco chloroplasts results in massive proliferation of the inner envelope membrane and upregulation of associated proteins. *Plant Cell*, **20**, 3405–3417.

Stern, D.B., Goldschmidt-Clermont, M. and Hanson, M.R. (2010) Chloroplast RNA metabolism. *Annu. Rev. Plant Biol.* **61**, 125–155.

Suzuki, Y. and Makino, A. (2013) Translational downregulation of RBCL is operative in the coordinated expression of Rubisco genes in senescent leaves in rice. *J. Exp. Bot.* **64**, 1145–1152.

Tabe, L.M., Wardley-Richardson, T., Ceriotti, A., Aryan, A., McNabb, W., Moore, A. and Higgins, T.J. (1995) A biotechnological approach to improving the nutritive value of alfalfa. *J. Anim. Sci.* **73**, 2752–2759.

Tangphatsornruang, S., Birch-Machin, I., Newell, C.A. and Gray, J.C. (2011) The effect of different 3′ untranslated regions on the accumulation and stability of transcripts of a gfp transgene in chloroplasts of transplastomic tobacco. *Plant Mol. Biol.* **76**, 385–396.

Tiller, N. and Bock, R. (2014) The translational apparatus of plastids and its role in plant development. *Mol. Plant*, **7**, 1105–1120.

Timmis, J.N., Ayliffe, M.A., Huang, C.Y. and Martin, W. (2004) Endosymbiotic gene transfer: organelle genomes forge eukaryotic chromosomes. *Nat. Rev. Genet.* **5**, 123–135.

Verma, D. and Daniell, H. (2007) Chloroplast vector systems for biotechnology applications. *Plant Physiol.* **145**, 1129–1143.

Verma, D., Moghimi, B., LoDuca, P.A., Singh, H.D., Hoffman, B.E., Herzog, R.W. and Daniell, H. (2010) Oral delivery of bioencapsulated coagulation factor IX prevents inhibitor formation and fatal anaphylaxis in hemophilia B mice. *Proc. Natl Acad. Sci. USA*, **107**, 7101–7106.

de Virgilio, M., De Marchis, F., Bellucci, M., Mainieri, D., Rossi, M., Benvenuto, E., Arcioni, S. and Vitale, A. (2008) The human immunodeficiency virus antigen Nef forms protein bodies in leaves of transgenic tobacco when fused to zeolin. *J. Exp. Bot.* **59**, 2815–2829.

Vitale, A., Bielli, A. and Ceriotti, A. (1995) The binding protein associates with monomeric phaseolin. *Plant Physiol.* **107**, 1411–1418.

Watson, J., Koya, V., Leppla, S.H. and Daniell, H. (2004) Expression of *Bacillus anthracis* protective antigen in transgenic chloroplasts of tobacco, a non-food/feed crop. *Vaccine*, **22**, 4374–4384.

Wirth, S., Segretin, M.E., Mentaberry, A. and Bravo-Almonacid, F. (2006) Accumulation of hEGF and hEGF-fusion proteins in chloroplast-transformed tobacco plants is higher in the dark than in the light. *J. Biotechnol.* **125**, 159–172.

Woodson, J.D. and Chory, J. (2008) Coordination of gene expression between organellar and nuclear genomes. *Nat. Rev. Genet.* **9**, 383–395.

Wostrikoff, K., Girard-Bascou, J., Wollman, F.A. and Choquet, Y. (2004) Biogenesis of PSI involves a cascade of translational autoregulation in the chloroplast of Chlamydomonas. *EMBO J.* **7**, 2696–2705.

Wostrikoff, K. and Stern, D. (2007) Rubisco large-subunit translation is autoregulated in response to its assembly state in tobacco chloroplasts. *Proc. Natl Acad. Sci. USA*, **10**, 6466–6471.

Wostrikoff, K., Clark, A., Sato, S., Clemente, T. and Stern, D. (2012) Ectopic expression of Rubisco subunits in maize mesophyll cells does not overcome barriers to cell type-specific accumulation. *Plant Physiol.* **160**, 419–432.

Yang, H., Gray, B.N., Ahner, B.A. and Hanson, M.R. (2013) Bacteriophage 5′ untranslated regions for control of plastid transgene expression. *Planta*, **237**, 517–527.

Ye, G.N., Hajdukiewicz, P.T., Broyles, D., Rodriguez, D., Xu, C.W., Nehra, N. and Staub, J.M. (2001) Plastid expressed 5-enolpyruvylshikimate-3-phosphate synthase genes provide high level glyphosate tolerance in tobacco. *Plant J.* **25**, 261–270.

Zoschke, R., Watkins, K.P. and Barkan, A. (2013) A rapid ribosome profiling method elucidates chloroplast ribosome behavior in vivo. *Plant Cell*, **25**, 2265–2275.

7

The dynamics of protein body formation in developing wheat grain

Katie L. Moore[1], Paola Tosi[2], Richard Palmer[3], Malcolm J. Hawkesford[3], Chris R.M Grovenor[4] and Peter R. Shewry[3,*]

[1]School of Materials, University of Manchester, Manchester, UK

[2]School of Agriculture Policy and Development, Reading University, Reading, UK

[3]Rothamsted Research, Harpenden, UK

[4]Department of Materials, University of Oxford, Oxford, UK

Summary

Wheat is a major source of protein in the diets of humans and livestock but we know little about the mechanisms that determine the patterns of protein synthesis in the developing endosperm. We have used a combination of enrichment with ^{15}N glutamine and NanoSIMS imaging to establish that the substrate required for protein synthesis is transported radially from its point of entrance in the endosperm cavity across the starchy endosperm tissues, before becoming concentrated in the cells immediately below the aleurone layer. This transport occurs continuously during grain development but may be slower in the later stages. Although older starchy endosperm cells tend to contain larger protein deposits formed by the fusion of small protein bodies, small highly enriched protein bodies may also be present in the same cells. This shows a continuous process of protein body initiation, in both older and younger starchy endosperm cells and in all regions of the tissue. Immunolabeling with specific antibodies shows that the patterns of enrichment are not related to the contents of gluten proteins in the protein bodies. In addition to providing new information on the dynamics of protein deposition, the study demonstrates the wider utility of NanoSIMS and isotope labelling for studying complex developmental processes in plant tissues.

*Correspondence
email peter.
shewry@rothamsted.ac.uk

Keywords: endosperm development, NanoSIMS, protein deposition, wheat.

Introduction

Wheat is one of the three major cereal crops which feed the human race and the major staple crop in temperate countries. Because of this, it is a major source of protein for human health. For example, bread alone provides over 10% of the daily protein intake in UK adults (Steer *et al.*, 2008). This contribution is more important in low-income groups and particularly important in some developing countries where wheat provides between 50 and 70% of the total calories (Cakmak, 2002). Despite this, we know little about the mechanisms that determine protein content and deposition and hence lack the tools required to improve this aspect of grain quality.

The major storage tissue in the cereal grain is the starchy endosperm, which comprises about 80% starch and about 10% protein. This is a highly organized and differentiated tissue, with significant gradients in cell composition, particularly protein that is concentrated in the outer few layers of starchy endosperm cells (called subaleurone cells) in all cereals, including wheat (Tosi *et al.*, 2011) and rice (Ohdaira *et al.*, 2011). These gradients will affect the recovery of protein in fractions produced by grain processing, such as milling of wheat to give white flour and polishing of rice. Both of these processes can result in significant loss of protein from the human diet if the outer layers of the starchy endosperm are removed with the bran.

Little is known about when the protein gradients in the starchy endosperm are established, or the mechanisms that determine them. However, it is known that the subaleurone cells of the cereal endosperm have a different origin from the central endosperm cells, being derived from the aleurone cells which retain their ability to divide both anticlinally and

periclinally up to about 12 days postanthesis (dpa). The inner cells from these divisions redifferentiate into subaleurone cells with only a single layer of aleurone cells being present in the mature wheat grain (Becraft and Yi, 2011). Hence, the accumulation of protein could be associated with the redifferentiation of the subaleurone cells into a new type of protein-rich storage tissue.

Secondary ion mass spectrometry (SIMS) has a number of advantages over conventional methods for imaging minerals in plant tissues, including high sensitivity (parts per million) and the ability to discriminate between isotopic forms of elements (Wilson *et al.*, 1989). It has been applied previously to plant tissues (Grignon *et al.*, 1997; Jauneau *et al.*, 1994), including determining cereal grain composition (Feeney *et al.*, 2003; Heard *et al.*, 2002; Moore *et al.*, 2010, 2012b). However, these studies were all of static systems, whereas we show here that SIMS can also be combined with isotopic labelling to study complex dynamic processes in developing tissues. We have therefore combined the *in vivo* incorporation of ^{15}N-labelled glutamine, the main transported form of nitrogen in wheat (Fisher and Macnicol, 1986), into developing caryopses with NanoSIMS analysis of tissue sections. The high lateral resolution of the NanoSIMS allowed the precise ratio of ^{15}N:^{14}N to be determined in individual protein bodies, providing the first description of the dynamics of nitrogen transport in relation to protein deposition.

Results

Developing wheat grains (caryopses) were fed ^{15}N-glutamine via the rachis (Figure S1) at two stages of development: at 10 dpa, corresponding to the start of grain filling, and at 20 dpa, corresponding to the middle of the grain-filling period. This treatment did not affect grain development, and caryopses were removed and analysed for ^{15}N enrichment at 6 h, 24 h and 7 days after the start of application. Analyses of whole caryopses confirmed that there was substantial enrichment with ^{15}N, which accounted for over 8% of the total N present at 6 h from the beginning of the feeding period at 10 dpa (when over 1 ml of the total 1.8 ml of ^{15}N supplied to each ear had been taken up). Enrichment remained high (over 10%) at 24 h (at 11 dpa) after the beginning of the feeding period but fell to about 5% after 7 days (at 17 dpa) due to dilution with ^{14}N from the normal uptake and transport of nitrogen from the soil. By contrast, transport from the rachis to the spike and into the caryopses was slower when the ^{15}N was fed at 20 dpa, enrichment being only about 1.8% after 6 h (when about 400 µl of solution had been taken up) and increasing to about 5% after 24 h. Enrichment decreased slightly, to about 4%, after 7 days, presumably due again to dilution with ^{14}N taken up from the soil (Table S1).

To study the incorporation of ^{15}N into proteins in the tissues, rather than into amino acids and other low-molecular-weight compounds, sections of developing caryopses from these times were fixed using paraformaldehyde and glutaraldehyde. This type of preservation, known as chemical fixation, adequately preserves elements that are covalently bound in large molecules, such as proteins, but will wash out diffusible ions, such as K and Na, and free amino acids that were not of interest in this study (Moore et al., 2012a). Transverse sections of the developing caryopses were prepared by ultramicrotomy and stained with toluidine blue to show protein distribution in the starchy endosperm (Figures 1 and S2). Serial sections, adjacent to those used for light microscopy, were prepared for NanoSIMS analysis to determine the patterns of protein enrichment.

NanoSIMS analyses showed little ^{15}N enrichment in either the maternal or endosperm tissues after 6 h feeding at either 10 or 20 dpa, although mass spectroscopy of the whole (un-fixed) caryopses showed 8% and 1.8% enrichment, respectively. Free glutamine would not have been preserved during sample preparation for microscopy, and it was therefore concluded that little or none of the ^{15}N transported to the endosperm had become incorporated into proteins at this time. The 6-hour

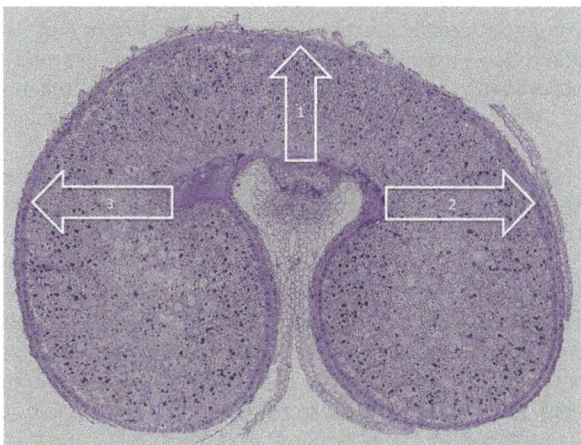

Figure 1 Transverse section of developing grain of wheat at 17 days postanthesis (dpa) stained with toluidine blue and annotated to show the positions and directions of the three transects selected for NanoSIMS analysis.

samples were therefore not studied further. By contrast, clear enrichment of endosperm proteins with ^{15}N was observed at 24 h and 7 days after feeding at both stages of development. Three transects of cells across the starchy endosperm were therefore selected for detailed study: one extending from the nucellar projection to the dorsal surface of the grain (transect 1) and two extending laterally across the lobes from the nucellar projection (transects 2 and 3) (Figure 1). Essentially similar results were obtained for these three transects so only data for transect 1 are reported here (see Figure S3 for a comparison of the three transects at three different time points).

Data for caryopses at 7 days after the start of feeding with ^{15}N glutamine are presented in Figure 2 (17 dpa) and Figure 3 (27 dpa) and data for 24 h after feeding (11 and 21 days) in Figures S4 and S5, respectively. The upper panel of Figure 2 shows a transect from a serial microtome section close to the NanoSIMS section, taken from a caryopsis at 17 dpa (7 days after feeding at 10 dpa) stained with toluidine blue to reveal protein. This shows the presence of small protein bodies deposited within vacuoles in all cell layers, from close to the transfer cells (area a) to the subaleurone cells (area e). A similar transect taken at 27 dpa (7 days after feeding at 20 dpa) (Figure 3, upper panel) reveals more extensive protein deposits, which vary in size and include both small newly initiated bodies and the fusion of protein bodies to give large irregular deposits.

The lower panels of Figures 2 and 3 show optical images which were used to select the areas where secondary ion NanoSIMS images have been acquired at high lateral resolution (areas a-e in Figure 2, a-j in Figure 3). The expanded images of these areas are shown as hue saturation intensity colour maps of the $^{15}N:^{14}N$ ratio with the level of ^{15}N enrichment being indicated by the colour scale at the bottom of each of figure. The NanoSIMS analysis shows clear differences in the extent of enrichment across the transects and between individual protein bodies within single cells. For example, Figure 2 shows greater enrichment of the protein bodies in the starchy endosperm cells closest to the transfer cells (i.e. close to the point of entry of the ^{15}N label, areas a and b), while Figure 3 shows high enrichment of large protein bodies in the outer endosperm cells (farthest from the transfer cells, areas i and j) but also of small protein bodies in the cells closest to the transfer cells (areas a and b).

It is clearly not possible to make a quantitative comparison of enrichment patterns by visual inspection alone. Transects from caryopses of each time stage (24 h and 7 days after labelling at 10 and 20 dpa) were therefore analysed in detail to measure the areas and enrichment levels of individual protein bodies, with over 6000 individual protein deposits being analysed in total. Examples of this analysis are displayed graphically in Figure 4, with individual protein bodies displayed as 'bubbles', the sizes of which correspond to their measured areas. The positions of the 'bubbles' along the x-axis corresponds to their location along the transect, while their degree of enrichment with ^{15}N is shown on the y-axis. Replicate analyses from different spikelets (Figure S6) show a high level of similarity.

The increase in the distance of the outermost protein bodies from the nucellar projection (see Figures 4, S3 and S6) is due to expansion of the endosperm during the period from 10 to 27 dpa, which corresponds under our controlled environment growth conditions to the point of maximum grain size, and may be due to cell division, cell expansion or a combination of these. In fact, counts of cells along transect 1 showed a small decrease in the number of endosperm cells after 10 dpa: from

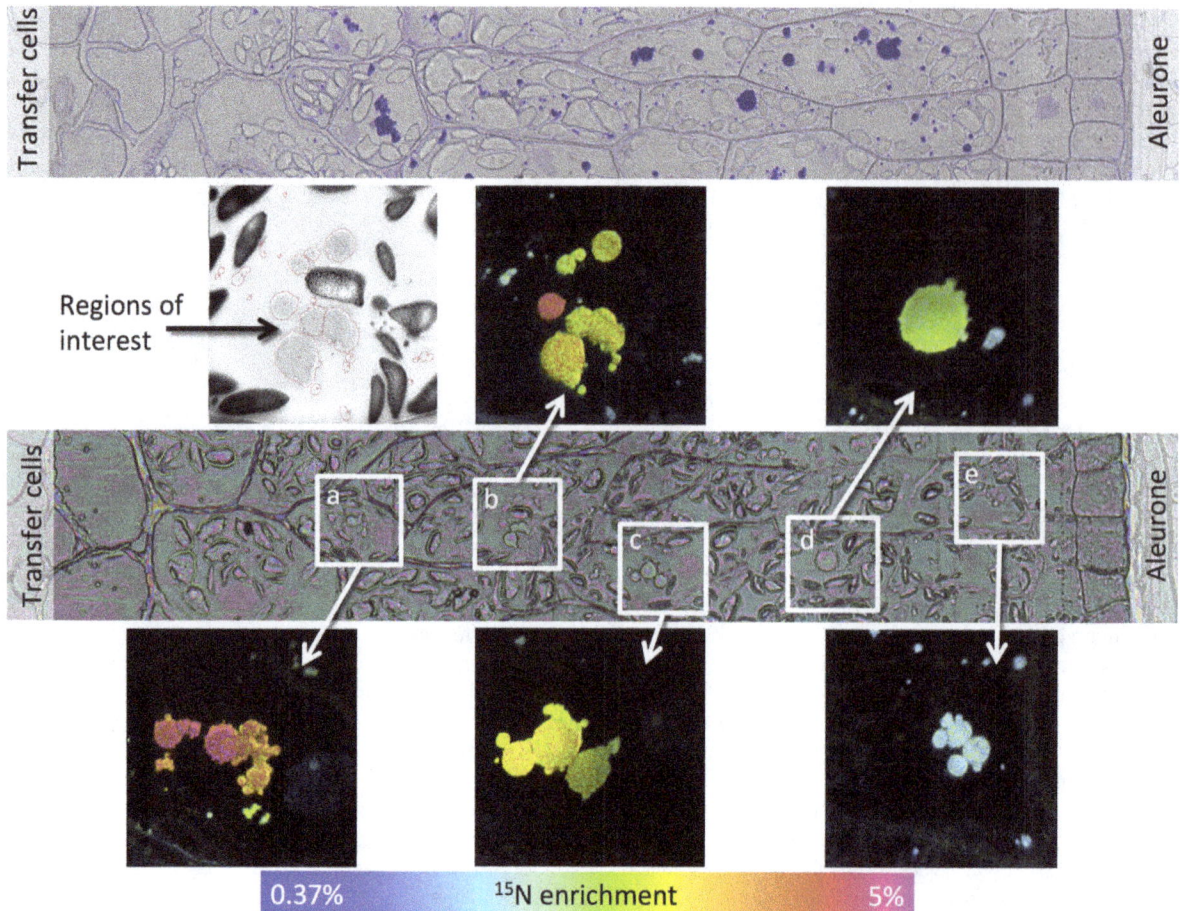

Figure 2 Analysis of transect 1 of a wheat starchy endosperm taken from a developing caryopsis at 17 dpa, after feeding ^{15}N glutamine at 10 dpa. The top panel shows a conventional light microscopy image after staining with toluidine blue to identify protein bodies. The central panel shows an optical image from a serial section which was used to select the areas where secondary ion NanoSIMS images were acquired at high lateral resolution. Areas a to e marked on this optical image are expanded in the boxes, and enrichment with ^{15}N is shown using a hue saturation intensity colour scale with the ^{15}N enrichment shown in the scale at the bottom. The grey scale image is a NanoSIMS secondary electron image from area b. Regions of interest, protein deposits, are outlined in red and indicate the regions from which the ^{15}N enrichment data were extracted.

11.7 +/− 1.2 at 11 dpa to 10.9 +/− 1.3 at 17 dpa, 10.8 +/ − 1.5 at 21 dpa and 10.8 +/− 0.8 at 27 dpa. This cell loss is consistent with previous studies (Gao *et al.*, 1992) and occurs despite continued anticlinal division of the aleurone cells until at least 14 dpa (Bechtel and Wilson, 2003).

The graphs demonstrate that the protein bodies present at 11 dpa were numerous but small, with enrichment increasing towards the outside of the grain at 24 h after labelling (Figure 4 a). At 17 days, the number of protein bodies had decreased and their average size increased massively, presumably due to fusion. However, small (presumably newly initiated) protein bodies were present in the same cells (Figure 4 b). The level of enrichment also decreased from the inner to outer cells at 17 days, which was probably due to dilution of the protein in the outer cells with ^{14}N. These results therefore show that the ^{15}N was rapidly transported into the caryopses and incorporated into small protein bodies (by 24 h) which subsequently grew in size, particularly in the outer layers of the starchy endosperm.

Analysis of the 21 dpa samples (i.e. at 24 h after labelling) showed large and small protein bodies, with the smaller bodies being more highly enriched in ^{15}N (Figure 4 c), indicating that they had been recently initiated. Differences were observed at

27 days (7 days after labelling). Firstly, whereas the central cells contained small highly enriched protein bodies, a similar level of enrichment was observed in the large protein bodies in the outer (most recently formed) subaleurone cells (Figure 4 d). This pattern may result from a decreased level of protein synthesis and deposition in the central cells combined with a continued high level of protein synthesis in the protein-rich subaleurone cells.

Differences in protein body morphology were also observed between the inner and outer starchy endosperm cells, particularly at 27 dpa (Figure 3). The large protein bodies in the inner cells appeared to be formed by the fusion of small protein bodies (Figure 3 panels a-h) and were irregular in shape (Figure 3 e, f, h), probably because they became squashed between expanding starch granules. By contrast, those in the subaleurone cells (which contained fewer starch granules) were more regular in shape and appeared to grow by direct accumulation of newly synthesized protein as well as fusion (Figure 4 i and j).

The major protein components stored in the starchy endosperm cells of wheat are prolamin storage proteins, which correspond to the gluten proteins that underpin the processing properties of wheat flour and dough. Gluten comprises over 50

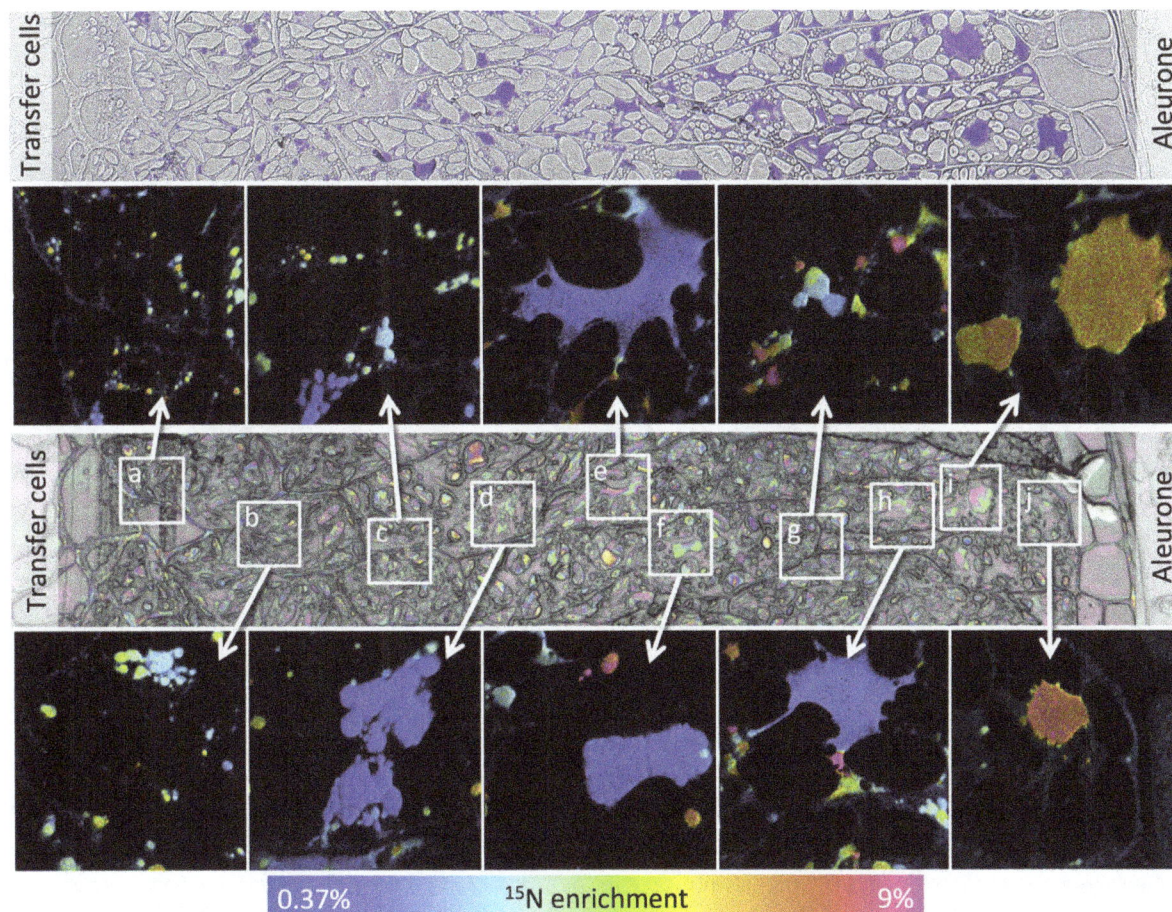

Figure 3 Analysis of transect 1 of a wheat starchy endosperm taken from a developing caryopsis at 27 dpa, after feeding ^{15}N glutamine at 20 dpa. The top panel shows a conventional light microscopy image after staining with toluidine blue to identify protein bodies. The central panel shows an optical image from a serial section which was used to select the areas where secondary ion NanoSIMS images were acquired at high lateral resolution. Areas a to j marked on this optical image are expanded in the boxes and enrichment with ^{15}N is shown using a hue saturation intensity colour scale with the ^{15}N enrichment shown in the scale at the bottom.

individual components, which are divided broadly into polymeric glutenins and monomeric gliadins. These two protein groups differ little in their timing of synthesis during grain development (Shewry et al., 2009), but it has been suggested that they are partially segregated into different types of protein body. In particular, the gliadins appear to be enriched in protein bodies of vacuolar origin while glutenins appear to accumulate directly within the lumen of the endoplasmic reticulum (Rubin et al., 1992; Tosi et al., 2009). It was not possible to discriminate between these two types of protein body in this study, although it is clear that the large aggregated structures present at later stages of development represent aggregates of small protein bodies within vacuoles. However, we did determine whether the differences in enrichment of protein bodies within cells was related to their protein composition by analysing serial sections by NanoSIMS and by immunomicroscopy using specific antibodies for two major classes of gluten protein (Figure 5). The monoclonal antibody IFR0610 (red, Figure 5 b) recognizes gliadins and low-molecular-weight subunits (Brett et al., 1999), which are classified as sulphur-rich prolamins, while the polyclonal antibody R2-HMW (green, Figure 5 e) recognizes epitopes on high-molecular-weight subunits of glutenin (Mills et al., 2000). Comparison of the labelling patterns with these two antibodies showed no relationship between protein composition and the degree of enrichment with [15]N (Figure 5).

Discussion

We have demonstrated that a combination of isotopic labelling and NanoSIMS imaging can be used to determine the spatial and temporal dynamics of protein deposition in the developing wheat endosperm. However, the interpretation is not straightforward as transport and deposition are accompanied by dilution with [14]N.

Firstly, we have demonstrated that ^{15}N provided as glutamine is transported radially across the developing starchy endosperm, from the transfer cells to the subaleurone cells. However, the rate of transport appears to be slower as the endosperm develops, with enrichment of the subaleurone cells being observed within 24 h when the labelling was carried out at 10 dpa, but not at 20 dpa. Clear differences in ^{15}N enrichment were also observed at 7 days after labelling compared with 24 h after labelling, for both developmental stages. Dilution with ^{14}N may account for the failure to observe enrichment of the subaleurone cells at 17 days, whereas the enrichment observed at 27 dpa may reflect lower dilution combined with a slower rate of ^{15}N transport.

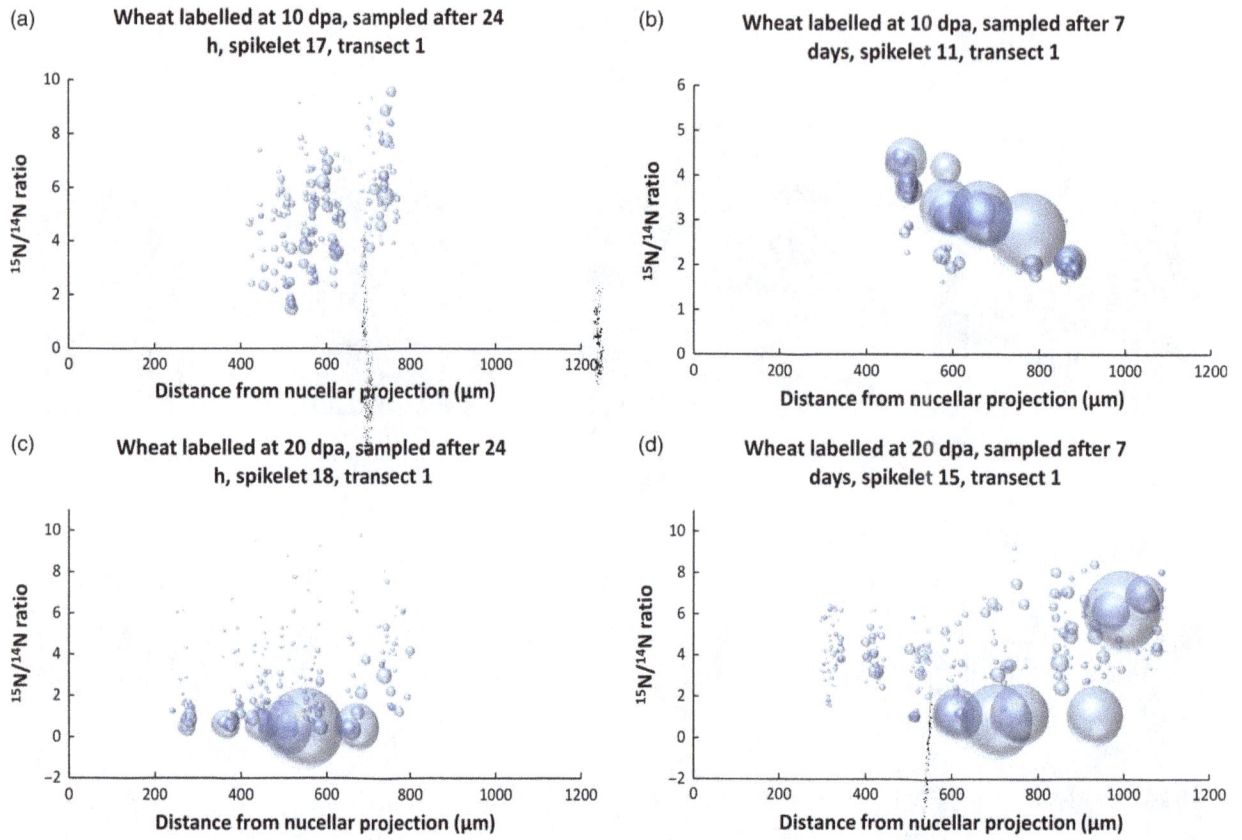

Figure 4 Graphical representation of the size and ^{15}N enrichment of protein bodies along transect 1 of starchy endosperm tissue after labelling at 10 dpa 24 h (a), 10 dpa 7 days (b) 20 dpa 24 h (c) and 20 dpa 7 days (d). Individual protein bodies are displayed as 'bubbles', which correspond in size to their measured areas. The positions of the protein bodies correspond to their locations along the transect (x-axis) and their degree of enrichment with ^{15}N (y-axis).

The studies described here do not provide information on the mechanisms that determine the patterns of protein accumulation but it can be speculated that the subaleurone cells, which have a different and more recent origin than the central starchy endosperm cells (originating from divisions in the aleurone cells late in development), have a higher requirement for amino acids as they are programmed to store high levels of protein (up to 40% dry weight). This strong sink activity may result in a gradient which effectively drives amino acid transport from the transfer cells across the central starchy endosperm to the subaleurone

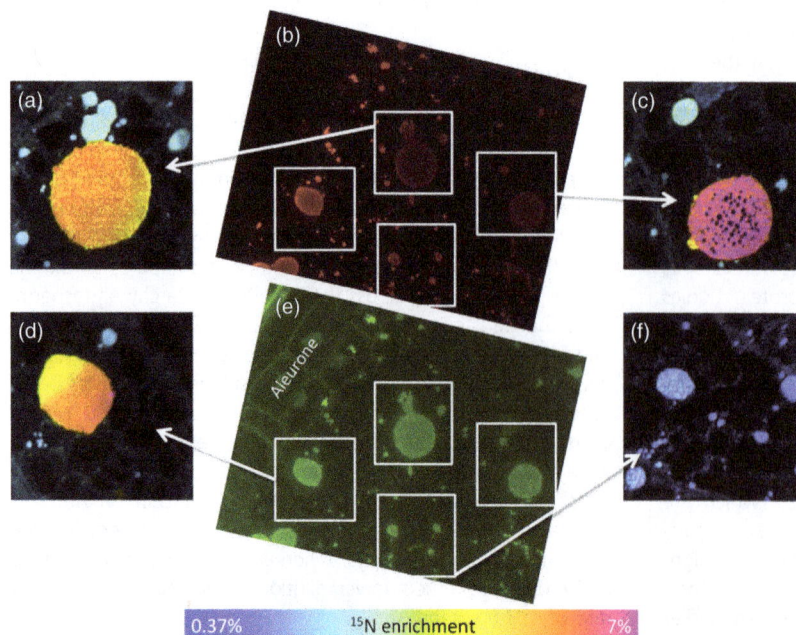

Figure 5 Comparison of ^{15}N enrichment of protein bodies in starchy endosperm cells at 17 dpa (7 days after labelling at 10 dpa) determined by NanoSIMS (panels a, c, d, f) with the contents of gluten proteins determined by immunolabeling with the monoclonal antibody IFR0610 which recognizes gliadins and low-molecular-weight subunits (red, panel b) and the polyclonal antibody R2-HMG which recognizes high-molecular-weight subunits of glutenin (green, panel e). The primary antibody binding in panel b was detected using an Alexa 568 anti-mouse conjugated antibody, while in panel e, an Alexa 488 anti-rabbit conjugated antibody was used. In panels a, c, d and f, enrichment with ^{15}N determined by NanoSIMS is displayed using a hue saturation intensity colour scale. Comparison of immunofluorescence and NanoSIMS images suggests that ^{15}N incorporation is independent of protein body composition, reflecting instead time of synthesis in relation to ^{15}N labelling.

cells, and this may explain why the large protein bodies in the subaleurone cells are highly enriched at 27 days.

Finally, we have shown that the larger protein deposits in older starchy endosperm cells are formed by the fusion of small protein bodies and that small protein bodies can also be observed in these same cells. Furthermore, some of these small protein bodies were highly enriched with ^{15}N at 21 and 27 dpa, indicating a continuous process of protein body initiation, in both older and younger starchy endosperm cells and in all regions of the tissue. This contrasts with the deposition of starch which shows two clear phases of granule initiation, with large A granules being initiated up to 5 dpa and small B granules between 9 and 14 dpa (Stone and Morell, 2009).

Hence, the analyses reported here clearly show that the nitrogen required for the synthesis of storage proteins in the wheat starchy endosperm is transported radially across the tissue from the groove and transfer cells, rather than around the circumference. Secondly, although a gradient in protein concentration is established, the initiation and expansion of protein bodies occur continuously in all cells, irrespective of their ontogeny or position.

This novel information has wider relevance to developmental processes in other cereals and is important if we wish to manipulate the pattern of protein accumulation to improve grain quality. It also demonstrates that combination of NanoSIMS and isotope labelling has potential for wider applications for analysing dynamic processes in other complex biological systems, such as seeds and lignified tissues in which the more widely used confocal imaging cannot be applied due to the opacity of the material.

Experimental procedures

Plant material and application of ^{15}N glutamine

Plants of bread wheat cv Cadenza were grown in controlled environment rooms at Rothamsted Research at 18°C day/15°C night temperature with a photoperiod of 16 h provided by banks of 400W hydrargyrum quartz iodide (HQI) lamps (Osram Ltd, UK) generating a light intensity of ~700 μmol/m^2/s photosynthetically active radiation (PAR) at the pot surface.

Ears at 10 dpa and 20 dpa were fed 1.6 ml of a 34 mM ^{15}N glutamine (L-Glutamine-(amine-^{15}N) 98 atom % ^{15}N, Sigma-Aldrich) solution via a glass capillary tube (Drummund Microcaps, Sigma-Aldrich) inserted into the rachis through a small incision between the first two basal spikelets, and immersed in an Eppendorf tube containing 1.8 ml of the labelled amino acid solution. Uptake of the solution by capillarity proceeded at different rates in different plants and was generally faster in ears at 10 dpa than at 20 dpa. To ensure that a minimum of 1.6 ml of ^{15}N-labelled glutamine solution was taken up by the ears within a 24-hour period, capillary tubes were regularly checked for blockages and re-inserted if necessary. The feasibility and efficiency of such a capillary feeding method was previously tested using a solution of aniline blue dye (0.5% w/v) (Figure S1). Dissection and analysis of the ears showed that the solution had reached all of the grains in all the spikelets after 4 h from the start of feeding. Because the feeding was carried out on intact plants, the spikes and caryopses developed normally and were harvested immediately after the 6-hour feeding period and then after 24 h and 7 days. Different ears were used for each labelling treatment and collection time point with caryopses from the 10-11th and 16-17th spikelets being used for analysis.

Preparation of grain sections

Transverse sections (approximately 1 mm thick) were cut in fixative (2.5% (w/v) paraformaldehyde, 0.5% (w/v) glutaraldehyde in 0.1M Sorenson's phosphate buffer, pH 7.2) from the middle of the grain. Sections were fixed at room temperature for 4 h and then rinsed three times in buffer before dehydration in an ethanol series immediately followed by infiltration with LR White resin (medium grade, TAAB L012) for several days. Resin-infiltrated samples were polymerized in polyethylene capsules (TAAB) at 55°C.

The embedded wheat grains were sectioned using a Reichert-Jung Ultracut ultramicrotome to a thickness of 1 micron, collected on drops of distilled water on silicon wafers and dried on a hot plate at 40 °C to stretch them flat. To prevent charging during NanoSIMS analysis, the samples and substrates were coated with 10 nm of platinum before loading into the NanoSIMS.

NanoSIMS

SIMS analysis was carried out with a NanoSIMS 50 instrument. A 16 keV Cs$^+$ beam with a current of 1.2-3.2 pA was scanned over the sample surface, and the bombardment resulted in the generation of negative secondary ions. These secondary ions were analysed by mass using a double focusing mass spectrometer. Detectors were aligned to detect ^{12}C$^-$, ^{16}O$^-$, ^{12}C^{14}N$^-$, ^{12}C^{15}N$^-$ and ^{31}P$^-$ with most detectors aligned using the sample except for the ^{31}P$^-$ which was aligned using a standard of GaP. To make quantitative comparisons between different samples and ensure that the analysed area was at steady state, Cs$^+$ ions were implanted into the surface to achieve a dose of 1×10^{17} ions cm^{-2}. Measurements of a control sample were acquired with an image dwell time of 20 ms, while images of the ^{15}N-labelled samples were acquired with a dwell time of 30 ms and at an image size of 50 ×× 50 μm and 256 by 256 pixels. Image J with the OpenMIMS plugin (Harvard, Cambridge, MA, USA) was used to generate isotope ratio images and extract quantitative data. Hue saturation intensity (HIS) colour maps were used to display the isotopic variation, the colour scale being set so that blue represented the natural background level of ^{15}N (0.37%) and pink the highest ratio found in the set of images from each transect.

For each of the four different time stages, two sections were examined taken from duplicate grains from the same ears. Three transects were taken from each section as shown in Figure 1. In total, 6025 individual protein bodies were measured for size and enrichment.

Light and immunomicroscopy

Semi-thin sections were cut using a Reichert-Jung Ultracut ultramicrotome, collected on drops of distilled water on multiwell slides coated with poly-L-lysine hydrobromide (Sigma) and dried on a hot plate at 40°C. Sections were stained with 0.01% (w/v) toluidine blue in 1% (w/v) sodium tetraborate, pH9. Immunofluorescence analysis was carried out as described in Tosi et al. (2011), using the antibodies R2–HMG rabbit polyclonal (specific for high-molecular-weight (HMW) glutenin subunits) and IFRN 0610 mouse monoclonal (recognizing epitopes present on gliadins and low-molecular-weight (LMW) glutenin subunits).

Bulk ^{15}N enrichment analysis by mass spectroscopy

Hand-dissected caryopses were frozen in liquid nitrogen and reduced to powder using a mortar and pestle. A known volume of 0.5 mg samples (± 0.1 mg) were analysed for total %N with a Vario Micro Elemental Analyser (Elementar Analysis Systems, Hanau, Germany) connected to an Isoprime 100 isotope ratio mass spectrometer (Isoprime Ltd, Cheadle Hume, UK) to determine ^{15}N enrichment, using commercial wheat flour as a control for the natural baseline level of ^{15}N.

Acknowledgements

Rothamsted Research receives strategic funding from the Biotechnological and Biological Sciences Research Council (BBSRC) as part of the 20:20 Wheat ® and Designing Seeds programmes. Katie Moore was supported by Engineering and Physical Sciences Research Council (EPSRC) grant EP/I026584/1. We are grateful to Dr Olivier Tranquet and Dr Sandra Denery-Papini (INRA, Nantes, France) for providing the R2-HMG antibody.

References

Bechtel, D.B. and Wilson, J.D. (2003) Amyloplast formation and starch granule development in hard red winter wheat. Cereal Chem. 80, 175–183.

Becraft, P.W. and Yi, G.B. (2011) Regulation of aleurone development in cereal grains. J. Exp. Bot. 62, 1669–1675.

Brett, G.M., Mills, E.N.C., Goodfellow, B.J., Fido, R.J., Tatham, A.S., Shewry, P.R. and Morgan, M.R.A. (1999) Epitope mapping studies of broad specificity monoclonal antibodies to cereal prolamins. J. Cereal Sci. 29, 117–128.

Cakmak, I. (2002) Plant nutrition research: priorities to meet human needs for food in sustainable ways. Plant Soil, 247, 3–24.

Feeney, K.A., Heard, P.J., Zhao, F.J. and Shewry, P.R. (2003) Determination of the distribution of sulphur in wheat starchy endosperm cells using secondary ion mass spectroscopy (SIMS) combined with isotope enhancement. J. Cereal Sci. 37, 311–318.

Fisher, D.B. and Macnicol, P.K. (1986) Amino-acid-composition along the transport pathway during grain filling in wheat. Plant Physiol. 82, 1019–1023.

Gao, X.P., Francis, D., Ormrod, J.C. and Bennett, M.D. (1992) Changes in cell number and cell-division activity during endosperm development in allohexaploid wheat, Triticum aestivum L. J. Exp. Bot. 43, 1603–1609.

Grignon, N., Halpern, S., Jeusset, J., Briançon, C. and Fragu, P. (1997) Localization of chemical elements and isotopes in the leaf of soybean (Glycine max) by secondary ion mass spectrometry microscopy: critical choice of sample preparation procedure. J. Microsc.-Oxford. 186, 51–66.

Heard, P.J., Feeney, K.A., Allen, G.C. and Shewry, P.R. (2002) Determination of the elemental composition of mature wheat grain using a modified secondary ion mass spectrometer (SIMS). Plant J. 30, 237–245.

Jauneau, A., Ripoll, C., Verdus, M.-C., Lefebvre, F., Demarty, M. and Thellier, M. (1994) Imaging the K, Mg, Na and Ca distributions in flax seeds using SIMS microscopy. Bot. Acta. 107, 81–89.

Mills, E.N.C., Field, J.M., Kauffman, J.A., Tatham, A.S., Shewry, P.R. and Morgan, M.R.A. (2000) Characterization of a monoclonal antibody specific for HMW subunits of glutenin and its use to investigate glutenin polymers. J. Agric. Food Chem. 48, 611–617.

Moore, K.L., Schröder, M., Lombi, E., Zhao, F.-J., McGrath, S.P., Hawkesford, M.J., Shewry, P.R. et al. (2010) NanoSIMS analysis of arsenic and selenium in cereal grain. New Phytol. 185, 434–445.

Moore, K.L., Lombi, E., Zhao, F.-J. and Grovenor, C.R.M. (2012a) Elemental imaging at the nanoscale: nanoSIMS and complementary techniques for element localisation in plants. Anal. Bioanal. Chem. 402, 3263–3273.

Moore, K.L., Zhao, F.-J., Gritsch, C.S., Tosi, P., Hawkesford, M.J., McGrath, S.P., Shewry, P.R. et al. (2012b) Localisation of iron in wheat grain using high resolution secondary ion mass spectrometry. J. Cereal Sci. 55, 183–187.

Ohdaira, Y., Masumura, T., Nakatsuka, N., Shigemitsu, T., Saito, Y. and Sasaki, R. (2011) Analysis of storage protein distribution in rice grain of seed-protein mutant cultivars by immunofluorescence microscopy. Plant Prod. Sci. 14, 219–228.

Rubin, R., Levanony, H. and Galili, G. (1992) Evidence for the presence of two different types of protein bodies in wheat endosperm. Plant Physiol. 99, 718–724.

Shewry, P.R., Underwood, C., Wan, Y., Lovegrove, A., Bhandari, D., Toole, G., Mills, E.N.C. et al. (2009) Storage product synthesis and accumulation in developing grains of wheat. J. Cereal Sci. 50, 106–112.

Steer, T., Thane, C., Stephen, A. and Jebb, S. (2008) Bread in the diet: consumption and contribution to nutrient intakes of British adults. Proc. Nutr. Soc. 67, E363.

Stone, B. and Morell, M.K. (2009) Carbohydrates. In Wheat: Chemistry and Technology (Khan, K. and Shewry, P.R., eds), pp. 299–362. St Paul, MN: AACC International.

Tosi, P., Parker, M., Gritsch, C.S., Carzaniga, R., Martin, B. and Shewry, P.R. (2009) Trafficking of storage proteins in developing grain of wheat. J. Exp. Bot. 60, 979–991.

Tosi, P., Sanchis Gritsch, C., He, J. and Shewry, P.R. (2011) Distribution of gluten proteins in bread wheat (Triticum aestivum) grain. Ann. Bot. 108, 23–35.

Wilson, R.G., Magee, C.W. and Stevie, F.A. (1989) Secondary Ion Mass Spectrometry: A Practical Handbook for Depth Profiling and Bulk Impurity Analysis. New York: Wiley.

8

A novel approach to identify genes that determine grain protein deviation in cereals

Ellen F. Mosleth[1,2], Yongfang Wan[2], Artem Lysenko[2], Gemma A. Chope[3], Simon P. Penson[3], Peter R. Shewry[2] and Malcolm J. Hawkesford[2]*

[1]Nofima AS, Ås, Norway
[2]Rothamsted Research, Harpenden, Hertfordshire, UK
[3]Cereals and Ingredients Processing, Campden BRI, Chipping Campden, Gloucestershire, UK

*Correspondence

email malcolm.hawkesford@rothamsted.ac.uk

Keywords: wheat, grain protein content, transcriptome, grain protein deviation, ANOVA–PCA, one-block means of scores regression.

Summary

Grain yield and protein content were determined for six wheat cultivars grown over 3 years at multiple sites and at multiple nitrogen (N) fertilizer inputs. Although grain protein content was negatively correlated with yield, some grain samples had higher protein contents than expected based on their yields, a trait referred to as grain protein deviation (GPD). We used novel statistical approaches to identify gene transcripts significantly related to GPD across environments. The yield and protein content were initially adjusted for nitrogen fertilizer inputs and then adjusted for yield (to remove the negative correlation with protein content), resulting in a parameter termed corrected GPD. Significant genetic variation in corrected GPD was observed for six cultivars grown over a range of environmental conditions (a total of 584 samples). Gene transcript profiles were determined in a subset of 161 samples of developing grain to identify transcripts contributing to GPD. Principal component analysis (PCA), analysis of variance (ANOVA) and means of scores regression (MSR) were used to identify individual principal components (PCs) correlating with GPD alone. Scores of the selected PCs, which were significantly related to GPD and protein content but not to the yield and significantly affected by cultivar, were identified as reflecting a multivariate pattern of gene expression related to genetic variation in GPD. Transcripts with consistent variation along the selected PCs were identified by an approach hereby called one-block means of scores regression (one-block MSR).

Introduction

Wheat is the most important food crop in temperate zones, with 713 million tonnes being produced globally in 2013 (http://faostat3.fao.org/faostat-gateway/go/to/home/E). It is also the most important crop in the UK, with up to 15 million tonnes being harvested annually and about 6 million tonnes milled for making bread and other food products. However, the yields of major crops, including wheat, are highly dependent on inputs, particularly of nitrogen fertilizer which is required for canopy development and carbon capture. Wheat production is particularly dependent on nitrogen availability as the quality for bread making is largely determined by the amount and composition of the grain storage proteins (see Shewry, 2007), and it may be necessary to apply additional nitrogen (i.e. above the optimum required for grain yield) in order to achieve an adequate content of grain protein for processing. Nitrogen is currently the major production cost for wheat farmers in the UK and Europe and may also have a significant environmental footprint when applied at high levels. Increases in cereal production must therefore be viewed against this economic and environmental background (Hawkesford, 2014).

Plant breeders have been highly successful in increasing wheat yields, by an average of about 1% a year in the UK (Mackay et al., 2011). However, increased yield is associated with lower protein concentration in grain (Barraclough et al., 2010) and the high protein content required for bread making (a minimum of 13% dry weight in the UK) means that modern bread-making cultivars require about 35 kg N/ha more than older cultivars. For example, Dampney et al. (2006) reported that six of 16 modern cultivars required >280 kg N/ha to achieve 13% dry weight protein, while four of 16 required >300 kg N/ha. The sustainability of such farming practices is now being questioned, in terms of economic returns, diffuse pollution and water framework compliance.

There is a well-established negative relation between grain yield and protein concentration (see i.e. Frey, 1951; Krapp et al., 2005; Lam et al., 1996; Simmonds, 1995) which reflects the inter-relationships between these traits. One hypothesis is that the negative correlation between grain yield and grain protein concentration results from the dilution of protein by carbohydrates (Acreche and Slafer, 2009). Another hypothesis is competition between carbon and nitrogen for energy (Munier-Jolain and Salon, 2005).

The negative relationship between yield and grain protein content is similar for most bread-making wheat cultivars when grown under the same conditions of nitrogen availability. However, some cultivars show reproducible deviation from this relationship, with high yield being combined with high grain protein content. This relationship has been called GPD (Monaghan et al., 2001) calculated as the residual from a regression analysis of grain yield on protein content. In some studies, GPD was calculated within each growth environment and compared across environments (i.e. Bogard et al., 2010;

Oury and Godin, 2007), whereas in other studies, environmental factors were incorporated into the regression (i.e. Monaghan et al., 2001).

It has been reported that GPD is under genetic control (Bogard et al., 2010; Oury et al., 2003). However, the analysis of GPD is complicated by the fact that both grain protein and grain yield have strong genotype–environmental interactions (Oury and Godin, 2007). Bogard et al. (2010) compared wheat grown under different conditions and showed that in most situations, GPD was correlated with postanthesis nitrogen uptake, but not with nitrogen remobilization, or with remobilization efficiency, although there was some variation between the different growth conditions. Uauy et al. (2006) also showed that Gpc-B1, a QTL associated with high contents of protein and minerals in wheat grain, encodes a transcription factor that controls nutrient remobilization from the leaves to the grain during senescence.

As much of the final grain nitrogen is accumulated in the plant before flowering and later mobilized to the grain (Barneix, 2007; Triboi and Triboi-Blondel, 2002), we hypothesize that genetic differences in GPD could directly or indirectly be reflected in differential expression of genes in the developing grain. We have therefore compared the expression patterns of gene transcripts in developing grain of six UK wheat cultivars grown in the field over three seasons at two different sites. This required the development of a novel statistical approach to dissociate differences in grain protein content and yield from the direct effects of nitrogen supply and from indirect effects related to yield and growth environment, in order to identify gene transcripts associated with GPD alone. We also suggest that this approach may have wider applicability in dissecting the transcriptional control of other complex phenotypic traits.

Results

Calculation of corrected grain protein deviation

Six UK cultivars were selected on the basis of differences in grain protein content: five high protein bread-making cultivars (Hereward, Marksman, Cordiale, Malacca and Xi19) and Istabraq which is a feed wheat cultivar known to have lower protein content. These cultivars were grown over three seasons (2008–2009, 2009–2010 and 2010–2011) at Rothamsted Research (Harpenden, UK) and at four other sites in the south-east of the UK, and at three N levels: 100 kg/ha as a 'low input' level, 200 kg/ha to reflect modern practice for bread-making wheats in the UK and 350 kg/ha as an extreme high input to achieve high grain protein (see Barraclough et al., 2010; Chope et al., 2014). The total number of samples was 594. Transcriptome data were determined for the experiments grown at Rothamsted Research and RAGT at three N levels in 2009 and 2010 and for one N level in 2011 giving a total of 161 samples (with one missing value). The trials grown at Rothamsted Research in 2009 and 2010 were used for feature extraction, while the remaining field trials were used to study the consistency of the expression of the selected genes across growth environments.

The yields at Rothamsted in 2009 ranged between 8.2 and 12.7 t/ha (at 85% dry matter), with grain %N ranging from 1.4 to 2.4. Istabraq had the highest yields and lowest %N which is consistent with the fact that it was the only feed cultivar. The yields in 2010 were substantially lower than in 2009, from 7.3 to 10.2 t/ha, with grain %N varying from 1.4 to 2.8. Both yield and grain %N were very responsive to N inputs in 2009, but yield was much less responsive in 2010, while %N remained very respon-

sive. Consequently, grain %N was highest at high N inputs in 2010. This may relate to the fact that 2010 had below average rainfall, with the exception of August which was very wet. In 2011, March to May also had exceptionally low rainfall, but this was followed by a relatively wet summer (summaries of temperature and rainfall for the three growth years are provided in Table S1). The yields of the samples where gene expression data are available ranged from 7.6 to 11.5 t/ha and grain %N from 1.6 to 3.2 (Table S2).

A negative relationship between grain %N and grain yield was observed within each year and at each N level, as shown in Figure 1 for the experiments at Rothamsted and RAGT where transcriptome data were available. The different cultivars are represented by different colours, and lines are drawn for the linear relationships between yield and grain %N at the different N inputs.

In order to quantify the extent of GPD, and to identify associated transcripts, novel statistical approaches were developed to dissociate effects on grain protein content from the direct effect of nitrogen and the indirect effect of yield, and to thereby relate transcript expression profiles to this trait alone.

The yields and grain %N contents of the samples grown in 2009 (Figure 2) and 2010 (Figure S1) were initially adjusted for the direct effects of N fertilization, with Figure 2a,b (and Figure S1a,b) showing the uncorrected data and Figure 2c,d (and Figure S1c,d) the data corrected for the impact of the applied N fertilizer. A second correction was then applied to remove the inverse relationship between grain %N content and yield, providing a measure of GPD called corrected GPD (Figure 3). Figure 3(a,b) therefore show grain %N content vs. grain yield for 2009 and 2010, respectively, where both the grain %N content and the yield have been corrected for the direct effect of N level (as illustrated in Figure 2 and Figure S1). Similarly, Figure 3(c,d) show grain %N contents for the same years after correction for yield (i.e. corrected GPD).

Whereas Figure 3(a,b) show the well-established negative correlation between grain %N content and yield, this is replaced by straight lines in Figure 3(c,d) with samples showing positive and negative GPD falling above and below these lines, respectively.

Analysis of variance (ANOVA) was performed to determine the effects of the design parameters on grain %N contents, grain yield and the corrected values (Table 1a). This showed significant effects of the cultivars on GPD as well as on the uncorrected and corrected values for grain %N content and yield. Whereas nitrogen level was significant for the uncorrected values, it was not for the corrected values, showing that the effect of N fertilization had been successfully removed. There were no significant interactions between cultivar (CV) and nitrogen fertilization for any of the parameters (results not shown).

The mean values for the cultivars within each site and year (Table 2) show that Hereward generally had high GPD, whereas Istabraq had low GPD, with Istabraq generally having higher yields than Hereward. This is also seen in Figure 3(c) where Hereward is generally is located to the left in the figure in the low-yield area and Istabraq to the right. To determine whether significant genetic variation in GPD existed in the absence of a relationship to genetic differences in yield, we also performed ANOVA without these two cultivars. This again showed a significant effect of cultivar for GPD (Table 1b). The mean values for yield and grain %N, both corrected for N (Figure 4), for the remaining four cultivars show significant differences in GPD that are not related to variation in grain yield. Malacca had lower

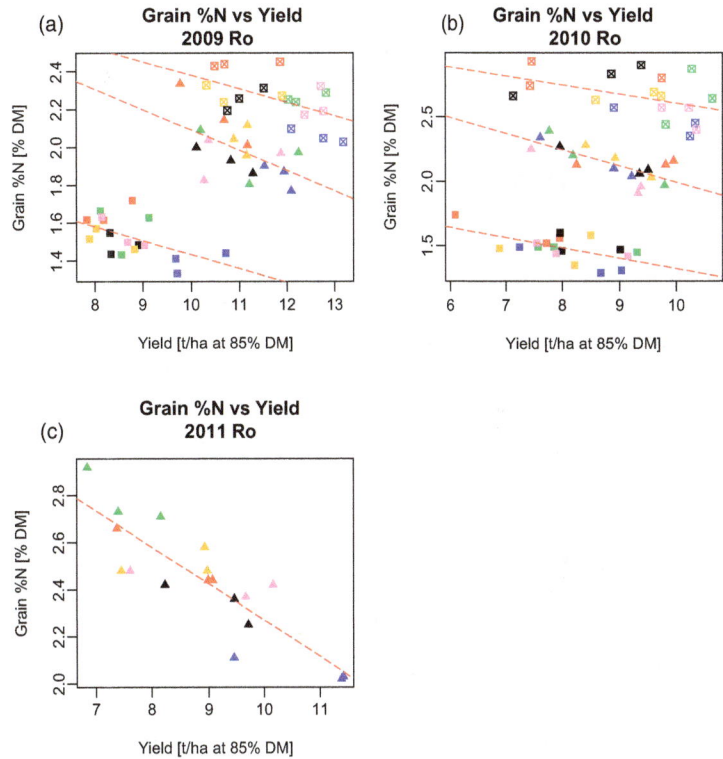

Figure 1 Raw-data plots of grain %N as a function of yield for (a) 2009, (b) 2010 and (c) 2011 at Rothamsted (Ro). In 2009 and 2010, there were three N levels: 100 kg/ha (filled squares), 200 kg/ha (triangles) and 250 kg/ha (open squares). Cultivars are colour-coded: Cordiale (green), Hereward (red), Istabraq (blue), Malacca (black), Marksman (yellow) and Xi19 (purple).

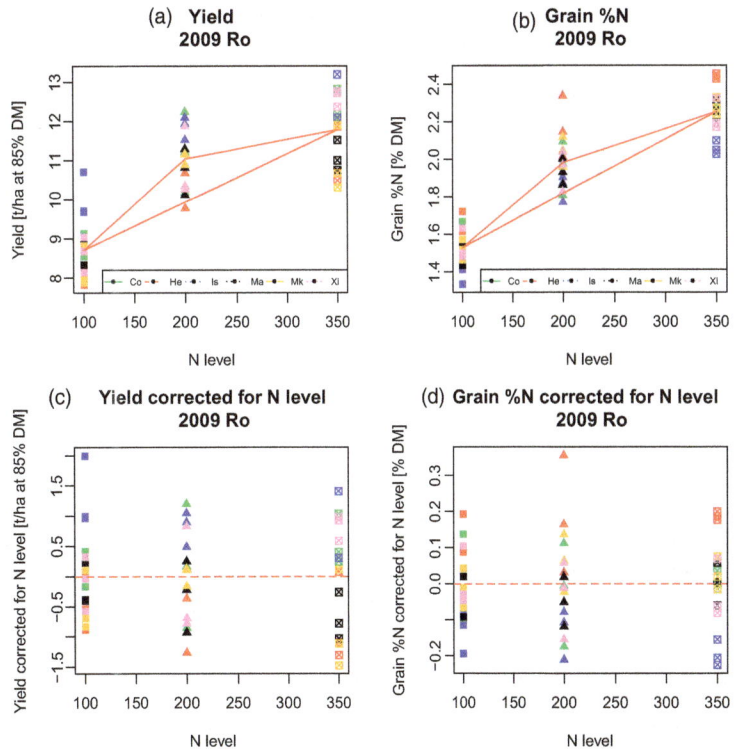

Figure 2 Correction of yield and grain %N content for their relation to N fertilisation (Yield ~N + N²) and (Grain %N ~N + N²) for wheat grown at Rothamsted in 2009. The x-axes of all plots are the N levels and the y-axes are as follows: (a) Yield and (b) Grain %N where the red lines are the linear and the quadratic effects. The deviation from the linear regression with N and N^2 is presented as: (c) Yield corrected for N level, and (d) Grain %N corrected for N level. N levels: 100 kg/ha (filled squares), 200 kg/ha (triangle) and 250 kg/ha (open squares). Cultivars are colour-coded: Cordiale (green), Hereward (red), Istabraq (blue), Malacca (black), Marksman (yellow) and Xi19 (purple).

values for both yield and grain %N, (corrected for N level), as well as for GPD compared with the other three cultivars, whereas Cordiale has the highest values for grain %N corrected for N level and GPD and higher Yield corrected for N level than Malacca.

The calculated GPDs (expressed as grain %N dry weight) for the six cultivars are summarized for samples grown at Rothamsted in 2009 in Figure 5(a) and in 2010 in Figure 5(b), which both show each cultivar at all N levels, and in Figure 5(c) which shows data for all sites and years at 200 kg N/ha. Figure 5(d) summarizes

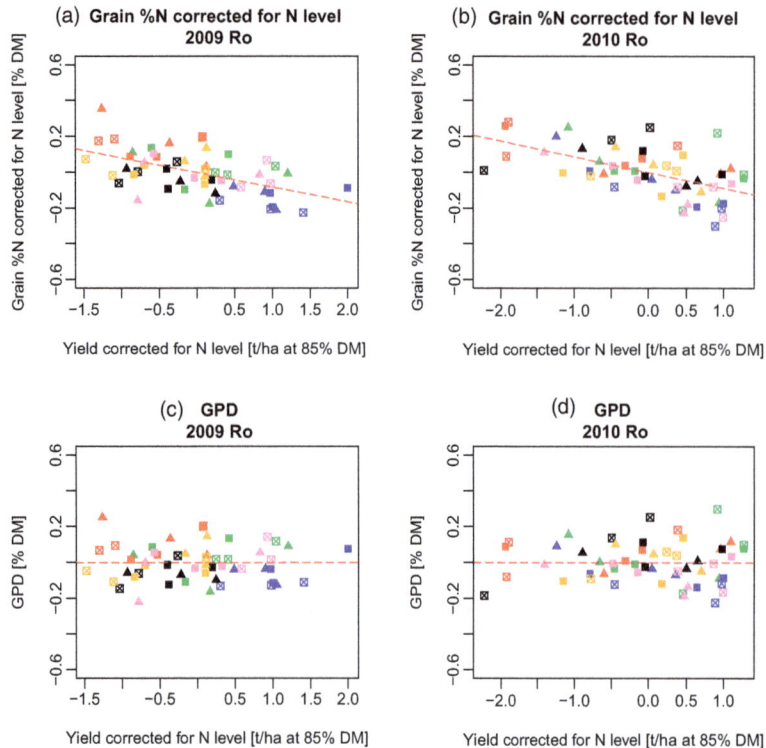

Figure 3 Correction of grain %N for yield for the experiment at Rothamsted (Ro). (a, b) Grain %N content vs. yield corrected for N level for 2009 and 2010, respectively. Cultivars are colour-coded: Cordiale (green), Hereward (red), Istabraq (blue), Malacca (black), Marksman (yellow) and Xi19 (purple). N levels: 100 kg/ha (filled squares), 200 kg/ha (triangle) and 350 kg/ha (open squares). (c, d) grain protein deviation (GPD) vs. yield corrected for N level for 2009 and 2010, with the data corrected for N level.

Table 1 *P*-values from ANOVA on the effect of the cultivar (CV) and N fertilization (N level) on the phenotypic characteristics; Yield and Protein both corrected for the effect of N level, and the double correction of protein to give GPD. The analyses were performed across years and sites

		Yield	Grain %N	Yield corrected for N level	Grain %N corrected for N level	GPD
(a) All data, 584 samples	N level	0.000	0.000	0.993	0.993	0.993
11 sites over 3 years	N level^2	0.020	0.000	0.937	0.937	0.937
3 N levels, 6 CV,	CV	0.000	0.000	0.000	0.000	0.000
3 biological replicates						
10 missing values						
(b) A subset of 391 samples	N level	0.001	0.000	0.993	0.993	0.993
11 sites over 3 years	N level^2	0.073	0.000	0.963	0.895	0.895
3 N levels, 4 CV (no Is or He),	CV	0.022	0.009	0.011	0.009	0.003
3 biological replicates						
5 missing values						

the results of the overall means for all of the trials at Rothamsted and RAGT over 3 years where the transcriptome data are available. Table 2 and Figure 4 shows results for all years and sites: Hereward shows positive GPD and Istabraq negative GPD in all years (2009–11), and Malacca being consistently lower than Hereward and Cordiale and higher than Istabraq in all data sets.

Identification of transcripts correlated with GPD

Principal component analysis (PCA) was used as the first step to identify gene transcripts related to GPD. The analysis was performed separately on the gene transcript profiles from the material grown at Rothamsted in 2009 and 2010. Consequently, there is no relationship between the principal components (PCs) identified for the 2 years.

The PCs were related to the design parameters (Table 3) and to phenotypic characters using the means of scores for the latter (Table 4). The same means of scores of the gene transcripts are

used here both towards the phenotypic characters, and as will be seen below, as an internal validation towards the transcriptional data used to generate the scores. The two data blocks of measured variables are thereby connected by their design represented by means over the biological replicates of a multivariate expression, whereas the biological variation within each block is kept for validation. This approach is hereby called means of scores regression (MSR).

The PCs which are significantly related to the cultivar (Table 3) and to GPD without affecting the grain yield (Table 4) reflect the expression of gene transcripts that underlie the genetic variation in GPD.

For 2009, both PC2 and PC7 are related to GDP (Table 4), with PC2 explaining 10.1% and PC7 1.4 % of the total variation (Table 3). PC7 has the strongest relationship to GDP (*P* < 0.001) with no relationship to grain yield (*P* = 0.85), whereas PC2 has *P*-values of *P* = 0.065 for grain yield and of *P* = 0.053 for yield

Table 2 Mean values of the six cultivars across all experiments in all years (in total 11 experiments over 3 years, 584 samples). The sites were Kw (KWS), Limagrain, Ra (RAGT), Ro (Rothamsted), Sy (Syngenta). The yield and grain %N in this tables were mean centred and scaled to unit variance prior the calculation of the corrected values. Corresponding data without standardization is provided in Table S2

	Co	He	Is	Ma	Mk	Xi
Yield corrected for N after mean centring and standardising to unit variance						
2009 Ro	0.26	−0.73	1.28	−0.50	−0.54	0.22
2010 Kw	0.57	−0.80	0.91	−0.54	−0.27	0.20
2010 Li	0.58	−1.15	0.95	−0.91	−0.16	0.69
2010 Ra	0.35	−1.37	0.82	−0.68	0.77	0.11
2010 Ro	0.31	−0.53	0.18	−0.19	−0.04	0.29
2010 Sy	−0.20	−1.65	1.41	0.03	−0.04	0.27
2011 Kw	−0.19	−1.25	1.60	−0.46	0.05	0.25
2011 Li	−0.70	−0.98	1.55	0.05	−0.20	0.29
2011 Ra	−0.54	−0.97	0.93	0.03	−0.48	1.02
2011 Ro	−1.22	−0.35	1.15	0.52	−0.35	0.25
2011 Sy	−0.35	−1.63	0.91	0.11	0.44	0.52
Grain %N corrected for N after mean centring and standardising to unit variance						
2009 Ro	0.09	1.38	−1.28	−0.25	0.21	−0.14
2010 Kw	0.23	1.00	−1.87	0.33	0.40	−0.13
2010 Li	0.51	0.97	−1.66	0.14	0.01	0.03
2010 Ra	0.33	1.32	−1.66	0.08	−0.09	0.01
2010 Ro	0.11	0.73	−0.70	0.43	0.06	−0.62
2010 Sy	0.53	1.49	−1.49	−0.09	−0.05	−0.24
2011 Kw	0.19	1.54	−1.43	−0.32	0.00	0.02
2011 Li	0.67	1.31	−1.26	−0.63	0.06	−0.16
2011 Ra	0.65	1.60	−1.22	−0.48	−0.08	−0.47
2011 Ro	1.53	0.39	−1.34	−0.63	0.10	−0.05
2011 Sy	0.16	1.81	−0.29	−0.79	−1.00	0.12
GPD calculated after mean centring and standardising to unit variance						
2009 Ro	0.27	1.17	−0.70	−0.62	−0.10	−0.03
2010 Kw	0.51	0.73	−1.64	0.11	0.32	−0.05
2010 Li	0.72	0.66	−1.44	−0.14	−0.04	0.24
2010 Ra	0.62	0.70	−1.45	−0.33	0.38	0.08
2010 Ro	0.34	0.52	−0.73	0.39	0.04	−0.56
2010 Sy	0.66	0.21	−0.57	−0.13	−0.16	−0.02
2011 Kw	0.07	0.92	−0.35	−0.99	0.05	0.30
2011 Li	0.22	0.87	−0.14	−0.91	−0.14	0.10
2011 Ra	0.38	1.29	−0.79	−0.64	−0.57	0.34
2011 Ro	0.95	0.19	−0.69	−0.36	−0.40	0.32
2011 Sy	−0.07	1.02	0.35	−0.92	−0.93	0.56

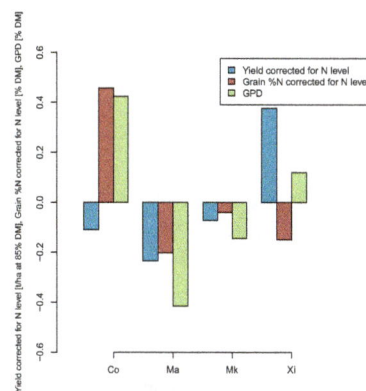

Figure 4 Yield and grain %N, both corrected for N, and corrected GPD, all mean centred and standardised to unit variance, for all data and four of the cultivars: Cordiale, Malacca, Marksman and Xi19. In total, 391 samples.

two figures show remarkable similarity when comparing PC7 with GPD for 2009 and PC2 and 3 with GPD for 2010, with Figure 5 showing the GDP values as means of the cultivars and Figure 6 showing the means of PCs from the PCA of the gene expression data selected to represent GDP. Figure 7 shows the relationship between GPD and PC7 as means of cultivars and N levels. There is a close relationship between PC7 and GPD for five of the six cultivars (omitting Xi19) with a correlation coefficient of $r = 0.86$, whereas Xi19 deviates from this. This is consistent with the genes underlying the selected PCs being responsible for, or correlated with, the variation in GPD in the cultivars Cordiale, Hereward, Istabraq, Malacca and Marksman, but not in Xi19.

The selected PCs which are shown in Table 4 to be related to GPD therefore reflect a multivariate pattern of gene expression. However, not all of these genes may be relevant to the traits that are reflected by the PCs. To identify a smaller number of candidate genes to be studied in more detail, we applied ANOVA to each of the selected PCs, using the means of the biological replicates of the scores as inputs and the gene expression values as the responses. Thus, we used the means of the scores obtained on the same data that were used to generate the PCA. We here call this approach one-block MSR.

For 2009, PC7 was of particular relevance as it was strongly related to GPD without any relation to yield or yield corrected for N level ($P < 0.001$) (Table 4). To limit the number of genes, we therefore focused on genes that were significant for PC7 in 2009, and at the same time significant for either PC2 or PC3 in 2010. This gave 959 transcripts as the best candidates for determining the corrected GPD.

The genes selected as significant by one-block MSR as having stable values for the selected PCs for the three different biological replicates were generally those with high or low loadings of the selected PCs (see Figure S2). To further reduce the number of potential candidate genes, we performed partial least-squares regression (PLSR) with Jackknife (see Figure S3). As shown in Figure 7, the cultivar Xi19 is clearly separated from the other cultivars in terms of the relationship between the transcriptome profiles reflected by the selected PCs and the corrected GPD.

The most interesting transcripts in terms of GPD were found to be 136 transcripts positively related to GPD by the PLS regression analysis (shown as pink-filled triangles in the Figure S3). A raw

corrected for N level (Table 4). Therefore, PC7 is the most relevant to GPD for 2009. For 2010, PC2 and PC3 are of most interest as they are both related to all protein parameters without any relationships to yield. These PCs accounted for 11.5% and 7.9%, respectively, of the variation in the transcriptome data (Table 4). The PCs above PC10 did not capture any information relevant to GPD. All of these PCs are significantly affected by the cultivar (Table 3).

The means of the selected scores for each cultivar for 2009 and 2010 are shown in Figure 6. As the directions of the scores and loading plots are arbitrary in PCA, the directions have been selected to facilitate the comparison with the GDP plots. These

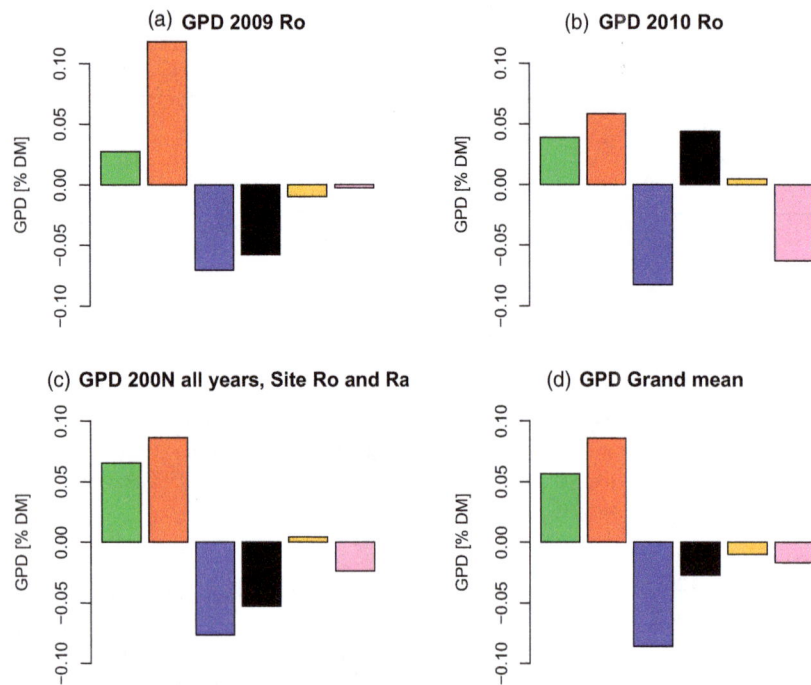

Figure 5 Mean value of GDP where protein content are corrected for N levels, and its relation with yield for (a) 2009 and (b) 2010, (c) All three growth years 2009, 2010 and 2011 at N level 200 kg/ha at Rothamsted and RAGT, corrected for yield and the effect of year and (d) all 161 data points summarised. Cultivars are colour-coded: Cordiale (green), Hereward (red), Istabraq (blue), Malacca (black), Marksman (yellow) and Xi19 (purple).

Table 3 Results of ANOVA (FDR-adjusted *P*-values) showing the effect of the design parameters (CV, linear and quadratic effects of N, and the interaction between N and CV) (input of the model) on the scores of PCA of the gene expression data (output of the model) for (a) 2009 [A total 53 samples: 1 site (Rothamsted), 3 N levels, 6 CV, 3 biological replicates, 1 missing value] and (b) 2010 [a total 54 samples: 1 site (Rothamsted), 3 N levels, 6 CV, 3 biological replicates]

	ExplVar	CV	N	N^2	CV*N
(a)					
PC1	33.5	0.003	0.414	0.451	0.943
PC2	10.1	0.002	0.603	0.832	0.859
PC3	8.9	0.005	0.808	0.712	0.943
PC4	3.4	0.001	0.838	0.712	0.645
PC5	2.3	0.000	0.273	0.712	0.271
PC6	1.3	0.001	0.414	0.656	0.672
PC7	1.4	0.000	0.669	0.739	0.271
PC8	1.1	0.004	0.371	0.522	0.705
PC9	1.2	0.001	0.808	0.522	0.943
PC10	1.4	0.000	0.002	0.246	0.432
(b)					
PC1	24.2	0.177	0.018	0.547	0.919
PC2	11.5	0.001	0.018	0.691	0.831
PC3	7.9	0.004	0.432	0.115	0.754
PC4	2.0	0.000	0.037	0.122	0.738
PC5	4.7	0.000	0.168	0.115	0.938
PC6	1.9	0.000	0.000	0.115	0.938
PC7	1.7	0.000	0.002	0.115	0.938
PC8	4.0	0.000	0.033	0.370	0.738
PC9	5.0	0.000	0.082	0.115	0.863
PC10	0.8	0.959	0.918	0.961	0.738

Table 4 Results of ANOVA (FDR-adjusted *P*-values) showing the effect of the scores of PCA of the gene expression data (input of the model) on the phenotypic characteristics (output of the model) for (a) 2009 [a total 53 samples: 1 site (Rothamsted), 3 N levels, 6 CV, 3 biological replicates, 1 missing value, and (b) 2010 (a total 54 samples: 1 site (Rothamsted), 3 N levels, 6 CV, 3 biological replicates]

	Yield	Protein	Yield corrected for N level	Grain%N corrected for N level	GPD
(a)					
Mean PC1	0.078	0.031	0.740	0.519	0.581
Mean PC2	0.065	0.991	0.053	0.005	0.043
Mean PC3	0.152	0.46	0.008	0.006	0.129
Mean PC4	0.736	0.682	0.139	0.911	0.421
Mean PC5	0.362	0.037	0.000	0.015	0.643
Mean PC6	0.357	0.195	0.823	0.572	0.415
Mean PC7	0.853	0.063	0.199	0.000	0.000
Mean PC8	0.071	0.034	0.431	0.636	0.285
Mean PC9	0.314	0.446	0.080	0.067	0.297
Mean PC10	0.000	0.000	0.447	0.088	0.010
(b)					
Mean PC1	0.001	0.000	0.293	0.392	0.756
Mean PC2	0.603	0.000	0.170	0.012	0.038
Mean PC3	0.942	0.021	0.285	0.030	0.062
Mean PC4	0.966	0.085	0.449	0.081	0.111
Mean PC5	0.143	0.255	0.369	0.837	0.708
Mean PC6	0.013	0.000	0.086	0.244	0.120
Mean PC7	0.313	0.005	0.641	0.491	0.608
Mean PC8	0.283	0.295	0.740	0.084	0.061
Mean PC9	0.026	0.245	0.136	0.087	0.301
Mean PC10	0.836	0.948	0.704	0.561	0.329

(a) **Scores of PCA 2009 PC2**

(b) **Scores of PCA 2009 PC7**

(c) **Scores of PCA 2010 PC2**

(d) **Scores of PCA 2010 PC3**

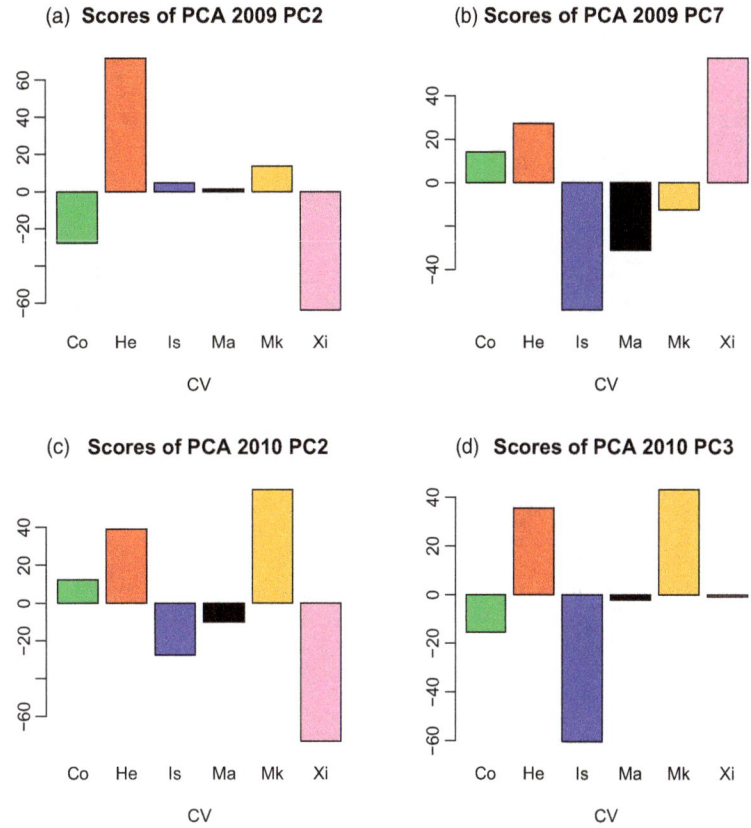

Figure 6 Mean values of the CVs for scores of selected PCs from PCA of the gene expression in 2009 (a,b) and 2010 (c,d). The PCs are selected as reflecting variation in GDP with no relation to the yield. (a,b) 2009: PC2 and PC7, respectively. (c,d) 2010: PC2 and PC3, respectively.

data plot of all these transcripts is given in Figure S4 where the gene transcripts are sorted according to the regression coefficient of the PLS regression.

The molecular functions of the 959 transcripts selected as significant for PC7 in 2009 and PC2 or PC3 in 2010 were predicted (Table S3) and assigned to functional groups (Table S4) by gene ontogeny (GO) analysis, based on sequence similarities with characterized genes from other plant species.

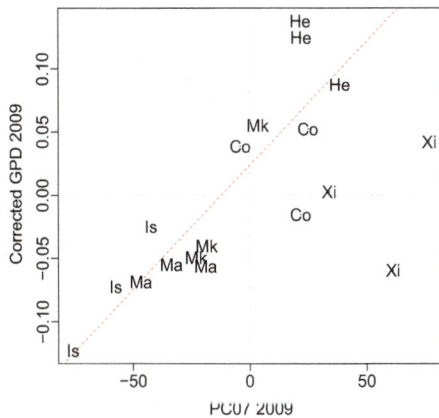

Figure 7 Plot of corrected GPD vs. PC7 for 2009 as means of the CVs and N level: N levels: 100 kg/ha (filled squares), 200 kg/ha (triangle) and 350 kg/ha (open squares). Cultivars are colour-coded: Co = Cordiale (green), He = Hereward (red), Is = Istabraq (blue), Ma = Malacca (black), Mk = Marksman (yellow) and Xi = Xi19 (purple).

To determine how the expression of selected genes was affected by genetic and environmental factors, ANOVA was applied for all samples where transcriptome data were available using the design factors as input and the expression profiles of the selected gene transcripts as output. Most transcripts displayed differences related to the genotype, with the year and N level also affecting a number of the transcripts. However, the relative importance of these factors differed. Four of the 136 gene transcripts identified as candidate genes positively related to GPD are displayed in Figure 8 for the five cultivars Cordiale, Hereward, Istabraq, Malacca and Marksman, for all 161 samples where transcriptome data were available. All four transcripts were significantly related to both the cultivar differences and to the year of growth, but the relative importance of these two factors differed. For transcripts Ta.8367.2.S1_a_at and Ta.14543.2. A1_at, the cultivar differences dominated with environmental factors having little impact, whereas the year of growth had a relatively larger impact on transcripts Ta.6968.1.S1_at and Ta.10471.1.S1_x_at. For Ta.8367.2.S1_a_at, the cultivar difference primarily resulted from lower expression of Istabraq vs. the remaining cultivars, whereas for Ta14543.2.A1_at showed more gradual variation in expression among the five cultivars. Similar expression profile for all 136 gene transcripts across all growth environments are shown in Figure S4.

Discussion

The identification of gene transcripts whose expression is significantly related to traits is a challenge in functional genomics, as the number of features can be high. For wheat, the Affymetrix arrays comprise approximately 60 000 features (gene probe sets

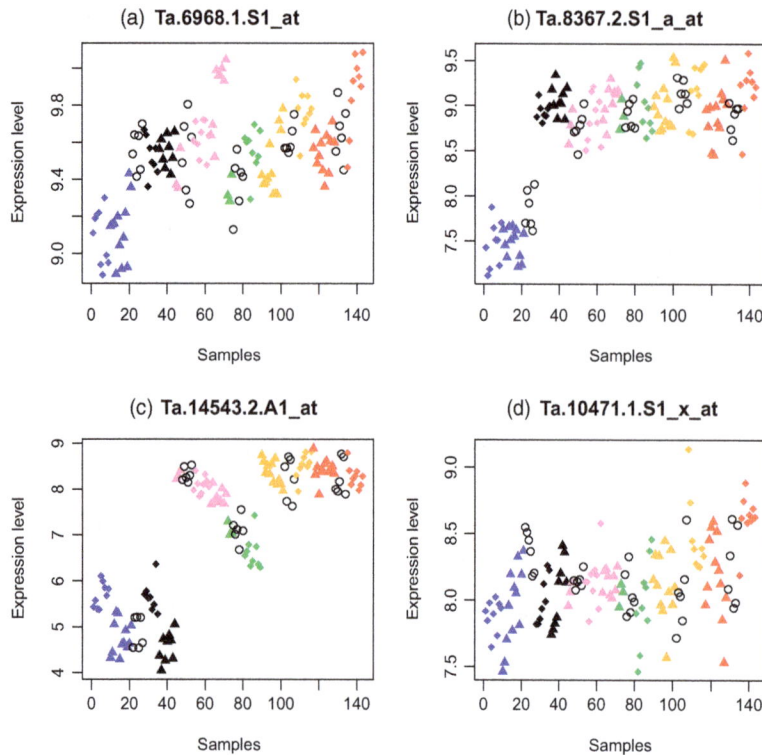

Figure 8 Examples of four genes (a-d) located in quadrant 4 of the PCA plot in Figure S3 and by PLS regression as being positively related to GPD. The colour codes indicate cultivars: Co=Cordiale (green), He=Hereward (red), Is=Istabraq (blue), Ma=Malacca (black), and Mk=Marksman (yellow), sorted by cultivar by their impact on GPD as found in regression analysis. The symbols indicate the year of growth: 2009 (filled squares): year 2010 (triangles) and 2011 (open circles), which to different degrees significantly affected the expression of these genes. N-level did not have significant impact on the expression of any of these four genes (not shown).

corresponding to slightly fewer unique transcripts), which may show separate or coordinated patterns of expression. The multivariate nature of functional genomics data must therefore be taken into account (Faergestad et al., 2009). One family of multivariate methods is based on the projection of the original variables onto new variables defined as linear combinations of the original variables [i.e. PCA (Hotelling, 1933a,b) and PLS regression (Wold et al., 1983)]. These methods present results as multivariate patterns, also called latent factors, which reflect the underlying phenomena that gives rise to the variation observed in the data. However, there is also a need for feature selection at the level of the observed variables (Lazar et al., 2012; Saeys et al., 2007). For the present study, yet another level of complexity had to be taken into account as we did not search for gene transcripts related to single phenotypic traits, but for those that resulted in the optimal deviation of two traits that are negatively related. Namely, genes that were associated with the highest protein content at a given yield (i.e. GPD). We therefore developed a novel methodology, to resolve the phenotypic characters and combine this with multivariate projection and feature extraction. After projection of the main information in the data onto latent factors (PCs), feature extraction was first performed at the level of the latent variation to identify genetically determined patterns of variation relating to the property of interest (here GPD alone). Feature extraction was then repeated within the selected latent factors to identify gene transcripts which consistently showed significant variation along the selected PCs. This is obtained by relating means of scores both to the corrected data for the phenotypic characteristics and to the gene transcripts used to generate the scores. Data from one site over two growth years was used to calculate GPD and identify genes related to the trait, and data for the second site and growth year were used in the validation of the consistency of the selected genes across different growth environments.

We suggest that this approach may have wider applicability in dissecting the transcriptional control of other complex traits. One of the strengths of the present approach is that we use methodologies well known to most biologists in the functional genomics area (PCA and ANOVA), combined in a novel way to solve complex problems.

Conclusions

We developed novel statistical approaches to identify transcripts whose expression in developing wheat caryopses is correlated with GPD at grain maturity. The transcripts that we identified probably represent both genes that control the trait and genes whose expression is affected as a consequence. In total, 136 gene transcripts were identified, and their behaviour was observed across different growth environments. Further work is required to identify the biological functions of these genes and identify those that can be exploited in for crop improvement. The availability of new rapid approaches for transcript analysis, such as next generation sequencing should facilitate the profiling of selected genes in large numbers of samples. It is necessary to identify and compare further material showing positive deviation from the negative relation between grain %N and yield, including larger collections of cultivars for association genetics or crosses between varieties differing in GPD for classical Mendelian analysis. These approaches should lead to the development of markers for breeders as well as providing information on the mechanisms controlling the trait. Although the cost of expression analysis has often limited the application of molecular approaches to crop improvement in the past, this is not likely to be the case in the future. Instead, the practical limitation is more likely to be the production of appropriate plant material and the availability of approaches to handle the complex data sets that are generated. As wheat yields can only be accurately determined in replicated field trials carried out in least three environments (years and/or sites), this not only requires appropriate facilities but also a long-term commitment.

Experimental procedures

Wheat material

Six UK cultivars (Istabraq, Hereward, Marksman, Condiale, Malacca and Xi 19) were grown over three seasons (2008–2009, 2009–2010 and 2010–2011) at Rothamsted Research and at four other sites in the south-east of the UK (RAGT, Ickleton, Cambridge; Limagrain, Woolpit, Suffolk; Syngenta, Whittlesford, Cambridge; KWS-UK, Thriplow, Hertfordshire) in 2009–2010 and 2010–2011 only. Three replicate plots were grown at three N levels: 100 kg/ha (N100), 200 kg/ha (N200) and 350 kg/ha (N350) (see Barraclough et al., 2010; Chope et al., 2014).

Developing heads (10 per plot) were tagged, and caryopses were harvested from the Rothamsted (2009, 2010 and 2011) and RAGT (2010 and 2011) sites at 21 days postanthesis (dpa), which represents the middle of grain filling when gene expression is at its highest (Wan et al., 2008). Gene expression was measured using Affymetrix wheat microarrays giving a total of 161 samples.

Yield and grain protein determination

Trials were performed as previously described (Barraclough et al., 2010; Chope et al., 2014). Yields are standardized to 85% dry matter, after determining moisture content of individual samples. DM total nitrogen was determined in mature grain using the Dumas combustion method (Dumas, 1831), using a CNS (carbon, nitrogen, sulphur) Combustion Analyser (Leco Corp., St. Paul, MN). Nitrogen is presented as % of 100% dry matter content (Grain %N).

Affymetrix Genechip® hybridization

Microarrays were used to profile transcriptome. A time point of 21 dpa was chosen as a key developmental stage (mid-grain filling) in which grain storage proteins were being synthesized. Ten years per plot (three replicate plots per treatment/variety) were tagged at anthesis, and around 100 caryopses per sample were taken from the mid-third of the year and were harvested 21 days later. Gene expression was determined by profiling RNA extracted from this material against a gene chip representing 61 313 probe sets equating to 55 052 transcripts. Data from the profiling are semiquantitative, giving a good indication of the relative levels of expression of all RNAs in the sample simultaneously. Data were collected for 3 years at Rothamsted and for 2010 and 2011 at the RAGT site, for the three N levels in 2009 and 2010, and for the 200 kg N/ha treatment in 2011. One sample from 2009 was omitted (Malacca in 2009 grown at 2003, replicate three) as this sample was not analysed.

Microarray data are submitted to Gene Expression Omnibus (GEO) (http://www.ncbi.nlm.nih.gov/geo/).

Data analysis

Measurements of the deviation from the negative relation between grain %N and yield were obtained by first adjusting yield and protein content for the effect of N level (N) and the second order effect of N (N^2). The corrected value of protein (Grain%N_corrN)

was further corrected for its negative relation to the corrected value of yield (Yield_corrN), giving corrected GPD as the residual. ANOVA was performed to investigate how the design factors: cultivars (CV), N level and the interaction between cultivar and N level, affected the phenotypic characters, and their corrected values using P-values were adjusted for multiple comparisons by false discovery rate using rotation test (Langsrud, 2005).

Principal component analysis was performed on the gene expression data. By PCA, the original data are decomposed into PCs where the PCs give multivariate patterns that describe in decreasing order the main variation in the data. All variables were centred and scaled to unit variance to allow gene transcript with small variation have the same impact on the model as gene transcript with large variation.

The mean values of each score were calculated across the three biological replicates to leave the variation in the phenotypic character across the biological replicates for validation. By this approach, we could identify multivariate pattern that might cause a positive deviation between grain %N and yield without negatively influencing the yield. The approach of using means of the scores for the regression is here called MSR.

To identify the individual gene transcripts that varied consistently along the relevant latent factors, a regression analysis was performed to relate means of the scores of the selected PCs to the gene expression data. This is an internal analysis performed within one block of data for validation of the consistency of each variable along relevant PCs. As above, we use MSR for this analysis, and we call this one-block MSR. The test was performed by rotation test (Langsrud, 2005) for correction of multiple comparisons by false discovery rate using rotation test.

To visualize positive vs. negative direction of the effects for the selected gene transcripts on GPD, and to further narrow down the number of candidate genes, PLSR was performed to relate the selected gene transcripts to GPD. The model was validated by Jackknife adapted to bilinear models (Martens and Martens, 2000). Gene transcripts significant for the selected PCs which showed a consistent pattern of variation related to the response for the different cross-validation segments and a positive relation to GPD were thereby selected.

For the significant genes positively related to GPD, a gene expression profile was made where the selected gene expression data are plotted for all the available data, also those not used for the selection of the gene transcripts. Whereas only growth year 2009 and 2010 at site Rothamsted were used to identify gene transcripts significantly related to GPD, all three growth years at both sites (in total 161 samples) were investigated for their behaviour over environmental conditions.

GO function analysis

GO functions for significantly over-expressed transcripts are found from the annotation offered by the B2G-FAR resource (Götz et al., 2011). B2G-FAR GO annotation is a broad-specificity data set derived from homology and protein domain annotation; it therefore included some erroneous annotations for functions that are not found in plants. To address this issue, the original annotation set was filtered and only plant-relevant terms where retained. The filtering was carried out based on a high-quality reference set of all plant-relevant terms that was created by taking a nonredundant union of all terms and their ancestor terms from manually annotated rice and Arabidopsis GO annotation sets. In total, about 11% of nonplant annotations were removed.

Acknowledgements

Rothamsted Research is funded by the Biotechnology and Biological Sciences Research Council (BBSRC). The work reported here was supported by the BBSRC Industrial Partnership Award BB/G022437 with support from the Home Grown Cereals Authority grant RD-2007-3409 'Sustainability of UK-grown wheat for breadmaking'. EFM was supported by grants from the Fund for Research Levy on Agricultural Products in Norway. Field trials at Rothamsted Research were supported by the Defra-funded WGIN project. Trials at KWS, Limagrain, RAGT and Syngenta were supported by the respective breeding companies. We are grateful to the reviewers of this paper for their constructive comments.

References

Acreche, M.M. and Slafer, G.A. (2009) Variation of grain nitrogen content in relation with grain yield in old and modern Spanish wheats grown under a wide range of agronomic conditions in a Mediterranean region. *J. Agric. Sci.* **147**, 657–667.

Barneix, A.J. (2007) Physiology and biochemistry of source-regulated protein accumulation in the wheat grain. *J. Plant Physiol.* **164**, 581–590.

Barraclough, P.B., Howarth, J.R., Jones, J., Lopez-Bellido, R., Parmar, S., Shepherd, C.E. and Hawkesford, M.J. (2010) Nitrogen efficiency of wheat: genotypic and environmental variation and prospects for improvement. *Eur. J. Agron.* **33**, 1–11.

Bogard, M., Allard, V., Brancourt-Hulmel, M., Huemez, E., Machet, J.-M., Jeuffroy, M.-H., Gate, P., Martre, P. and Le Gouis, J. (2010) Deviation from the grain protein concentration–grain yield negative relationship is highly correlated to post-anthesis N uptake in winter wheat. *J. Exp. Bot.* **61**, 4303–4312.

Chope, G.A., Wan, Y., Penson, S.P., Bhandari, D.G., Powers, S.J., Shewry, P.R. and Hawkesford, M.J. (2014) Effects of genotype, season, and nitrogen nutrition on gene expression and protein accumulation in wheat grain. *J. Agric. Food Chem.* **62**, 4399–4407.

Dampney, P.M.R., Edwards, A. and Dyer, C.J. (2006) Managing nitrogen applications to new Group 1 and 2 wheat varieties. *HGCA Proj Rep* No. 400.

Dumas, J.B.A. (1831) Procedes de l'analyse organique. *Ann. Chim. Phys.* **2**, 198–213.

Faergestad, E.M., Langsrud, Ø., Høy, M., Hollung, K., Sæbø, S., Liland, K.H., Kohler, A., Gidskehaug, L., Almergren, J., Anderssen, E. and Martens, H. (2009) Analysis of megavariate data in functional genomics. In *Comprehensive Chemometrics*, vol. **4** (Brown, S., Tauler, R. and Walczak, R., eds), pp. 221–278. Oxford: Elsevier.

Götz, S., Arnold, R., Sebastián-León, P., Martín-Rodríguez, S., Tischler, P., Jehl, M.-A., Dopazo, J., Rattei, T. and Conesa, A. (2011) B2G-FAR, a species-centered GO annotation repository. *Bioinformatics*, **27**, 919.

Hawkesford, M.J. (2014) Reducing the reliance on nitrogen fertiliser for wheat production. *J. Cereal Sci.* **59**, 276–283.

Hotelling, H. (1933a) Analysis of a complex of statistical variables into principal components. *J. Edu. Psychol.* **24**, 417–441.

Hotelling, H. (1933b) Analysis of a complex of statistical variables into principal components. *J. Edu. Psychol.* **24**, 498–520.

Krapp, A., Saliba-Colombani, V. and Daniel-Vedele, F. (2005) Analysis of C and N metabolisms and of C/N interactions using quantitative genetics. *Photosynth. Res.* **83**, 251–263.

Lam, H.-M., Coschigano, K.T., Oliviera, I.C., Melo-Oliviera, R. and Coruzzi, G. (1996) The molecular-genetics of nitrogen assimiliation into amino acids in higher plants. *Annu. Rev. Plant Phys.* **47**, 569–593.

Langsrud, Ø. (2005) Rotation test. *Stat. Comput.* **15**, 53–60.

Lazar, C., Taminau, J., Meganck, S., Steenhoff, D., Coletta, A., Molter, C., de Schaetzen, V., Duque, R., Bersini, H. and Nowé, A. (2012) A survey on filter techniques for feature selection in gene expression microarray analysis. *IEEE/ACM. Trans. Comput. Biol. Bioinform.* **9**, 1106–1119.

Mackay, I., Horwell, A., Garner, J., White, J. and Philpott, H. (2011) Reanalyses of the historical series of UK variety trials to quantify the contributions of genetic and environmental factors to trends and variability in yield over time. *Theor. Appl. Genet.* **122**, 225–238.

Martens, H. and Martens, M. (2000) Modified Jack-knife estimation of parameter uncertainty in bilinear modelling by partial least squares regression (PLSR). *Food Qual. Prefer.* **11**, 5–16.

Monaghan, J.M., Snape, J.W., Chojecki, J.S. and Kettlewell, P.S. (2001) The use of grain protein deviation for identifying wheat cultivars with high grain protein concentration and yield. *Euphytica*, **122**, 309–317.

Munier-Jolain, N.G. and Salon, C. (2005) Are the carbon costs of seed production related to the quantitative and qualitative performance? An appraisal of legumes and other crops. *Plant, Cell Environ.* **28**, 1388.

Oury, F.X. and Godin, C. (2007) Yield and grain protein concentration in bread wheat: how to use the negative relationship between the two characters to identify favourable genotypes? *Euphytica*, **157**, 45–57.

Oury, F.X., Be'rard, P. and Brancourt-Hulmel, M. (2003) Yield and grain protein concentration in bread wheat: a review and a study of multi-annual data from a French breeding program. *J. Genet. Breed.* **57**, 59–68.

Saeys, Y., Inza, I. and Larrañaga, P. (2007) A review of feature selection techniques in bioinformatics. *Bioinformatics*, **23**, 2507–2517.

Shewry, P.R. (2007) Improving the protein content and composition of cereal grain. *J. Cereal Sci.* **46**, 239–250.

Simmonds, N.W. (1995) The relation between yield and protein in cereal grain. *J. Sci. Food Agric.* **67**, 309.

Triboi, E. and Triboi-Blondel, A.-M. (2002) Productivity and grain or seed composition: a new approach to an old problem. *Eur. J. Agron.* **16**, 163–186.

Uauy, C., Distelfeld, A. and Fahima, T. (2006) A NAC gene regulating senescence improves grain protein, zinc, and iron content in wheat. *Science* **314**, 1298–1301.

Wan, Y., Poole, R.L., Huttly, A.K., Toscano-Underwood, C., Feeney, K., Welham, S., Gooding, M.J., Mills, E.N.C., Edwards, K.J., Shewry, P.R. and Mitchell, R.A.C. (2008) Transcriptome analysis of grain development in hexaploid wheat. *BMC Genomics*, **9**, 121.

Wold, S., Martens, H. and Wold, H. (1983) The multivariate calibration problem in chemistry solved by the PLS method. In: *Proc. Conf. Matrix Pencils. Lecture Notes in Mathematics.* (Ruhe, A. and Kågstrom, B., eds), pp. 286–293. Springer: Heidelberg.

9

Stable expression of silencing-suppressor protein enhances the performance and longevity of an engineered metabolic pathway

Fatima Naim[1,2,*,†], Pushkar Shrestha[1], Surinder P. Singh[1], Peter M. Waterhouse[1,2,3,†] and Craig C. Wood[1,*]

[1]CSIRO Agriculture, Canberra, ACT, Australia
[2]School of Biological Sciences, The University of Sydney, Sydney, NSW, Australia
[3]School of Molecular Bioscience, The University of Sydney, Sydney, NSW, Australia

*Correspondence

email craig.wood@csiro.au

email fatima.naim@qut.edu.au
†Present address: The Centre for Tropical Crops and Biocommodities, Queensland University of Technology, Brisbane, Qld, Australia.

Keywords: metabolic engineering, viral silencing-suppressor proteins, long chain polyunsaturated fatty acid, transgene longevity.

Summary

Transgenic engineering of plants is important in both basic and applied research. However, the expression of a transgene can dwindle over time as the plant's small (s)RNA-guided silencing pathways shut it down. The silencing pathways have evolved as antiviral defence mechanisms, and viruses have co-evolved viral silencing-suppressor proteins (VSPs) to block them. Therefore, VSPs have been routinely used alongside desired transgene constructs to enhance their expression in transient assays. However, constitutive, stable expression of a VSP in a plant usually causes pronounced developmental abnormalities, as their actions interfere with endogenous microRNA-regulated processes, and has largely precluded the use of VSPs as an aid to stable transgene expression. In an attempt to avoid the deleterious effects but obtain the enhancing effect, a number of different VSPs were expressed exclusively in the seeds of *Arabidopsis thaliana* alongside a three-step transgenic pathway for the synthesis of arachidonic acid (AA), an ω-6 long chain polyunsaturated fatty acid. Results from independent transgenic events, maintained for four generations, showed that the VSP-AA-transformed plants were developmentally normal, apart from minor phenotypes at the cotyledon stage, and could produce 40% more AA than plants transformed with the AA transgene cassette alone. Intriguingly, a geminivirus VSP, V2, was constitutively expressed without causing developmental defects, as it acts on the siRNA amplification step that is not part of the miRNA pathway, and gave strong transgene enhancement. These results demonstrate that VSP expression can be used to protect and enhance stable transgene performance and has significant biotechnological application.

Introduction

The introduction of complex transgenic pathways specifically expressed in oilseeds to generate modified oils has long been a target of plant genetic engineers, producing a range of nutritional, pharmaceutical and industrial oils (Cahoon and Kinney, 2005; Nykiforuk *et al.*, 2012; Petrie *et al.*, 2012; Vanhercke *et al.*, 2013). The process of generating elite transgenic events in many crops is arduous (Saunders and Lomonossoff, 2013) and from the large population of transgenic events produced in the initial tissue culture stage, only a small subpopulation give high levels of transgene performance that are stable in subsequent generations (Hagan *et al.*, 2003). It has been suggested that high transgene expression triggers silencing pathways and that initial highly expressing events soon become incapacitated (Lindbo *et al.*, 1993; Saunders and Lomonossoff, 2013; Schubert *et al.*, 2004). It is clear that this silencing operates at both the transcriptional and post-transcriptional level and is guided by small RNAs (sRNAs) (Hagan *et al.*, 2003; Schubert *et al.*, 2004).

Plant viruses have evolved ways to evade or negate the host's silencing mechanisms. They deploy a range of strategies, including the expression of viral silencing-suppressor proteins (VSPs). These proteins interfere with silencing pathway components, sometimes targeting a number of them simultaneously (Ding and Voinnet, 2007; Incarbone and Dunoyer, 2013). They therefore have great potential as enhancers of transgene expression.

However, normal plant growth and responses to the environment are also dependent on sRNA-guided processes, so the generic inhibition or ablation of sRNAs to prevent gene silencing can have developmental and defence-response consequences. We sought to find a VSP, or a way of applying it, that enhanced stable transgene expression without deleterious side effects. In our study, we examined three VSPs: p19, V2 and PO[PE].

Tombusvirus p19 is a small protein encoded by *Tomato bushy stunt virus* (TBSV) (a *positive-strand RNA virus*). It was one of the first VSPs used in transient leaf assay systems to inhibit the post-transcriptional gene silencing of transgenes (Voinnet *et al.*, 2003). This small (19 kDa) and soluble protein has been extensively studied, including the resolution of crystal structures of p19 bound to sRNA. In conjunction with *in vitro* and *in vivo* experiments (Silhavy *et al.*, 2002; Ye *et al.*, 2003), the crystal structures show that p19 forms a homodimer specifically with 21 nt sRNA duplexes that contain 2 nt 3' overhangs (Lakatos *et al.*, 2004; Vargason *et al.*, 2003; Ye *et al.*, 2003). Constitutive expression of p19 with *Cauliflower mosaic virus* (CaMV) 35S promoter causes major developmental defects in *A. thaliana*, including flower abnormalities and leaf serration (Dunoyer *et al.*, 2004). The V2 VSP is encoded by the *Tomato yellow leaf curl virus* (TYLCV), a single-stranded DNA begomovirus. In the Israeli isolate of TYLCV, V2 is well characterized and mutation of it results in loss of viral infection in *Nicotiana benthamiana* and tomato (Wartig *et al.*, 1997). V2 is involved in systemic movement of the virus (Wartig *et al.*, 1997) and also allows overexpression of GFP in transient assays (Zrachya *et al.*, 2007). The direct interaction between V2 and SGS3 has been proposed to disrupt the RDR6-driven production of sRNA against TYLCV (Glick *et al.*, 2008). However, Fukunaga and Doudna (2009) reported that V2 shares RNA binding selectivity with SGS3 and that it outcompetes SGS3 in binding double-stranded RNA structures that have 5′ overhangs. Lastly, P0 proteins are encoded by poleroviruses and enamoviruses, which are in the *Luteoviridae* family of plant viruses. P0^{PE} is encoded by the first open reading frame of enamovirus *Pea enation mosaic virus-1* (PEMV-1). Fusaro *et al.* (2012) reported that P0^{PE} acts as an F-Box protein and that its constitutive expression in *A. thaliana* causes major developmental defects. Polerovirus P0 and enamovirus P0^{PE} proteins have been shown to inhibit production of secondary sRNA through destabilization of AGO1 (Bortolamiol *et al.*, 2007; Fusaro *et al.*, 2012). These three VSPs are encoded by phylogenetically unrelated viruses and have different modes of action, yet they all display the common feature of enhancing transgene expression in transient assay expression systems (Fusaro *et al.*, 2012; Incarbone and Dunoyer, 2013; Voinnet *et al.*, 2003; Zrachya *et al.*, 2007).

The trait that we sought to enhance by judicious use of a VSP was the production of arachidonic acid (AA), an ω-6 long chain polyunsaturated fatty acid (LCPUFA) required for brain development in infants and the precursor for a number of mammalian signalling compounds (Spychalla *et al.*, 1997). It is made of a 20-carbon chain with four double bonds and synthesized in some micro-organisms from linoleic acid by a 3-step pathway involving an elongase and two desaturases. Plants do not contain AA; however, introducing transgenes for these enzymes into *A. thaliana* has produced lines that accumulate up to 20% AA in their seed oil profiles (Petrie *et al.*, 2012). The promoters used in driving transgene expression in seeds are chosen from pathways that are exclusively active in seed protein and oil storage, a period of development distinct from early embryogenesis (Belmonte *et al.*, 2013; Le *et al.*, 2010). Here, we investigate whether VSP expression can be tolerated and can enhance transgene activity in plants during late seed development.

Results

Effects of expressing VSP on development and seed oil composition in *Arabidopsis thaliana*

To test whether seed-specific expression of a VSP gives localized enhanced transgene expression without causing developmental defects, the napin FP1 promoter from *Brassica napus* was investigated. FP1 has been reported to give high level of expression throughout the oil storage and protein synthesis stages of seed development but not during critical events of embryogenesis (Le *et al.*, 2010). First, a construct with GFP under the control of the FP1 promoter was made (Figure 1) and transformed into *A. thaliana*. Then the promoter was placed upstream of the coding region of one of three different VSPs (V2, p19 and P0^{PE}; Figure 1a) and these constructs transformed into *A. thaliana*. A constitutive promoter, CaMV 35S, was also used to drive V2. Visual observation under blue light, and Western blot analysis (Figure S1), of T2 FP1:GFP lines confirmed that the GFP expression was restricted to the seed; it was not observed in roots, cotyledons, true leaves or shoot apical meristems. The pFP1:P0^{PE} construct gave relatively few transgenic lines compared to the numbers obtained using pFP1:V2 and pFP1:p19 constructs; the FP1:P0^{PE} and FP1:p19 lines displayed altered cotyledon

phenotypes (Figure 1b). Despite this, the transgenic lines from all three constructs grew vigorously and were fertile.

The composition and levels of lipids in 50-seed samples from three independent T3 lines per genotype were quantified (Figure 1c & 1d). FP1:P0^{PE} and FP1:V2 lines had slightly reduced total lipid levels compared to wild-type Col-0, but these were no more reduced than in FP1:GFP. However, FP1:p19 seeds had lower total lipid levels than those in FP1:GFP plants, suggesting that p19 expression has an adverse effect on seed oil biosynthesis (Figure 1c). Examining the effects on the canonical *A. thaliana* seed oil pathway revealed that the oleic acid (18:1) levels in FP1:GFP, FP1:p19 and FP1:V2 were decreased by ~4%, whereas the linoleic acid (18:2) levels in FP1:GFP and FP1:V2 increased by ~6% and in FP1:P0^{PE} and FP1:p19 by ~4%. The relative percentage of linolenic acid (18:3) decreased by ~2% in all lines (Figure 1d).

Overall, the expression of the VSPs, in the absence of a transgenic oil modification pathway, caused little or no developmental defects and only minor changes in seed oil composition and content.

An intergenerational study of populations of transgenic *Arabidopsis thaliana* expressing AA and VSP in seed

To provide a three-transgene metabolic pathway, with which to evaluate the effects of VSP expression, the single AA pathway cassette of Petrie *et al.* (2012) was utilized. This contains the fatty acid biosynthesis genes *Isochrysis galbana* Δ9-elongase (*IgΔ9E*), *Pavlova salina* Δ8-desaturase (*PsΔ8D*) and *Pavlova salina* Δ5-desaturase (*PsΔ5D*) (Figure 2a). *IgΔ9E* is driven by the *A. thaliana* FAE1 promoter and *PsΔ8D* and *PsΔ5D* are both driven by the FP1 promoter (Figure 2b). The FP1:GFP:Nos was inserted to the original triple gene construct, then the cassette containing these and the NPTII gene was excised and inserted into pFP1:p19 and pFP1:V2 to generate pAA-p19 and pAA-V2, (Figure 2b). P0^{PE} was excluded in this part of the study due to the low number of transformants obtained earlier. The GFP:AA intermediate construct was retained, termed 'pNo-VSP' and subsequently used as control. All of these T-DNA binary constructs were transformed into *A. thaliana* genotype MC49, which is a *fad3/fae1* double mutant containing high levels of 18:2 in seed oil (Zhou *et al.*, 2006). The 18:2 fatty acid is the starting substrate for the introduced transgenic pathway.

Fifteen independent T1 lines (Table 1) were randomly selected, for each AA construct, from a large number of primary transformants, and followed for four generations. Segregation analysis was performed for T2 (Figure S2) and T3 generations. This analysis showed that the segregation pattern was relatively similar in all transgenic populations, that the vast majority of the VSP expressing lines are not multiloci insertion events (Figure S2), and that by T3 11/15 No-VSP, 13/15 AA-p19, and 11/15 AA-V2 lines were homozygous (Figure S3).

In each generation, the metabolic profiles of seed samples were measured for 15 individual progeny (Figures 3 and 4). The expression of either V2 or p19 improved the population median for AA in T5 and one elite event, AA-p19-26 (Figure S3), produced 39–41% AA in T3, T4 and T5 seeds. The V2 and p19 lines with high levels of AA, had low levels of the pathway intermediates, 20:2 and 20:3 (Figure 3). This is consistent with efficient metabolic flux of 18:2 through to AA and indicates that all three transgene-encoded enzymes were working efficiently. The poorly performing lines generally accumulated higher levels of 20:2, indicating that transgenic constraints were occurring at

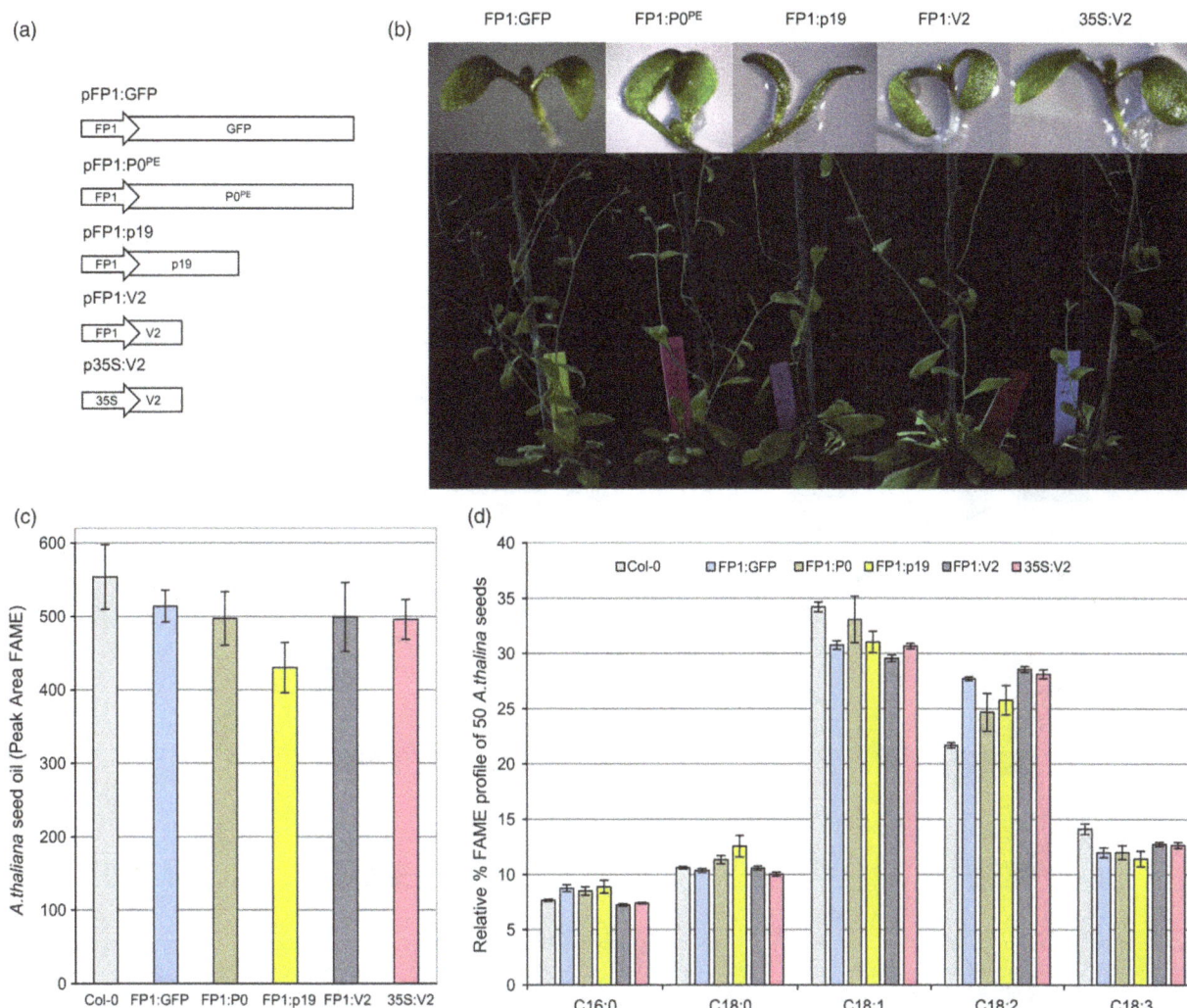

Figure 1 Phenotypic and oil profile of transgenic *Arabidopsis thaliana* lines expressing various VSPs. (a) Schematic of T-DNA binary plasmids designed for stable expression of the VSPs (P0PE p19 and V2) in *A. thaliana*. Genes driven by the *Brassica napus* truncated napin FP1 promoter and the CaMV 35S promoter as indicated. For comparison purposes, a control construct encoding GFP driven by the FP1 promoter was also designed. (b) The phenotypic changes in cotyledons of *A. thaliana* stably transformed with constructs outlined in (a). Representative mature plants expressing the respective constructs outlined in (a). Images presented are of stably transformed *A. thaliana* in the third generation.(c–d) Total oil and relative percentages of the major fatty acids extracted as fatty acid methyl esters (FAME) from seeds of Col-0 and T3 transgenic *A. thaliana*. Data presented are from 3 independent events for each plasmid described in Figure 1a. Error bars are the standard error of the mean.

the first desaturation step, IgΔ8D. For all VSP-transgenic populations the overall trend was for an increase in the median level of AA production over generations 2, 3 and 4. The No-VSP lines stabilized for AA production in the T4 and T5 generations, whereas AA-p19 or AA-V2 events continued to improve in overall production of AA in the T5 generation (Figure 3).

Segregation analysis shows that by T3 most of the lines are homozygous for the transgene cassette and strongly suggests that the jump of AA level between T2 and T3 is due to the lines transitioning from the hemizygous to homozygous state. From T3 onwards the AA levels of most No-VSP lines decline, indicative of transgene silencing whereas most of the AA-p19 and AA-V2 lines maintain their elevated levels.

All data for each transgenic population at the T5 seed stage is shown in Figure 4. The T5 transgenic populations of AA-V2 and AA-p19 show a median of 15% and 16% AA, respectively, whereas the No-VSP population shows a median of ~11% AA. The data for the populations also revealed that the first and third

quartiles were more compact for the AA-V2 or AA-p19 events, compared to a wider spread in No-VSP events.

The effect of high levels of AA on total seed oil and seed germination and seedling vigour

Various phenotypic characteristics were measured in seeds containing varying amounts of AA. In early generations, it was observed that seeds with high levels of AA were smaller than wild-type MC49 seeds. This was investigated further by quantifying the seed oil content of independent events with levels of AA ranging from 0 to 38% as a percentage of total fatty acids (Figure 5). The nontransgenic MC49, FP1:GFP expressing MC49 and AA-V2-03 (null segregant) seeds contained 33%, 32% and 34% oil, respectively, indicating that the transgenesis process had not affected the seed oil profiles. Seeds containing up to ~15% AA had negligible reductions in seed oil content, however at higher levels of AA, the oil content was reduced (Figure 5a). This trade-off between AA and oil content was most pronounced in a

(a)   **ω-6 pathway**

18:2$^{\Delta 9,12}$ linoleic acid (LA)

↓ IgΔ9E, elongase

20:2$^{\Delta 11,14}$ eicosadienoic acid (EDA)

↓ PsΔ8D, desaturase

20:3$^{\Delta 8,11,14}$ dihomo-γ-linolenic acid (DGLA)

↓ PsΔ5D, desaturase

20:4$^{\Delta 5,8,11,14}$ arachidonic acid (AA)

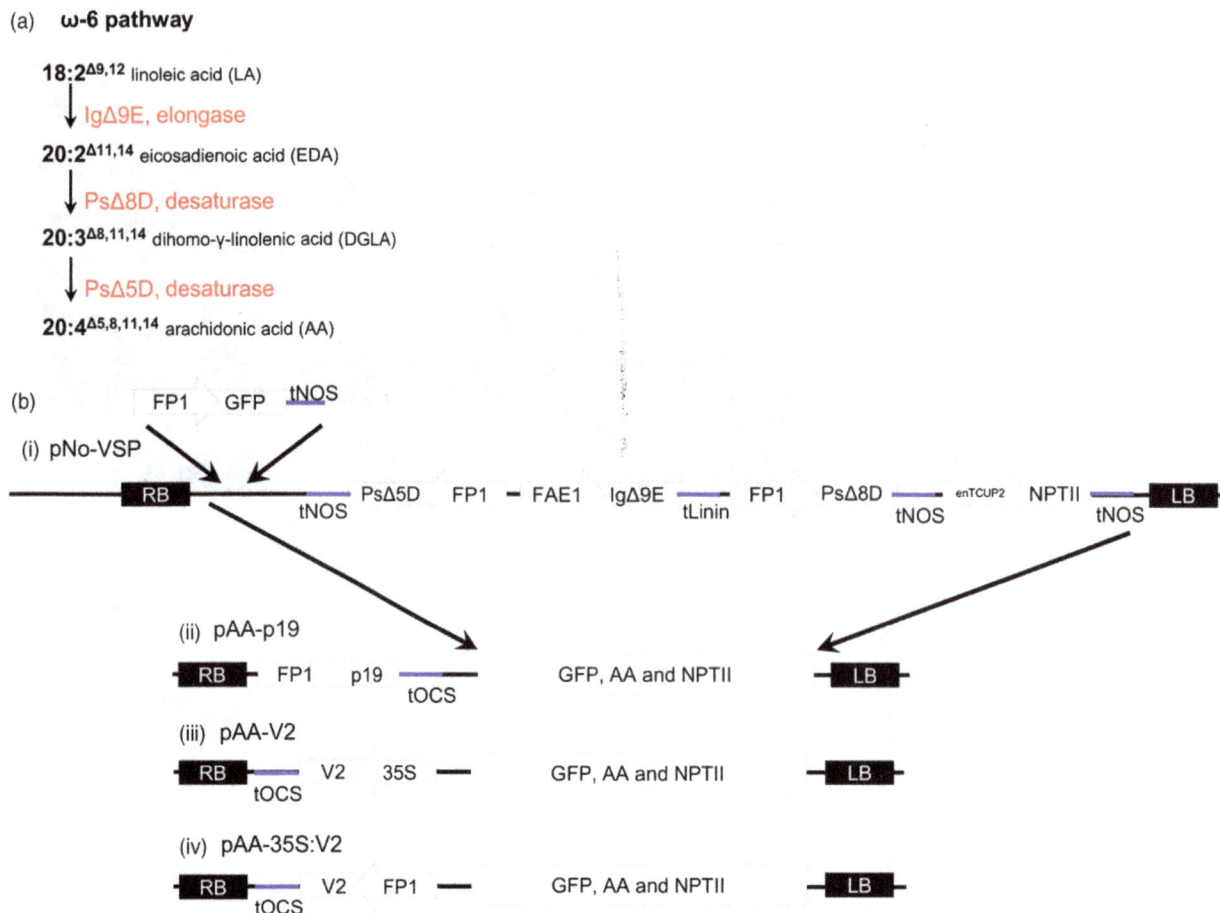

Figure 2 Design of constructs containing a three-gene pathway for biosynthesis of AA (20:4) in seed oil. (a) A schematic of enzymes involved in the biosynthesis of 20:4 via the ω-6 pathway using 18:2 (LA) as the starting substrate. Abbreviations are as follows: IgΔ9E, *Isochrysis galbana* Δ9-elongase; PsΔ8D, *Pavlova salina* Δ8-desaturase; PsΔ5D, *Pavlova salina* Δ5-desaturase. (b) A schematic of the construct maps designed for stable transformation of *Arabidopsis thaliana*. (i) The construct pNo-VSP consisted of the *IgΔ9E* driven by the *A. thaliana* FAE1 promoter, *PsΔ8D* and *PsΔ5D* and *GFP* driven by the FP1 promoter. (ii–iv) Addition of a VSP to pNo-VSP generated pAA-p19, pAA-V2 and pAA-35S:V2. Abbreviations are as follows: RB, right border; tOCS, *Agrobacterium tumefaciens* octopine synthase terminator; 35S, *Cauliflower mosaic virus* 35S promoter; GFP, green fluorescent protein reporter; tNOS, *A. tumefaciens* nopaline synthase terminator (polyadenylation signal); tLinin, *Linum usitatissimum* conlinin2 terminator; enTCUP2, enhanced tobacco cryptic promoter 2; NPTII, neomycin phosphotransferase II selectable marker; LB, left border.

Table 1 Summary of the transgenic *Arabidopsis thaliana* population generated with the AA constructs

Plasmid	Independent T1 events	Events taken to T5
pNo-VSP	24	15
pAA-p19	39	15
pAA-V2	26	15
pAA-35S:V2	8	5

The table summarizes the generation of independent transgenic events for each plasmid. *Arabidopsis thaliana* was transformed with these plasmids and a large number of independent events were generated. Each data point for parent and descendent was preserved and graphed, generating box-whisker plots in Figure 3 and progeny plots in Figure S3.

line (AA-p19-26) containing ~38% AA that resulted in a 70% reduction in oil content (Figure 5a).

Slower germination, reduced seedling vigour and small seed phenotypes were observed in a number of AA lines, independent of the presence or absence of a VSP but correlated with LCPUFA

content. The seedlings of wild-type MC49 (0% AA) and AA-p19-13 (11% AA) developed normally, however AA-p19-26 (38%) grew more slowly (Figure 5b) and produced seed with 50% viability. This observation was recorded in all later generations as well (data not presented).

The effect of constitutive expression of V2 on plant development and AA pathway

Transgenic plants constitutively expressing VSPs, including p19, and P0, show severe development abnormalities (Dunoyer *et al.*, 2004; Fusaro *et al.*, 2012) but there have been no reports of plants with constitutive transgenic expression of V2. We have shown previously that V2 expression has the rare attribute of simultaneously allowing hairpin RNA-directed silencing and enhancing overexpression of transgenes in transient Agro-infiltration assays of *Nicotiana benthamiana* leaves (Naim *et al.*, 2012). Therefore, we investigated the effect of constitutive expression of V2 on plant development and on the transgenic oil modification pathway. Wild-type Col-0 and MC49 *A. thaliana* were transformed with p35S:V2 and pAA-35S:V2, respectively

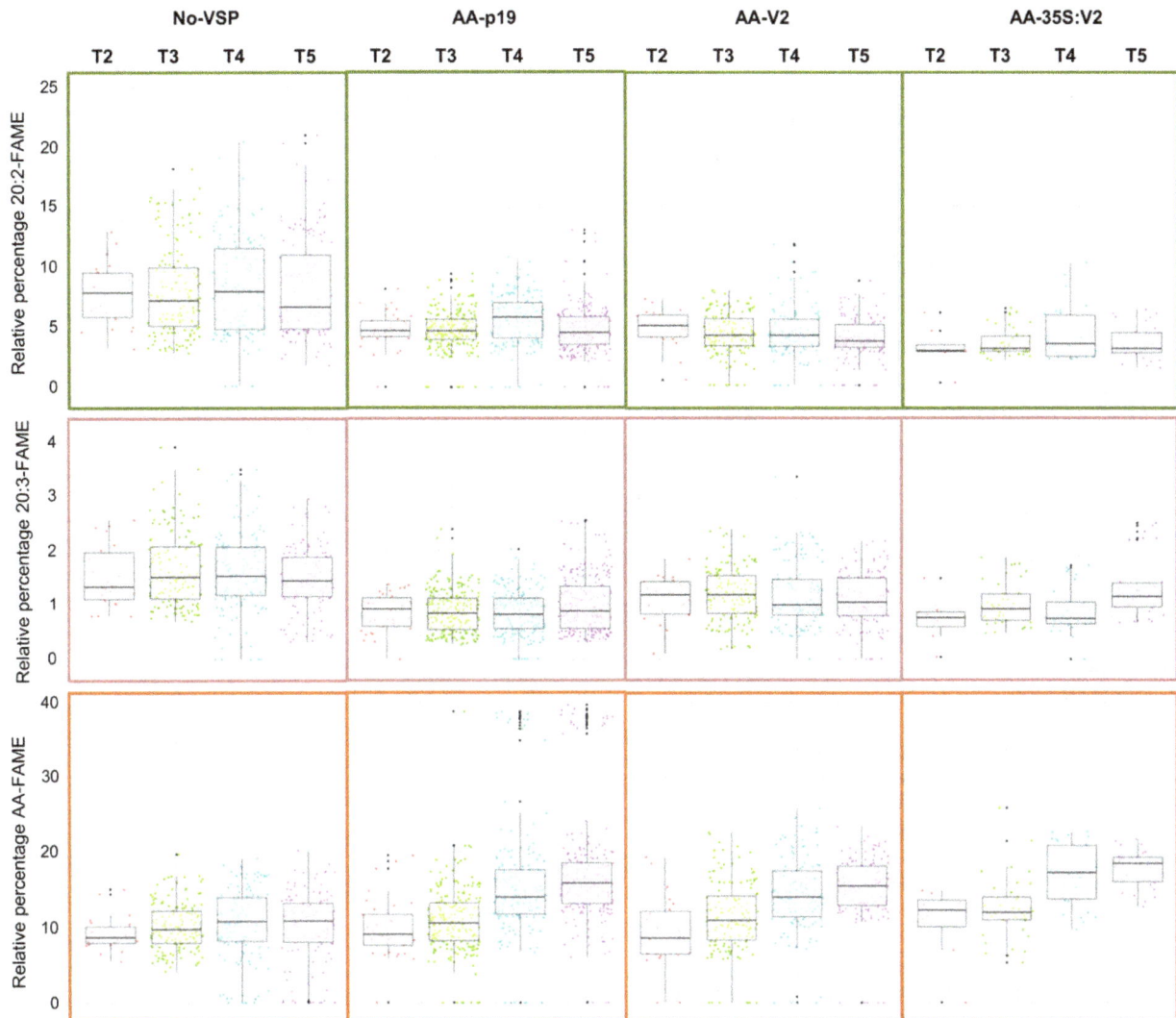

Figure 3 Relative percentages of 20:2-FAME, 20:3-FAME and AA-FAME extracted from T2–T5 seeds of transgenic *A. thaliana*. Box-whisker plot analysis of a large population of transgenic *A. thaliana* expressing the constructs described in Figure 2. Relative percentages of 20:2, 20:3 and AA shown and dot points refer to independent transformation events in T2 and their respective progenies in T3–T5. Points outside of the whiskers are not included in the calculation of median or quartiles.

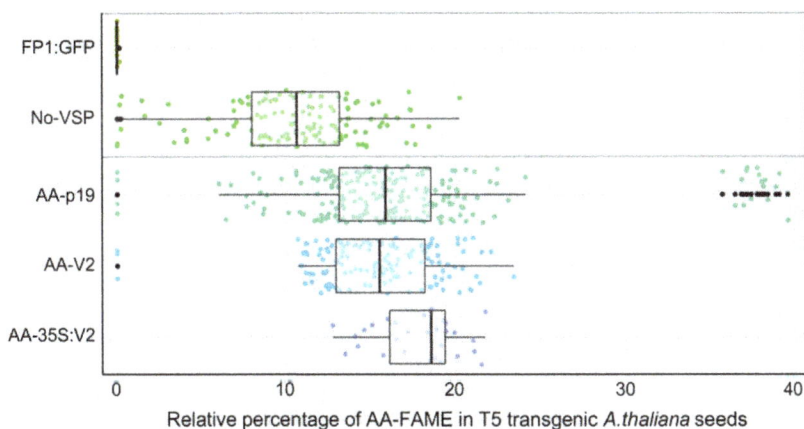

Figure 4 Relative percentages of AA-FAME extracted from T5 seeds of transgenic *Arabidopsis thaliana*. Box-whisker plot analysis of a T5 population of transgenic *A. thaliana* expressing the constructs described in Figure 2. Relative percentage of AA measured in T5 seeds and each line represented as a dot point. FP1:GFP is the negative control of MC49 transformed with pFP1: GFP plasmid.

(Figures 1b and 2b) and transgenic populations followed for four generations. The 35S:V2 lines had no obvious phenotypic abnormality or changes in oil lipid profile (Figure 1b–d) and the AA-35S:V2 population produced AA at ~19% in the T5 seed (Figures 3 and 4). Although fewer lines were obtained with pAA-35S:V2 (Table 1).

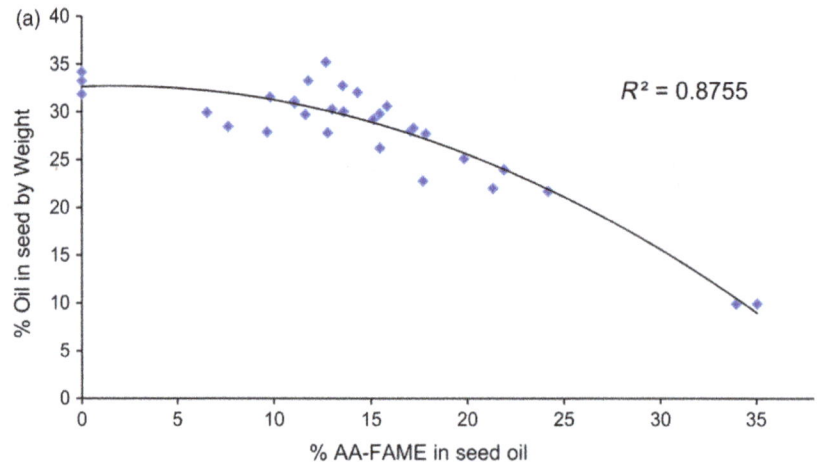

(a) $R^2 = 0.8755$

% Oil in seed by Weight

% AA-FAME in seed oil

(b)

Figure 5 The effect of high levels of AA on the accumulation of oil (TAG) in the seed and seedling vigour. (a) Total oil extracted from T4 transgenic *A. thaliana* lines expressing pAA-p19. Independent lines analysed were selected based on the varied levels of AA to determine a relationship between seed oil and relative percentage of AA. Oil also extracted from seeds of wild-type MC49 and MC49 expressing FP1:GFP as controls. (b) The phenotypic comparison between AA-p19-26, AA-p19-13 and MC49 seedlings. The relative sizes and leaf shapes of AA-p19-26 seedlings (~38% AA) compared to AA-p19-13 (11% AA) and MC49 seedlings. Seedlings photographed 3 weeks after imbibition.

Discussion

The ubiquitous expression of VSPs (e.g. driven by the CaMV 35S promoter) usually leads to abnormal plant development and embryo lethality (Dunoyer et al., 2004; Fusaro et al., 2012) due to their effects on endogenous miRNA-regulated functions. To avoid this, but capture VSP enhancement of transgene expression, a previous study has developed a mutated form of p19 that retains some silencing-suppressor activity yet does not affect plant development (Saxena et al., 2011). Our alternative approach has been to investigate the use of fully functional VSPs expressed only in tissues that are important for seed oil production but not development. Using the FP1 promoter to drive p19 expression greatly enhanced the efficacy and longevity of the oil biosynthesis transgene cassette, as measured by the production of AA. Its seed-specific expression pattern almost completely avoided the developmental problems encountered when constitutively expressed (Dunoyer et al., 2004). We conclude that the FP1 promoter is active in the cotyledon tissue, causing both the modified seed oil enhancement and the unusual cotyledon phenotype of seedlings, but not in the meristem, thus avoiding disruption of sRNA driven regulatory pathways in the shoot apex.

The seed-specific expression of V2 or p19 not only improved the transgenic biosynthesis of AA but also maintained this effect for four generations. A possible mechanism for the initial VSP enhancement might have been by favouring the selection of multiple transgene insertion lines (which may otherwise have been rapidly silenced), especially as *A. thaliana* transformation is

by floral dip and selection at the germinating seed stage. However, the range in transgene locus number, as judged from segregation ratios, was very similar for VSP and no-VSP lines (Figure S2). This suggests that the VSP is having an enhancing effect through prevention of silencing of predominantly single loci rather than by increasing the number of transgene loci per plant.

The continued increase in AA content from T2 to T5 in the populations of VSP lines, but lack of increase in the no-VSP lines, seems to result from the plants' responses to transgene homozygosity. In the No-VSP lines, the transgenes becoming homozygous trigger their own silencing, resulting in a reduction of AA biosynthesis, whereas the gains in transgene expression from homozygosity are protected in VSP plants. As the population of a VSP line becomes 100% homozygous, the cumulative effect of doubling transgene loci is reflected in increasing AA content.

Interestingly, the best performing lines had low levels of intermediates whereas the poor performing lines showed accumulation of 20:2. This suggests that the PsΔ8D transgene is most sensitive to silencing or is the bottleneck step of the pathway. While fatty acid elongation reactions are generally considered to occur in the CoA pool (Fehling and Mukherjee, 1991), desaturation reactions, such as PsΔ8D and PsΔ5D, are typically considered to operate on the phosphatidylcholine (PC) pool. In metabolic pathways that are sequential, such as for the conversion of 18:2 to AA, there is an implied movement of acyl chains between different lipid head groups in subcellular compartments. In *A. thaliana* the LCPUFA intermediates may be preferentially channelled towards AA. At the upper most extreme of metabolic capacity recorded in this study, one event was capable of ~40%

AA, a level considerably higher than previously reported (Petrie et al., 2012). However, high levels of LCPUFA in A. thaliana resulted in reduced oil content and reduced seedling vigour. The production of exotic fatty acids in seeds often results in compromised seed oil content (van Erp et al., 2011; Nykiforuk et al., 2012) most likely due to the inefficient flux of new fatty acids from their site of synthesis into storage oils and altered properties of the membranes in the developing seed (Thomæus et al., 2001). Previous reports show that the production of a novel fatty acid, ricinoleic acid, in A. thaliana resulted in poor oil content (van Erp et al., 2011). It has been demonstrated that the reduced oil content could be recovered with the inclusion of more specialized oil accumulation enzymes specific for transfer of ricinoleic acid into seed oil (van Erp et al., 2015). Similar strategies could be employed to recover the total oil content in plants generated in this study that produce very high levels of AA.

The seed-specific expression of V2 or p19 could easily be combined with transgenic pathways to produce high levels of other biocommodities, especially those that do not have an adverse effect on seed germination. Bio-products such as recombinant antibodies and vaccines have been produced in leaf, tuber, and seeds (Bortesi et al., 2009; Magnusdottir et al., 2013; Schünmann et al., 2002) at sufficient levels for commercial production. Using our approach of expressing the bio-product transgene and a VSP in seeds may significantly further enhance the bio-product yield. Moreover, because V2 can be expressed without side effects in the whole plant this has the potential to enhance the other nonseed transgenic expression systems. A great concern when taking a transgenic trait into the field for commercial production is that the trait will be lost over subsequent generations. It takes at least 7–10 generations from production of the primary transformant to the first generation for commercial sale. Our results suggest that inclusion of VSP expression is preventing a decline in gene expression over generations. Such surety of performance is likely to be highly prized in agribusiness.

Experimental procedures

Molecular cloning – Preparation of constructs

Each VSP was cloned under the expression control of the FP1 (Stalberg et al., 1993) and OCS3' transcription termination/polyadenylation region with the exception of 35S:V2 where the V2 coding region was under the control of the CaMV 35S promoter. The fatty acid biosynthesis genes IgΔ9E, PsΔ8D and PsΔ5D and the bacterial selection marker were obtained on a single DNA fragment from pJP3010 (Petrie et al., 2012) by digestion with PmeI and AvrII giving rise to a 9560 bp fragment. The three fatty acid biosynthesis genes on this fragment were oriented and spaced in the same manner as in pJP107 (Petrie et al., 2012). Binary plasmids of pFP1:p19, pFP1:V2 and p35S:V2 were digested with SwaI and AvrII. The 9560 bp of pJP3010 was ligated to the ~8000 bp fragment of the VSP containing plasmids (T4 DNA Ligase – Promega, Sydney, NSW, Australia) and electroporated into DH5α. This gave rise to three plasmids namely pAA-p19, pAA-35:V2 and pAA-V2. The pAA-p19 plasmid was digested with AhdI to drop out the FP1:p19 cassette. The backbone was re-cirularized to form the pNo-VSP plasmid. Plasmids comprised an NPTII selectable marker gene within the T-DNA and adjacent to the left border and an RK2 origin of replication for maintenance of the plasmids in Agrobacterium tumefaciens. The newly prepared vectors were electroporated

into GV3101 strain of A. tumefaciens for stable transformation of A. thaliana.

Transformation of Arabidopsis thaliana with the VSP and AA constructs

To transform A. thaliana, plants were treated by the floral dip method (Clough and Bent, 1998). The treated plants were grown to maturity and T1 seeds harvested from them were plated on MS media (Murashige and Skoog, 1962) containing kanamycin. Screening for GFP expression in the seed was also used as a visual marker for selection of T1 seeds. The seedlings that survived on MS media containing 50 mg/L of kanamycin or which were obtained from GFP-positive seeds were transferred to soil and grown to maturity for T2 seeds.

Segregation analysis and plant growth conditions

Segregation analysis was also carried out for each individual event by growing 60-70 seeds on MS media containing kanamycin (Figure S2). Six to ten positive seedlings from each event were transferred to soil and grown to maturity. Plants were grown with an 8/16 h (light/dark) photoperiod in 22 °C. The A. thaliana plants generated during this study were grown in the same PC2 growth room with unchanged conditions throughout the two-year period.

Lipid analyses

Approximately 100 pooled seeds of transgenic A. thaliana were taken from each transformed plant for the determination of seed fatty acid composition. FAME (fatty acid methyl esters) was prepared with 750 μL of MeOH:HCl:DCM (10:1:1) containing 2 μg of internal standard (17:0 free fatty acid; NU-CHECK PREP, Inc.). Samples were heated at 80 °C for 2 h. After cooling, 300 μL of MQ water and 500 μL of Hexane:DCM (4:1) were added to each sample, which was then vortexed and layers allowed to separate. In most cases, it was not necessary to centrifuge. The top organic layer was extracted into a clean vial and evaporated under N_2 gas, and FAME was dissolved in 200 μL of hexane. Gas chromatography analysis performed as described previously (Wood et al., 2009). Each run included a freshly diluted 411 standard (Nu-Check Prep Inc, Elysian, MN, USA) used for normalization of FAME profiles and fatty acid composition analysed as described previously (Wood et al., 2009).

For the quantification of oil, seeds were cleaned to remove debris and desiccated for 24 h. Approximately 5 mg of seeds was weighed using a precision balance (Mettler Toledo) and transferred to 2 mL glass vial. The addition of a known amount of internal standard (heptadecanoin; Nu-Check Prep Inc, Elysian, MN, USA) and seed mass were used to calculate amount of oil/mg of seed. FAME was prepared by incubating the seed samples in 0.7 mL of 1N Methanolic-HCl (Supelco) for 2 h at 80 °C. After incubation, the vials were cooled to room temperature followed by addition of 0.35 mL of 0.9% NaCl and 0.1 mL hexane and mixed in a shaker (Vibramax) for 10 min. The vials were centrifuged at 1700 g for 5 min and the upper phase containing FAME was transferred to a new vial and analysed by GC. Oil content (TAG) in the seeds was calculated as the sum of glycerol- and fatty acyl moieties using a relation:%TAG by weight = 100 × ((41 × total mol FAME / 3) + (total g FAME − (15 × total mol FAME))) / g seed dry weight, where 41 and 15 are molecular weights of glycerol moiety and methyl group, respectively.

FW10-23 is the elite T3 control line generated by Petrie *et al.* (2012). Therefore, to monitor deviations in sample preparation and the sensitivity of analytical instruments, FAME was extracted from seeds of FW10-23 and analysed in every batch of transgenic *A. thaliana* during the two-year period. Statistical analyses were performed on the FAME profile of FW10-23 collected during data collection period and there were no statistically significant differences.

Western blot analysis

Total proteins were separated on a 12% SDS-PAGE gel, transferred to Immobilon P membrane (Millipore, Bedford, MA) and detected by chemiluminescence. GFP was detected with anti-GFP monoclonal antibody (Clontech, 1:5000, Clayton, Victoria, Australia) followed by anti-mouse Ig HRP conjugate (Promega, 1:5000).

Acknowledgements

The authors thank CSIRO OCE scholarship fund for support of this PhD project, Dr James Petrie for providing the published AA transgenic line and construct, Drs Srinivas Belide and Xue-Rong Zhou for technical advice, Cheryl Blundell, Anne Mackenzie, Lucy Yuan, Luch Hac for excellent technical support and Dr Alex Whan for generation of progeny plots. The authors declare that there are no conflicts of interest.

References

Belmonte, M.F., Kirkbride, R.C., Stone, S.L., Pelletier, J.M., Bui, A.Q., Yeung, E.C., Hashimoto, M. et al. (2013) Comprehensive developmental profiles of gene activity in regions and subregions of the Arabidopsis seed. *Proc. Natl Acad. Sci. USA*, **110**, E435–E444.

Bortesi, L., Rossato, M., Schuster, F., Raven, N., Stadlmann, J., Avesani, L., Falorni, A. et al. (2009) Viral and murine interleukin-10 are correctly processed and retain their biological activity when produced in tobacco. *BMC Biotechnol.* **9**, 1–13.

Bortolamiol, D., Pazhouhandeh, M., Marrocco, K., Genschik, P. and Ziegler-Graff, V. (2007) The Polerovirus F box protein P0 targets ARGONAUTE1 to suppress RNA silencing. *Curr. Biol.* **17**, 1615–1621.

Cahoon, E.B. and Kinney, A.J. (2005) The production of vegetable oils with novel properties: Using genomic tools to probe and manipulate plant fatty acid metabolism. *Eur. J. Lipid Sci. Technol.* **107**, 239–243.

Clough, S. and Bent, A. (1998) Floral dip: a simplified method for Agrobacterium-mediated transformation of Arabidopsis thaliana. *Plant J.* **16**, 735–743.

Ding, S.W. and Voinnet, O. (2007) Antiviral immunity directed by small RNAs. *Cell*, **130**, 413–426.

Dunoyer, P., Lecellier, C.-H., Parizotto, E.A., Himber, C. and Voinnet, O. (2004) Probing the microRNA and small interfering RNA pathways with virus-encoded suppressors of RNA silencing. *The Plant Cell Online*, **16**, 1235–1250.

van Erp, H., Bates, P.D., Burgal, J., Shockey, J. and Browse, J. (2011) Castor phospholipid:diacylglycerol acyltransferase facilitates efficient metabolism of hydroxy fatty acids in transgenic Arabidopsis. *Plant Physiol.* **155**, 683–693.

van Erp, H., Shockey, J., Zhang, M., Adhikari, N.D. and Browse, J. (2015) Reducing isozyme competition increases target fatty acid accumulation in seed triacylglycerols of transgenic Arabidopsis. *Plant Physiol.* **168**, 36–46.

Fehling, E. and Mukherjee, K.D. (1991) Acyl-CoA elongase from a higher plant (Lunaria annua): metabolic intermediates of very-long-chain acyl-CoA products and substrate specificity. *Biochim. Biophys. Acta*, **1082**, 239–246.

Fukunaga, R. and Doudna, J.A. (2009) dsRNA with 5′ overhangs contributes to endogenous and antiviral RNA silencing pathways in plants. *EMBO J.* **28**, 545–555.

Fusaro, A.F., Correa, R.L., Nakasugi, K., Jackson, C., Kawchuk, L., Vaslin, M.F.S. and Waterhouse, P.M. (2012) The Enamovirus P0 protein is a silencing

suppressor which inhibits local and systemic RNA silencing through AGO1 degradation. *Virology*, **426**, 178–187.

Glick, E., Zrachya, A., Levy, Y., Mett, A., Gidoni, D., Belausov, E., Citovsky, V. et al. (2008) Interaction with host SGS3 is required for suppression of RNA silencing by tomato yellow leaf curl virus V2 protein. *Proc. Natl Acad. Sci. USA*, **105**, 157–161.

Hagan, N.D., Spencer, D., Moore, A.E. and Higgins, T.J.V. (2003) Changes in methylation during progressive transcriptional silencing in transgenic subterranean clover. *Plant Biotechnol. J.* **1**, 479–490.

Incarbone, M. and Dunoyer, P. (2013) RNA silencing and its suppression: novel insights from in planta analyses. *Trends Plant Sci.* **18**, 382–392.

Lakatos, L., Szittya, G., Silhavy, D. and Burgyan, J. (2004) Molecular mechanism of RNA silencing suppression mediated by p19 protein of tombusviruses. *EMBO J.* **23**, 876–884.

Le, B.H., Cheng, C., Bui, A.Q., Wagmaister, J.A., Henry, K.F., Pelletier, J., Kwong, L. et al. (2010) Global analysis of gene activity during Arabidopsis seed development and identification of seed-specific transcription factors. *Proc. Natl Acad. Sci. USA*, **107**, 8063–8070.

Lindbo, J.A., Silva-Rosales, L., Proebsting, W.M. and Dougherty, W.G. (1993) Induction of a highly specific antiviral state in transgenic plants: implications for regulation of gene expression and virus resistance. *Plant Cell*, **5**, 1749–1759.

Magnusdottir, A., Vidarsson, H., Björnsson, J.M. and Örvar, B.L. (2013) Barley grains for the production of endotoxin-free growth factors. *Trends Biotechnol.* **31**, 572–580.

Murashige, T. and Skoog, F. (1962) A revised medium for rapid growth and bio assays with Tobacco tissue cultures. *Physiol. Plant.* **15**, 473–497.

Naim, F., Nakasugi, K., Crowhurst, R.N., Hilario, E., Zwart, A.B., Hellens, R.P., Taylor, J.M. et al. (2012) Advanced engineering of lipid metabolism in Nicotiana benthamiana using a draft genome and the V2 viral silencing-suppressor protein. *PLoS ONE*, **7**, e52717.

Nykiforuk, C., Shewmaker, C., Harry, I., Yurchenko, O., Zhang, M., Reed, C., Oinam, G. et al. (2012) High level accumulation of gamma linolenic acid (C18:3Δ6.9,12 cis) in transgenic safflower (Carthamus tinctorius) seeds. *Transgenic Res.* **21**, 367–381.

Petrie, J.R., Shrestha, P., Belide, S., Mansour, M.P., Liu, Q., Horne, J., Nichols, P.D. et al. (2012) Transgenic production of arachidonic acid in oilseeds. *Transgenic Res.* **21**, 139–147.

Saunders, K. and Lomonossoff, G.P. (2013) Exploiting plant virus-derived components to achieve in planta expression and for templates for synthetic biology applications. *New Phytol.* **200**, 16–26.

Saxena, P., Hsieh, Y.-C., Alvarado, V.Y., Sainsbury, F., Saunders, K., Lomonossoff, G.P. and Scholthof, H.B. (2011) Improved foreign gene expression in plants using a virus-encoded suppressor of RNA silencing modified to be developmentally harmless. *Plant Biotechnol. J.* **9**, 703–712.

Schubert, D., Lechtenberg, B., Forsbach, A., Gils, M., Bahadur, S. and Schmidt, R. (2004) Silencing in Arabidopsis T-DNA transformants: the predominant role of a gene-specific RNA sensing mechanism versus position effects. *Plant Cell*, **16**, 2561–2572.

Schünmann, P.D., Coia, G. and Waterhouse, P. (2002) Biopharming the SimpliRED™ HIV diagnostic reagent in barley, potato and tobacco. *Mol. Breeding*, **9**, 113–121.

Silhavy, D., Molnar, A., Lucioli, A., Szittya, G., Hornyik, C., Tavazza, M. and Burgyan, J. (2002) A viral protein suppresses RNA silencing and binds silencing-generated, 21- to 25-nucleotide double-stranded RNAs. *EMBO J.* **21**, 3070–3080.

Spychalla, J.P., Kinney, A.J. and Browse, J. (1997) Identification of an animal omega-3 fatty acid desaturase by heterologous expression in Arabidopsis. *Proc. Natl Acad. Sci. USA*, **94**, 1142–1147.

Stalberg, K., Ellerstrom, M., Josefsson, L.G. and Rask, L. (1993) Deletion analysis of a 2S seed storage protein promoter of Brassica napus in transgenic tobacco. *Plant Mol. Biol.* **23**, 671–683.

Thomæus, S., Carlsson, A.S. and Stymne, S. (2001) Distribution of fatty acids in polar and neutral lipids during seed development in Arabidopsis thaliana genetically engineered to produce acetylenic, epoxy and hydroxy fatty acids. *Plant Sci.* **161**, 997–1003.

Vanhercke, T., Wood, C.C., Stymne, S., Singh, S.P. and Green, A.G. (2013) Metabolic engineering of plant oils and waxes for use as industrial feedstocks. *Plant Biotechnol. J.* **11**, 197–210.

Vargason, J.M., Szittya, G., Burgyán, J. and Hall, T.M.T. (2003) Size selective recognition of siRNA by an RNA silencing suppressor. *Cell*, **115**, 799–811.

Voinnet, O., Rivas, S., Mestre, P. and Baulcombe, D. (2003) An enhanced transient expression system in plants based on suppression of gene silencing by the p19 protein of tomato bushy stunt virus. *Plant J.* **33**, 949–956.

Wartig, L., Kheyr-Pour, A., Noris, E., De Kouchkovsky, F., Jouanneau, F., Gronenborn, B. and Jupin, I. (1997) Genetic analysis of the monopartite Tomato yellow leaf curl geminivirus: Roles of V1, V2, and C2 ORFs in viral pathogenesis. *Virology*, **228**, 132–140.

Wood, C.C., Petrie, J.R., Shrestha, P., Mansour, M.P., Nichols, P.D., Green, A.G. and Singh, S.P. (2009) A leaf-based assay using interchangeable design principles to rapidly assemble multistep recombinant pathways. *Plant Biotechnol. J.* **7**, 914–924.

Ye, K., Malinina, L. and Patel, D.J. (2003) Recognition of small interfering RNA by a viral suppressor of RNA silencing. *Nature*, **426**, 874–878.

Zhou, X., Singh, S., Liu, Q. and Green, A. (2006) Combined transgenic expression of Δ12-desaturase and Δ12-epoxygenase in high linoleic acid seeds leads to increased accumulation of vernolic acid. *Funct. Plant Biol.* **33**, 585–592.

Zrachya, A., Glick, E., Levy, Y., Arazi, T., Citovsky, V. and Gafni, Y. (2007) Suppressor of RNA silencing encoded by Tomato yellow leaf curl virus-Israel. *Virology*, **358**, 159–165.

10

Leaf proteome rebalancing in *Nicotiana benthamiana* for upstream enrichment of a transiently expressed recombinant protein

Stéphanie Robert[1], Marie-Claire Goulet[1], Marc-André D'Aoust[2], Frank Sainsbury[1,3] and Dominique Michaud[1,*]

[1]*Centre de recherche et d'innovation sur les végétaux, Pavillon Envirotron, Université Laval, Québec, QC, Canada*
[2]*Medicago Inc., Québec, QC, Canada*
[3]*The University of Queensland, Australian Institute for Bioengineering and Nanotechnology, St Lucia, QLD, Australia*

*Correspondence

email dominique.michaud@fsaa.ulaval.ca

Keywords: agroinfiltration, methyl jasmonate, *Nicotiana benthamiana*, leaf proteome, molecular farming, recombinant protein yield, transient expression.

Summary

A key factor influencing the yield of biopharmaceuticals in plants is the ratio of recombinant to host proteins in crude extracts. Postextraction procedures have been devised to enrich recombinant proteins before purification. Here, we assessed the potential of methyl jasmonate (MeJA) as a generic trigger of recombinant protein enrichment in *Nicotiana benthamiana* leaves before harvesting. Previous studies have reported a significant rebalancing of the leaf proteome via the jasmonate signalling pathway, associated with ribulose 1,5-bisphosphate carboxylase oxygenase (RuBisCO) depletion and the up-regulation of stress-related proteins. As expected, leaf proteome alterations were observed 7 days post-MeJA treatment, associated with lowered RuBisCO pools and the induction of stress-inducible proteins such as protease inhibitors, thionins and chitinases. Leaf infiltration with the *Agrobacterium tumefaciens* bacterial vector 24 h post-MeJA treatment induced a strong accumulation of pathogenesis-related proteins after 6 days, along with a near-complete reversal of MeJA-mediated stress protein up-regulation. RuBisCO pools were partly restored upon infiltration, but most of the depletion effect observed in noninfiltrated plants was maintained over six more days, to give crude protein samples with 50% less RuBisCO than untreated tissue. These changes were associated with net levels reaching 425 µg/g leaf tissue for the blood-typing monoclonal antibody C5-1 expressed in MeJA-treated leaves, compared to less than 200 µg/g in untreated leaves. Our data confirm overall the ability of MeJA to trigger RuBisCO depletion and recombinant protein enrichment in *N. benthamiana* leaves, estimated here for C5-1 at more than 2-fold relative to host proteins.

Introduction

The potential of plants as bio-factories for clinically useful recombinant proteins has been confirmed in recent years with the approval of a first plant-made biopharmaceutical for human therapy, the demonstrated feasibility of readily producing therapeutic antibodies in plants during the recent Ebola virus outbreak, and an increasing number of plant-made proteins undergoing Phase I or Phase II clinical trials (Maxmen, 2012; Merlin *et al.*, 2014; Sack *et al.*, 2015). Plants show several advantages for recombinant protein expression, including the ability to perform mammalian-like post-translational maturation of protein backbones, low infrastructure costs compared to classical systems based on industrial-scale fermenters and no inherent biosafety issues regarding product contamination with microbial toxins or human pathogens (Fischer *et al.*, 2013; Stoger *et al.*, 2014). Much progress has been made over the past decade to decipher basic processes underlying transgene expression in plants, and to develop molecular tools for high-yield protein expression in different plant platforms (Makhzoum *et al.*, 2014; Streatfield, 2007). Significant advances have also been made to understand the plant cellular machinery driving protein biosynthesis, maturation and protease processing, useful in designing strategies to modulate these processes and introducing novel cellular functions for recombinant protein design *in planta* (Benchabane *et al.*, 2008; Faye *et al.*, 2005; Gomord *et al.*, 2010; Pillay *et al.*, 2014).

Current research efforts to further establish the potential of plants as suitable protein production hosts include the elaboration of predictive models linking culture conditions and host plant growth parameters with protein expression rate, recovery operations and net protein yield (Buyel and Fischer, 2012; Robert *et al.*, 2013). Current efforts also include the development of customized procedures for an efficient primary recovery and purification of the expressed biopharmaceuticals (reviewed in Wilken and Nikolov, 2012). Plant hosts contain numerous contaminants, including fibres, oils, proteases, phenolics and highly abundant proteins, that must be neutralized or removed from crude samples upon tissue disruption to avoid recombinant protein loss, denaturation or proteolytic processing (Benchabane *et al.*, 2008; Menkhaus *et al.*, 2004). The downstream processing of recombinant proteins, from tissue disruption to the isolation of a suitable purified product, may account for as much as 80% of total production costs and eventually represent a key determinant in the commercial success or failure of any plant-made protein intended for clinical use (Fischer *et al.*, 2012).

An important factor influencing the overall efficiency of downstream processing is the ratio of recombinant to host (native) proteins, which strongly impacts purification yield and contamination of the final product with host proteins or their proteolytic fragments (Wilken and Nikolov, 2012). Significant progress has been made towards the prevention of protein degradation in crude extracts or culture media (Baur et al., 2005; Benchabane et al., 2009; Huang et al., 2009; Komarnytsky et al., 2006; Mandal et al., 2014; Rivard et al., 2006), and towards the optimization of recovery schemes for protein enrichment (e.g. Wilken and Nikolov, 2006; Hassan et al., 2008; Buyel and Fischer, 2014). Approaches to enrich recombinant proteins in plant extracts often rely on the choice of a proper extraction buffer pH or ionic strength, in such a way as to preferentially remove host proteins and facilitate further purification (e.g. Balasubramaniam et al., 2003; Farinas et al., 2005; Woodard et al., 2009; Zhang et al., 2005; Zhong et al., 2007). Ammonium sulphate precipitation and extraction at low pH are particularly useful in green tissue extracts to precipitate contaminants such as cellular debris and photosynthetic pigments. These procedures are also useful to precipitate the highly abundant enzyme ribulose 1,5-bisphosphate carboxylase oxygenase (RuBisCO) (Bendandi et al., 2010; Lai et al., 2010; Peckham et al., 2006; Woodard et al., 2009), which represents up to 50% of total soluble proteins (TSP) in leaves (Ellis, 1979) and often complicates the preparation of a highly purified protein product (Buyel et al., 2013; Gaeda et al., 2007; Ross and Zhang, 2010). On the other hand, ammonium sulphate precipitation is hardly adaptable to large-scale production schemes and low pH extraction is not compatible with most pH-sensitive protein products.

In this study, we explored the potential of methyl jasmonate (MeJA), a volatile derivative of the stress hormone jasmonic acid (Okada et al., 2015), to trigger a generic enrichment of recombinant proteins in planta, before tissue harvesting and further processing. Studies have reported strong down-regulating effects for MeJA and jasmonic acid-inducing signals such as wounding, herbivory or insect oral secretions on the transcription of photosynthesis-related genes (Bilgin et al., 2010; Duceppe et al., 2012; Hermsmeier et al., 2001; Jung et al., 2007; Lawrence et al., 2008; Zubo et al., 2011). These suppressing effects are typically associated with a concomitant accumulation of stress-related proteins and carbon-metabolizing enzymes in leaves, leading overall to strongly depleted pools of RuBisCO and a significant rebalancing of the leaf proteome (Duceppe et al., 2012; Giri et al., 2006; Mahajan et al., 2014; Ullmann-Zeunert et al., 2013; Wei et al., 2009). Here, we assessed whether MeJA can reproduce similar proteome alterations in leaves of the protein expression host Nicotiana benthamiana (Leuzinger et al., 2013), and whether this could positively impact the specific and relative yields of a mammalian monoclonal antibody transiently expressed in leaf tissue.

Results and discussion

MeJA induces leaf proteome rebalancing in N. benthamiana

Forty-two-day-old N. benthamiana plants were treated with 0.5, 1 or 2 mM MeJA to determine the effects of jasmonic acid signalling on the overall protein profile in leaf tissue (Figure 1). On a fresh weight basis, mature leaves of MeJA-treated plants had their TSP contents reduced by 20%–30% after 7 days compared to nontreated control leaves, depending on the elicitor dose (ANOVA; $P < 0.001$) (Figure 1a). As observed with other Solana-

ceae (Duceppe et al., 2012; Lawrence et al., 2008; Ullmann-Zeunert et al., 2013), MeJA treatment caused a strong depletion of RuBisCO large and small subunit pools in leaf extracts (Figure 1b), estimated at 40% less than in controls for the 0.5 mM MeJA dose to more than 90% less for the 2 mM dose (ANOVA; $P < 0.05$ for RbcL; $P < 0.001$ for RbcS) (Figure 1c). RuBisCO depletion was counterbalanced by the up-regulation of several proteins, notably in the ~30-kDa, ~25-kDa and ~6-kDa molecular mass ranges as visualized on Coomassie blue-stained gels following SDS-PAGE (Figure 1b, boxes A, B and C). A shotgun proteomic analysis was conducted to identify the most abundant proteins in these mass ranges, based on a spectral count analysis of tandem mass spectrometry (MS/MS) peptides obtained from trypsin digests of protein bands in boxes A, B and C of Figure 1b (Table 1). As expected, most identified proteins were typical MeJA-inducible stress-related (or defence) proteins, notably including thionins, chitinases, Ser protease inhibitors of the Kunitz and proteinase inhibitor II protein families, and stress-related enzymes such as superoxide dismutase, carbonic anhydrase and thioredoxin peroxidase (see Table S1 for details on MS/MS peptide sequences).

A number of plants were infiltrated 24 h post-MeJA treatment with Agrobacterium tumefaciens cells harbouring an 'empty' pCAMBIA2300 vector (Goulet et al., 2012), to assess whether MeJA-mediated alterations of the leaf proteome could be maintained over the usual 6- to 7-days period left following bacterial infection for recombinant protein expression (Figure 1). Agroinfiltration was shown previously to trigger the active secretion of PR proteins—mostly PR-2 (ß-glucanases) and PR-3 (chitinase) proteins—in the N. benthamiana leaf apoplast, presumably involving the pathogen-inducible salicylic acid signalling pathway (Goulet et al., 2010). Accordingly, three protein bands in the mass range of 25- to 33-kDa expected for PR-2 and PR-3 proteins were strongly up-regulated in agroinfiltrated leaves, regardless of the MeJA dose applied (Figure 1b). Not surprisingly given the antagonistic interactions between the jasmonate and salicylate signalling pathways (Derksen et al., 2013), PR protein induction in infiltrated leaves was associated with a strong reversal of the MeJA-mediated stress protein inductions detected in uninfected plants, resulting in faint protein signals in gel areas corresponding to boxes A, B and C 6 days postinfection (Figure 1b). RuBisCO depletion was also tempered in infected leaves, but both the large and small subunits were still found at less than 50% of their original content, relative to untreated plants, in leaves treated with 1 or 2 mM MeJA (Figure 1b,c). Together, these data demonstrate the effectiveness of MeJA as a potent, pre-infection trigger of RuBisCO depletion in N. benthamiana leaves, potentially useful in reducing RuBisCO load and increasing recombinant protein-relative content in crude protein extracts prior to postrecovery enrichment and purification.

MeJA has little effect on the N. benthamiana–A. tumefaciens interaction

Data showing a near-complete reversal of MeJA up-regulating effects on defence proteins following agroinfiltration suggested a limited influence of jasmonate treatment on both the plant's ability to mount a PR protein-based defence response to bacterial infection, and the eventual ability of the bacterium to transfect plant cells and persist normally into leaf tissue. To further confirm these conclusions, we looked closer at the plant–bacterium interaction using immunoblots for PR-2 proteins as a reference for PR protein induction in leaves (Goulet et al., 2010) and mRNA

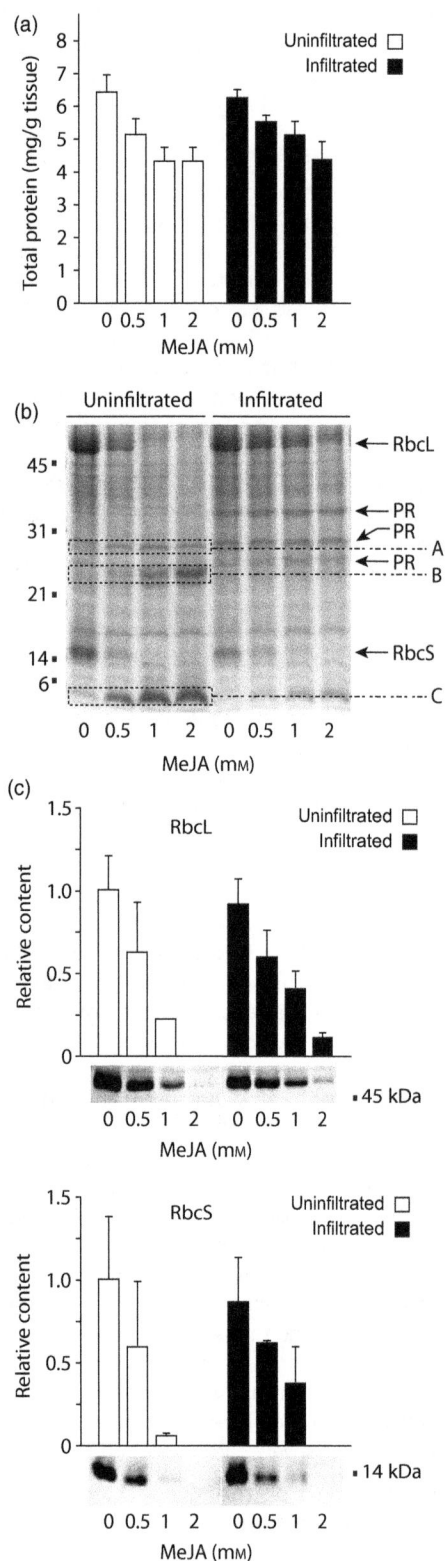

(a)

(b)

(c)

Figure 1 Total soluble proteins (TSP), 1-D proteome profile and RuBisCO subunit levels in agroinfiltrated and control *N. benthamiana* leaves treated with the stress elicitor MeJA 24 h before infection. (a) TSP on a fresh weight basis in leaves treated with 0, 0.5, 1 or 2 mM MeJA. (b) Coomassie blue-stained protein profile in control and MeJA-treated leaves following SDS-PAGE. Numbers on the left correspond to commercial molecular mass markers (kDa). PR, pathogenesis-related proteins up-regulated in agroinfiltrated leaves; RbcL and RbcS, RuBisCO large and small subunit, respectively; A, B and C boxes, gel areas containing MeJA-inducible proteins in uninfiltrated plants (*see* Table 1 for protein identities). (c) Relative amounts of RuBisCO large (RbcL) and small (RbsS) subunits in MeJA-treated leaves as determined by densitometry following immunodetection with appropriate antibodies. Protein signals on the immunoblots were quantified using the Phoretix 2-D Expression software v. 2005 (NonLinear USA, Durham NC) after scanning nitrocellulose membranes with a Microtek ScanMaker II digitalizer (Microtek Laboratory, Torrance, CA). Data are expressed as relative contents compared to nontreated controls assigned an arbitrary value of 1.0. Each bar on panels (a) and (c) is the mean of three independent (leaf replicate) values ± SE. Infiltrated leaves were transfected with *A. tumefaciens* cells 24 h post-MeJA treatment. All plant samples were harvested 7 days post-MeJA treatment (i.e. 6 days postagroinfiltration for the transfected plants).

Table 1 Stress-related proteins up-regulated in *Nicotiana benthamiana* leaves 7 days post-MeJA treatment[1,2]

Protein	Accession No.[3]	Species	No. spectral counts
Box A			
Acidic endochitinase	P17514	*N. tabacum*	34
Carbonic anhydrase	A4D0J8	*N. benthamiana*	30
Kunitz-type protease inhibitor	A9UF61	*N. alata*	22
Proteasome subunit ß type-6	Q9XG77	*N. tabacum*	14
Superoxide dismutase	P22302	*N. plumbaginifolia*	12
Chaperonin 21	Q9M5A8	*Solanum lycopersicum*	12
Box B			
Kunitz-type protease inhibitor	A9UF61	*N. alata*	40
Thioredoxin peroxidase	Q8RVF8	*N. tabacum*	18
Pathogenesis-related protein R	P13046	*N. tabacum*	18
Proteasome subunit ß type-6	P93395	*N. tabacum*	16
Superoxide dismutase [Fe]	P22302	*N. plumbaginifolia*	14
Box C			
Flower-specific thionin	B2BLV8	*N. tabacum*	39
Trypsin proteinase inhibitor	Q1WL50	*N. benthamiana*	11

[1]These proteins correspond to the most abundant protein species identified by LC-MS/MS in boxes A, B and C of Figure 1b. Protein identifications based on a minimal spectral count of 10 spectra are included in the list.

[2]Additional information on MS/MS sequences is available in Table S1.

[3]Accession numbers from the National Center for Biotechnology Information/GenBank database (http://www.ncbi.nlm.nih.gov).

transcript numbers of two agrobacterial virulence genes regulated by salicylate as markers for the transfection process (Yuan *et al.*, 2007) (Figure 2). Several studies have shown strong antagonistic effects for salicylic acid or functional homologues on jasmonate signalling, and divergent regulatory patterns for salicylate-inducible PR proteins and jasmonate-inducible defence

proteins upon salicylate or MeJA treatment (Thaler *et al.*, 2012). In line with the Coomassie blue-stained protein profiles above (*see* Figure 1b), MeJA sprayed at 0.5 or 1 mM had no significant effect on the expression of a constitutively expressed 33-kDa pathogen-inducible PR-2 protein in noninfiltrated leaves (ANOVA; $P > 0.05$) (Figure 2a). In agreement with Goulet et al. (2010) reporting a strong secretion of PR proteins upon agroinfiltration, this protein was strongly up-regulated in agroinfiltrated leaves ($P < 0.001$), independent of the MeJA dose ($P > 0.05$).

Salicylic acid is known to attenuate agroinfection in leaves (Anand et al., 2008; Veena et al., 2003; Yuan et al., 2007) via a down-regulation of the bacterium vir regulon affecting virulence gene expression and T-DNA integration into host cells (Yuan et al., 2007). Plants defective in salicylic acid have been shown to be more susceptible to the pathogen, while plants overproducing this metabolite showed increased recalcitrance to infection (Yuan et al., 2007). Bacterial counts and real-time RT-PCR assays were here performed to compare Agrobacterium cell numbers and mRNA transcript pools of virulence proteins in infiltrated leaves, with or without MeJA treatment, to look for a possible salicylate-repressing effect of the jasmonate derivative facilitating Agrobacterium growth and virulence genes expression (Figure 2b,c). Similar numbers of bacteria were retrieved from the apoplast of control and MeJA-treated leaves, as inferred from bacterial counts of 10 to 100 million colony-forming units (CFU) per mL of apoplast extract up to 2 days postinfiltration, to less than a million CFU after 4 or 6 days (Figure 2b). DNA coding sequences for VirB1 and VirE1, two virulence proteins involved in T-DNA translocation into recipient host cells and subsequent integration into the nucleus, respectively (Lacroix et al., 2006), were used as salicylate-responsive bacterial markers for RT-PCR assays. VirB1 expression was negatively altered in MeJA-treated leaves and no positive effect on transcription was observed for either gene 7 days post-MeJA treatment despite the natural antagonistic effect of jasmonates on salicylate signalling (Figure 2c). These data confirm overall the onset of a strong defence response to bacterial infection in MeJA-treated and control leaves upon agroinfiltration, and no positive impact of MeJA pretreatment on

both this response and the bacterium's potency for gene transfection.

MeJA also has little effect on protease activities in transfected leaves

Enzymatic assays were carried out with synthetic peptide substrates to investigate the effect of MeJA treatment on protease activity in leaf crude extracts (Figure 3). Endogenous proteases have a strong impact on recombinant protein yield in plant systems given their direct role in protein turnover either in planta during expression or ex planta upon tissue disruption for protein recovery (Benchabane et al., 2008). Protease profiles are influenced by different developmental or environmental factors in N. benthamiana, including leaf age, agrobacterial infection and recombinant protein expression (Robert et al., 2013). Keeping in mind the complex crosstalk between defence-related signalling pathways (Robert-Seilaniantz et al., 2011), the importance of secreted proteases upon pathogenic infection (Figueiredo et al., 2014; Hörger and van der Hoorn, 2013; Höwing et al., 2014; Ramirez et al., 2013) and the MeJA-mediated overexpression of protease inhibitors that could influence the activity of endogenous proteases in crude extracts (see Figure 1b and Table 1), we assessed the impact of MeJA treatment on major endoprotease activities in control and agroinfiltrated leaves.

Protease activities measured in crude extracts represent net values reflecting both the relative abundance of protease and protease inhibitor molecules in the extraction medium upon tissue disruption, and the inhibitory specificity of the released inhibitors towards endogenous proteases (Benchabane et al., 2009). Cathepsin L-like Cys protease, trypsin-like Ser protease and cathepsin D/E-like Asp protease activities were assayed in crude extracts of control and MeJA-treated leaves harvested 7 days post-MeJA treatment to document the basic long-term effect of jasmonate signalling on leaf protease profiles. The volatile elicitor had no impact on cathepsin L-like activity for the three tested

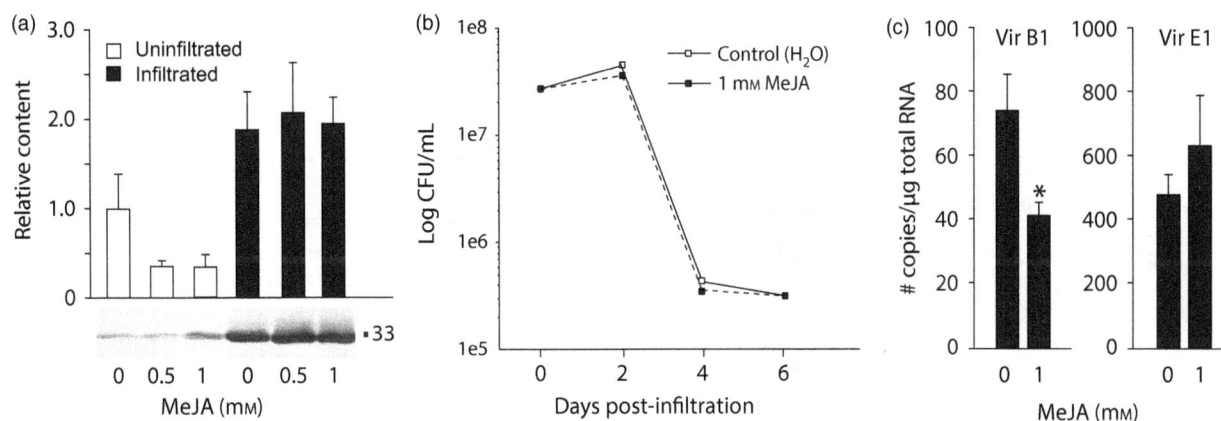

Figure 2 Effect of MeJA treatment on PR-2 protein accumulation, agrobacteria numbers and transcript numbers of the two A. tumefaciens virulence proteins VirB1 and VirE1 in N. benthamiana leaves. (a) Effects of MeJA treatment on accumulation of the 33-kDa, pathogen-inducible PR-2 protein in control and agroinfiltrated leaves as assayed by densitometry following immunodetection. Immunoblot signals were quantified using the Phoretix 2-D Expression software v. 2005 (NonLinear USA) after scanning nitrocellulose membranes with a Microtek ScanMaker II digitalizer (Microtek Laboratory). Data are the mean of three leaf replicates ± SE. An arbitrary value of 1.0 was assigned to PR-2 level in control healthy leaves. (b) Bacteria retrieved from N. benthamiana leaves 0, 2, 4 or 6 days postagroinfiltration. Data are expressed as log numbers of colony-forming units (CFU) on agar plates, and each point is the mean of five independent (leaf replicate) values. (c) mRNA transcript numbers for VirB1 and VirE1 in agroinfiltrated leaves treated with 0 or 1 mM MeJA, as assayed by real-time RT-PCR with appropriate DNA primers. Each value is the mean of five biological (leaf replicate) values ± SE. Asterisk indicates a significantly lower value for VirB1 transcripts in MeJA-treated leaves compared to control leaves (Student's t-test; $P < 0.05$).

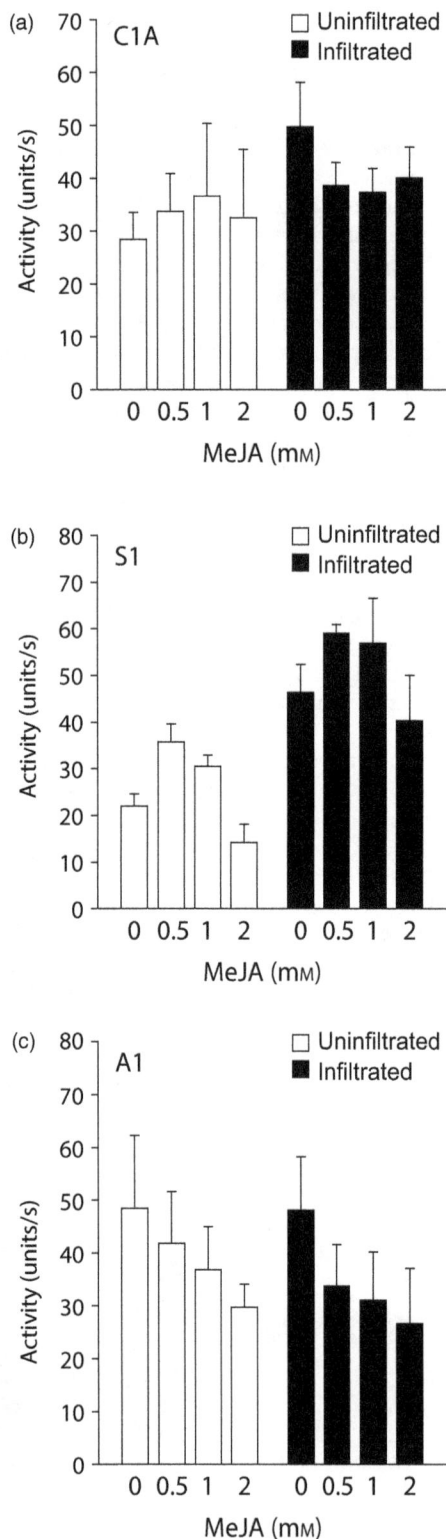

Figure 3 Protease activities in crude protein extracts of control and agroinfiltrated *N. benthamiana* leaves treated with 0, 0.5, 1 or 2 mM MeJA 24 h before infiltration. Protease assays were conducted *in vitro* using fluorigenic peptide substrates specific to cathepsin L-like (C1A) Cys proteases (a), trypsin-like (S1) Ser proteases (b) and cathepsin D/E-like (A1) Asp proteases (c). Leaf samples were harvested 7 days post-MeJA treatment. Each bar is the mean of three independent (leaf replicate) values ± SE.

doses (ANOVA; $P > 0.05$) (Figure 3a), while trypsin-like enzymes showed increased activity for the 0.5 and 1 mM doses ($P < 0.001$) (Figure 3b) despite a concomitant up-regulation of Ser protease inhibitors in leaves (*see* Table 1). By contrast, cathepsin D/E-like activity showed a dose-dependent decrease 7 days post-MeJA treatment in uninfiltrated plants ($P < 0.001$), which was also observed in agroinfiltrated leaves (Figure 3c). As for PR proteins above, cathepsin L-like ($P < 0.05$) and trypsin-like ($P < 0.001$) activities were up-regulated in infiltrated leaves 6 days postinfection, independent of MeJA pretreatment (Figure 3a,b). Together, our data suggest a possible role for Cys and Ser proteases upon bacterial infection and the onset of specific expression patterns for these enzymes in agroinfiltrated leaves, independent of MeJA-mediated protease inhibitor inductions. Our data also point to the establishment of a 'consolidated', rebalanced proteome in MeJA-treated leaves 6 days postinfiltration presenting, on the one hand, a significantly depleted pool of RuBisCO subunits; and, on the other hand, strongly increased amounts of pathogen-inducible PR proteins apparently substituting depleted pools of MeJA-inducible defence proteins accumulated temporarily in leaves before agroinfiltration.

MeJA treatment increases the accumulation of a transiently expressed recombinant protein

Agroinfiltration assays were conducted to determine the impact of MeJA treatment on the expression and steady-state levels of a clinically useful recombinant protein transiently expressed in leaf tissue (Figure 4). The human blood-typing monoclonal antibody C5-1 (Khoudi *et al.*, 1999) was chosen as a model given the wealth of information available about the expression, maturation and proteolytic processing of this protein in plant systems (Bardor *et al.*, 2003; D'Aoust *et al.*, 2009; Goulet *et al.*, 2012; Khoudi *et al.*, 1999; Robert *et al.*, 2013; Sainsbury *et al.*, 2008; Vézina *et al.*, 2009). The light and heavy chains of C5-1 co-expressed in *N. benthamiana* leaves are detected on immunoblots as a high molecular weight, multiband protein pattern following SDS-PAGE in nonreducing conditions, including a ~150-kDa, fully assembled version of the antibody and a number of smaller, yet active, fragments (Goulet *et al.*, 2012; Robert *et al.*, 2013). Accordingly, a major protein band of about 150 kDa was immunodetected using anti-IgG primary antibodies, along with the expected smaller bands (Figure 4a). Visually similar protein band patterns were observed in leaf extracts of MeJA-treated plants compared to control plants, except for the band signals being apparently stronger.

A quantitative enzyme-linked immunosorbent assay (ELISA) was performed to confirm the apparent positive effect of MeJA on antibody accumulation, and to define a possible dose–curve relation for the jasmonate-mediated response (Figure 4b). Significantly higher amounts of antibody were measured in MeJA-treated leaves, to reach steady-state levels about 1.5 to 2.5 times the levels measured in control plants (ANOVA; $P < 0.05$). The up-regulating effect of MeJA followed a quadratic curve, with a maximum yield value of ~425 μg/g leaf tissue measured at 1 mM MeJA compared to ~325 μg/g leaf tissue at 2 mM MeJA or less than 200 μg/g in control leaves (Figure 4b). The effect of MeJA was also confirmed on a protein-relative basis to give a net yield of 70 ng antibody/μg soluble proteins (or ~7% TSP) in leaves treated with 1 mM MeJA, about twice the yield obtained with control leaves ($P < 0.05$) (Figure 4b). As inferred from real-time RT-PCR data for plants sprayed with 1 mM MeJA, higher C5-1 yields in MeJA-treated leaves were associated with higher

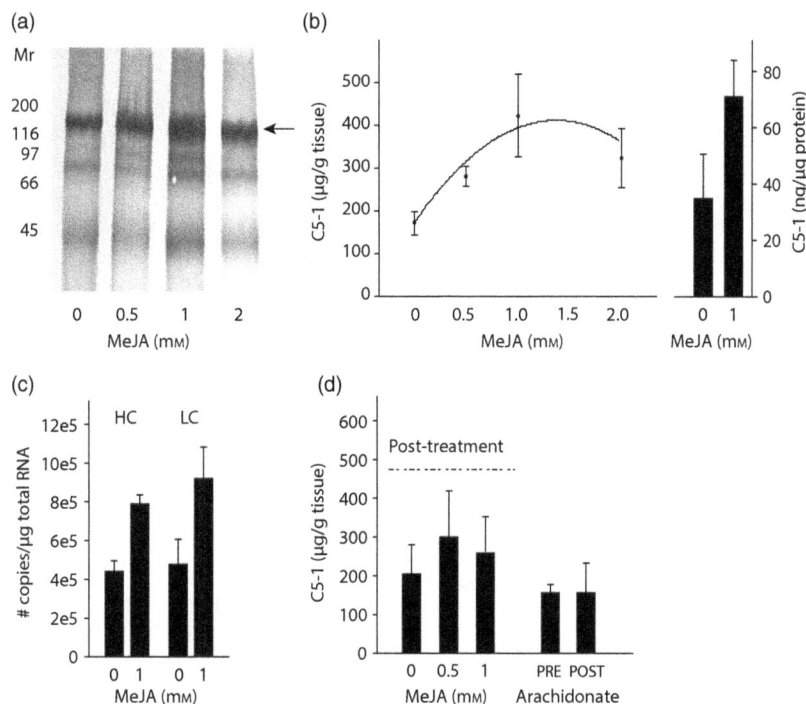

Figure 4 C5-1 antibody yield and expression in agroinfiltrated *N. benthamiana* leaves treated with 0, 0.5, 1 or 2 mM MeJA. (a) C5-1 heavy and light chain full (arrow) and partial complexes immunodetected following SDS-PAGE in nonreducing conditions. Mr, commercial molecular mass markers. (b) ELISA-assayed C5-1 in MeJA-treated leaves. Data are expressed on a weight basis (quadratic curve; r^2=0.885) or on a relative basis compared to total soluble proteins (histogram). (c) mRNA transcripts for C5-1 light (LC) and heavy (HC) chains in leaves treated with 0 or 1 mM MeJA, as assayed by real-time RT-PCR. (d) ELISA-assayed C5-1 in leaves treated with 0, 0.5 or 1 mM MeJA 24 h after infiltration or with 1 mM arachidonic acid 24 h before or after infiltration. Each bar or point on panels (b), (c) and (d) is the mean of five independent (leaf replicate) values ± SE.

numbers of mRNA transcripts for both the light and heavy chains, about twofold the numbers measured in control leaf extracts (Figure 4c).

Unlike Jung *et al.* (2014) recently reporting increased expression of two plant cell wall-degrading enzymes in detached sunflower leaves infiltrated with a MeJA-supplemented agrobacterial solution, no yield increase was observed when applying MeJA at the moment of infiltration (not shown). Similarly, no yield increase was observed when spraying the same elicitor 24 h postinfiltration (Figure 4d) or when treating leaves 24 h before infiltration with arachidonic acid, a functional analogue of salicylic acid (Girard *et al.*, 2007) (Figure 4d). Induction of the 33-kDa PR-2 protein in leaves transfected with the empty pCAMBIA2300 vector (*see* Figure 2a) was similar in leaves transfected with the antibody-encoding vector (Figure S1), suggesting a limited, if not null, impact of C5-1 antibody expression on the host plant's response to agroinfiltration. Our observations suggest overall the practical usefulness of pre-infiltration MeJA treatment to boost recombinant protein expression in *N. benthamiana* leaves, via a yet to be understood transcription-promoting effect of the jasmonate pathway.

Conclusions

Our first goal in this study was to evaluate the potential of MeJA to induce a re-organization, or rebalancing, of the *N. benthamiana* leaf proteome useful in enriching transiently expressed recombinant proteins *in planta* via a concomitant down-regulation of RuBisCO. Several studies have reported the strong down-regulating impact of jasmonate signalling on RuBisCO expression

and accumulation in leaves, parallel to the onset of an active defence response involving the biosynthesis of various defensive compounds such as antimicrobial enzymes and antidigestive defence proteins (e.g. Bilgin *et al.*, 2010; Duceppe *et al.*, 2012; Ullmann-Zeunert *et al.*, 2013). In line with these reports, MeJA treatment induced a strong depletion of RuBisCO large and small subunit pools in *N. benthamiana* leaves under our conditions, initially compensated by increased levels of jasmonate-inducible defence proteins including thionins, Ser protease inhibitors and antimicrobial hydrolases. The proteome rebalancing effect of MeJA on jasmonate-inducible defence proteins was readily compromised by a strong antibacterial response upon agroinfiltration, but the down-regulating effect observed on RuBisCO in noninfiltrated plants was maintained for the most part in infected plants over the usual 6- to 7-day incubation period allocated for protein expression. MeJA treatment generated a RuBisCO-depleted cellular environment in agroinfiltrated *N. benthamiana* leaves, allowing for an effective enrichment of heterologous proteins *in situ*, before their downstream recovery. Compared to nontreated plants, MeJA treatment resulted in an almost fivefold enrichment factor for the transiently expressed C5-1 antibody relative to RuBisCO, derived from a greater than twofold depletion of RuBisCO and a twofold increase of C5-1 mRNA transcripts leading to a twofold increase in recombinant antibody levels on a fresh weight basis.

The mechanism driving the positive impact of MeJA treatment on C5-1 expression remains to be elucidated but given the well established constitutive expression pattern of the CaMV 35S viral promoter in plant cells (Potenza *et al.*, 2004), the positive effect on transcription is likely to be nonspecific. A simple explanation for the improved accumulation of C5-1 on a leaf weight basis would be a low 'sink pressure' on amino acid pools towards RuBisCO biosynthesis in RuBisCO-depleted leaves, and a resulting increased availability of metabolite and cellular resources for the production of less abundant (e.g. recombinant) proteins. MeJA was shown recently to promote an elevated expression of ribosomal genes in leaves, presumably useful in keeping the protein biosynthesis machinery active in cells responding to the jasmonate pathway (Noir et al., 2013). This observation, along with recently reported data indicating that nitrogen resources not invested in RuBisCO biosynthesis in RuBisCO-silenced plants are reallocated in part to the expression of other soluble proteins (Stanton et al., 2013), supports the idea of a nonspecific, free amino acid pool-related effect for MeJA. Studies have described the positive impact of silencing the expression of highly abundant endogenous proteins in storage organs to make more amino acids available and to promote recombinant protein deposition (Goossens et al., 1999; Kim et al., 2012; Schmidt and Herman, 2008; Shigemitsu et al., 2012). Work is underway to test the hypothesis of a similar nonspecific, metabolite resource-related effect of MeJA on CaMV 35S promoter-driven transgenes in N. benthamiana leaves. Work is also underway to confirm the general applicability of MeJA as a trigger of recombinant protein enrichment in plant systems, using as models different protein candidates and expression platforms. Questions currently being addressed include the impact of transfection procedures on MeJA-mediated effects, the impact of leaf proteome alterations on yield and maturation of recombinant proteins targeted to different cell locations, and the eventual metabolic interference effects of biologically active proteins expressed in an altered physiological context (Badri et al., 2009).

Experimental procedures

Plant tissue sampling and elicitor treatments

Forty-two-day-old N. benthamiana plants grown in greenhouse were used for the experiments. Each plant was sprayed evenly with 50 mL of 0.5 mM, 1 mM or 2 mM MeJA in water containing 0.1% (v/v) Triton X-100 (Sigma-Aldrich, Mississauga ON, Canada) or with 50 mL of 1 mM arachidonic acid (Girard et al., 2007) in the same solvent, transferred to a Conviron growth chamber (Conviron, Winnipeg MB, Canada) and let to incubate for 7 days before leaf tissue harvesting. Plants sprayed with 50 mL of 0.1% (v/v) Triton X-100 in water were used in parallel as negative controls to avoid confounding effects due to experimental conditions. Two or three 1-cm² leaf discs were harvested from the third main stem leaf (see Robert et al., 2013) after 7 days, as source material for protein and RNA extraction. The leaf samples were frozen immediately in liquid nitrogen and stored at −80 °C until protein or RNA extraction. Three to five biological (plant) replicates were used for each treatment, to allow for statistical treatment of the data.

Gene constructs and leaf agroinfiltration

Two gene vectors were used for the agroinfiltration assays: an engineered pcambia2300 vector encoding the light and heavy chains of C5-1 fused to an N-terminal signal peptide sequence for cellular secretion (see Goulet et al., 2012 for details on gene construct); and the original ('empty') pcambia2300 binary vector (CAMBIA, Canberra, Australia) as a control for agroinfection. The

two binary vectors were electroporated into A. tumefaciens strain LBA4404, and the cultures were maintained in lysogeny broth (LB) medium supplemented with 50 µg/mL kanamycin and 50 µg/mL rifampicin. For infiltration, bacteria were grown to stable phase at 28 °C to an OD_{600} of 1.0 and collected by centrifugation at 2000 g. The bacterial pellets were resuspended in 10 mM 2-[N-morpholino] ethanesulfonic acid] (MES) buffer, pH 5.8, containing 100 µM acetosyringone and 10 mM $MgCl_2$. Leaf infiltration was performed using a needle-less syringe as described earlier (D'Aoust et al., 2009), after mixing the C5-1 antibody (or empty vector) agrobacterial suspension with an equal volume of bacterial suspension carrying a binary vector for the gene silencing suppressor p19 (Voinnet et al., 2003). Infiltrated tissue for molecular characterization was collected six days postinfection (i.e. seven days post-MeJA or arachidonate treatment), unless otherwise indicated. Three to five biological (plant) replicates were used for each treatment to allow for statistical analyses.

Protein extraction, SDS-PAGE and immunoblotting

Whole leaf proteins were extracted from three 1-cm² leaf discs in 400 µL of ice-cold 50 mM Tris-HCl, pH 7.5, containing 150 mM NaCl, by disrupting the leaf samples with tungsten carbide beads (Qiagen, Mississauga ON, Canada) in a Mini-Beadbeater apparatus (BioSpec, Bartlesville OK). The cOmplete protease inhibitor cocktail (Roche, Laval QC, Canada) was added in the extraction buffer before tissue disruption as specified by the supplier, except for those extracts dedicated to protease activity monitoring. Crude leaf extracts were clarified by centrifugation at 20 000 g for 30 min at 4 °C, and total soluble proteins assayed according to Bradford (1976) with bovine serum albumin as a protein standard (Sigma-Aldrich). Electrophoretic separation of the proteins was performed by 10% (w/v) SDS-PAGE in reducing conditions (Laemmli, 1970), unless otherwise indicated. The resolved proteins were stained with Coomassie blue to visualize protein band patterns, or electrotransferred onto nitrocellulose membranes for immunodetection with appropriate primary and secondary antibodies. RuBisCO small subunit was detected using polyclonal IgG raised in rabbit against a synthetic small subunit peptide (Agrisera, Vännäs, Sweden) and alkaline phosphatase-conjugated goat anti-rabbit IgG as secondary antibodies (Sigma-Aldrich). RuBisCO large subunit was detected using polyclonal IgY raised in chicken against a synthetic large subunit peptide (Agrisera) and alkaline phosphatase-conjugated goat anti-chicken IgY secondary antibodies (Sigma-Aldrich). The 33-kDa pathogen-inducible PR-2 protein was detected using rabbit polyclonal IgG primary antibodies (Agrisera) and alkaline phosphatase-conjugated goat anti-rabbit IgG secondary antibodies (Sigma-Aldrich). C5-1 light and heavy chains were detected with alkaline phosphatase-conjugated goat anti-mouse IgG antibodies (Sigma-Aldrich).

Mass spectrometry

Leaf proteins for MS identification (corresponding to protein gel areas in boxes A, B and C of Figure 1b) were excised manually from the gels, put in 100 µL of Milli-Q water, and sent to the Eastern Québec Proteomics Center (Centre de Recherche du CHUL, Québec QC, Canada) for ion trap MS/MS analysis. In-gel protein digestion with sequencing grade trypsin (Promega, Madison WI) was performed in a MassPrep liquid handling robot (Waters, Milford MA) according to the provider's instructions. Peptide samples were resolved by online reversed-phase nanoscale capillary liquid chromatography and analysed by electrospray mass spectrometry using a Thermo Surveyor MS pump connected to an LTQ linear ion trap mass spectrometer equipped with a nanoelectrospray ion source (ThermoFisher, San Jose, CA). Peptide aliquots of 10 µL were loaded onto a 75-µm internal diameter BioBasic C18 picofrit column (New Objective, Woburn, MA). The peptides were eluted along a water–acetonitrile/0.1 (v/v) formic acid gradient, at a flow rate of 200 nL/min obtained by flow splitting. Mass spectra were acquired under the data-dependent acquisition mode, using the Xcalibur software, v. 2.0. Each full MS scan (from 400 to 2000 m/z) was followed by MS/MS scans of the seven most intense precursor ions using collision-induced dissociation. The relative collisional fragmentation energy was set at 35, and the dynamic exclusion function enabled with a 30-s exclusion duration.

Protein identification

MS/MS spectral data were extracted using the ExtractMSn utility of Thermo's Bioworks application package, with no charge state deconvolution or deisotoping. Protein identities were assessed using the Mascot program, v. 2.3.02 (Matrix Science, London, UK) and the Open Source software X! Tandem (Craig et al., 2004). Both softwares were set up to search a custom database containing all known protein sequences of Solanaceae species (Taxonomy ID: 4070, for 39 896 proteins), A. tumefaciens protein sequences (12 554 proteins) and data sequences of protein contaminants commonly found in trypsin digests. The database was searched with a fragment ion mass tolerance of 0.50 Da and a parent ion tolerance of 2.0 Da. The iodoacetamide derivative of Cys residues was specified in both Mascot and X! Tandem as a fixed modification; citrullination of Arg residues and oxidation of Met residues were specified as variable modifications. MS/MS-based peptide and protein identifications were validated using Scaffold, v. 3.4.9 (Proteome Software, Portland, OR). Identifications were accepted if they included at least four peptides and could be established at greater than 95% probability as specified by the Peptide Prophet algorithm (Keller et al., 2002). Protein probabilities were assigned by the Protein Prophet algorithm as described by Nesvizhskii et al. (2003). Proteins that contained similar peptides and could not be differentiated based on MS/MS data were grouped to satisfy the principles of parsimony.

Real-time RT-PCR

mRNA transcripts for C5-1 and A. tumefaciens virulence proteins VirB1 and VirE1 were assayed by real-time RT-PCR using an ABI PRISM 7500 Fast real-time PCR apparatus, system version 2.0.1 (Applied Biosystems, Carlsbad, CA). Total RNA was extracted from two 1-cm^2 leaf discs collected on the third main stem leaf (Robert et al., 2013) using the Qiagen RNeasy plant mini kit (Qiagen), following the supplier's instructions. RNA samples were treated with DNase I (Roche Diagnostics) to remove contaminant DNA and assessed for quality and quantity using a Nanodrop® ND-1000 spectrophotometer (NanoDrop Technologies, Wilmington DE). First-strand cDNA was synthesized from 500 ng of total RNA using four units of Omniscript reverse transcriptase (Qiagen) and 1 µM of oligo-dT(15) nucleotides (Roche). PCR mixtures contained 10 µL of Fast SYBR Green PCR Master Mix (Applied Biosystems), 2 µL of cDNA template, and 2.5 µL each of appropriate forward and reverse primers at 625 nM final concentration (Table S2). A no-template mixture control was included in each 96-well plate. Amplification rounds consisted of a 20-s denaturation step at 95 °C, followed by 40 two-step cycles

of 3 s at 95 °C and 30 s at 60 °C. A dissociation curve analysis was performed after amplification with the SYBR Green Master Mix, and the cycle threshold of each sample was then compared to a DNA standard curve designed for each pair of primers. Standard curves were generated with 2 µL of cDNA template following the NEB Taq polymerase routine protocol (New England Biolabs, Pickering ON, Canada). Amplification products were purified using the Illustra GFX kit (GE Healthcare), and DNA standard curves were devised with serial dilutions of the purified PCR products in nuclease-free water (from 10^7 to 10^2 copies per µL). C_t data were plotted against the corresponding number of transcript copies. All amplifications were carried out in duplicate.

Bacterial counts

Bacterial loads in leaves were determined by the counting of CFU on LB agar plates coated with bacteria recovered at different moments from the third main stem leaf. Each replicate consisted of two 1-cm^2 leaf discs collected 0, 2, 4 or 6 days postagroinfiltration. The leaf discs were homogenized in 10 mM MES buffer, pH 5.8, containing 10 mM MgCl$_2$ in the BioSpec Mini-Beadbeater (see Protein extraction, SDS-PAGE and immunoblotting, above). The resulting suspensions were dilution-plated on LB medium supplemented with kanamycin (50 mg/mL) and incubated at 28 °C until CFU counting after 2 days.

Protease assays

Protease activities were assayed by the monitoring of substrate hydrolysis progress curves (Goulet et al., 2012) using the following synthetic fluorigenic substrates (Peptides International, Louisville, KY): Z-Phe–Arg–methylcoumarin (MCA) for cathepsin L-like C1A Cys proteases, Z-Arg-MCA for trypsin-like S1 Ser proteases, and MOCAc-Gly–Lys–Pro–Ile–Leu–Phe–Phe–Arg–Leu–Lys(Dnp)-D-Arg-NH$_2$ for cathepsin D/E-like A1 Asp proteases. Substrate hydrolysis by the leaf extract proteases (36 ng leaf protein per µL of reaction mixture) was allowed to proceed at 25 °C in 50 mM MES, pH 5.8, containing 10 mM L-Cys for the cathepsin L substrate. Protease activities were monitored using a Fluostar Galaxy microplate fluorometer (BMG, Offenburg, Germany) with excitation and emission filters of 360 and 450 nm, respectively, for the MCA substrates; or of 340 and 400 nm, respectively, for the MOCAc substrate. Three independent (biological) replicates were used for each assay.

C5-1 quantification

ELISA plates for C5-1 antibody quantification (Becton Dickinson, Mississauga ON, Canada) were coated with 3.75 µg/mL goat anti-mouse heavy chain-specific IgG1 (Sigma-Aldrich) in 50 mM carbonate buffer (pH 9.0) at 4 °C for 16–18 h. The plates were washed three times in 10 mM phosphate-buffered saline containing 0.1% (v/v) Tween-20 (PBS-T), blocked through a 1-h incubation at 37 °C in 1% (w/v) casein in phosphate-buffered saline (PBS) (Pierce, Rockford IL), and washed three times in PBS-T. A standard curve was generated for each plate with 0, 4, 8, 16, 24, 32, 40 and 60 ng/mL of purified mouse IgG1 (Sigma-Aldrich). All dilutions (controls and samples) were performed in a control extract obtained from leaf tissue infiltrated with a mock inoculum so that any matrix effect was eliminated. The plates were incubated with protein samples for 1 h at 37 °C, washed three times in PBS-T and then incubated with peroxidase-conjugated goat anti-mouse IgG (H+L) antibodies (0.02 µg/mL in blocking solution) (Jackson ImmunoResearch) for 1 h at 37 °C. After additional washes with PBS-T, the plates were incubated with the

3,3',5,5'-tetramethylbenzidine Sure Blue peroxidase substrate (KPL, Gaithersburg, MD). The reaction was stopped by the addition of 1 N HCl and the absorbance was read at 450 nm. Three independent (biological) replicates were used for each assay.

Statistical analyses

Statistical analyses were performed using the SAS program v. 9.1 (SAS Institute, Cary, NC). ANOVA and PROC GLM procedures were used to compare C5-1 yields among treatments, densitometry data and protease activities. Contrast calculations were made when the ANOVA was significant using an α value threshold of 5%. Student t-tests were performed to compare mRNA transcript numbers for *A. tumefaciens* virulence proteins and C5-1 chains in control and MeJA-treated plants.

Acknowledgements

We thank Charles Goulet and Philippe Jutras for fruitful discussions while preparing this manuscript, and Émilie Desfossés-Foucault for help with and useful advice on the qPCR assays.

References

Anand, A., Uppalapati, S.R., Ryu, C.M., Allen, S.N., Kang, L., Tang, Y.H. and Mysore, K.S. (2008) Salicylic acid and systemic acquired resistance play a role in attenuating crown gall disease caused by *Agrobacterium tumefaciens*. *Plant Physiol.* **146**, 703–715.

Badri, M.A., Rivard, D., Coenen, K. and Michaud, D. (2009) Unintended molecular interactions in transgenic plants expressing clinically-useful proteins–The case of bovine aprotinin travelling the potato leaf cell secretory pathway. *Proteomics*, **9**, 746–756.

Balasubramaniam, D., Wilkinson, C., Van Cott, K. and Zhang, C. (2003) Tobacco protein separation by aqueous two-phase extraction. *J. Chromatogr. A*, **989**, 119–129.

Bardor, M., Loutelier-Bourhis, C., Paccalet, T., Cosette, P., Fitchette, A.C., Vézina, L.-P., Trepanier, S., Dargis, M., Lemieux, R., Lange, C., Faye, L. and Lerouge, P. (2003) Monoclonal C5-1 antibody produced in transgenic alfalfa plants exhibits a N-glycosylation that is homogenous and suitable for glyco-engineering into human-compatible structures. *Plant Biotechnol. J.* **1**, 451–462.

Baur, A., Reski, R. and Gorr, G. (2005) Enhanced recovery of a secreted recombinant human growth factor using stabilizing additives and by co-expression of human serum albumin in the moss *Physcomitrella patens*. *Plant Biotechnol. J.* **3**, 331–340.

Benchabane, M., Goulet, C., Rivard, D., Faye, L., Gomord, V. and Michaud, D. (2008) Preventing unintended proteolysis in plant protein biofactories. *Plant Biotechnol. J.* **6**, 633–648.

Benchabane, M., Rivard, D., Girard, C. and Michaud, D. (2009) Companion protease inhibitors to protect recombinant proteins in transgenic plant extracts. *Methods Mol. Biol.* **483**, 265–273.

Bendandi, M., Marillonnet, S., Kandzia, R., Thieme, F., Nickstadt, A., Herz S. Fröde, R., Inogés, S., Lopez-Diaz de Cerio, A., Soria, E., Villanueva, H., Vancanneyt, G., McCormick, A., Tusé, D., Lenz, J., Butler-Ransohoff, J.E., Klimyuk, V. and Gleba, Y. (2010) Rapid, high-yield production in plants of individualized idiotype vaccines for non-Hodgkin's lymphoma. *Ann. Oncol.*, **21**, 2420–2427.

Bilgin, D.D., Zavala, J.A., Zhu, J., Clough, S.J., Ort, D.R. and DeLucia, E.H. (2010) Biotic stress globally downregulates photosynthesis genes. *Plant, Cell Environ.* **33**, 1597–1613.

Bradford, M.M. (1976) A rapid and sensitive method for the quantitation of microgram quantities of protein utilizing the principle of protein-dye binding. *Anal. Biochem.* **72**, 248–254.

Buyel, J.F. and Fischer, R. (2012) Predictive models for transient protein expression in tobacco (*Nicotiana tabacum* L.) can optimize process time, yield, and downstream costs. *Biotechnol. Bioeng.* **109**, 2575–2588.

Buyel, J.F. and Fischer, R. (2014) Flocculation increases the efficacy of depth filtration during the downstream processing of recombinant pharmaceutical proteins produced in tobacco. *Plant Biotechnol. J.* **12**, 240–252.

Buyel, J.F., Woo, J.A., Cramer, S.M. and Fischer, R. (2013) The use of quantitative structure–activity relationship models to develop optimized processes for the removal of tobacco host cell proteins during biopharmaceutical production. *J. Chromatogr. A*, **1322**, 18–28.

Craig, R., Cortens, J.P. and Beavis, R.C. (2004) Open source system for analyzing, validating, and storing protein identification data. *J. Proteome Res.* **3**, 1234–1242.

D'Aoust, M.-A., Lavoie, P.-O., Belles-Isles, J., Bechtold, N., Martel, M. and Vézina, L.-P. (2009) Transient expression of antibodies in plants using syringe agroinfiltration. *Methods Mol. Biol.* **483**, 41–50.

Derksen, H., Rampitsch, C. and Daayf, F. (2013) Signaling cross-talk in plant disease resistance. *Plant Sci.* **207**, 79–87.

Duceppe, M.-O., Cloutier, C. and Michaud, D. (2012) Wounding, insect chewing and phloem sap feeding differentially alter the leaf proteome of potato, *Solanum tuberosum* L.. *Proteome Sci.* **10**, 73.

Ellis, R.J. (1979) Most abundant protein in the world. *Trends Biochem. Sci.* **4**, 241–244.

Farinas, C.S., Leite, A. and Miranda, E.A. (2005) Aqueous extraction of recombinant human proinsulin from transgenic maize endosperm. *Biotechnol. Prog.* **21**, 1466–1471.

Faye, L., Boulaflous, A., Benchabane, M., Gomord, V. and Michaud, D. (2005) Protein modifications in the plant secretory pathway: current status and practical implications in molecular pharming. *Vaccine*, **23**, 1770–1778.

Figueiredo, A., Monteiro, F. and Sabastiana, M. (2014) Subtilisin-like proteases in plant–pathogen recognition and immune priming: a perspective. *Front. Plant Sci.* **5**, 739.

Fischer, R., Schillberg, S., Hellwig, S., Twyman, R.M. and Drossard, J. (2012) GMP issues for recombinant plant-derived pharmaceutical proteins. *Biotechnol. Adv.* **30**, 434–439.

Fischer, R., Schillberg, S., Buyel, J.F. and Twyman, R.M. (2013) Commercial aspects of pharmaceutical protein production in plants. *Curr. Pharm. Des.* **19**, 5471–5479.

Gaeda, D., Valdés, R., Escobar, A., Ares, D.M., Torres, E., Blanco, R., Ferro, W., Dorta, D., Gonzalez, M., Aleman, M.R., Padilla, S., Gomez, L., del Castillo, N., Mendoza, O., Urquiza, D., Soria, Y., Brito, J., Leyva, A., Borroto, C. and Gavilondo, J.V. (2007) Detection of Rubisco and mycotoxins as potential contaminants of a plantibody against the hepatitis B surface antigen purified from tobacco. *Biologicals*, **35**, 309–315.

Girard, C., Rivard, D., Kiggundu, A., Kunert, K., Gleddie, S.C., Cloutier, C. and Michaud, D. (2007) A multicomponent, elicitor-inducible cystatin complex in tomato, *Solanum lycopersicum*. *New Phytol.* **173**, 841–851.

Giri, A.P., Wunsche, H., Mitra, S., Zavala, J.A., Muck, A., Svatos, A. and Baldwin, I.T. (2006) Molecular interactions between the specialist herbivore *Manduca sexta* (Lepidoptera, Sphingidae) and its natural host *Nicotiana attenuata*. VII. Changes in the plant's proteome. *Plant Physiol.* **142**, 1621–1641.

Gomord, V., Fitchette, A.C., Menu-Bouaouiche, L., Saint-Jore-Dupas, C., Plasson, C., Michaud, D. and Faye, L. (2010) Plant-specific glycosylation patterns in the context of therapeutic protein production. *Plant Biotechnol. J.* **8**, 564–587.

Goossens, A., Van Montagu, M. and Angenon, G. (1999) Co-introduction of an antisense gene for an endogenous seed storage protein can increase expression of a transgene in *Arabidopsis thaliana* seeds. *FEBS Lett.* **456**, 160–164.

Goulet, C., Goulet, C., Goulet, M.-C. and Michaud, D. (2010) 2-DE proteome maps for the leaf apoplast of *Nicotiana benthamiana*. *Proteomics*, **10**, 2536–2544.

Goulet, C., Khalf, M., Sainsbury, F., D'Aoust, M.-A. and Michaud, D. (2012) A protease activity-depleted environment for heterologous proteins migrating towards the leaf cell apoplast. *Plant Biotechnol. J.* **10**, 83–94.

Hassan, S., Van Dolleweerd, C.J., Ioakeimidis, F., Keshavarz-Moore, E. and Ma, J.K.-C. (2008) Considerations for extraction of monoclonal antibodies targeted to different subcellular compartments in transgenic tobacco plants. *Plant Biotechnol. J.* **6**, 733–748.

Hermsmeier, D., Schittko, U. and Baldwin, I.T. (2001) Molecular interactions between the specialist herbivore *Manduca sexta* (Lepidoptera, Sphingidae) and its natural host *Nicotiana attenuata*. I. Large-scale changes in the accumulation of growth- and defense-related plant mRNAs. *Plant Physiol.* **125**, 683–700.

Hörger, A.C. and van der Hoorn, R.A.L. (2013) The structural basis of specific protease–inhibitor interactions at the plant–pathogen interface. *Curr. Opin. Struct. Biol.* **23**, 842–850.

Höwing, T., Huesmann, C., Hoefle, C., Nagel, M.-K., Isono, E., Hückelhoven, R. and Gietl, E. (2014) Endoplasmic reticulum KDEL-tailed cysteine endopeptidase 1 of *Arabidopsis* (AtCEP1) is involved in pathogen defense. *Front. Plant Sci.* **5**, 58.

Huang, T.-K., Plesha, M.A., Falk, B.W., Dandekar, A.M. and McDonald, K.A. (2009) Bioreactor strategies for improving production yield and functionality of a recombinant human protein in transgenic tobacco cell cultures. *Biotechnol. Bioeng.*, **102**, 508–520.

Jung, C., Lyou, S.H., Yeu, S., Kim, M.A., Rhee, S., Kim, M., Lee, J.S., Choi, Y.D. and Cheong, J.J. (2007) Microarray-based screening of jasmonate-responsive genes in *Arabidopsis thaliana*. *Plant Cell Rep.* **26**, 1053–1063.

Jung, S.-K., Lindenmuth, B.E., McDonald, K.A., Hwang, M.S., Nguyen Bui, M.Q., Falk, B.W., Uratsu, S.L., Phu, M.L. and Dandekar, A.M. (2014) *Agrobacterium tumefaciens* mediated transient expression of plant cell wall-degrading enzymes in detached sunflower leaves. *Biotechnol. Prog.* **30**, 905–915.

Keller, A., Nesvizhskii, A.I., Kolker, E. and Aebersold, R. (2002) Empirical statistical model to estimate the accuracy of peptide identifications made by MS/MS and database search. *Anal. Chem.* **74**, 5383–5392.

Khoudi, H., Laberge, S., Ferullo, J.M., Bazin, R., Darveau, A., Castonguay, Y., Allard, G., Lemieux, R. and Vézina, L.-P. (1999) Production of a diagnostic monoclonal antibody in perennial alfalfa plants. *Biotechnol. Bioeng.* **64**, 135–143.

Kim, Y.-M., Lee, J.-Y., Lee, T., Lee, Y.-H., Kim, S.-H., Kang, S.-H., Yoon, U.-H., Ha, S.-H. and Lim, S.-H. (2012) The suppression of the glutelin storage protein gene in transgenic rice seeds results in a higher yield of recombinant protein. *Plant Biotechnol. Rep.* **6**, 347–353.

Komarnytsky, S., Borisjuk, N., Yakoby, N., Garvey, A. and Raskin, I. (2006) Cosecretion of protease inhibitor stabilizes antibodies produced by plant roots. *Plant Physiol.* **141**, 1185–1193.

Lacroix, B., Li, J., Tzfira, T. and Citovsky, V. (2006) Will you let me use your nucleus? How *Agrobacterium* gets its T-DNA expressed in the host plant cell. *Can. J. Physiol. Pharmacol.* **84**, 333–345.

Laemmli, U.K. (1970) Cleavage of structural proteins during the assembly of the head of bacteriophage T4. *Nature* **227**, 680–685.

Lai, H., Engle, M., Fuchs, A., Keller, K., Johnson, S., Gorlatov, S., Diamond, M.S. and Chen, Q. (2010) Monoclonal antibody produced in plants efficiently treats West Nile virus infection in mice. *Proc. Natl Acad. Sci. USA*, **107**, 2419–2424.

Lawrence, S.D., Novak, N.G., Ju, C.J. and Cooke, J.E. (2008) Potato, *Solanum tuberosum*, defense against Colorado potato beetle, *Leptinotarsa decemlineata* (Say): microarray gene expression profiling of potato by Colorado potato beetle regurgitant treatment of wounded leaves. *J. Chem. Ecol.* **34**, 1013–1025.

Leuzinger, K., Dent, M., Hurtado, J., Stahnke, J., Lai, H., Zhou, X. and Chen, Q. (2013) Efficient agroinfiltration of plants for high-level transient expression of recombinant proteins. *J. Vis. Exp.* **77**, e50521.

Mahajan, N.S., Mishra, M., Tamhane, V.A., Gupta, V.S. and Giri, A.P. (2014) Stress inducible proteomic changes in *Capsicum annuum* leaves. *Plant Physiol. Biochem.* **74**, 212–217.

Makhzoum, A., Benyammi, R., Koustafa, K. and Trémouillaux-Guiller, J. (2014) Recent advances on host plants and expression cassettes' structure and function in plant molecular pharming. *BioDrugs*, **28**, 145–159.

Mandal, M.K., Fischer, R., Schillberg, S. and Schiermeyer, A. (2014) Inhibition of protease activity by antisense RNA improves recombinant protein production in *Nicotiana tabacum* cv. Bright Yellow 2 (BY-2) suspension cells. *Biotechnol. J.* **9**, 1065–1073.

Maxmen, A. (2012) Drug-making plant blooms. *Nature*, **485**, 160.

Menkhaus, T.J., Bai, Y., Zhang, C.M., Nikolov, Z.L. and Glatz, C.E. (2004) Considerations for the recovery of recombinant proteins from plants. *Biotechnol. Prog.* **20**, 1001–1014.

Merlin, M., Gecchele, E., Capaldi, S., Pezzotti, M. and Avesani, L. (2014) Comparative evaluation of recombinant protein production in different biofactories: The green perspective. *Biomed. Res. Int.* **2014**, 136419.

Nesvizhskii, A.I., Keller, A., Kolker, E. and Aebersold, R. (2003) A statistical model for identifying proteins by tandem mass spectrometry. *Anal. Chem.* **75**, 4646–4658.

Noir, S., Bömer, M., Takahashi, N., Ishida, T., Tsui, T.-L., Balbi, V., Shanahan, H., Sugimoto, K. and Devoto, A. (2013) Jasmonate controls leaf growth by repressing cell proliferation and the onset of endoreduplication while maintaining a potential stand-by mode. *Plant Physiol.* **161**, 1930–1951.

Okada, K., Abe, H. and Arimura, G. (2015) Jasmonates induce both defense responses and communication in monocotyledonous and dicotyledonous plants. *Plant Cell Physiol.* **56**, 16–27.

Peckham, G.D., Bugos, R.C. and Su, W.W. (2006) Purification of GFP fusion proteins from transgenic plant cell cultures. *Protein Expr. Purif.* **49**, 183–189.

Pillay, P., Schlüter, U., van Wyk, S., Kunert, K.J. and Vorster, B.J. (2014) Proteolysis of recombinant proteins in bioengineered plant cells. *Bioengineered*, **5**(1), 1–6.

Potenza, C., Aleman, L. and Sengupta-Gopalan, C. (2004) Targeting transgene expression in research, agricultural, and environmental applications: Promoters used in plant transformation. *In Vitro Cell Dev. Biol.* **40**, 1–22.

Ramirez, V., Lopez, A., Mauch-Mani, B., Gil, M.J. and Vera, P. (2013) An extracellular subtilase switch for immune priming in Arabidopsis. *PLoS Pathog.* **9**, e1003445.

Rivard, D., Anguenot, R., Brunelle, F., Le, V.Q., Vézina, L.-P., Trépanier, S. and Michaud, D. (2006) An in-built proteinase inhibitor system for the protection of recombinant proteins recovered from transgenic plants. *Plant Biotechnol. J.* **4**, 359–368.

Robert, S., Khalf, M., Goulet, M.-C., D'Aoust, M.-A., Sainsbury, F. and Michaud, D. (2013) Protection of recombinant mammalian antibodies from development-dependent proteolysis in leaves of *Nicotiana benthamiana*. *PLoS ONE*, **8**, e70203.

Robert-Seilaniantz, A., Grant, M. and Jones, J.D. (2011) Hormone crosstalk in plant disease and defence: more than just jasmonate-salicylate antagonism. *Annu. Rev. Phytopathol.* **49**, 317–343.

Ross, K.C. and Zhang, C. (2010) Separation of recombinant ß-glucuronidase from transgenic tobacco by aqueous two-phase extraction. *Biochem. Eng. J.* **49**, 343–350.

Sack, M., Hofbauer, A., Fischer, R. and Stoger, E. (2015) The increasing value of plant-made proteins. *Curr. Opin. Biotechnol.* **32**, 163–170.

Sainsbury, F., Lavoie, P.O., D'Aoust, M.-A., Vézina, L.-P. and Lomonossoff, G.P. (2008) Expression of multiple proteins using full-length and deleted versions of cowpea mosaic virus RNA-2. *Plant Biotechnol. J.* **6**, 82–92.

Schmidt, M.A. and Herman, E.M. (2008) Proteome rebalancing in soybean seeds can be exploited to enhance foreign protein accumulation. *Plant Biotechnol. J.* **6**, 832–842.

Shigemitsu, T., Ozaki, S., Saito, Y., Kuroda, M., Morita, S., Satoh, S. and Masumura, T. (2012) Production of human growth hormone in transgenic rice seeds: co-introduction of RNA interference cassette for suppressing the gene expression of endogenous storage proteins. *Plant Cell Rep.* **31**, 539–549.

Stanton, M., Ullmann-Zeunert, L., Wielsch, N., Bartram, S., Svatos, A., Baldwin, I.T. and Groten, K. (2013) Silencing ribulose-1,5-bisphosphate carboxylase/oxygenase expression does not disrupt nitrogen allocation to defense after simulated herbivory in *Nicotiana attenuata*. *Plant Signal. Behav.* **8**, e27570.

Stoger, E., Fischer, R., Moloney, M. and Ma, J.K.-C. (2014) Plant molecular pharming for the treatment of chronic and infectious diseases. *Annu. Rev. Plant Biol.* **65**, 743–768.

Streatfield, S.J. (2007) Approaches to achieve high-level heterologous protein production in plants. *Plant Biotechnol. J.* **5**, 2–16.

Thaler, J.S., Humphrey, P.T. and Whiteman, N.K. (2012) Evolution of jasmonate and salicylate signal crosstalk. *Trends Plant Sci.* **17**, 260–270.

Ullmann-Zeunert, L., Stanton, M.A., Wielsch, N., Bartram, S., Hummert, C., Svatos, A., Baldwin, I.T. and Groten, K. (2013) Quantification of growth–defense trade-offs in a common currency: nitrogen required for phenolamide biosynthesis is not derived from ribulose-1,5-bisphosphate carboxylase/oxygenase turnover. *Plant J.* **75**, 417–429.

Veena, Jiang, H., Doerge, R.W. and Gelvin, S.B. (2003) Transfer of T-DNA and Vir proteins to plant cells by *Agrobacterium tumefaciens* induces expression of host genes involved in mediating transformation and suppresses host defense gene expression. *Plant J.* **35**, 219–226.

Vézina, L.-P., Faye, L., Lerouge, P., D'Aoust, M.-A., Marquet-Blouin, E., Burel, C., Lavoie, P.-O., Bardor, M. and Gomord, V. (2009) Transient co-expression for fast and high-yield production of antibodies with human-like *N*-glycans in plants. *Plant Biotechnol. J.* **7**, 442–455.

Voinnet, O., Rivas, S., Mestre, P. and Baulcombe, D. (2003) An enhanced transient expression system in plants based on suppression of gene silencing by the p19 protein of tomato bushy stunt virus. *Plant J.* **33**, 949–956.

Wei, Z., Hu, W., Lin, Q., Cheng, X., Tong, M., Zhu, L., Chen, R. and He, G. (2009) Understanding rice plant resistance to the Brown Planthopper (*Nilaparvata lugens*): a proteomic approach. *Proteomics*, **9**, 2798–2808.

Wilken, L.R. and Nikolov, Z.L. (2006) Factors influencing recombinant human lysozyme extraction and cation exchange adsorption. *Biotechnol. Prog.* **22**, 745–752.

Wilken, L.R. and Nikolov, Z.L. (2012) Recovery and purification of plant-made recombinant proteins. *Biotechnol. Adv.* **30**, 419–433.

Woodard, S.L., Wilken, L.R., Barros, G.O.F., White, S.G. and Nikolov, Z.L. (2009) Evaluation of monoclonal antibody and phenolic extraction from transgenic

Lemna for purification process development. *Biotechnol. Bioeng.* **104**, 562–571.

Yuan, Z.-C., Edlind, M.P., Liu, P., Saenkham, P., Banta, L.M., Wise, A.A., Roznone, E., Binns, A.N., Kerr, K. and Nester, E.W. (2007) The plant signal salicylic acid shuts down expression of the *vir* regulon and activates quormone-quenching genes in *Agrobacterium*. *Proc. Natl Acad. Sci.* **104**, 11790–11795.

Zhang, C., Lillie, R., Cotter, J. and Vaughan, D. (2005) Lysozyme purification from tobacco extract by polyelectrolyte precipitation. *J. Chromatogr. A*, **1069**, 107–112.

Zhong, Q., Xu, L., Zhang, C. and Glatz, C. (2007) Purification of recombinant aprotinin from transgenic corn germ fraction using ion exchange and hydrophobic interaction chromatography. *Appl. Microbiol. Biotechnol.* **76**, 607–613.

Zubo, Y.O., Yamburenko, M.V., Kusnetsov, V.V. and Börner, T. (2011) Methyl jasmonate, gibberellic acid, and auxin affect transcription and transcript accumulation of chloroplast genes in barley. *J. Plant Physiol.* **168**, 1335–1344.

11

Protein body formation in leaves of *Nicotiana benthamiana*: a concentration-dependent mechanism influenced by the presence of fusion tags

Reza Saberianfar[1,2], Jussi J. Joensuu[3], Andrew J. Conley[3] and Rima Menassa*[1,2]

[1]*Department of Biology, Western University, London, ON, Canada*
[2]*Southern Crop Protection and Food Research Centre, Agriculture and Agri-Food Canada, London, ON, Canada*
[3]*VTT Technical Research Centre of Finland, Espoo, Finland*

*Correspondence

email rima.menassa@agr.gc.ca

Keywords: protein body, PB, protein body formation, elastin-like polypeptides, hydrophobin, *Nicotiana benthamiana*, transient expression.

Summary

Protein bodies (PBs) are endoplasmic reticulum (ER) derived organelles originally found in seeds whose function is to accumulate seed storage proteins. It has been shown that PB formation is not limited to seeds and green fluorescent protein (GFP) fused to either elastin-like polypeptide (ELP) or hydrophobin (HFBI) fusion tags induce the formation of PBs in leaves of *N. benthamiana*. In this study, we compared the ELP- and HFBI-induced PBs and showed that ELP-induced PBs are larger than HFBI-induced PBs. The size of ELP- and HFBI-induced PBs increased over time along with the accumulation levels of their fused protein. Our results show that PB formation is a concentration-dependent mechanism in which proteins accumulating at levels higher than 0.2% of total soluble protein are capable of inducing PBs *in vivo*. Our results show that the presence of fusion tags is not necessary for the formation of PBs, but affects the distribution pattern and size of PBs. This was confirmed by PBs induced by fluorescent proteins as well as fungal xylanases. We noticed that in the process of PB formation, secretory and ER-resident molecules are passively sequestered into the lumen of PBs. We propose to use this property of PBs as a tool to increase the accumulation levels of erythropoietin and human interleukin-10 by co-expression with PB-inducing proteins.

Introduction

During the past two decades, plants have been extensively studied as efficient green bioreactors for production of recombinant proteins (Egelkrout *et al.*, 2012). In 2012, the first plant-produced pharmaceutical protein was approved by the US Food and Drug Administration (Maxmen, 2012). Several host plants and production systems including cell culture, seed- and leaf-based systems have been used for producing recombinant proteins (Stoger *et al.*, 2014).

Seeds are naturally suited to produce and store high levels of proteins. High protein content, low protease activity and low water content make seeds very attractive for producing and accumulating target proteins (Benchabane *et al.*, 2008). Seeds have developed specialized protein storage organelles such as protein bodies (PBs), protein storage vacuoles (PSVs) and oil bodies (Khan *et al.*, 2012). Over the years, several plant-made pharmaceuticals and industrial recombinant proteins have been produced in plant seeds (Lau and Sun, 2009). However, seed-based expression systems face issues related to biocontainment due to the possibility of gene leakage into the environment through distribution of seed or pollen. As well, relatively long periods of time are required to generate transgenic plant lines and screen them for high expressing lines (Ma and Wang, 2012).

Unlike seed-based systems, leaves can be harvested before flowering therefore minimizing the risk of gene leakage to the environment. In addition, leaves produce high biomass yield along with high soluble protein content. Among leaf-based expression systems, two approaches have produced high levels of recombi-

nant proteins, chloroplast transformation (Bock, 2014; Chebolu and Daniell, 2009) and transient expression (Komarova *et al.*, 2010). However, although leaf-based expression systems have advantages for biocontainment, the production yield of recombinant proteins is often low due to proteolytic degradation of foreign proteins (Benchabane *et al.*, 2008; Fischer *et al.*, 2004). Several fusion tags such as Zera®, elastin-like polypeptides (ELPs) and hydrophobins (HFBs) have been used to improve accumulation levels of recombinant fusion partners in leaves (Conley *et al.*, 2011). The use of these fusion tags was also shown to facilitate the process of recombinant protein purification (Conley *et al.*, 2011; Khan *et al.*, 2012; Torrent *et al.*, 2009b). All three fusion tags were hypothesized to improve accumulation of recombinant proteins by inducing the formation of novel organelles resembling seed PBs, which protect recombinant proteins from proteolytic degradation. Indeed, these PBs are surrounded by a membrane studded with ribosomes, contain soluble proteins and are thought to bud off the ER and to be stored in the cytosol (Conley *et al.*, 2009b). This feature differentiates PBs from inclusion bodies found in *E. coli* which contain insoluble, partially folded proteins (Baneyx and Mujacic, 2004), and from inclusion bodies and protein crystals found in chloroplasts (Fernandez-San Millan *et al.*, 2003; de Cosa *et al.*, 2001).

Zera® is a proline-rich region at the N-terminal region of a maize storage protein known as γ-zein (Llop-Tous *et al.*, 2010). Proteins fused to Zera® were found in the form of PBs in a wide range of host cells including mammalian, insect, fungi and plants (Torrent *et al.*, 2009b). Zera-fused proteins accumulate in the endoplasmic reticulum (ER) independent of the conventional ER retrieval signal (HDEL/KDEL), and it is thought that the interaction of Zera® with the ER membrane induces the budding of PBs from the ER. The Zera-induced PBs can be purified by gradient density centrifugation (Torrent *et al.*, 2009a).

ELPs are synthetic biopolymers of repeating pentapeptides, Val-Pro-Gly-Xaa-Gly, similar to repetitive pentapeptides of the mammalian protein elastin, in which the guest amino acid (Xaa) can be any amino acid except proline. ELP tags containing 100–120 pentapeptide repeats can be used for the rapid, nonchromatographic purification method called inverse transition cycling (ITC) (Urry and Parker, 2002). Because of this property, ELPs have been investigated in recombinant protein production in plants (Floss *et al.*, 2010). In chloroplasts, a 121-pentapeptide-repeat ELP did not accumulate to expected high levels (Guda *et al.*, 2000), while a 100-repeat ELP increased the accumulation of the 2G12 antibody by 2.5 to 3 fold up to 1% of total soluble protein both in seeds and in leaves when retrieved to the ER (Floss *et al.*, 2009). The difference in accumulation between the two systems may be due to a limitation in the amino acid or tRNA pool in the chloroplast (Guda *et al.*, 2000). An ELP size library expression experiment targeting ELP fusions to the ER showed that smaller ELPs (5–20 repeats) accumulate to higher levels than larger ELPs (80–240 repeats) (Conley *et al.*, 2009a). Subsequently, an ER-retrieved GFP-ELP$_{28}$ fusion was shown to induce the formation of PBs as well as boost accumulation levels of GFP both in transient and in transgenic expression (Conley *et al.*, 2009b; Gutierrez *et al.*, 2013).

Hydrophobins are surface-active proteins originally found in fungi (Wessels, 1997). Hydrophobins share eight conserved cysteine residues responsible for the formation of four intramolecular disulphide bridges which maintain the structural stability of the molecule. The rigid core enables hydrophobins to display a group of hydrophobic amino acids on their surface forming a 'hydrophobic patch' responsible for their amphiphilic nature (Hakanpää *et al.*, 2004). Therefore, hydrophobins can self-assemble at hydrophilic–hydrophobic interfaces, a property that can be exploited for purification using a surfactant-based aqueous two-phase separation system (ATPS) (Linder *et al.*, 2001). Hydrophobin-I (HFBI) fusion has been shown to increase the accumulation levels of GFP, and to induce PB formation *in vivo* similar to ELP by transient and transgenic expression (Gutierrez *et al.*, 2013; Joensuu *et al.*, 2010).

Interestingly, fusion of recombinant proteins to ELP and HFBI does not always impact the accumulation levels of their fused protein in the same way. For instance, Phan *et al.* (2014) showed that ELP fusion to hemagglutinin increases accumulation levels, while HFBI fusion does not significantly impact hemagglutinin levels.

In this study, we compared ELP and HFBI fusion tags in the process of PB formation. We showed that PB formation is not exclusively promoted by the fusion tags and that protein accumulation level is a critical element involved in this process. Our results indicate that PB formation is not a selective mechanism and proteins targeted to the secretory pathway can be passively sequestered into PBs. We have used this property of PBs as a novel approach to increase the accumulation levels of erythropoietin and human interleukin-10.

Results

Elastin-like polypeptides induce larger protein bodies than hydrophobin-I

To compare the effects of ELP and HFBI fusion tags on PB formation and accumulation levels of their fused proteins, constructs (Conley *et al.*, 2009b; Joensuu *et al.*, 2010) targeting GFP to the endomembrane system (Figure 1) were co-agroinfiltrated into *N. benthamiana* with a construct containing the suppressor of gene silencing, p19, from *Cymbidium ringspot* virus (CymRSV) (Silhavy *et al.*, 2002). Previous independent reports had shown that GFP fusion with ELP or HFBI increased the accumulation levels of the GFP and gave rise to clusters of PBs with variable distribution patterns (Conley *et al.*, 2009b; Joensuu *et al.*, 2010). To compare PBs induced by ELP or HFBI and consequent GFP accumulation levels, the sizes of PBs induced by GFP-ELP and GFP-HFBI were measured, and the amount of GFP produced in every treatment was quantified 3, 4 and 5 days postinfiltration (dpi) (Figure 2). To obtain accurate measurements of all PBs within a cell, protoplasts were isolated from the infiltrated leaf tissue and used for imaging by confocal microscopy. Z-stack assemblies of 1.5 to 2μm slices spanning entire protoplasts and covering approximately 70 to 80 μm were used for three-dimensional visualization. With this technique, PBs within the cell were visualized and their detailed information was analysed using the Imaris software (Figure 2a,b) (Movie S1).

Day to day comparison of ELP- and HFBI-induced PBs suggests that ELP-induced PBs were significantly larger in size than HFBI-induced PBs (Figure 2c). At 3 and 4 dpi, 75% of the ELP-induced PBs were larger than 1.7 μm in diameter with the median of

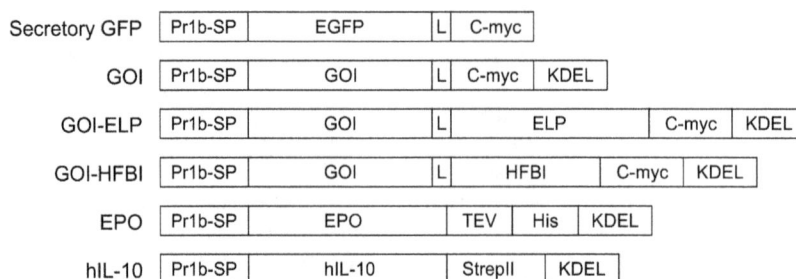

Secretory GFP	Pr1b-SP	EGFP	L	C-myc			
GOI	Pr1b-SP	GOI	L	C-myc	KDEL		
GOI-ELP	Pr1b-SP	GOI	L	ELP	C-myc	KDEL	
GOI-HFBI	Pr1b-SP	GOI	L	HFBI	C-myc	KDEL	
EPO	Pr1b-SP	EPO	TEV	His	KDEL		
hIL-10	Pr1b-SP	hIL-10	StrepII	KDEL			

Figure 1 Schematic representation of constructs used for *Agrobacterium*-mediated transient expression in *N. benthamiana* leaves. Pr1b, secretory signal peptide from the tobacco pathogenesis-related protein 1; EGFP, enhanced green fluorescent protein; GOI, gene of interest; L, linker (GGGS)$_3$; TEV, tobacco etch virus protease recognition site; c-myc, detection tag; KDEL, endoplasmic reticulum retrieval tetra peptide; ELP, elastin-like polypeptide; HFBI, hydrophobin-I; EPO, human erythropoietin; HIS, (His)$_6$ purification tag; hIL-10, human interleukin-10; Strep II, purification tag. In GOI, GOIs include EGFP, xyn10A, or xyn11A. In GOI-ELP, GOIs include EGFP or RFP. In GOI-HFBI, GOIs include EGFP, RFP, or xyn11A. Schematic not drawn to scale.

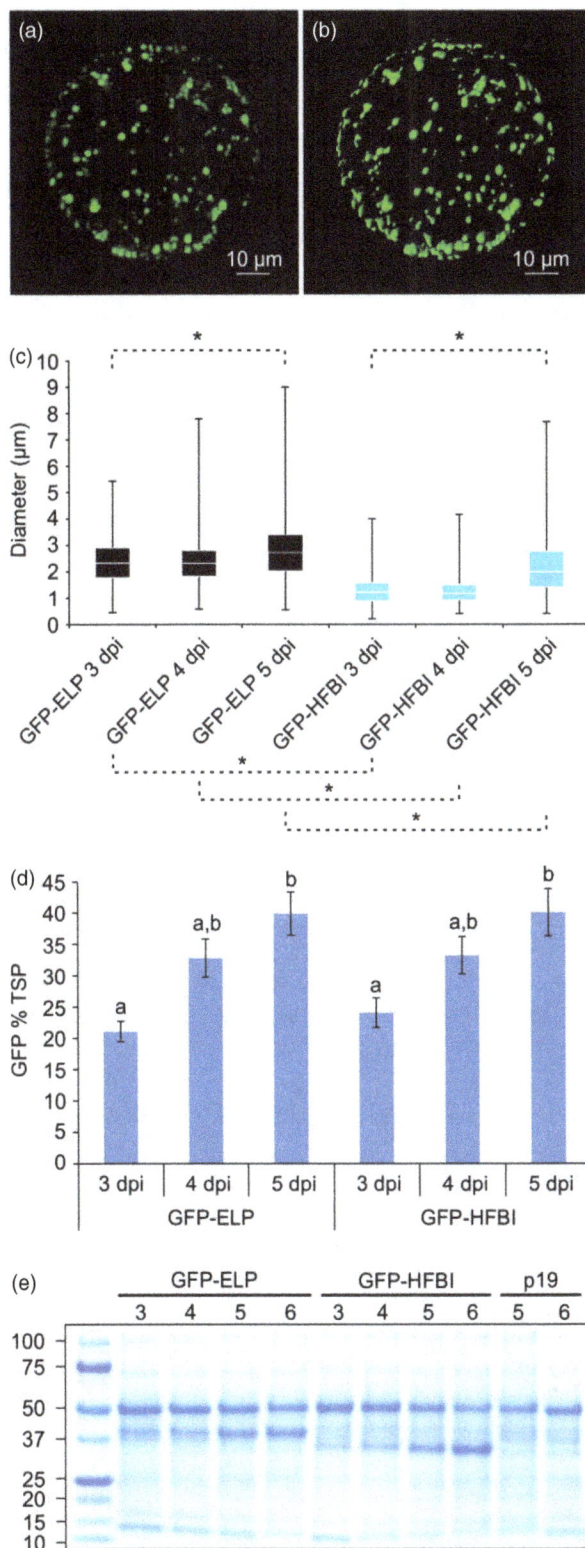

Figure 2 Protein body size and GFP accumulation in protoplasts expressing GFP-ELP and GFP-HFBI increase simultaneously. (a) 2D representation of the z-stacks acquired from a protoplast expressing GFP-ELP at 4 dpi. (b) 3D-rendered z-stack image of the protoplast shown in (a). (c) Size distribution pattern of GFP-ELP- and GFP-HFBI-induced PBs based on their respective diameters (µm) from 3 to 5 days postinfiltration (dpi). The box represents the middle 50% of the protein body size distribution, and upper and lower 'whiskers' represent the entire spread of the data acquired from 10 protoplasts per day per treatment; dotted lines indicate pairwise significant differences at $*P < 0.0001$ using Kruskal–Wallis one-way ANOVA. (d) GFP accumulation levels as % TSP. Four biological replicates (each replicate containing four leaf discs of the infiltrated tissue from a different plant) were used for fluorometry. Columns not denoted with the same letter are significantly different from each other ($P < 0.05$) using one-way ANOVA. Error bars represent the standard error of the mean. (e) Time course comparison of ELP and GFP-HFBI accumulation levels by 4%–15% continuous gradient SDS–PAGE analysis. Total soluble protein (10 µg/lane) of *N. benthamiana* leaf extracts infiltrated with GFP-ELP, GFP-HFBI and p19 (as control), at 3, 4, 5 and 6 dpi (pooled from 6 individual plants) were loaded and detected by GelCode staining.

located at 2 µm. The maximum size of GFP-ELP-induced PBs increased from 5.5 µm at 3 dpi to 9 µm at 5 dpi, while the maximum size of HFBI-induced PBs increased from 4 µm at 3 dpi to 7.7 µm at 5 dpi.

We found that PB size and protein accumulation levels increase simultaneously, as the size increase of ELP- or HFBI-induced PBs was accompanied with an increase in GFP accumulation from 21% and 24% of TSP at 3 dpi for GFP-ELP and GFP-HFBI, respectively to 40% of TSP at 5 dpi for both proteins (Figure 2d). Time course comparison of GFP-ELP or GFP-HFBI accumulation levels by SDS–PAGE and GelCode Blue staining also showed an increase in the amount of both proteins from 3 to 6 dpi (Figure 2e), with the recombinant protein band intensity similar to that of the large subunit of RuBisCo on day 5 post-infiltration, and even higher on day 6 with GFP-HFBI.

Protein body formation is a concentration-dependent mechanism

Previous studies have shown that by addition of fusion tags, the accumulation levels of recombinant proteins can increase (Conley *et al.*, 2009a; Joensuu *et al.*, 2010; Kaldis *et al.*, 2013; Torrent *et al.*, 2009a). Conley *et al.* (2009b) have also shown that by co-agroinfiltrating GFP-ELP and p19, a suppressor of post-transcriptional gene silencing (Silhavy *et al.*, 2002), recombinant protein levels increase and PB formation is induced in a higher percentage of cells. Surprisingly, using the same experimental set-up with unfused GFP, PBs were found in over 40% of the examined cells, suggesting a potential role for increased protein accumulation in the process of PB formation (Conley *et al.*, 2009b).

To investigate the effect of recombinant protein accumulation on PB formation, we conducted an experiment with a gradient of GFP levels. The GFP gradient was achieved by co-infiltrating a dilution series matrix of *Agrobacterium* containing the respective GFP constructs at an OD at A_{600} of 0.015 to 0.05 and the p19 suppressor of post-transcriptional gene silencing at an OD at A_{600} of 0.01 to 0.05. As expected, GFP accumulation levels increased with increasing *Agrobacterium* optical density (data not shown).

To compare the effects of fusion tags on PB formation at similar concentrations, we divided the treatments based on their respective accumulation levels into low (~0.2% of TSP) and high

approximately 2.4 µm, while 75% of the HFBI-induced PBs were smaller than 1.7 µm in diameter with the median of approximately 1.2 µm. At 5 dpi, the PBs were bigger in both the ELP and the HFBI fusions, with 75% of the ELP-induced PBs larger than 2 µm in diameter with the median of 2.7 µm, while 50% of the HFBI-induced PBs were also within this range with the median

(>6.5% of TSP) categories (Figure 3a) and looked for the effect of protein concentration on PB formation in each category (Figure 3b–g). At low accumulation levels, ER-targeted GFP highlighted the typical ER network pattern with minuscule PBs appearing along the ER network (Figure 3b). At a similar accumulation level of around 0.2% of TSP, both GFP-ELP and GFP-HFBI produced PBs. GFP-ELP-induced PBs were larger in size (in some cases reaching 5 μm) compared to GFP-HFBI-induced PBs, although the number of GFP-HFBI-induced PBs was larger (Figure 3c,d).

On the other hand, PBs were present in all treatments when accumulation levels were high (>6.5% of TSP), irrespective of the presence or absence of the fusion tags (Figure 3e–g). As shown in Figure 3e, ER-targeted GFP induced PBs with sizes ranging from very small (~0.5 μm) to very large (~6 μm). At similar accumulation levels, GFP-ELP and GFP-HFBI induced higher numbers of homogenously sized PBs that were clustered together (Figure 3f,g). GFP-ELP generally induced PBs that were larger than GFP-HFBI.

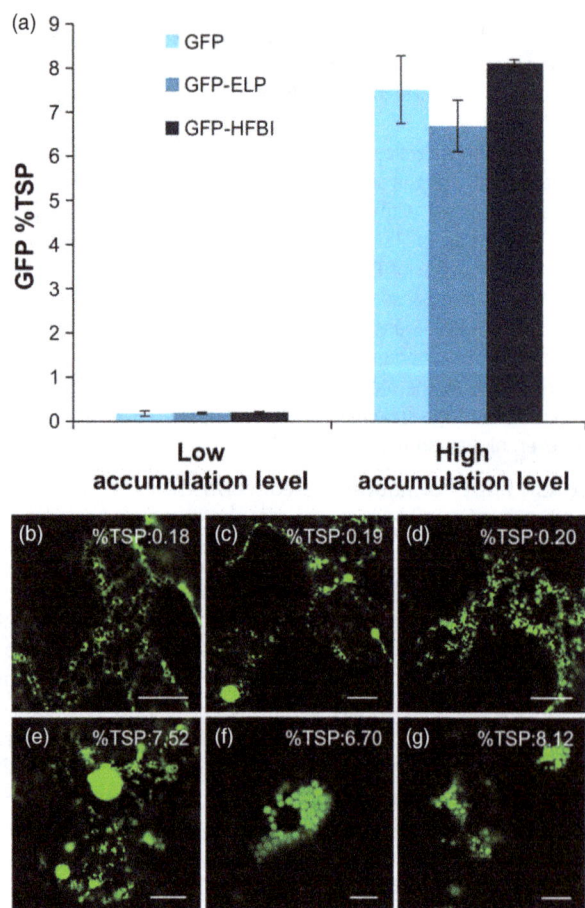

Figure 3 High recombinant protein concentration leads to protein body formation. (a) Leaf tissue samples were divided into low and high expressing groups and monitored for PB formation. Error bars represent the standard deviation of the mean from technical replicates quantified by dot blots. (b–d) PB formation in leaf tissues with low accumulation levels (~0.2% of TSP). Leaf tissue was infiltrated with GFP, GFP-ELP and GFP-HFBI in (b), (c) and (d), respectively. (e–g) PB formation in leaves with high accumulation levels (>6.5% TSP). Leaf tissue was infiltrated with GFP, GFP-ELP and GFP-HFBI in (e), (f) and (g), respectively. Bar, 10 μm.

The appearance of very small PBs at 0.2% of TSP in the unfused GFP treatment, and their increase in number and size at higher accumulation levels, indicates that levels of GFP, and possibly any other protein accumulating at high levels, may play a major role in PB biogenesis. The different patterns of PB appearance, that is their clustering and size uniformity may be due to the physico-chemical properties of the ELP and HFBI tags.

Secretory or ER-targeted proteins are sequestered passively into protein bodies

A proteomics study of the composition of Zera-induced PBs has shown the presence of a broad range of proteins other than Zera, including secretory pathway proteins, in PBs (Joseph et al., 2012). Conley et al. (2009b) also showed that the ER-resident chaperone BiP is found in GFP-ELP-induced protein bodies. Therefore, we hypothesize that during the formation of PBs, secretory or ER-resident proteins can be sequestered passively into the core of the PBs. To understand if this is the case, we co-expressed secretory GFP or ER-targeted GFP along with RFP-HFBI and RFP-ELP. All of the treatments were performed in the presence of p19 to ensure high accumulation levels.

Secretory GFP highlighted the apoplastic space between the cells without formation of PBs (Figure 4a), while ER-targeted GFP induced the formation of very small PBs along the ER network (Figure 4b), and as expected, both RFP-HFBI and RFP-ELP gave rise to clusters of homogeneous PBs with RFP-ELP PBs being larger in size (Figure 4c,d). Co-expression of secretory GFP and RFP-HFBI (Figure 4e–g) or RFP-ELP (Figure 4h–j) resulted in the sequestration of secretory GFP into HFBI- or ELP- induced PBs, shown by colocalization of both green and red fluorescent proteins into the same PBs. Co-expression of ER-targeted GFP with RFP-HFBI and RFP-ELP also resulted in the sequestration of GFP into PBs induced by HFBI and ELP (Figure 4k–p). In all of the co-expression treatments, the typical HFBI- or ELP-induced clustered patterns of PBs were observed (Figure 4e–p). Our results indicate that proteins targeted to the secretory pathway, or retrieved to the ER (Figure 1) can be sequestered into HFBI- or ELP-induced PBs.

Fungal xylanases induce the formation of protein bodies

To find out whether PB formation is limited to GFP or occurs upon overexpression of other proteins, we examined the possibility of PB formation with two other recombinant proteins accumulating at or higher than 0.2% of TSP. Two fungal xylanases from *Aspergillus niger* known as xyn11A (Nordberg et al., 2014) with and without the HFBI tag, and xyn10A (Nordberg et al., 2014) were selected and targeted to the ER with the ER retrieval tetrapeptide KDEL (Figure 1). Accumulation levels of the three recombinant proteins were quantified by Western blot analysis at 4 and 6 dpi (Figure 5a,b). Xyn11A-HFBI accumulated to 2.1% of TSP at 6 dpi, almost 10-fold higher than the unfused enzyme. Based on our results with GFP, PB formation was expected in all three groups. The subcellular localization of xyn11A, xyn11A-HFBI and xyn10A was determined by the whole-mount immunolocalization of the infiltrated leaf tissue using anti-c-myc monoclonal mouse antibody as the primary antibody and donkey anti-mouse IgG antibody fused to Alexa Fluor® 488 dye as the secondary antibody, followed by confocal microscopy analysis. As shown in Figure 5c and e, xyn11A and xyn10A induced the formation of dispersed PBs irregular in size, whereas higher accumulating xyn11A-HFBI induced numerous clustered PBs homogeneous in size (Figure 5d). This property of xyn11A-HFBI resembles the typical pattern of HFBI-induced PBs previously observed with GFP-

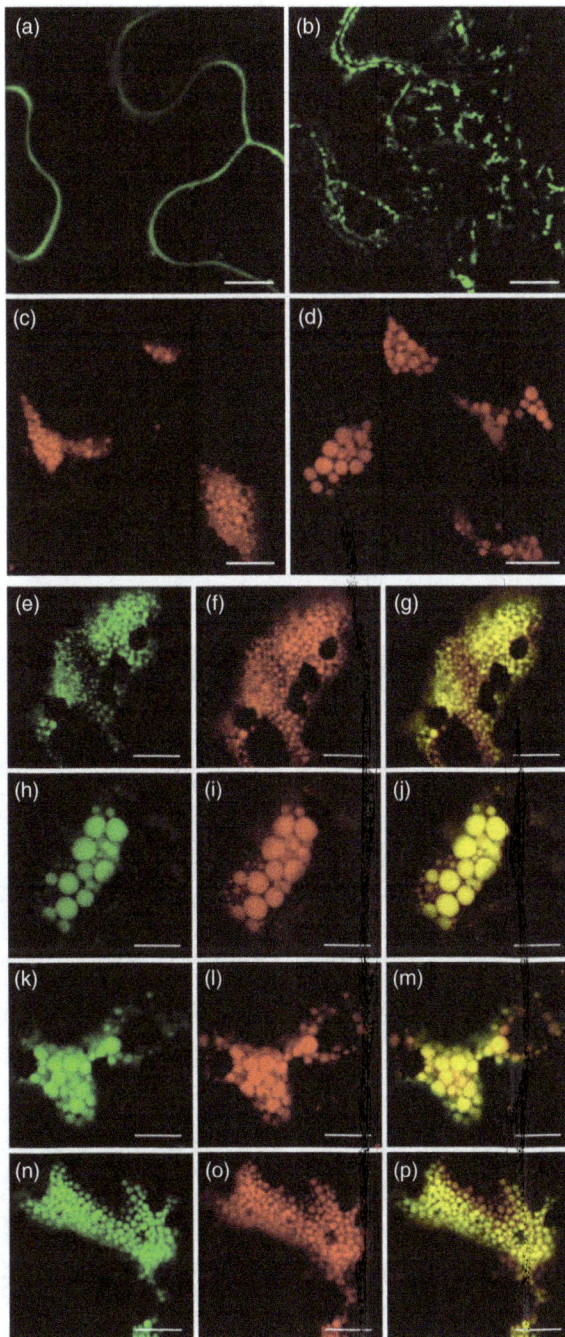

Figure 4 Secretory and ER-targeted GFP are sequestered into RFP-HFBI- and RFP-ELP-induced protein bodies. (a) Secretory GFP localizes to the apoplast. (b) ER-targeted GFP induces the formation of small PBs along the ER. (c) ER-targeted RFP-HFBI induces the formation of clusters of small PBs. (d) ER-targeted RFP-ELP induces the formation of large PB clusters. (e–g) Co-expression of Sec-GFP and RFP-HFBI results in sequestration of secretory GFP into RFP-HFBI-induced PBs. (h–j) Co-expression of Sec-GFP and RFP-ELP results in sequestration of secretory GFP into RFP-ELP-induced PBs. (k–m) Co-expression of GFP and RFP-HFBI results in co-localization of both proteins into the same PBs. (n–p) Co-expression of GFP and RFP-ELP results in co-localization of both proteins into the same PBs. All images were acquired in sequential mode. Bar, 10 μm.

HFBI. According to these results, we conclude that PB formation is not limited to GFP or its fusion partners and can be seen with other recombinant proteins as well.

As xyn11A and xyn11A-HFBI induce the formation of PBs, and since we observed that secretory GFP is sequestered in RFP-HFBI-induced PBs, we asked whether secretory GFP would be also sequestered in xyn11A-induced PBs. If so, simple co-localization of secretory GFP would provide a tool for visualizing PBs induced by nonfluorescent proteins without having to conduct immunolocalization, a lengthy and arduous process. We found that co-expression of xyn11A and xyn11A-HFBI with sec-GFP indeed allowed us to visualize the xyn11A-induced PBs. As shown in Figure 6a, secretory GFP was localized to the apoplastic space of *N. benthamiana* leaf epidermal cells. Co-expression of secretory GFP and xyn11A resulted in sequestration of secretory GFP into xyn11A-induced PBs (Figure 6b). These large PBs were found in all of the cells producing PBs in the infiltrated tissue (Table 1). Similarly, co-expression of the secretory GFP and xyn11A-HFBI resulted in sequestration of secretory GFP into xyn11A-HFBI-

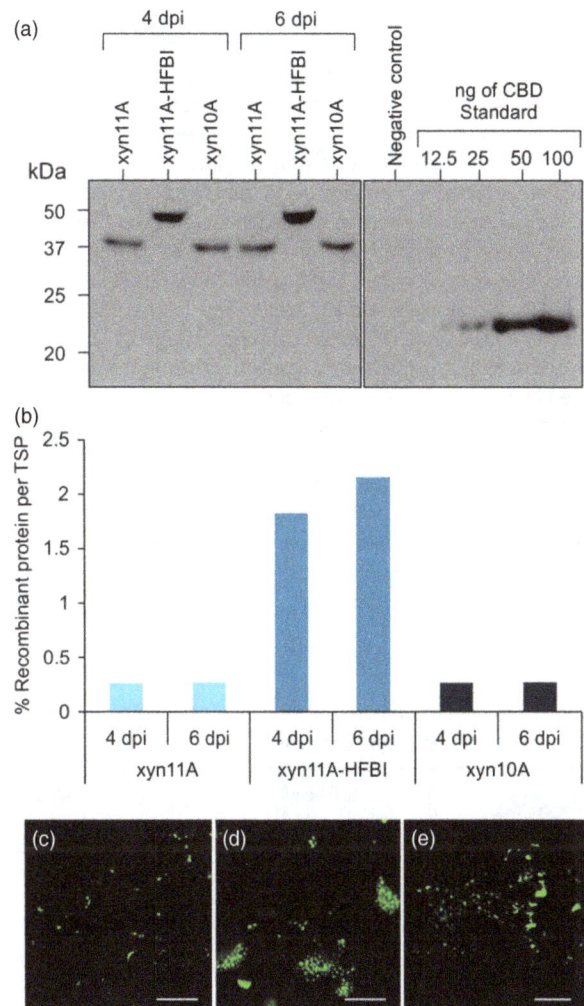

Figure 5 Fungal xylanases accumulate in PBs. Xylanases xyn11A, xyn11A-HFBI and xyn10A were co-expressed with p19 in *N. benthamiana* leaves. Leaf discs from four individual plants at 4 and 6 days postinfiltration (dpi). (a and b) Equal amounts of TSP (3 μg) was loaded per lane. TSP from p19-infiltrated *N. benthamiana* tissue was used as negative control. Known amounts of a c-myc-tagged protein were used as reference (12.5–100 ng). (a) Western blot analysis. (b) Quantification of (a) by densitometry. (c–e) PBs were visualized by confocal microscopy after whole-mount immunolocalization. Bar, 10 μm.

induced PBs. As expected, these PBs were all small and homogenous in size and formed clusters (Figure 6c). This pattern was consistently observed in all of the cells forming PBs (Table 1).

Also, as we observed the sequestration of ER-targeted GFP into RFP-HFBI-induced PBs, we examined the possibility of sequestration of ER-targeted GFP into xyn11A-induced PBs. Even though ER-targeted GFP induced the formation of minuscule PBs along the ER network (Figure 6d), its co-expression with xyn11A resulted in the appearance of large PBs (Figure 6e). Co-expression of GFP-ER and xyn11A-HFBI gave rise to clusters of homogenously-sized small PBs along with a number of large PBs (Figure 6f) (Table 1). Our results suggest that in the process of PB formation, proteins present in the ER, such as proteins on their way to secretion (secretory GFP) or ER-retrieved proteins (GFP-KDEL) are sequestered into PBs.

PB induction can be used as a tool to increase accumulation levels of valuable proteins

Our results indicate that proteins accumulating at levels higher than 0.2% of TSP are capable of PB formation *in vivo*, and that proteins targeted to the secretory pathway or retrieved to the ER using a KDEL retrieval signal can be sequestered into PBs. Based on these observations, and because we believe that PBs may protect recombinant proteins from turnover, we hypothesized that we can use PB formation as a strategy to help increase the accumulation levels of other recombinant proteins that do not accumulate at high levels. We first tested this hypothesis by co-expressing erythropoietin (EPO) with other proteins already shown to accumulate at high levels and induce PBs such as GFP, GFP-ELP and GFP-HFBI (Lee *et al.*, 2012). EPO is a glycoprotein best known for its regulatory roles in the production of red blood cells and its applications in the treatment of anaemia caused by renal failure, chemotherapy and AIDS (Lee *et al.*, 2012).

EPO transiently co-expressed with p19 in *N. benthamiana* leaves accumulated at slightly above 0.04% (±0.01%) of TSP, almost a fivefold increase compared to the very low transient expression of the same EPO construct without p19 using binary vectors (Figure 7a) (Conley *et al.*, 2009a,c). Therefore, p19 was

used in all further experiments to obtain high accumulation levels of the recombinant proteins. Using the same construct as Conley *et al.* (2009c), co-expression of EPO in this work with GFP-ELP and GFP-HFBI, EPO accumulation levels was enhanced significantly by twofold up to 0.09% (±0.02%) of TSP with GFP-ELP, and more than threefold up to 0.14% (±0.01%) of TSP with GFP-HFBI (Figure 7a). Co-expression of EPO with GFP showed an increase in the mean EPO accumulation, but not a significant difference from EPO expressed alone (Figure 7a).

To further validate the idea that accumulation of valuable unfused proteins can be increased by co-expression with a PB-inducing fusion, we co-expressed the human cytokine interleukin-10 (hIL-10) with GFP-HFBI. Human IL-10 is an anti-inflammatory cytokine with multiple roles in the regulation of the immune responses (Denes *et al.*, 2010). Several previous attempts to express unfused hIL-10 in plants resulted in low accumulation levels including expression in transgenic tobacco (accumulating at 0.0055% of TSP) (Menassa *et al.*, 2001), BY-2 cell suspension cultures (accumulating at 0.046% of TSP) (Kaldis *et al.*, 2013), and transient expression in *N. benthamiana* (accu-

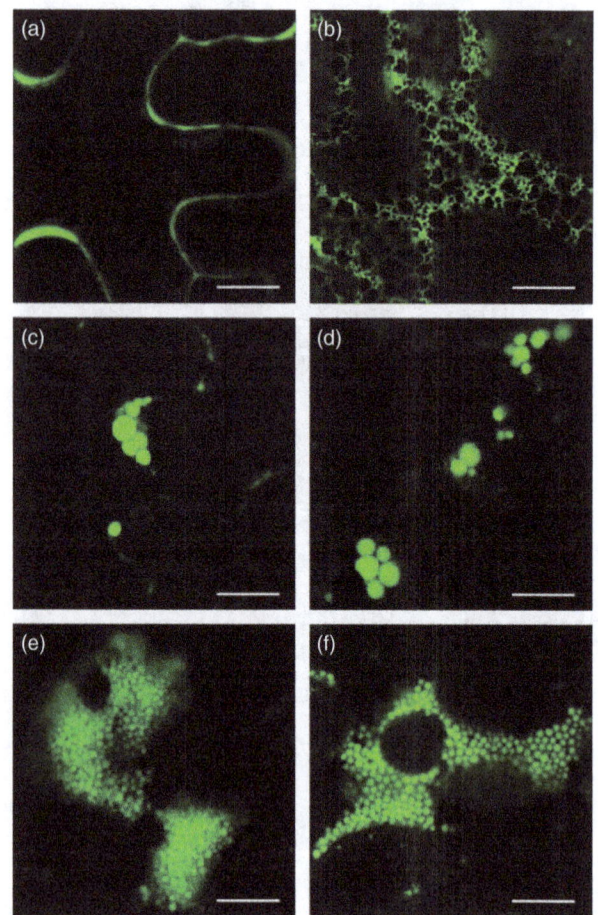

Figure 6 Secretory and ER-retrieved GFP localize to protein bodies induced by fungal xylanases. (a) Secretory GFP highlights the apoplast of *N. benthamiana* epidermal cells. (b-c) Co-expression of secretory GFP with xyn11A (b) and xyn11A-HFBI (c) results in localization of secretory GFP to PBs. (d) ER-targeted GFP highlights the ER network and induces the formation of very small PBs along the ER. (e–f) Co-expression of GFP with xyn11A (e) and xyn11A-HFBI (f) results in localization of GFP to PB clusters. Bar, 10 μm.

Table 1 Occurrence and size distribution of PBs in cells transiently transformed with different constructs

Treatments[†]	Protein body		Protein body size	
	Presence (%)	Absence[‡] (%)	Small[§] (%)	Large[¶] (%)
SecGFP	0	100	0	0
SecGFP / xyn11A	61	39	0	100
SecGFP / xyn11A-HFBI	48	52	100	0
GFP	71	29	90	10
GFP / xyn11A	90	10	43	57
GFP / xyn11A-HFBI	87	13	51	49

[†]To avoid any bias, the presence or absence of PBs in 100 cells in four biological replicates (25 cells per replicate) were evaluated per treatment with confocal microscopy at 4 dpi.
[‡]No PBs were observed in these cells.
[§]Maximum PB diameter was <1.0 μm.
[¶]Maximum PB diameter was >1.0 μm.

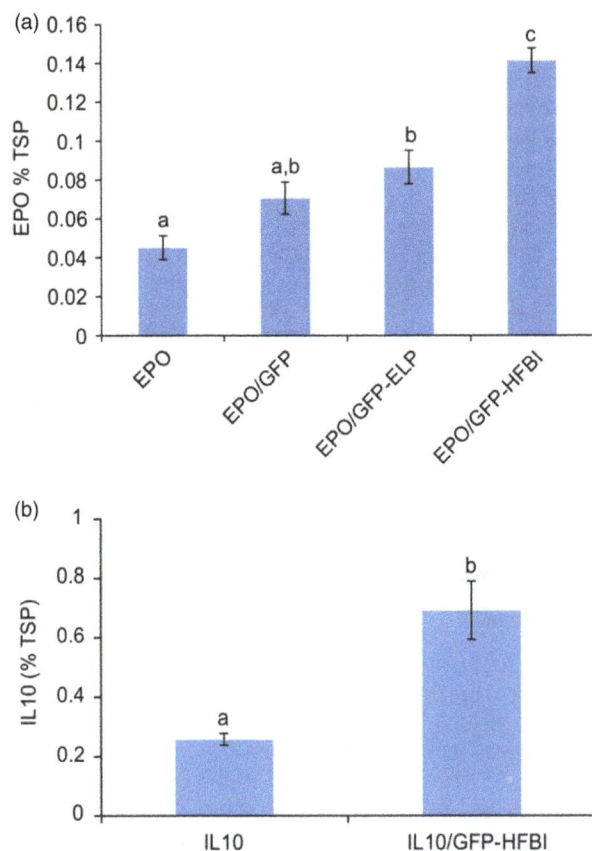

Figure 7 Co-expression of EPO and IL-10 with PB-inducing proteins increases EPO and IL-10 accumulation. (a) Co-expression of erythropoietin (EPO) with GFP, GFP-ELP and GFP-HFBI. ELISA was used to quantify EPO accumulation levels in agroinfiltrated *N. benthamiana* leaves. Columns denoted with different letters are significantly different ($P < 0.05$) using one-way ANOVA. (b) Co-expression of interleukin-10 (IL-10) with GFP-HFBI. IL-10 accumulation levels were determined by ELISA. Each column represents the mean value of eight biological replicates. Columns denoted with different letters are significantly different ($P < 0.002$) using the student *t*-test. The error bars represent the standard error of the mean.

mulating at approximately 0.05% of TSP) (Conley *et al.*, 2009a). Transient co-expression of hIL-10 with p19 increased hIL-10 accumulation levels to 0.26% ($\pm$0.05) of TSP. Upon co-expression of IL-10 with GFP-HFBI and p19, accumulation levels of IL-10 were significantly increased almost threefold, up to 0.69% ($\pm$0.27) of TSP (Figure 7b); this is approximately 125-fold higher than transgenic tobacco plants (Menassa *et al.*, 2001), and 15-fold higher than BY-2 cell culture (Kaldis *et al.*, 2013) or *N. benthamiana* transient expression in the absence of p19 (Conley *et al.*, 2009a).

Discussion

Protein body size and recombinant protein accumulation increase simultaneously

Characterization of PBs induced by GFP-ELP and GFP-HFBI revealed a clear size difference in PBs induced by the fusion tags in *N. benthamiana* leaf protoplasts. We have shown that ELP-induced PBs are larger than HFBI-induced PBs. Comparison of our data with the sizes of Zera-induced PBs (Llop-Tous *et al.*, 2010) suggests that both ELP and HFBI induce PBs larger than Zera. In

this study, we used a novel technique by combining confocal imaging and 3D analysis to accurately measure the sizes of all of the PBs inside the protoplasts. The results presented here confirm that the size of ELP- and HFBI-induced PBs increases over time as well as their associated amount of fused recombinant protein. Previous studies by Llop-Tous *et al.*, in 2010 indicated that the size of Zera-eCFP-induced PBs increases over time from 2 to 10 dpi. This property of Zera-induced PBs was attributed to the continuous biosynthesis of Zera-eCFP polypeptides by the ER active ribosomes (Llop-Tous *et al.*, 2010). Joseph *et al.* (2012) showed that Zera-DsRed PBs were surrounded with a typical ER membrane decorated with ribosomes. Similarly, GFP-ELP- and GFP-HFBI-induced PBs were shown to be surrounded with membranes studded with ribosomes (Conley *et al.*, 2009b; Joensuu *et al.*, 2010). Our results, in accordance with previous observations, suggest a possible role for ribosomes attached to membranes surrounding PBs in the increase in sizes of GFP-ELP- and GFP-HFBI-induced PBs from 3 to 5 dpi and the simultaneous increase in the amount of GFP during this time period. Meanwhile, in a series of fluorescent recovery after photobleaching (FRAP) experiments, Conley *et al.* (2009b) have shown recovery of GFP-ELP into photobleached PBs (within 5 min) which suggests the possibility of trafficking of GFP-ELP from surrounding PBs into the photobleached PB. Although protein trafficking between PBs or between the ER and the PBs needs further confirmation, it can still be considered a potential factor responsible for increasing the sizes of GFP-ELP- and GFP-HFBI-induced PBs.

Role of protein accumulation level in protein body formation

The occurrence of PBs in unfused ER-retrieved GFP and the increase in PB size at higher GFP accumulation levels suggests a direct role of protein accumulation in PB formation. This observation is consistent with our previous study in which a threshold level of protein accumulation was observed at 0.2% of TSP in transgenic tobacco plants expressing GFP, GFP-ELP and GFP-HFBI (Gutierrez *et al.*, 2013). Below 0.2% of TSP, no PBs were observed, while above 0.2% of TSP PBs were observed in almost all plants, regardless of the fusion tag. We confirmed the formation of PBs with fungal xylanases xyn11A and xyn10A, both of which reached accumulations slightly above 0.2% of TSP. Addition of the HFBI fusion tag to xyn11A not only increased accumulation levels by 10-fold, but also produced more PBs with the typical small and clustered HFBI-associated PB pattern. This pattern was previously observed with GFP-HFBI-induced PBs in independent studies (Gutierrez *et al.*, 2013; Joensuu *et al.*, 2010). Although we have shown that PBs form regardless of the availability of fusion tags above the threshold level, the distribution pattern and sizes of PBs were affected by the fusion tags, with obvious clustering and with larger PBs produced by ELP as compared to HFBI.

Protein body formation, a new tool in increasing recombinant protein accumulation

A proteomics study of Zera-DsRed-induced PBs has previously shown that aside from Zera-DsRed, which composed the majority of PB content, other ER-resident and secretory proteins were found in PBs (Joseph *et al.*, 2012). In an independent study, the co-expression of ER-resident molecular chaperone binding protein (BiP) fused with CFP (BiP-CFP-KDEL) and ER-targeted YFP-ELP (YFP-ELP-KDEL) showed the sequestration of the ER chaperone

BiP into the YFP-ELP-induced PBs (Conley *et al.*, 2009b). Furthermore, co-immunoprecipitation analysis of these two proteins suggested that BiP does not specifically interact with the ELP (Conley *et al.*, 2009b). This might have been due to the passive sequestration of BiP into the YFP-ELP-induced PBs. The reason why PBs contain a majority of recombinant proteins may be due to high localized accumulation of the recombinant protein that triggers PB formation and is therefore trapped in the PBs as a mechanism to relieve ER stress in the cell. Our results confirmed the passive sequestration of ER-retrieved and secretory proteins into PBs. This was shown separately by sequestration of secretory GFP and ER-retrieved GFP into RFP-HFBI-, RFP-ELP-, xyn11A-HFBI- and xyn11A-induced PBs.

Based on our observations of passive sequestration of ER-resident proteins into PBs, we designed a series of experiments to study the potential of PB induction as a new approach to increase recombinant protein accumulation in plants. We chose EPO and hIL-10 both of which are highly valuable pharmaceutical proteins that accumulate at low levels when expressed in plants. Although recent reports have shown accumulation levels of EPO fusions to high levels, EPOFc to 0.18% of TSP and EPO-ELD to approximately 2% of TSP using viral-based magnICON® vectors (Castilho *et al.*, 2013; Jez *et al.*, 2013), the highest accumulation level achieved in plants using binary vectors approached 0.01% of TSP in transient expression and up to 0.05% of TSP in transgenic tobacco (Conley *et al.*, 2009c). ER-retained, asialylated EPO was previously shown to bind its receptor efficiently and was hypothesized to provide tissue protection without inducing red blood cell production (Conley *et al.*, 2009c). Here, we have shown that by co-expressing ER-retained EPO with a PB-inducing construct and p19, we could increase EPO accumulation levels up to 0.14% of TSP, 14-fold higher than the previously reported transient expression levels using binary vectors.

Human IL-10 was previously expressed in several plant systems including transgenic tobacco, transgenic BY-2 cell lines and transiently in *N. benthamiana* and in all cases accumulated below 0.05% of TSP (Conley *et al.*, 2009a; Kaldis *et al.*, 2013; Menassa *et al.*, 2001). Unfused, ER-retained hIL-10 was shown to be biologically active in cell assays, and therapeutic in a mouse model of colitis when administered orally (Menassa *et al.*, 2001, 2007). By co-expressing ER-retained hIL-10 and GFP-HFBI, we obtained hIL-10 accumulation of 0.69% TSP on average which is approximately a 14-fold increase from the highest previously reported hIL-10 accumulation level. This increase in accumulation levels of ER-targeted EPO or hIL-10 might be due to nonselective sequestration of ER-targeted proteins into PBs. We hypothesize that by co-expressing low-accumulating valuable proteins such as EPO and hIL-10 with high accumulating PB-inducing proteins, both recombinant proteins are sequestered in PBs. It has already been proposed that PBs protect recombinant proteins from physiological turnover within the cell (Conley *et al.*, 2011; Khan *et al.*, 2012; Müntz and Shutov, 2002), which might consequently explain the increased accumulation levels of EPO and hIL-10 in this work. It is worth noting that hIL-10-ELP fusion expressed in BY-2 cell lines were shown to accumulate to 3% of TSP; however, the activity of the fused hIL-10 was significantly reduced when fused to ELP (Kaldis *et al.*, 2013). With our proposed co-expression technique, the need for addition of fusion tags that might affect the proper folding and activity of their fused partner is eliminated. Nevertheless, proteins that require complex glycans for activity such as antibodies may not be fully functional using this strategy.

In conclusion, the main objectives of this study were to understand the role of fusion tags and recombinant protein accumulation levels in the process of PB formation, and to investigate the potential utility of PBs as a tool to facilitate recombinant protein production in leaves of *N. benthamiana*. We have shown that the ELP fusion tag induces PBs larger in size compared to HFBI and that the PB size correlates with protein accumulation levels. We have also shown that the presence of ELP or HFBI fusion tags is not necessary for PB formation, which occurs with several ER-targeted proteins, but that ELP and HFBI affect the distribution patterns and size of PBs. We have also shown that PB formation is not limited to the conventional fluorescent proteins (e.g. GFP or RFP) as fungal xylanases also induce PBs. In addition, our results indicate that in the process of PB formation, secretory and ER-resident proteins can passively integrate into the lumen of PBs. We have used this property of PBs as a tool to increase accumulation levels of EPO and hIL-10 by co-expression with PB-inducing proteins.

Experimental procedures

Construct design and cloning

Secretory GFP, GFP, GFP-ELP, GFP-HFBI and EPO plant expression vectors were previously published (Conley *et al.*, 2009b,c; Joensuu *et al.*, 2010). The coding sequence of RFP was PCR amplified using primers RS1 (5′GGGGACAAGTTTGTACAAAAAAGCAGG CTTGATG-GCCTCCTCCGAGGACGT3′) and RS2 (5′GGGGACCA CTTTGTACAAGAAAGCTGGGTCGGCGCCGGTG-GAGTGGC3′). The PCR construct was then inserted into the Gateway® donor vector pDONR/Zeo™ (Invitrogen, Carlsbad, CA). The integrity of the PCR construct was confirmed with sequencing analysis. Using Gateway technology, RFP was recombined into the pCamGate-ER, pCaMGate-HFBI-ER and pCaMGate-ELP-ER (Figure 1) expression vectors (Conley *et al.*, in preparation) under the control of a double-enhanced cauliflower mosaic virus (CaMV) 35S promoter (Kay *et al.*, 1987; Wu *et al.*, 2001), a tCUP translational enhancer (Malik *et al.*, 2002) and a nopaline synthase (nos) terminator (Bevan *et al.*, 1983). The final expression constructs were then transformed into *Agrobacterium tumefaciens* strain EHA105. The xyn11A and xyn11A-HFBI plant expression vectors were obtained from Ruoyu Yan (Yan, 2011).

Nicotiana benthamiana growth and maintenance

Nicotiana benthamiana plants were grown for 7 weeks and used for transient expression. Plants were grown in a growth chamber at 22 °C with a 16-h photoperiod at a light intensity of 110 μmol/m^2/s. Plants were always watered with the water soluble fertilizer (N: P: K = 20: 8: 20) at 0.25 g/L (Plant Products, Brampton, ON, Canada).

Transient expression in *Nicotiana benthamiana* plants

Agrobacterium tumefaciens cultures were grown to an optical density at 600 nm (OD$_{600}$) of 0.5–0.8, and collected by centrifugation at 1000 g for 30 min. The pellets were then resuspended in Agro-infiltration solution (3.2 g/L Gamborg's B5 medium and vitamins, 20 g/L sucrose, 10 mM MES pH 5.6, 200 μM 4′-Hydroxy-3′,5′-dimethoxyacetophenone) to a final OD$_{600}$ of 0.30, followed by incubation at room temperature (21°C) with gentle agitation for 1 h. The suspension was then used for infiltration of the abaxial leaf epidermis through the stomata of *N. benthamiana* plants with a 1 mL syringe (Kapila *et al.*, 1997).

Tissue sampling and protein extraction

Nicotiana benthamiana leaf samples were collected at 4–6 dpi depending on the experiment. Four leaf discs were collected from at least three biological replicates per sample. Total soluble protein was extracted and quantified as previously described Conley *et al.* (2009a).

Protoplast preparation

Protoplasts were prepared as described by Sheen (Sheen, 2002). Briefly, 1 g of infiltrated leaf tissue was cut into strips of 0.5–1 mm, covered with enzyme solution (1.5% cellulase R10 (216016; Yakult Pharmaceutical, Tokyo, Japan), 0.4% macerozyme® R10 (202051; Yakult Pharmaceutical), 0.4 M mannitol, 20 mM KCl, 20 mM MES (pH 5.7), 10 mM CaCl$_2$, 5 mM β-mercaptoethanol and 0.1% BSA (Sigma A-6793 Saint Louis, MO, USA)) and incubated overnight in the dark at 21 °C. Protoplasts were filtered with a 100-μm mesh, and centrifuged for 10 min at 80 × ***g*** at 4 °C. The supernatant was removed, replaced with an equal volume of cold washing and incubation solution (WI) (0.5 M mannitol, 4 mM MES (pH 5.7) and 20 mM KCl) and centrifuged again for 10 min at 80 × ***g*** at 4 °C. The solution was then transferred gently onto flotation medium (0.6 M sucrose, 10 mM MES (pH 5.7) and 20 mM KCl) to remove debris and centrifuged for 10 min at 80 × ***g*** at 4 °C. Protoplasts were removed from the top layer of the flotation medium and transferred to cold WI solution.

Confocal microscopy and image analysis

To visualize protoplasts, z-stack confocal images of protoplasts were acquired with a Leica TCS SP2 confocal laser scanning inverted microscope (Leica Microsystems, Wetzlar, Germany) equipped with a 63× water immersion objective. The surpass module of Imaris® ×64 software (version 7.6.1; Bitplane Scientific Software, Bitplane, Zurich, Switzerland) was used to generate 3D reconstructed images and videos. To measure the size and volume of PBs, the surface module of Imaris software was used to detect separate PBs.

To visualize leaf samples, the abaxial leaf epidermis was imaged with a Leica TCS SP2 CLSM. For the imaging of GFP and chlorophyll autofluorescence, excitation with a 488-nm argon laser was used and fluorescence was detected at 500–525 nm and 630–690 nm, respectively. The excitation for RFP was 543 nm and its emission was recorded at 553–630 nm. Collected images were analysed with the Leica Application Suite for Advanced Fluorescence (LAS AF, version 2.3.5) (Leica Microsystem). To avoid cross-talk between the fluorescent channels, images were acquired in the sequential mode at all times.

To visualize whole-mounted tissue, 488-nm argon laser was used to detect Alexa Fluor® 488 dye, and green fluorescence was detected at 495 to 525 nm.

Recombinant protein quantification

GFP quantification was performed with either fluorometry or immunodot blot as described by Gutierrez *et al.* (2013). For Western blot analysis of xylanases, 3 μg of total soluble protein was resolved by 10% sodium dodecylsulphate–polyacrylamide gel electrophoresis (SDS–PAGE) and transferred to PVDF membrane. The recombinant proteins were detected with the primary mouse anti-c-myc monoclonal antibody (GenScript, Cat. No. A00864) diluted 1 : 5000 in blocking solution and HRP-conjugated goat anti-mouse IgG secondary antibody diluted 1 : 3000 (Biorad, Cat. No. 170-6516). Membranes were visualized with the enhanced chemiluminescence (ECL) detection kit (GE Healthcare, Mississauga, ON, Canada) as described by the manufacturer. Quantification was performed using TotalLab TL 100 software (Nonlinear Dynamics, Durham, NC) with intensity analysis of specific bands, with known amounts of a c-myc-tagged standard protein as a standard.

Quantification of prEPO from *N. benthamiana* leaf extracts was performed by sandwich ELISA as described previously Conley *et al.* (2009c). At least four independent biological samples were quantified per treatment. Leaf tissue infiltrated with p19 was used as control.

The concentration of plant recombinant IL-10 was determined by comparison with a human IL-10 standard curve (BD Biosciences, Mississauga, Canada) using indirect ELISA as described by Kaldis *et al.* (2013).

Immunocytochemistry

Whole-mounting of *N. benthamiana* leaves was performed as described by Sauer *et al.* (2006). Briefly, pieces of infiltrated and noninfiltrated leaves were fixed in 4% paraformaldehyde (Electron Microscopy Sciences, Hatfield, PA) in PBS supplemented with 0.1% Triton™ X-100 (Sigma, Cat. No. T-9284), and permeabilized in Driselase (Sigma, Cat. No. D-9515) and IGEPAL® (Sigma, Cat. No. CA-630) plus 10% dimethyl sulfoxide (DMSO) (Sigma, Cat. No. D8418). For immunocytochemistry, samples were preblocked in 3% BSA (Fisher Scientific, Cat. No. BP1600-100), incubated with c-myc monoclonal mouse antibody (1:1000 dilution) as the primary antibody and donkey anti-mouse IgG antibody conjugated to Alexa Fluor® 488 dye (1 : 130 dilution) as the secondary antibody (Life Technologies, Cat. No. A-21202). Confocal microscopy was performed as described above.

Statistical analysis

Statistical analysis of the protein body size comparison experiment was performed with Prism 6 (GraphPad Software, GraphPad Inc., La Jolla, CA). The Kolmogorov–Smirnov test was used to confirm the normal distribution of data. As the data were not normally distributed, the Kruskal–Wallis non-parametric test of variance was used to assess whether there were statistical differences between the median values of different treatments. Other statistical analyses were performed with SPSS Statistics v 22.0 for Windows (IBM, Armonk, NY). Depending on the number of comparisons involved, a student *t*-test or one-way ANOVA test was performed followed by Tukey–Kramer's test to find means significantly different from one another. A post hoc test was then performed to analyse for significance of mean difference at 95% confidence intervals (the level of statistical difference was defined as $P < 0.05$).

Acknowledgements

We thank Angelo Kaldis, Hong Zhu and Tuuli Teikari for technical assistance, Danny Poinapen for help with statistical analyses, and Alex Molnar for assistance with preparation of figures. This research was supported by Agriculture and Agri-Food Canada A-base project 1725 and Eurostars Hydropro 5320 project grants.

References

Baneyx, F. and Mujacic, M. (2004) Recombinant protein folding and misfolding in Escherichia coli. *Nat. Biotechnol.* **22**, 1399–1408.

Benchabane, M., Goulet, C., Rivard, D., Faye, L., Gomord, V. and Michaud, D. (2008) Preventing unintended proteolysis in plant protein biofactories. *Plant Biotechnol. J.* **6**, 633–648.

Bevan, M., Barnes, W.M. and Chilton, M.D. (1983) Structure and transcription of the nopaline synthase gene region of T-DNA. *Nucleic Acids Res.* **11**, 369–385.

Bock, R. (2014) Genetic engineering of the chloroplast: novel tools and new applications. *Curr. Opin. Biotechnol.* **26**, 7–13.

Castilho, A., Neumann, L., Gattinger, P., Strasser, R., Vorauer-Uhl, K., Sterovsky, T., Altmann, F. and Steinkellner, H. (2013) Generation of biologically active multi-sialylated recombinant human EPOFc in plants. *PLoS ONE*, **8**, e54836.

Chebolu, S. and Daniell, H. (2009) Chloroplast-derived vaccine antigens and biopharmaceuticals: expression, folding, assembly and functionality. *Curr. Top. Microbiol. Immunol.* **332**, 33–54.

Conley, A.J., Joensuu, J.J., Jevnikar, A.M., Menassa, R. and Brandle, J.E. (2009a) Optimization of elastin-like polypeptide fusions for expression and purification of recombinant proteins in plants. *Biotechnol. Bioeng.* **103**, 562–573.

Conley, A.J., Joensuu, J.J., Menassa, R. and Brandle, J.E. (2009b) Induction of protein body formation in plant leaves by elastin-like polypeptide fusions. *BMC Biol.* **7**, 48.

Conley, A.J., Mohib, K., Jevnikar, A.M. and Brandle, J.E. (2009c) Plant recombinant erythropoietin attenuates inflammatory kidney cell injury. *Plant Biotechnol. J.* **7**, 183–199.

Conley, A.J., Joensuu, J.J., Richman, A. and Menassa, R. (2011) Protein body-inducing fusions for high-level production and purification of recombinant proteins in plants. *Plant Biotechnol. J.* **9**, 419–433.

De Cosa, B., Moar, W., Lee, S.B., Miller, M. and Daniell, H. (2001) Overexpression of the Bt cry2Aa2 operon in chloroplasts leads to formation of insecticidal crystals. *Nat. Biotechnol.* **19**, 71–74.

Denes, B., Fodor, I. and Langridge, W.H. (2010) Autoantigens plus interleukin-10 suppress diabetes autoimmunity. *Diabetes Technol. Ther.* **12**, 649–661.

Egelkrout, E., Rajan, V. and Howard, J.A. (2012) Overproduction of recombinant proteins in plants. *Plant Sci.* **184**, 83–101.

Fernandez-San Millan, A., Mingeo-Castel, A.M., Miller, M. and Daniell, H. (2003) A chloroplast transgenic approach to hyper-express and purify human serum albumin, a protein highly susceptible to proteolytic degradation. *Plant Biotechnol. J.* **1**, 71–79.

Fischer, R., Stoger, E., Schillberg, S., Christou, P. and Twyman, R.M. (2004) Plant-based production of biopharmaceuticals. *Curr. Opin. Plant Biol.* **7**, 152–158.

Floss, D.M., Sack, M., Arcalis, E., Stadlmann, J., Quendler, H., Rademacher, T., Stoger, E., Scheller, J., Fischer, R. and Conrad, U. (2009) Influence of elastin-like peptide fusions on the quantity and quality of a tobacco-derived human immunodeficiency virus-neutralizing antibody. *Plant Biotechnol. J.* **7**, 899–913.

Floss, D.M., Schallau, K., Rose-John, S., Conrad, U. and Scheller, J. (2010) Elastin-like polypeptides revolutionize recombinant protein expression and their biomedical application. *Trends Biotechnol.* **28**, 37–45.

Guda, C., Lee, S.B. and Daniell, H. (2000) Stable expression of a biodegradable protein-based polymer in tobacco chloroplasts. *Plant Cell Rep.* **19**, 257–262.

Gutierrez, S.P., Saberianfar, R., Kohalmi, S.E. and Menassa, R. (2013) Protein body formation in stable transgenic tobacco expressing elastin-like polypeptide and hydrophobin fusion proteins. *BMC Biotechnol.* **13**, 40.

Hakanpää, J., Paananen, A., Askolin, S., Nakari-Setälä, T., Parkkinen, T., Penttilä, M., Linder, M.B. and Rouvinen, J. (2004) Atomic resolution structure of the HFBII hydrophobin, a self-assembling amphiphile. *J. Biol. Chem.* **279**, 534–539.

Jez, J., Castilho, A., Grass, J., Vorauer-Uhl, K., Sterovsky, T., Altmann, F. and Steinkellner, H. (2013) Expression of functionally active sialylated human erythropoietin in plants. *Biotechnol. J.* **8**, 371–382.

Joensuu, J.J., Conley, A.J., Lienemann, M., Brandle, J.E., Linder, M.B. and Menassa, R. (2010) Hydrophobin fusions for high-level transient protein expression and purification in *Nicotiana benthamiana. Plant Physiol.* **152**, 622–633.

Joseph, M., Ludevid, M.D., Torrent, M., Rofidal, V., Tauzin, M., Rossignol, M. and Peltier, J.B. (2012) Proteomic characterisation of endoplasmic reticulum-derived protein bodies in tobacco leaves. *BMC Plant Biol.* **12**, 36.

Kaldis, A., Ahmad, A., Reid, A., McGarvey, B., Brandle, J., Ma, S., Jevnikar, A., Kohalmi, S.E. and Menassa, R. (2013) High-level production of human interleukin-10 fusions in tobacco cell suspension cultures. *Plant Biotechnol. J.* **11**, 535–545.

Kapila, J., De Rycke, R., Van Montagu, M. and Angenon, G. (1997) An Agrobacterium-mediated transient gene expression system for intact leaves. *Plant Sci.* **122**, 101–108.

Kay, R., Chan, A., Daly, M. and McPherson, J. (1987) Duplication of CaMV 35S promoter sequences creates a strong enhancer for plant genes. *Science* **236**, 1299–1302.

Khan, I., Twyman, R.M., Arcalis, E. and Stoger, E. (2012) Using storage organelles for the accumulation and encapsulation of recombinant proteins. *Biotechnol. J.* **7**, 1099–1108.

Komarova, T.V., Baschieri, S., Donini, M., Marusic, C., Benvenuto, E. and Dorokhov, Y.L. (2010) Transient expression systems for plant-derived biopharmaceuticals. *Expert Rev. Vaccines*, **9**, 859–876.

Lau, O.S. and Sun, S.S. (2009) Plant seeds as bioreactors for recombinant protein production. *Biotechnol. Adv.* **27**, 1015–1022.

Lee, J.S., Ha, T.K., Lee, S.J. and Lee, G.M. (2012) Current state and perspectives on erythropoietin production. *Appl. Microbiol. Biotechnol.* **95**, 1405–1416.

Linder, M., Selber, K., Nakari-Setala, T., Qiao, M., Kula, M.R. and Penttila, M. (2001) The hydrophobins HFBI and HFBII from *Trichoderma reesei* showing efficient interactions with nonionic surfactants in aqueous two-phase systems. *Biomacromolecules*, **2**, 511–517.

Llop-Tous, I., Madurga, S., Giralt, E., Marzabal, P., Torrent, M. and Ludevid, M.D. (2010) Relevant elements of a Maize γ-zein domain involved in protein body biogenesis. *J. Biol. Chem.* **285**, 35633–35644.

Ma, S. and Wang, A. (2012) Molecular farming in plants: an overview. In *Molecular Farming in Plants: Recent Advances and Future Prospects* (2012 ed), pp. 1–20. Dordrecht, Netherlands: Springer.

Malik, K., Wu, K., Li, X.-Q., Martin-Heller, T., Hu, M., Foster, E., Tian, L., Wang, C., Ward, K. and Jordan, M. (2002) A constitutive gene expression system derived from the tCUP cryptic promoter elements. *Theor. Appl. Genet.* **105**, 505–514.

Maxmen, A. (2012) Drug-making plant blooms. *Nature*, **485**, 160.

Menassa, R., Nguyen, V., Jevnikar, A. and Brandle, J. (2001) A self-contained system for the field production of plant recombinant interleukin-10. *Mol. Breed.* **8**, 177–185.

Menassa, R., Du, C., Yin, Z.Q., Ma, S., Poussier, P., Brandle, J. and Jevnikar, A.M. (2007) Therapeutic effectiveness of orally administered transgenic low-alkaloid tobacco expressing human interleukin-10 in a mouse model of colitis. *Plant Biotechnol. J.* **5**, 50–59.

Müntz, K. and Shutov, A.D. (2002) Legumains and their functions in plants. *Trends Plant Sci.* **7**, 340–344.

Nordberg, H., Cantor, M., Dusheyko, S., Hua, S., Poliakov, A., Shabalov, I., Smirnova, T., Grigoriev, I.V. and Dubchak, I. (2014) The genome portal of the Department of Energy Joint Genome Institute: 2014 updates. *Nucleic Acids Res.* **42**, D26–D31.

Phan, H.T., Hause, B., Hause, G., Arcalis, E., Stoger, E., Maresch, D., Altmann, F., Joensuu, J. and Conrad, U. (2014) Influence of elastin-like polypeptide and hydrophobin on recombinant hemagglutinin accumulations in transgenic tobacco plants. *PLoS ONE*, **9**, e99347.

Sauer, M., Paciorek, T., Benková, E. and Friml, J. (2006) Immunocytochemical techniques for whole-mount in situ protein localization in plants. *Nat. Protoc.* **1**, 98–103.

Sheen, J. (2002) *A transient expression assay using Arabidopsis mesophyll protoplasts.*

Silhavy, D., Molnar, A., Lucioli, A., Szittya, G., Hornyik, C., Tavazza, M. and Burgyan, J. (2002) A viral protein suppresses RNA silencing and binds silencing-generated, 21- to 25-nucleotide double-stranded RNAs. *EMBO J.* **21**, 3070–3080.

Stoger, E., Fischer, R., Moloney, M. and Ma, J.K. (2014) Plant molecular pharming for the treatment of chronic and infectious diseases. *Annu. Rev. Plant Biol.* **65**, 743–768.

Torrent, M., Llompart, B., Lasserre-Ramassamy, S., Llop-Tous, I., Bastida, M., Marzabal, P., Westerholm-Parvinen, A., Saloheimo, M., Heifetz, P.B. and Ludevid, M.D. (2009a) Eukaryotic protein production in designed storage organelles. *BMC Biol.* **7**, 5.

Torrent, M., Llop-Tous, I. and Ludevid, M.D. (2009b) Protein body induction: a new tool to produce and recover recombinant proteins in plants. *Methods Mol. Biol.* **483**, 193–208.

Urry, D.W. and Parker, T.M. (2002) Mechanics of elastin: molecular mechanism of biological elasticity and its relationship to contraction. *J. Muscle Res. Cell Motil.* **23**, 543–559.

Wessels, J.G.H. (1997) Hydrophobins: proteins that change the nature of the fungal surface. *Adv. Microb. Physiol.*, **38**, 1–45.

Wu, K., Malik, K., Tian, L., Hu, M., Martin, T., Foster, E., Brown, D. and Miki, B. (2001) Enhancers and core promoter elements are essential for the activity of a cryptic gene activation sequence from tobacco, tCUP. *Mol. Genet. Genomics*, **265**, 763–770.

Yan, R. (2011) Expression of active fungal xylanases in *N. benthamiana* for hemicellulose degradation. In *Biology Department* p. 72. Department of Biology: University of Western Ontario.

12

Chromodomain, Helicase and DNA-binding CHD1 protein, CHR5, are involved in establishing active chromatin state of seed maturation genes

Yuan Shen[1], Martine Devic[2], Loïc Lepiniec[3] and Dao-Xiu Zhou[1,*]

[1]Université Paris-sud 11, Institut de Biologie des Plantes, CNRS, UMR8618, Saclay Plant Science, Orsay, France
[2]Régulation Epigénétique et Développement de la Graine, ERL 3500 CNRS-IRD UMR DIADE, IRD centre de Montpellier, Montpellier, France
[3]INRA, Institut Jean-Pierre Bourgin, Saclay Plant Science, Versailles, France

*Correspondence

email dao-xiu.zhou@u-psud.fr

Keywords: chromatin remodeling, seed gene expression, CHD protein.

Summary

Chromatin modification and remodelling are the basis for epigenetic regulation of gene expression. *LEAFY COTYLEDON 1* (*LEC1*), *LEAFY COTYLEDON 2* (*LEC2*), *ABSCISIC ACID-INSENSITIVE 3* (*ABI3*) and *FUSCA3* (*FUS3*) are key regulators of embryo development and are repressed after seed maturation. The chromatin remodelling CHD3 protein PICKLE (PKL) is involved in the epigenetic silencing of the genes. However, the chromatin mechanism that establishes the active state of these genes during early embryo development is not clear. We show that the Arabidopsis CHD1-related gene, *CHR5*, is activated during embryo development. Mutation of the gene reduced expression of *LEC1*, *ABI3* and *FUS3* in developing embryo and accumulation of seed storage proteins. Analysis of double mutants revealed an antagonistic function between *CHR5* and *PKL* in embryo gene expression and seed storage protein accumulation, which likely acted on the promoter region of the genes. CHR5 was shown to be associated with the promoters of *ABI3* and *FUS3* and to be required to reduce nucleosome occupancy near the transcriptional start site. The results suggest that CHR5 is involved in establishing the active state of embryo regulatory genes by reducing nucleosomal barrier, which may be exploited to enhance seed protein production.

Introduction

Chromatin remodelling involves altering histone–DNA contacts, sliding nucleosomes and exchanging or removing histones and entire nucleosomes (Narlikar et al., 2013). Chromatin remodellers, which are ubiquitous in eukaryotic cells (Flaus et al., 2006), are compositionally and functionally diverse, but they all share the presence of a subunit that belongs to the SNF2-like family of ATPases, among which are the chromodomain, helicase/ATPase and DNA-binding domain (CHD) proteins (Murawska and Brehm, 2011). Besides *Saccharomyces cerevisiae* that has only one CHD protein (Chd1), other higher eukaryotes have several CHD members that are divided into subfamily I (CHD1, CHD2), subfamily II (CHD3-5) and subfamily III (CHD6-9) (Murawska and Brehm, 2011). The two chromodomains in CHD1 proteins are essential for recognition of di- or trimethylated lysine 4 of histone H3 (H3K4me2/3) (Flanagan et al., 2005; Sims et al., 2005). CHD1 proteins are implicated in gene expression at different levels. Recent results suggest that CHD1 proteins have both chromatin assembly and remodelling functions to direct the positioning of nucleosomes required for gene transcription (Gkikopoulos et al., 2011; Torigoe et al., 2013). In addition to the double chromodomains and the helicase/ATPase, CHD3 proteins contain one or two plant homoeodomain (PHD) fingers at the N-terminus. CHD3 proteins in *Drosophila melanogaster* and mammalian cells are the central components of the nucleosome remodelling and histone deacetylase (NuRD) or Mi-2 complexes regulating transcriptional repression (Ramirez and Hagman, 2009).

Several CHD3-like proteins are identified in plants (Hu et al., 2012). The *Arabidopsis* CHD3-like protein PICKLE (PKL) was initially found to repress embryonic traits after seed germination (Ogas et al., 1997). PKL functions as a transcriptional repressor of embryo identity genes such as *LEAFY COTYLEDON 1* (*LEC1*), *LEAFY COTYLEDON 2* (*LEC2*), *ABSCISIC ACID-INSENSITIVE 3* (*ABI3*) and *FUSCA3* (*FUS3*) in seedlings (Aichinger et al., 2009; Ogas et al., 1999; Zhang et al., 2008, 2012). In contrast to animal CHD3 proteins, PKL is found to promote H3K27me3 over target genes rather than histone deacetylation (Zhang et al., 2008). Recent data revealed that the rice (*Oryza sativa*) CHD3 protein, CHR729, is a bifunctional chromatin regulator that recognizes and modulates H3K4 and H3K27 methylation over repressed tissue-specific genes (Hu et al., 2012). By contrast, only one CHD1-like protein gene is found in the genome of *Arabidopsis* or rice (Hu et al., 2012). The function of CHD1-like genes in plant gene expression and development is not studied.

LEC1 and the B3-domain ABI3, FUS3 and LEC2 (referred to as AFL) are key transcriptional regulators of zygotic embryo development (Giraudat et al., 1992; Lotan et al., 1998; Stone et al., 2001). These factors activate the seed maturation gene expression programme in a complex network (Kagaya et al., 2005; Monke et al., 2012; Santos-Mendoza et al., 2008; Suzuki and McCarty, 2008; Wang and Perry, 2013). The genes are expressed specifically during embryo development and epigenetically repressed after seed germination primarily by PRC2-mediated H3K27me3 (Aichinger et al., 2009; Berger et al., 2011; Makarevich et al., 2006). *Arabidopsis* PRC2 mutants with reduced

H3K27me3 display derepression of *LEC1* and *AFL* genes and embryonic traits in seedlings (Bouyer *et al.*, 2011; Chanvivattana *et al.*, 2004; Schubert *et al.*, 2006). Recent data indicate that the E3 H2A monoubiquitin ligase activity of the polycomb repressive complex 1 (PRC1) that recognizes and binds to H3K27me3 is also required for the postgermination repression of the genes (Bratzel *et al.*, 2010; Yang *et al.*, 2013).

Besides the epigenetic repression of the embryo identity or seed maturation genes after seed germination, little is known about the chromatin mechanism that establishes the active state of these genes during embryogenesis and seed development. In this work, we show that the *Arabidopsis* CHD1-like protein gene, known as *CHR5*, is expressed during embryo development and seed maturation and is directly involved in the activation of *ABI3* and *FUS3* expression. In addition, we show that CHR5 and PKL have antagonistic function in *ABI3* and *FUS3* expression in developing seeds. CHR5 binds to the promoter of *ABI3* and *FUS3* and reduces nucleosome occupancy near the transcriptional start site, whereas PKL represses the expression of *ABI3* and *FUS3* genes during seed maturation likely by promoting H3K27me3. The antagonistic function of CHR5 and PKL on embryo gene expression by acting on different aspects of chromatin modifications suggests important but complex roles of chromatin in the regulation of seed maturation gene expression.

Results

CHR5 is expressed during late embryogenesis

Analysis of microarray data suggested that *CHR5* is highly expressed in developing and mature seeds, shoot apex and floral organs (http://bbc.botany.utoronto.ca/efp/cgi-bin/efpWeb.cgi?

primaryGene=AT2G13370&modelInput=Absolute). To confirm the data, we produced >10 independent transgenic plants expressing the GUS reporter gene under the control of a 2-kb promoter region of *CHR5*. Most of the transgenic lines displayed a similar GUS staining pattern that was detected in seeds, seedlings, flower buds and floral organs such as carpel, stigma, stamen and pollen (Figure S1). In young seedlings, GUS staining was mostly detected in root tip, shoot apex and vasculature (Figure S1). In developing embryo, GUS expression started at about late globular-triangular stages and gradually increased till the mature stage of embryogenesis (Figure 1a). GUS staining could be also detected in endosperm during the early stages of seed development (Figures 1a and S1). To further study the *CHR5* expression in developing seeds, we analysed *CHR5* transcript levels by RT-qPCR using mRNA isolated from siliques from different stages as defined at https://www.genomforschung.uni-bielefeld.de/GF-research/AtGenExpress-SeedsSiliques.html. Stages 1–4 correspond to the zygote to globular stages of embryo development. The *CHR5* transcript was low before stage 4 (globular stage) (Figure 1b). After stage 4, the *CHR5* mRNA level increased gradually till the embryo maturation stage (Figure 1b). This confirmed the GUS staining data and indicated that *CHR5* is activated during embryo development, which roughly corresponded to that of the late *AFL* genes (i.e. *FUS3* and *ABI3*) (Figure 1b) (Santos-Mendoza *et al.*, 2008). These observations suggested that *CHR5* might play a role in late embryo development and seed maturation.

chr5 mutants characterization

To study CHR5 function in gene expression and plant development, we characterized *Arabidopsis* T-DNA insertion lines of the

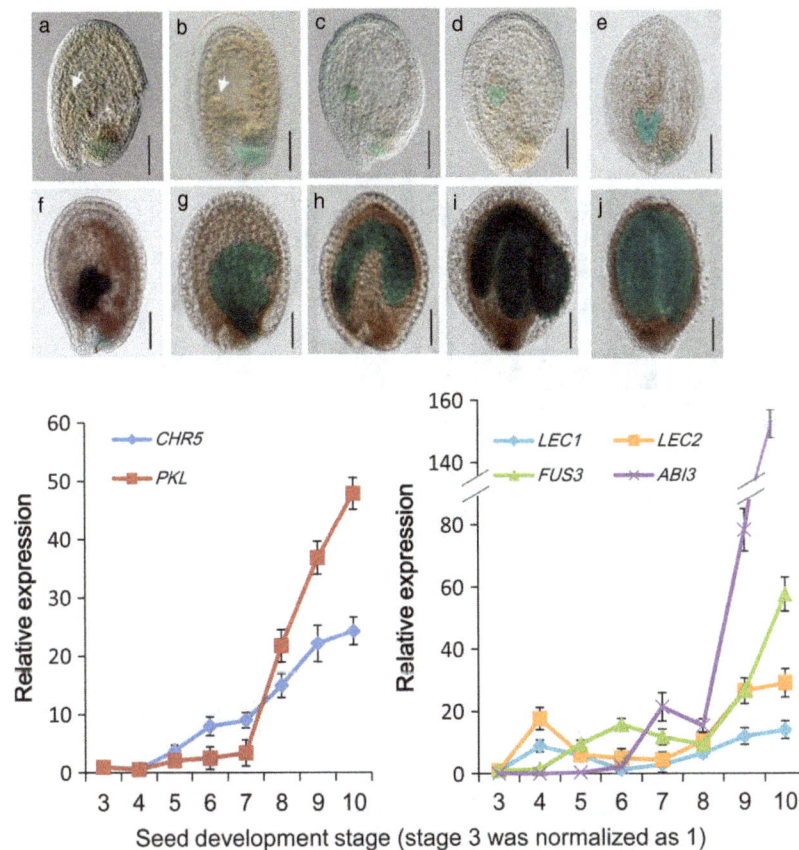

Figure 1 Expression pattern of CHR5 in developing seed. Upper panel: Developing seeds of *Arabidopsis* plants transformed by *pCHR5-GUS* were stained for GUS activity. Photographs of representative seeds at different stages of embryo development (a–j) are shown. White arrow heads indicate unstained embryo at the early stages. Scale bars = 0.1 μm. Lower panel: Left part: RT-qPCR detection of *CHR5* and *PKL* transcripts during the different stages of silique development as defined previously (https://www.genomforschung.uni-bielefeld.de/GF-research/AtGenExpress-SeedsSiliques.html). Right part: RT-qPCR detection of transcripts of *LEC1* and *AFL* genes from the same samples as in left part. Relative transcript levels are shown with that of stage 3 (mid-globular) set as 1.

gene in Columbia-0 (Col-0) and Wassilewskija (Ws) backgrounds (Figure S2). The insertions interrupted the production of the full-length transcript of the gene (Figure S2). The mutants did not display any visible morphological defect, except a weak long hypocotyl phenotype in different light conditions (Figure S3).

Because CHR5 was expressed during embryo development, we examined whether the chr5 mutations affected expression of the embryo regulatory genes in developing siliques (at stage 6 or mid-torpedo stage) by RT-qPCR. Two mutant alleles, chr5-1 and chr5-6, were selected for the analysis. In the mutants, the mRNA levels of LEC1, FUS3 and ABI3 were significantly reduced compared to wild type (Student's t-test, P-value < 0.01), while that of LEC2 was not clearly affected (Figure 2a). In addition, the expression of seed storage protein genes (i.e. 2S2, 7S1, OLE1 and CRA1), which are downstream targets of LEC1 and AFL (Kagaya et al., 2005), was also reduced (Figure 2a). By contrast, the expression of PIL5, a PHYTOCHROME-INTERACTING FACTOR3-LIKE gene that inhibits seed germination (Oh et al., 2004), was not changed in the mutants (Figure 2a). To confirm the data, we analysed the seed storage protein accumulations by SDS-PAGE according to previous description (Finkelstein and Somerville, 1990), in which 50 seeds per genotype were used for protein extraction; one-fifth of the extraction was loaded for the analysis. The analysis revealed a decrease of 2S albumin and 12S globulin proteins in mature seeds of the mutants compared to wild type (Col-0) (Figure 2b). More pronounced decrease was also detected in abi3 and fus3 loss-of-function mutants (Figure 2b).

To confirm the effect of chr5 mutation on seed gene expression and storage protein accumulation, we made complementation tests by transforming chr5-6 plants with the pCHR5-CHR5-HA fusion construct under the control of the 2.0-kb promoter region of CHR5. RT-qPCR analysis of CHR5 transcript confirmed the complementation of the mutation (Figure S4). The production of the fusion protein in the complementation plants was detected by Western blotting using the anti-HA antibody (Figure S4). The expression of LEC1, FUS3, ABI3 and the seed storage genes in the complementation plants was increased to wild-type levels (Figure 2a). The levels of seed storage proteins in chr5-6 were restored in the complementation plants (Figure 2b), indicating that the decrease of storage proteins was caused by the mutation of CHR5. These data suggested that CHR5 was involved in the gene expression of embryo and seed maturation genes in Arabidopsis.

Antagonistic function between CHR5 and PKL in seed maturation gene expression

The pickle (pkl) mutation derepresses embryo regulatory genes in young seedlings with the production of embryonic traits in the primary root of seedlings, accumulation of seed storage reserves and formation of somatic embryos (Ogas et al., 1997). However, this pickle root phenotype only occurs with a low penetrance (Ogas et al., 1997). To study the functional interaction between CHR5 and PKL, we generated chr5-6 pkl1 double mutants and examined the seedling root phenotype. The overall seedling phenotype of pkl was reduced in the double mutant (Figure 3a). The penetrance of the embryonic root in pkl1 (about 15%), revealed by fat red dye staining of fatty acids present in embryonic roots, was significantly (Student's t-test, P-value < 0.01) decreased to about 5% in the double mutants (Figure 3a,b). Similar results were obtained in the chr5-2 pkl1 double mutant (Figure S5). Ectopic expression of LEC1, ABI3 and FUS3 in pkl seedlings or seedling roots was suppressed in the double mutants (Figure 3c,d), suggesting antagonistic actions of the two CHD proteins on the expression of LEC1, ABI3 and FUS3. No clear decrease of LEC2 ectopic expression in the double mutant siliques (Figure 3e) corroborated the observation that CHR5 was not or weakly involved in the activation of the gene (Figure 2a).

Figure 2 chr5 mutations reduced the mRNA levels of embryo maturation genes and seed storage protein accumulation. (a) transcript levels of LEC1, AFL and the indicated seed storage genes in siliques harvested at stage 6 (mid-torpedo) from wild type (Col-0) chr5-1, chr5-6 and the complementation (pCHR5-CHR5-HA) plants. Bars = means ± SD from three biological repeats. *Significance of difference (Student's t-test, P-value < 0.01). (b) SDS-PAGE analysis of 5 mature seed proteins from the wild type (Col-0), mutants (chr5-6, fus3, abi3, pkl and chr5-6 pkl) and chr5-6 complementation plants. ImageJ: National Institutes of Health (NIH), USA was used to measure the density of a 12S peptide band (red arrow) with a higher weight protein band (black arrow). The calculated red/black ratios are indicated.

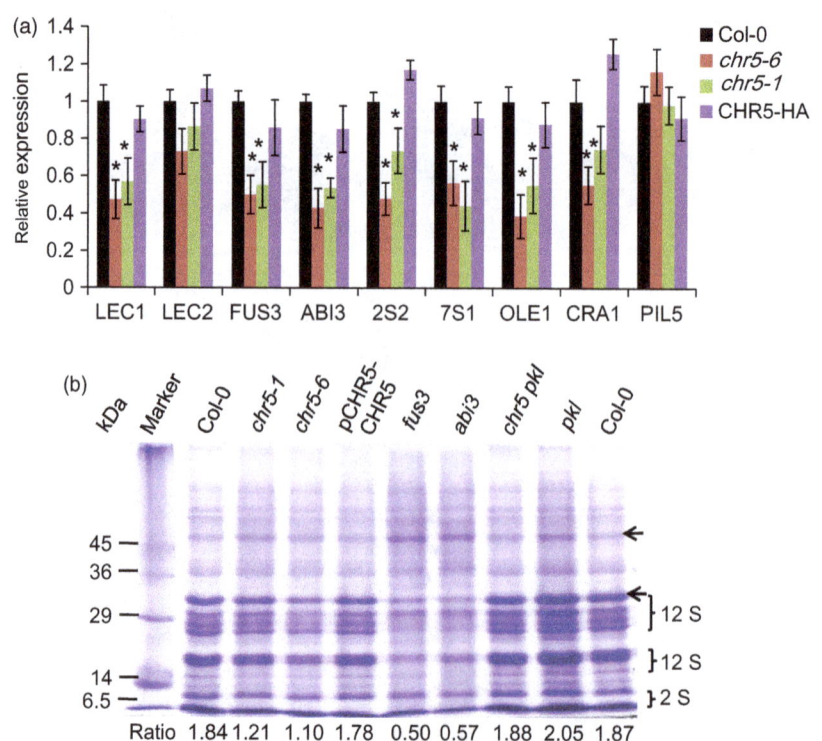

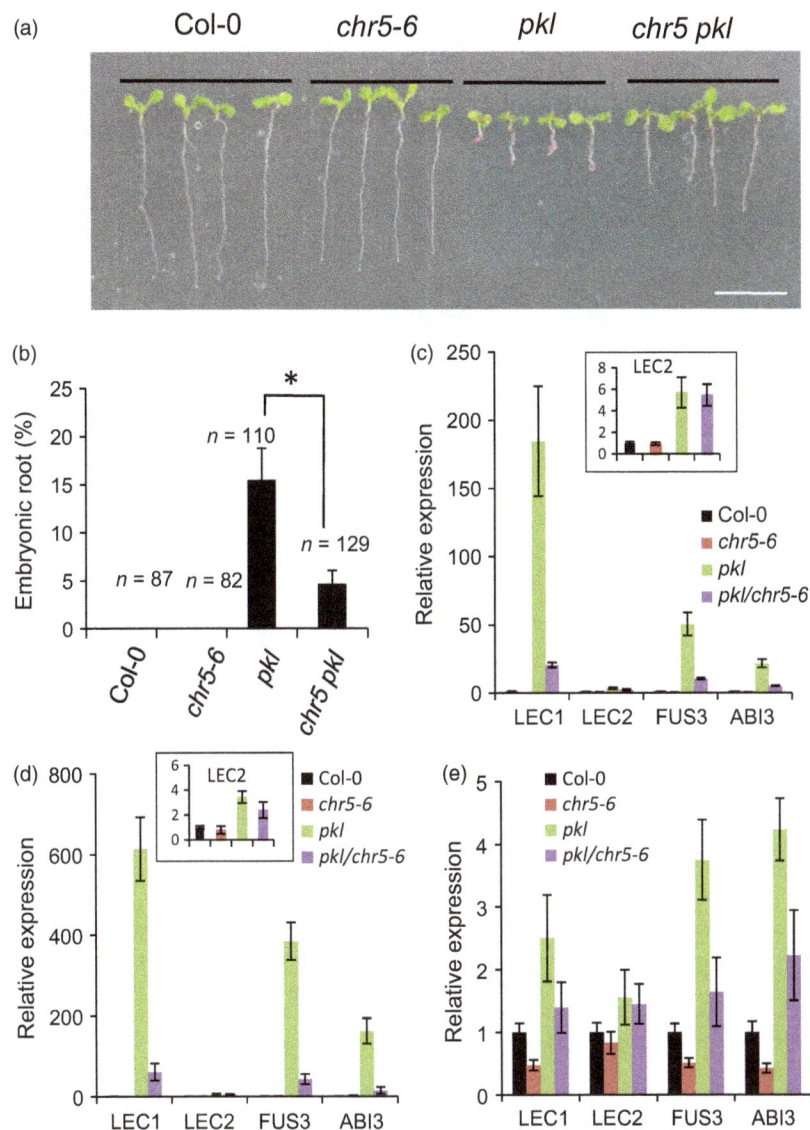

Figure 3 CHR5 and PKL have antagonistic function in the expression of embryo regulatory genes. (a) chr5 pkl double mutation reduced the overall seedling phenotype of pkl. Seedling phenotype of wild type (Col-0), chr5-6, pkl and chr5 pkl double mutants stained by fat red dye. Bar = 5 mm. (b) Embryonic root penetrance of the 4 genotypes calculated from 82 to 129 plants. Asterisks indicate the significance of difference by Student's t-test (P-value < 0.01). (c–e) Relative expression of LEC1 and AFL genes in 5-day-old seedlings (c), Five-day-old seedling roots (d) and siliques at stage 6 (mid-torpedo) (e), with the levels in wild type set as 1. Bars = means ± SD from three biological repeats.

In pkl siliques, the expression levels of LEC1, ABI3 and FUS3 were higher than in wild type (Figure 3e), indicating that PKL may also repress these genes during embryo and seed development. This was in agreement with gradual activation of the PKL gene from stage 4 (early heart stage) and sharply increased after stage 7 (late torpedo stage) of silique development (Figure 1b). In addition, some increases of seed storage proteins could be observed in pkl compared to wild type (Figure 2b). In the siliques of the chr5 pkl double mutants, the expression of LEC1, ABI3 and FUS3 was lower than that in pkl (Figure 3e), supporting the idea that CHR5 and PKL have an opposite function to regulate the seed maturation gene expression programme.

CHR5 and PKL acted on the promoter of ABI3 and FUS3

We next studied whether chr5 and pkl mutations affected the promoter activity and the expression pattern of the embryo regulatory genes during embryo development. Because of the similar expression pattern of CHR5 with ABI3 and FUS3, we chose the promoter region of ABI3 and FUS3 (with 5'-UTR) to control GUS expression in transgenic wild type, chr5-6 and pkl single and double mutant plants. A 4-kb promoter region of ABI3 and a 2.2-kb promoter region of FUS3 were used. Several (6–11) indepen-

dent transgenic lines for each construct in each genotype were characterized. GUS staining revealed that ABI3 and FUS3 promoters started to be active in the embryo during the late globular or early heart stages in the wild type (Figure 4a), which corresponded to the expression profile of the genes (Santos-Mendoza et al., 2008) indicating that the used promoter regions contained necessary elements for the expression patterns. We also noticed that the FUS3 promoter was active in the suspensor in the early stages of embryogenesis. The chr5 and pkl single and double mutations did not clearly alter the expression pattern during embryo development, except the suspensor expression of FUS3-GUS was not detected in chr5 background (Figure 4a). Quantification of the GUS transcripts in siliques (at stage 6 or mid-torpedo stage) of several independent transgenic plants indicated that the promoter activity of both genes was significantly reduced in chr5-6, but increased in pkl background compared to wild type (Student's t-test, P-value < 0.01) (Figure 4b). In the double mutants, the promoter activity was lower than that in pkl (Figure 4a,b). These results confirmed the above data (Figure 3) and revealed that CHR5 and PKL act antagonistically to regulate the promoter activity of ABI3 and FUS3 in developing embryo.

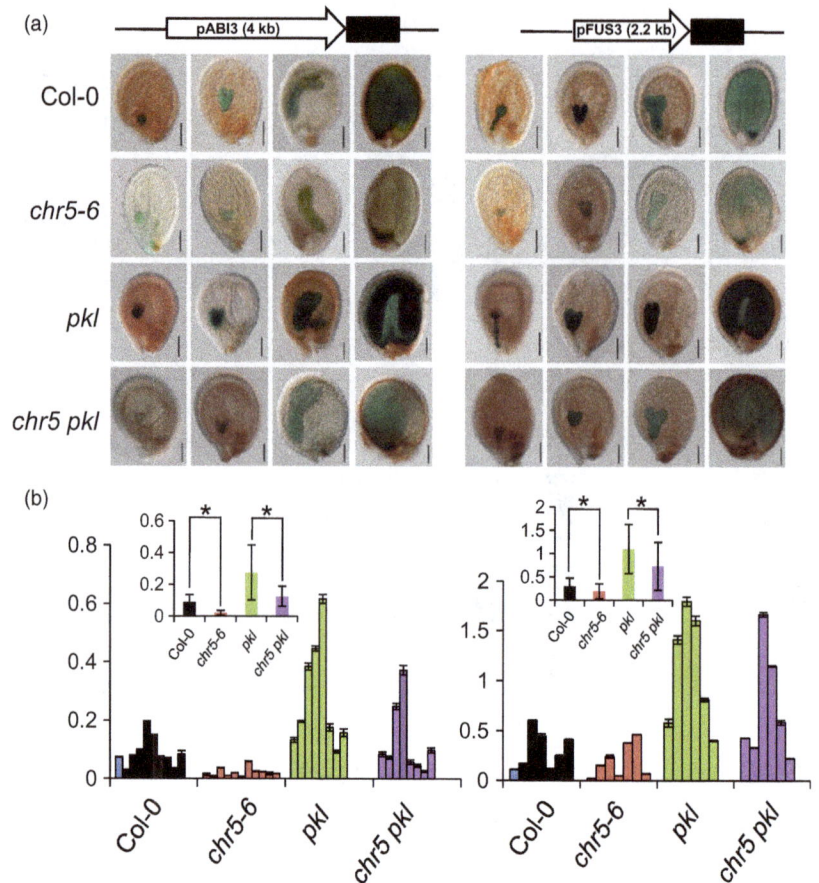

Figure 4 Promoter activity of ABI3 and FUS3 in wild type (Col-0), *chr5-6*, *pkl* and *chr5-6 pkl* double mutants. (a) GUS staining of developing seeds of plants transformed by *pABI3-GUS* (left panels) and *pFUS3-GUS* right panels. (b) Quantification of GUS transcripts (relative to *ACTIN2* mRNA) in siliques at stage 6 (mid-torpedo) from 6 to 10 independent transgenic lines per genotype for each construct. Insets: means ± SD are shown. Asterisks indicate significance of differences (Student's *t*-test, *P*-value < 0.01).

CHR5 binds directly to the promoter region of *ABI3* and *FUS3*

To examine whether CHR5 was directly associated with the embryo regulatory genes, we isolated chromatin from silique tissues of *chr5-6* plants complemented by *pCHR5-CHR5-HA* (Figure S4). The chromatin fragments were immunoprecipitated with anti-HA and analysed by qPCR using primer sets corresponding to several regions of *AFL* and *LEC1* loci and to *ACTIN2* gene and Ta3 transposon as controls (Figure 5). The analysis revealed that only the regions upstream to or near the transcription start site (TSS), not the gene body regions, of *ABI3* and *FUS3* were clearly enriched in the precipitated fractions (Figure 5), indicating that CHR5 was associated with the promoter and TSS of *ABI3* and *FUS3*. No association with *LEC1* or *LEC2* was detected, suggesting that CHR5 was not directly involved in their expression at this stage. However, it is not excluded that CHR5 may act at an earlier stage to establish a transcriptionally permissive state of the genes.

Chromatin modifications of *LEC1* and *AFL* loci in *chr5* and *pkl* mutants

To study whether *chr5* mutations affected histone modification on *LEC1* and *AFL* genes during seed development, chromatin fragments isolated from wild type and *chr5-6* developing siliques (at stage 6, mid-torpedo) were immunoprecipitated by antibodies of H3K4me3 and H3K27me3 and analysed by qPCR using primer sets corresponding to the 5′-UTR region of *FUS3*, *ABI3*, *LEC1*, *LEC2* and, as a control, *PIL5*. The analysis revealed that *LEC2* and *PIL5* displayed relatively higher levels of H3K4me3 compared to

the other genes (Figure 6a). However, there was no clear difference observed between the mutant and wild type. By contrast, relatively higher levels of H3K27me3 were detected on *FUS3*, *ABI3* and *LEC2* and *LEC1* compared to the control genes *RBCS* and *PIL5* (Figure 6b). The *chr5* mutation led to a near twofold increase of H3K27me3 on *LEC1* and *AFL* genes, but not on the control genes (Figure 6b). These data suggested that CHR5 was not involved in modulating H3K4me3 for *LEC1* and *AFL* activation during seed development. Because *LEC1* and *LEC2* were not directly bound by CHR5 at this stage, their increased H3K27me3 levels in *chr5* mutant background may arise from an indirect effect of the mutation, or from a direct effect but at a developmental stage prior to the one sampled for our enrichment analysis.

We also analysed histone methylation in *chr5* and *pkl* seedlings. The results revealed that the amount of H3K4me3 on *LEC1*, *LEC2*, *FUS3* and *ABI3* was not clearly altered in *chr5-6*, *pkl* and *chr5 pkl* compared to wild type, except some increase on *LEC2* was observed in *pkl* plants (Figure S6). This corroborated the data obtained in developing seeds/siliques (Figure 6). In *pkl* seedlings that ectopically express *LEC1* and *AFL* genes, H3K27me3 was reduced on *LEC1* and *LEC2*, but not on *ABI3* and *FUS3* (Figure S6), confirming previous observations (Aichinger *et al.*, 2009; Zhang *et al.*, 2012). Unlike in developing seeds, the *chr5* mutation did not lead to any clear change of H3K27me3 on the seed maturation genes in seedlings. In addition, the decrease of H3K27me3 on *LEC1* and *LEC2* in *pkl* was maintained in the *chr5 pkl* double mutants, suggesting that the antagonistic function of *CHR5* and *PKL* on the expression of the genes in seedlings may depend on different aspects of chromatin modification.

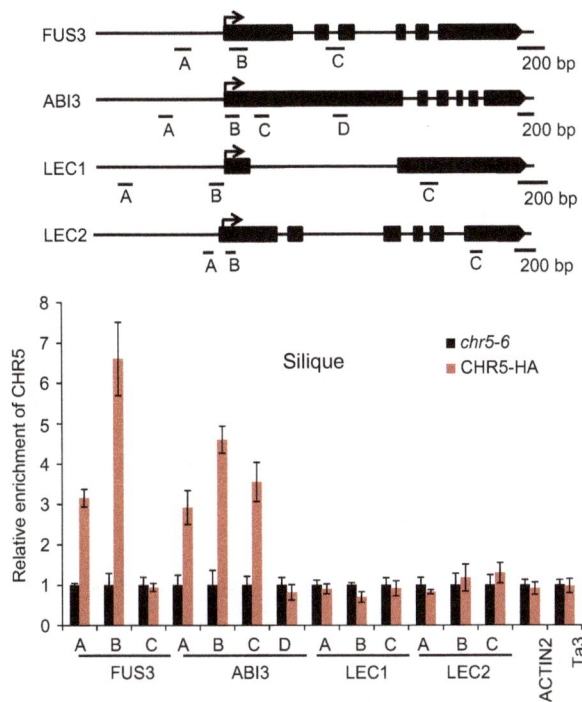

Figure 5 CHR5 is enriched near the transcriptional start site of *ABI3* and *FUS3*. Chromatin isolated respectively from siliques (at stage 6, mid-torpedo) of *chr5-6* mutant and its complementation plants with *pCHR5-CHR5-HA* (CHR5-HA) was immunoprecipitated with anti-HA antibody, and the amount of precipitated DNA was measured by quantitative PCR using 3–4 primer sets (A–D) as indicated on the *LEC1* and *AFL* genes (top). Transcriptional start sites of the four genes are indicated by arrows. Exons are represented by black boxes and untranslated regions by grey. Levels in noncomplemented mutant (*chr5-6*) were set to 1 after normalization to the levels of input DNA. *ACTIN2* and *Ta3* DNA were tested as controls. Bars = mean values ± SD from three biological repeats.

CHR5 may modulate nucleosome occupancy on *FUS3* promoter

Recent results have shown that CHD1 protein is involved in nucleosome positioning and turnover and thus regulates transcription rate (Zentner *et al.*, 2013). To study whether CHR5 was involved in chromatin structure of target genes, we next examined nucleosome positioning and occupancy at the *FUS3* TSS region in chromatin isolated from siliques (stage 6, mid-torpedo) using high-resolution micrococcal nuclease (MNase) mapping (Chodavarapu *et al.*, 2010; Han *et al.*, 2012). We identified one nucleosome upstream of the likely nucleosome-free region (NFR) between −182 and −17 relative to TSS and two nucleosomes downstream of the NFR of *FUS3*. NFR that is located just upstream of the TSS is commonly found in eukaryotic promoters. A typical nucleosome protects about 140–150 bp of genomic DNA from MNase digestion (Yen *et al.*, 2012), as was the case for the three nucleosomes near the TSS of *FUS3*. The upstream one was roughly located between −465 and −265, the two downstream ones protected from −40 to +97 and from 122 to 259 relative to the TSS (Figure 7). In *chr5* mutants, we observed reproducibly a moderate but significant increase (Student's *t*-test, *P*-value < 0.01) in nucleosome occupancy at the three positions in *FUS3* compared to wild type (Figure 7). However, there was no clear difference of nucleosome occupancy observed in the *LEC2*

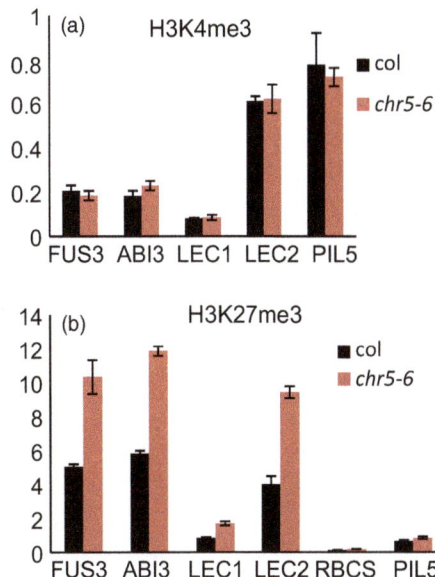

Figure 6 Histone methylation on *LEC1*, *AFL* and *PIL5* genes in wild type and *chr5* mutants. (a) H3K4me3 on the five genes in developing siliques (at stage 6, mid-torpedo) of wild type and *chr5-6*. (b) H3K27me3 on the five genes and RBCS (as a control) in developing siliques (at mid-torpedo stage) of wild type and *chr5-6*. *Y*-axis: relative levels of H3K4me3 or H3K27me3 to that on the *ACTIN2* locus. Bars = mean values ± SD from three biological repeats.

promoter region in siliques of the wild type and mutant plants (Figure S7). Considering that *FUS3* is expressed only in developing seeds and that the mixture of silique tissues at stage 6 was used for the analysis, the observed differences of nucleosome occupancy between wild type and the mutants were likely underestimated. In seedling tissues, the relative nucleosome occupancy on *FUS3* (especially in the NFR) was higher than in silique tissues, and there was no clear difference between the wild type and the mutants (Figure 7). Thus, CHR5 was likely to be required to reduce nucleosome occupancy near the NFR of *FUS3* in developing embryo, which might contribute to increased expression of the gene.

Discussion

In this work, we have shown that the *Arabidopsis* CHD1 gene *CHR5* is involved in embryo regulatory gene expression and seed storage protein accumulation. In addition to embryo, CHR5 is also expressed in the endosperm. As CHR5-targeted seed-specific genes are expressed in both embryo and endosperm, it is possible that CHR5 may be involved in the expression of these genes in endosperm as well. The observations that *chr5* mutations reduced the levels, but not patterns of *AFL* expression in developing embryo, suggest that CHR5 may be mostly involved in modulating gene expression levels instead of defining expression pattern of the key regulatory genes. The *CHR5* expression pattern is similar to that of *ABI3* and *FUS3* during embryo development (Figure 1) (Santos-Mendoza *et al.*, 2008), although the expression of *ABI3* and *FUS3* was detected in earlier stages (Le *et al.*, 2010; Raissig *et al.*, 2013). This partial overlap between CHR5 and *ABI3/FUS3* expression during embryo development suggests that CHR5 may be involved in establishing a more favourable

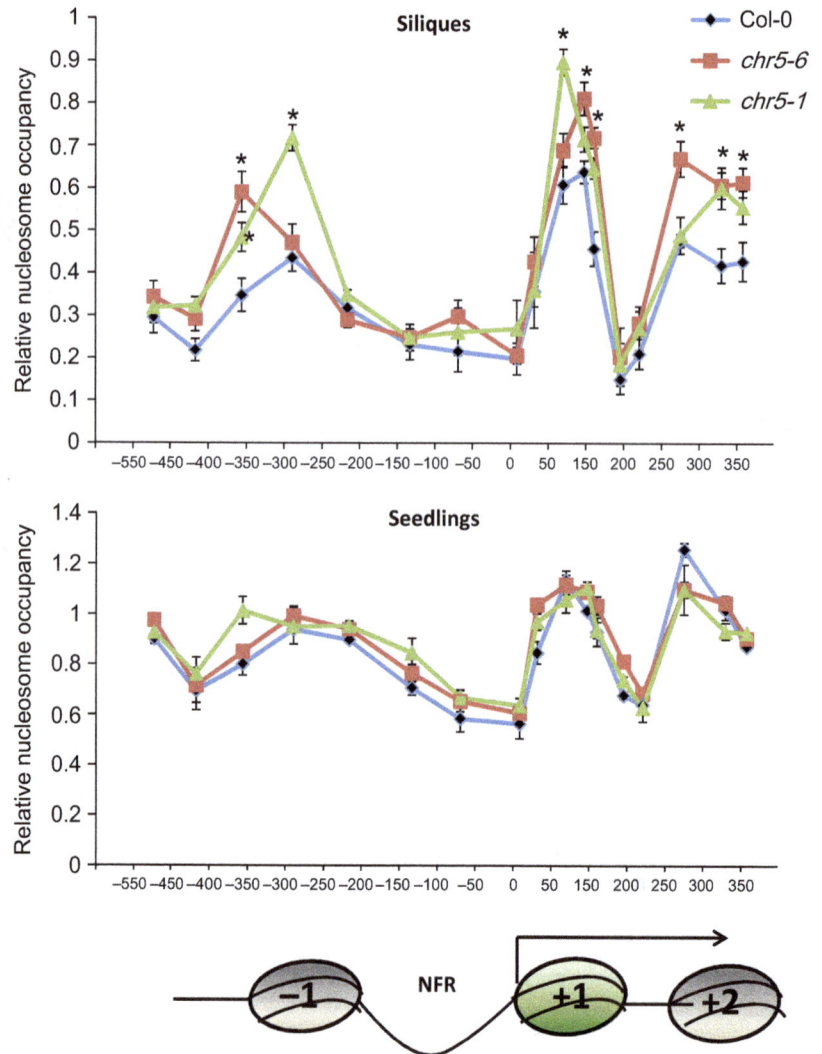

Figure 7 Effect of *chr5* mutations on nucleosome occupancy near the transcriptional start site of *FUS3*. Chromatin isolated from siliques (stage 6, mid-torpedo) or seedlings of wild type (Col-0), *chr5-1* and *chr5-6* was digested by micrococcal nuclease and then analysed by tiled primer qPCR to monitor nucleosome positioning and occupancy at the 5′-end of the *FUS3* locus. The fractions of input were calculated as $2^{-\Delta Ct}$ ($2^{-[Ct(mono)-Ct(gDNA)]}$) using undigested genomic DNA followed by normalization over that of gypsy-like retrotransposon locus (at −73 position) for each sample. NFR, nucleosome-free region. Bars = mean values ± SD from three biological repeats. Significance of differences between wild type and the mutants is indicated by asterisks (Student's *t*-test, $P < 0.01$).

chromatin environment to increase the expression levels of the genes during embryo growth instead of initiating the transcriptional activation process. This is consistent with normal embryo development in the *chr5* mutants. The high accumulation of *CHR5* transcripts in mature embryo/seed and the effects of the mutation of the gene on seed storage protein accumulation suggest that, in addition to *LEC1*, *AFL* and the tested seed storage genes, CHR5 may regulate many other seed maturation genes. Nevertheless, unlike *lec1*, *abi3* or *fus3* loss-of-function mutants that display a wrinkled seed phenotype with accumulation of anthocyanin in *lec1* and *fus3* cotyledons or chlorophyll in *lec1* and *abi3*, the *chr5* mutant seeds germinated normally and did not display any visible abnormality. Possibly, the reduction of *ABI3*, *FUS3*, *LEC1* expression and storage protein accumulation in *chr5* mutants was not sufficient to lead to a seed phenotype. In addition, the expression of *LEC2* was not affected by *chr5* mutations at mid-torpedo stage, which may attenuate the effect of *chr5* mutations in seed gene expression, as it has been shown that LEC2 up-regulate *FUS3* and, together with FUS3 and ABI3, to up-regulate *At2S3* expression (Kroj *et al.*, 2003).

In addition to developing embryo/endosperm, *CHR5* is also expressed in many rapidly growing organs/tissues, suggesting that the gene may be involved in other plant developmental processes. However, loss-of-function mutations of the gene did

not produce any severe developmental defects. Because several other chromatin remodelling activities are found in Arabidopsis, functional redundancy between CHR5 and other chromatin remodelling factors may help explain partly the mild developmental phenotype observed in *chr5* mutants.

Nucleosomes are organized into uniformly spaced arrays at the 5′-end of the genes, which starts with the +1 nucleosome at around the TSS (Jiang and Pugh, 2009). A '−1' nucleosome located on the upstream in the promoter is positioned to potentially control access to gene regulatory sequences. It is suggested that enrichment of CHD1 proteins within the NFR is to position nucleosomes within the promoter or around the TSS to facilitate gene transcription and also to position nucleosomes within the coding regions of genes, which are aligned with respect to TSS to prevent cryptic transcription within gene bodies by suppressing histone turnover (Gkikopoulos *et al.*, 2011; Hennig *et al.*, 2012; Smolle *et al.*, 2012; Zentner *et al.*, 2013). Our data showing that a relatively higher nucleosome occupancy of *FUS3* in seedling compared to silique tissues and that *chr5* mutations led to a relatively higher occupancy of the nucleosomes located near the NFR of *FUS3* in silique tissues (Figure 7), suggest that CHR5 may have a function to reduce the nucleosomal barrier to stimulate *FUS3* gene transcription during embryo development. This is in agreement with the finding that in *S. cerevisiae* with the deletion of Chd1 nucleosomes near the NFR display a relatively higher occupancy genomewide (Gkikopoulos et al., 2011). Possibly, the reduction of nucleosome occupancy in the promoter and near the TSS by the binding of CHD1 proteins within NFR may facilitate the functioning of regulatory sequence such as transcription factor binding and transcription initiation complex recruitment to stimulate gene transcription. It remains to know whether CHR5-regulated nucleosome occupancy near the TSS of *FUS3* affects the binding of transcription factors including LEC1 and AFL proteins to the gene.

Materials and methods

Plant material and growth

The *Arabidopsis thaliana* genetic resources used in this study were mostly in the Columbia ecotype, except the *chr5-3* was from the Wassilewskija ecotype. Mutant lines *chr5-1* (Sail_504_D01), *chr5-2* (Salk_020296), *chr5-3* (Flag_130A16), *chr5-4* (GABI_773A12), *chr5-5* (Sail_1259_B05) and *chr5-6* (Salk_046838) were obtained from the Nottingham *Arabidopsis* Stock Center and ABRC collections. The *pkl* mutant used in this study was the *pkl-1* allele as described (Ogas et al., 1997). The *abi3* mutant (Salk_023411) and *fus3* mutant (GABI_612E06) were used. The T-DNA insertion was confirmed by PCR using the primers described in the Table S1. Double mutants *chr5-6 pkl* and *chr5-2 pkl* were obtained by genetic crosses between *chr5-6 and chr5-2* with *pkl*. Seeds were surface-sterilized and plated on 0.5× Murashige and Skoog medium. After stratification at 4 °C for 2 days, plants were grown in a growth room under a long-day photoperiod (16 h light/8 h dark) at 22 °C.

Constructs and transformation

For the histochemical GUS assay, the 2-kb promoter of *CHR5* was amplified from WT DNA using proCHR5-F and proCHR5-R and then cloned into the pPR97 vector. For the complementation experiment, the full-length cDNA without a stop codon was amplified from total cDNA of Col-0 using primers comCHR5-F and comCHR5-R. *CHR5* promoter and cDNA were inserted into the binary vector pFA1300, which was modified based on pCAMBIA1300 (CAMBIA) and contained 2× HA tag. The *pCHR5-GUS* and *pCHR5-CHR5-HA* constructs were introduced into Col-0 and *chr5* mutants via Agrobacterium-mediated transformation by the floral dip method. The *pABI3:GUS* and *pFUS3:GUS* were made using a PCR-based gateway system. The promoter and 5′-UTR of 4 kb for *ABI3* and 2.2 kb for *FUS3* were inserted as translational fusion with the *uidA* gene into the vector pGWB553.

Gene expression analysis

Definition of stages for seed and silique development was according to previous description (Kleindt et al., 2010). Total RNA of siliques was extracted as described (Meng and Feldman, 2010). The crude RNA was further purified via the clean-up protocol of the RNeasy plant RNA isolation kit (Qiagen, Hilden, Germany) according to the manufacturer's protocol instead of TRIzol reagent. Two micrograms of total RNA was treated with DNase I (Promega, Madison, WI) and transcribed into cDNA by ImPromII reverse transcriptase (Promega). Real-time PCR was performed with the LightCycler® 480 SYBR Green I Master (Roche, Roche Diagnostics GmbH: Mannheim, Germany; Roche Diagnostics: Indianapolis, IN) following the manufacturer's instructions. Primers are listed in the Table S1. Three biological replicates and three technical repeats were performed for each sample, and the expression level was normalized with that of *ACTIN2*.

ChIP assay

Chromatin was isolated from siliques and dissected seeds after cross-linking proteins and DNA with 1% formaldehyde for 1 h under vacuum and termination of the reaction with glycine. Chromatin was fragmented to 200–1000 bp by sonication, and ChIP was performed using the following antibodies: anti-H3H4me3 (Millipore, 07–473; Temecula, CA), anti-H3K27me3 (Millipore, 07–449; Temecula, CA), anti-H3 (Abcam, ab1791: Cambridge, UK), anti-H3ac (Millipore, 06–599; Temecula, CA) and anti-HA antibody (Sigma, H6908: St. Louis, MO).

Histochemical GUS and lipid staining

GUS staining was performed as previously described (Bertrand et al., 2003). Briefly, plant siliques were fixed with 90% acetone on ice for 30 min and were washed with staining buffer (0.2% Triton X-100, 5 mM potassium ferrocyanide, 5 mM potassium ferricyanide, 100 mM NaH$_2$PO$_4$ and 100 mM Na$_2$HPO$_4$ pH 7.2). The dissected seeds from the fixed siliques were immersed in GUS staining solution with 1 mM X-Gluc and placed under vacuum for 1 h. After incubation at 37 °C overnight, the staining solution was removed and samples were cleared by sequential changes of 75% and 95% (v/v) ethanol. Lipid staining with fat red 7B was carried out as described previously (Ogas et al., 1997). Whole seedlings were incubated for 1 h in filtered fat red solution (0.5% fat red 7B in 60% isopropanol), washed three times with water and analysed under a dissecting microscope.

Seed protein analysis

Seed storage protein analysis was performed according to procedures described in (Finkelstein and Somerville, 1990). Seed protein was extracted from 50 seeds/genotype by grinding mature seeds in an ice-cold motor with 20 µL/mg seed of extraction buffer (100 mM Tris–HCl, pH 8.0, 0.5% SDS, 10% glycerol and 2% β-mercaptoethanol). Extracts were boiled for 3 min and centrifuged. The gel was loaded on a seed number basis, and the proteins were resolved by 12% SDS-PAGE. Proteins were visualized by Coomassie blue staining.

MNase assay

Siliques were harvested in liquid nitrogen after cross-linking in 1% formaldehyde. Nuclei and chromatin were isolated as described (Chodavarapu et al., 2010) with the following changes. The isolated chromatin was digested with 0.1 units/µL of micrococcal nuclease (Takara, Dalian, China) for 10 min in digestion buffer at 37 °C. Mononucleosomes were excised from 1.5% agarose gels and purified using a gel purification kit (Macherey-Nagel, Düren, Germany). The purified DNA was quantified using a NanoDrop-1000 spectrophotometer. Two nanograms of purified DNA was used for qPCR to monitor nucleosome occupancy. The fraction of input was calculated as $2^{-\Delta Ct}$ ($2^{-[Ct(mono)-Ct(gDNA)]}$) using undigested genomic DNA followed by normalization over that of gypsy-like retrotransposon locus (at −73 position) for each sample according to previous description (Gevry et al., 2009). The tiled primer sets used for real-time PCR are listed in the Table S1.

Acknowledgements

We thank F. Barneche for help in H2Bub detection, B. Dubreucq and M. Miquel for helpful discussion, and G.

Barthole for providing *fus3* and *abi3* mutants. This work was supported by the Agence National de Recherche (ANR 2010-BLAN-1238 CERES). Y.S. was supported by a PhD fellowship from the Chinese Scholar Council. The authors declare that there is no conflict of interest.

References

Aichinger, E., Villar, C.B.R., Farrona, S., Reyes, J.C., Hennig, L. and Kohler, C. (2009) CHD3 proteins and polycomb group proteins antagonistically determine cell identity in Arabidopsis. *PLoS Genet.* **5**, e1000605.

Berger, N., Dubreucq, B., Roudier, F., Dubos, C. and Lepiniec, L. (2011) Transcriptional regulation of Arabidopsis LEAFY COTYLEDON2 involves RLE, a cis-element that regulates trimethylation of histone H3 at lysine-27. *Plant Cell*, **23**, 4065–4078.

Bertrand, C., Bergounioux, C., Domenichini, S., Delarue, M. and Zhou, D.X. (2003) Arabidopsis histone acetyltransferase AtGCN5 regulates the floral meristem activity through the WUSCHEL/AGAMOUS pathway. *J. Biol. Chem.* **278**, 28246–28251.

Bouyer, D., Roudier, F., Heese, M., Andersen, E.D., Gey, D., Nowack, M.K., Goodrich, J., Renou, J.P., Grini, P.E., Colot, V. and Schnittger, A. (2011) Polycomb repressive complex 2 controls the embryo-to-seedling phase transition. *PLoS Genet.* **7**, e1002014.

Bratzel, F., Lopez-Torrejon, G., Koch, M., Del Pozo, J.C. and Calonje, M. (2010) Keeping cell identity in Arabidopsis requires PRC1 RING-finger homologs that catalyze H2A monoubiquitination. *Curr. Biol.* **20**, 1853–1859.

Chanvivattana, Y., Bishopp, A., Schubert, D., Stock, C., Moon, Y.H., Sung, Z.R. and Goodrich, J. (2004) Interaction of Polycomb-group proteins controlling flowering in Arabidopsis. *Development*, **131**, 5263–5276.

Chodavarapu, R.K., Feng, S., Bernatavichute, Y.V., Chen, P.Y., Stroud, H., Yu, Y., Hetzel, J.A., Kuo, F., Kim, J., Cokus, S.J., Casero, D., Bernal, M., Huijser, P., Clark, A.T., Kramer, U., Merchant, S.S., Zhang, X., Jacobsen, S.E. and Pellegrini, M. (2010) Relationship between nucleosome positioning and DNA methylation. *Nature*, **466**, 388–392.

Flanagan, J.F., Mi, L.Z., Chruszcz, M., Cymborowski, M., Clines, K.L., Kim, Y., Minor, W., Rastinejad, F. and Khorasanizadeh, S. (2005) Double chromodomains cooperate to recognize the methylated histone H3 tail. *Nature*, **438**, 1181–1185.

Flaus, A., Martin, D.M., Barton, G.J. and Owen-Hughes, T. (2006) Identification of multiple distinct Snf2 subfamilies with conserved structural motifs. *Nucleic Acids Res.* **34**, 2887–2905.

Gevry, N., Svotelis, A., Larochelle, M. and Gaudreau, L. (2009) Nucleosome mapping. *Methods Mol. Biol.* **543**, 281–291.

Giraudat, J., Hauge, B.M., Valon, C., Smalle, J., Parcy, F. and Goodman, H.M. (1992) Isolation of the Arabidopsis ABI3 gene by positional cloning. *Plant Cell*, **4**, 1251–1261.

Gkikopoulos, T., Schofield, P., Singh, V., Pinskaya, M., Mellor, J., Smolle, M., Workman, J.L., Barton, G.J. and Owen-Hughes, T. (2011) A role for Snf2-related nucleosome-spacing enzymes in genome-wide nucleosome organization. *Science*, **333**, 1758–1760.

Han, S.K., Sang, Y., Rodrigues, A., Wu, M.F., Rodriguez, P.L. and Wagner, D. (2012) The SWI2/SNF2 chromatin remodeling ATPase BRAHMA represses abscisic acid responses in the absence of the stress stimulus in Arabidopsis. *Plant Cell*, **24**, 4892–4906.

Hennig, B.P., Bendrin, K., Zhou, Y. and Fischer, T. (2012) Chd1 chromatin remodelers maintain nucleosome organization and repress cryptic transcription. *EMBO Rep.* **13**, 997–1003.

Hu, Y., Liu, D., Zhong, X., Zhang, C., Zhang, Q. and Zhou, D.X. (2012) CHD3 protein recognizes and regulates methylated histone H3 lysines 4 and 27 over a subset of targets in the rice genome. *Proc. Natl Acad. Sci. USA*, **109**, 5773–5778.

Jiang, C. and Pugh, B.F. (2009) Nucleosome positioning and gene regulation: advances through genomics. *Nat. Rev. Genet.* **10**, 161–172.

Kagaya, Y., Toyoshima, R., Okuda, R., Usui, H., Yamamoto, A. and Hattori, T. (2005) LEAFY COTYLEDON1 controls seed storage protein genes through its regulation of FUSCA3 and ABSCISIC ACID INSENSITIVE3. *Plant Cell Physiol.* **46**, 399–406.

Kleindt, C.K., Stracke, R., Mehrtens, F. and Weisshaar, B. (2010) Expression analysis of flavonoid biosynthesis genes during Arabidopsis thaliana silique and seed development with a primary focus on the proanthocyanidin biosynthetic pathway. *BMC Res Notes*, **3**, 255.

Kroj, T., Savino, G., Valon, C., Giraudat, J. and Parcy, F. (2003) Regulation of storage protein gene expression in Arabidopsis. *Development*, **130**, 6065–6073.

Le, B.H., Cheng, C., Bui, A.Q., Wagmaister, J.A., Henry, K.F., Pelletier, J., Kwong, L., Belmonte, M., Kirkbride, R., Horvath, S., Drews, G.N., Fischer, R.L., Okamuro, J.K., Harada, J.J. and Goldberg, R.B. (2010) Global analysis of gene activity during Arabidopsis seed development and identification of seed-specific transcription factors. *Proc. Natl Acad. Sci. USA*, **107**, 8063–8070.

Lotan, T., Ohto, M., Yee, K.M., West, M.A., Lo, R., Kwong, R.W., Yamagishi, K., Fischer, R.L., Goldberg, R.B. and Harada, J.J. (1998) Arabidopsis LEAFY COTYLEDON1 is sufficient to induce embryo development in vegetative cells. *Cell*, **93**, 1195–1205.

Makarevich, G., Leroy, O., Akinci, U., Schubert, D., Clarenz, O., Goodrich, J., Grossniklaus, U. and Kohler, C. (2006) Different Polycomb group complexes regulate common target genes in Arabidopsis. *EMBO Rep.* **7**, 947–952.

Meng, L. and Feldman, L. (2010) A rapid TRIzol-based two-step method for DNA-free RNA extraction from Arabidopsis siliques and dry seeds. *Biotechnol. J.* **5**, 183–186.

Monke, G., Seifert, M., Keilwagen, J., Mohr, M., Grosse, I., Hahnel, U., Junker, A., Weisshaar, B., Conrad, U., Baumlein, H. and Altschmied, L. (2012) Toward the identification and regulation of the Arabidopsis thaliana ABI3 regulon. *Nucleic Acids Res.* **40**, 8240–8254.

Murawska, M. and Brehm, A. (2011) CHD chromatin remodelers and the transcription cycle. *Transcription*, **2**, 244–253.

Narlikar, G.J., Sundaramoorthy, R. and Owen-Hughes, T. (2013) Mechanisms and functions of ATP-dependent chromatin-remodeling enzymes. *Cell*, **154**, 490–503.

Ogas, J., Cheng, J.C., Sung, Z.R. and Somerville, C. (1997) Cellular differentiation regulated by gibberellin in the Arabidopsis thaliana pickle mutant. *Science*, **277**, 91–94.

Ogas, J., Kaufmann, S., Henderson, J. and Somerville, C. (1999) PICKLE is a CHD3 chromatin-remodeling factor that regulates the transition from embryonic to vegetative development in Arabidopsis. *Proc. Natl Acad. Sci. USA*, **96**, 13839–13844.

Oh, E., Kim, J., Park, E., Kim, J.I., Kang, C. and Choi, G. (2004) PIL5, a phytochrome-interacting basic helix-loop-helix protein, is a key negative regulator of seed germination in Arabidopsis thaliana. *Plant Cell*, **16**, 3045–3058.

Raissig, M.T., Bemer, M., Baroux, C. and Grossniklaus, U. (2013) Genomic imprinting in the Arabidopsis embryo is partly regulated by PRC2. *PLoS Genet.* **9**, e1003862.

Ramirez, J. and Hagman, J. (2009) The Mi-2/NuRD complex: a critical epigenetic regulator of hematopoietic development, differentiation and cancer. *Epigenetics*, **4**, 532–536.

Santos-Mendoza, M., Dubreucq, B., Baud, S., Parcy, F., Caboche, M. and Lepiniec, L. (2008) Deciphering gene regulatory networks that control seed development and maturation in Arabidopsis. *Plant J.* **54**, 608–620.

Schubert, D., Primavesi, L., Bishopp, A., Roberts, G., Doonan, J., Jenuwein, T. and Goodrich, J. (2006) Silencing by plant Polycomb-group genes requires dispersed trimethylation of histone H3 at lysine 27. *EMBO J.* **25**, 4638–4649.

Sims, R.J. 3rd, Millhouse, S., Chen, C.F., Lewis, B.A., Erdjument-Bromage, H., Tempst, P., Manley, J.L. and Reinberg, D. (2005) Recognition of trimethylated histone H3 lysine 4 facilitates the recruitment of transcription postinitiation factors and pre-mRNA splicing. *Mol. Cell*, **28**, 665–676.

Smolle, M., Venkatesh, S., Gogol, M.M., Li, H., Zhang, Y., Florens, L., Washburn, M.P. and Workman, J.L. (2012) Chromatin remodelers Isw1 and Chd1 maintain chromatin structure during transcription by preventing histone exchange. *Nat. Struct. Mol. Biol.* **19**, 884–892.

Stone, S.L., Kwong, L.W., Yee, K.M., Pelletier, J., Lepiniec, L., Fischer, R.L., Goldberg, R.B. and Harada, J.J. (2001) LEAFY COTYLEDON2 encodes a B3 domain transcription factor that induces embryo development. *Proc. Natl Acad. Sci. USA*, **98**, 11806–11811.

Suzuki, M. and McCarty, D.R. (2008) Functional symmetry of the B3 network controlling seed development. *Curr. Opin. Plant Biol.* **11**, 548–553.

Torigoe, S.E., Patel, A., Khuong, M.T., Bowman, G.D. and Kadonaga, J.T. (2013) ATP-dependent chromatin assembly is functionally distinct from chromatin remodeling. *Elife*, **2**, e00863.

Wang, F. and Perry, S.E. (2013) Identification of direct targets of FUSCA3, a key regulator of Arabidopsis seed development. *Plant Physiol.* **161**, 1251–1264.

Yang, C., Bratzel, F., Hohmann, N., Koch, M., Turck, F. and Calonje, M. (2013) VAL- and AtBMI1-mediated H2Aub initiate the switch from embryonic to postgerminative growth in Arabidopsis. *Curr. Biol.* **23**, 1324–1329.

Yen, K., Vinayachandran, V., Batta, K., Koerber, R.T. and Pugh, B.F. (2012) Genome-wide nucleosome specificity and directionality of chromatin remodelers. *Cell*, **149**, 1461–1473.

Zentner, G.E., Tsukiyama, T. and Henikoff, S. (2013) ISWI and CHD chromatin remodelers bind promoters but act in gene bodies. *PLoS Genet.* **9**, e1003317.

Zhang, H., Rider, S.D., Henderson, J.T., Fountain, M., Chuang, K., Kandachar, V., Simons, A., Edenberg, H.J., Romero-Severson, J., Muir, W.M. and Ogas, J. (2008) The CHD3 remodeler PICKLE promotes trimethylation of histone H3 lysine 27. *J. Biol. Chem.* **283**, 22637–22648.

Zhang, H., Bishop, B., Ringenberg, W., Muir, W.M. and Ogas, J. (2012) The CHD3 remodeler PICKLE associates with genes enriched for trimethylation of histone H3 Lysine 27. *Plant Physiol.* **159**, 418–432.

13

Reduced paucimannosidic *N*-glycan formation by suppression of a specific β-hexosaminidase from *Nicotiana benthamiana*

Yun-Ji Shin[1], Alexandra Castilho[1], Martina Dicker[1], Flavio Sádio[1], Ulrike Vavra[1], Clemens Grünwald-Gruber[2], Tae-Ho Kwon[3], Friedrich Altmann[2], Herta Steinkellner[1] and Richard Strasser[1,*]

[1]*Department of Applied Genetics and Cell Biology, University of Natural Resources and Life Sciences, Vienna, Austria*
[2]*Department of Chemistry, University of Natural Resources and Life Sciences, Vienna, Austria*
[3]*NBM Inc., Wanju-gun, Jeollabuk-do, Korea*

*Correspondence
email richard.
strasser@boku.ac.at

Keywords: α1-antitrypsin, glyco-engineering, *N*-glycosylation, *Nicotiana benthamiana*, plant-made pharmaceuticals.

Summary

Plants are attractive hosts for the production of recombinant glycoproteins for therapeutic use. Recent advances in glyco-engineering facilitate the elimination of nonmammalian-type glycosylation and introduction of missing pathways for customized *N*-glycan formation. However, some therapeutically relevant recombinant glycoproteins exhibit unwanted truncated (paucimannosidic) *N*-glycans that lack GlcNAc residues at the nonreducing terminal end. These paucimannosidic *N*-glycans increase product heterogeneity and may affect the biological function of the recombinant drugs. Here, we identified two enzymes, β-hexosaminidases (HEXOs) that account for the formation of paucimannosidic *N*-glycans in *Nicotiana benthamiana*, a widely used expression host for recombinant proteins. Subcellular localization studies showed that HEXO1 is a vacuolar protein and HEXO3 is mainly located at the plasma membrane in *N. benthamiana* leaf epidermal cells. Both enzymes are functional and can complement the corresponding HEXO-deficient *Arabidopsis thaliana* mutants. *In planta* expression of HEXO3 demonstrated that core α1,3-fucose enhances the trimming of GlcNAc residues from the Fc domain of human IgG. Finally, using RNA interference, we show that suppression of HEXO3 expression can be applied to increase the amounts of complex *N*-glycans on plant-produced human α1-antitrypsin.

Introduction

The majority of therapeutic proteins including monoclonal antibodies, hormones and lysosomal enzymes are glycoproteins. For many glycoprotein drugs, a defined glycan structure is required for optimal efficacy. As a consequence, glycans from recombinant glycoproteins are considered as critical quality attributes by industry (Reusch and Tejada, 2015) and current manufacturing platforms are converted into systems with controllable glycosylation (Yang *et al.*, 2015a). Recent progress in glyco-engineering of plants has shown that *Nicotiana benthamiana* is highly suitable for the production of recombinant glycoproteins with tailor-made *N*- and *O*-glycan structures (Steinkellner and Castilho, 2015; Strasser *et al.*, 2014). In particular, the glyco-engineered ΔXT/FT mutant that exhibits stable down-regulation of the plant enzymes β1,2-xylosyl- and core α1,3-fucosyltransferase has been used for the transient expression of different therapeutically relevant glycoproteins with custom-made glycosylation (Castilho *et al.*, 2014; Dicker *et al.*, 2016; Jez *et al.*, 2013; Loos *et al.*, 2014, 2015; Schneider *et al.*, 2014; Strasser *et al.*, 2008, 2009; Wilbers *et al.*, 2016). Notably, the ΔXT/FT plants are used to manufacture ZMapp the experimental antibody cocktail for treatment of acute Ebola virus infections (Qiu *et al.*, 2014) and for the production of other monoclonal antibodies against infectious diseases that are currently under development (Loos *et al.*, 2015; Zeitlin *et al.*, 2016).

Complex *N*-glycan formation is initiated in the *cis*/medial Golgi by *N*-acetylglucosaminyltransferase I (GnTI) which transfers a single GlcNAc residue to $Man_5GlcNAc_2$ (Strasser *et al.*, 1999). The presence of this terminal GlcNAc is a prerequisite for all further modifications including mannose trimming by Golgi-α-mannosidase II, transfer of a second terminal GlcNAc residue by *N*-acetylglucosaminyltransferase II (GnTII), β1,2-xylosylation and core α1,3-fucosylation (Strasser, 2016). Commonly, these Golgi processing reactions result in the formation of complex *N*-glycans with two terminal GlcNAc residues at the nonreducing end (GnGnXF: $GlcNAc_2XylFucMan_3GlcNAc_2$ in wild-type or GnGn: $GlcNAc_2Man_3GlcNAc_2$ in ΔXT/FT plants) (Figure 1a). Such complex *N*-glycans are, for example, the predominant structures on plant-produced monoclonal antibodies. However, for other recombinant glycoproteins expression in *N. benthamiana* leaves resulted in the generation of *N*-glycans with a considerable amount of truncated oligosaccharide structures (Castilho *et al.*, 2014; Dicker *et al.*, 2016; Dirnberger *et al.*, 2001). These so-called paucimannosidic *N*-glycans (MMXF: $Man_3XylFucGlcNAc_2$ in wild-type) expose terminal mannose residues at the nonreducing end. Under physiological conditions, paucimannosidic *N*-glycans are quite rare on mammalian glycoproteins, but these truncated structures can be considerably increased in certain environments like cancer tissues (Schachter, 2009; Sethi *et al.*, 2015). Moreover, exposed mannose residues on glycoproteins can accelerate their turnover by receptor-mediated clearance from the blood (Yang *et al.*, 2015b). For biotechnological production of most secreted recombinant glycoproteins, it is therefore relevant to prevent the formation of paucimannosidic *N*-glycans.

As the majority of Golgi-mediated *N*-glycan processing steps are strictly dependent on the GlcNAc residue that is transferred by GnTI, the occurrence of paucimannosidic MMXF *N*-glycans can only be explained by enzymatic removal of one or two terminal GlcNAc residues (Figure 1a). The site for this trimming reaction is most likely in a post-Golgi compartment or in the apoplast. Plant β-hexosaminidases (HEXOs) are the class of enzymes that can cleave off terminal GlcNAc residues from complex *N*-glycans

(a)

(b)

Figure 1 (a) Schematic illustration of proposed HEXO activity on complex *N*-glycans. The symbols for the monosaccharides in the illustration are drawn according to the nomenclature from the Consortium for Functional Glycomics. (b) Schematic representation of used expression vectors. LB: left border; Pnos: nopaline synthase gene promoter; Hyg: hygromycin B phosphotransferase gene; Kan: neomycin phosphotransferase 2 gene; Tnos: nopaline synthase gene terminator; P35S: cauliflower mosaic virus 35S gene promoter; HEXO1: *Nicotiana benthamiana* HEXO1 open reading frame (ORF); HEXO3: *N. benthamiana* HEXO3 ORF; Sec Fc: signal peptide from α-glucosidase II fused to the Fc domain of human IgG1; GFP: green fluorescent protein; mRFP: monomeric red fluorescent protein; XYLT: *Arabidopsis thaliana* β1,2-xylosyltransferase ORF; HA: hemagglutinin tag; S: HEXO3 RNAi sequence in sense orientation; XTI2: intron 2 from *A. thaliana* XYLT; AS: HEXO3 RNAi sequence in antisense orientation; g7: agrobacterium gene 7 terminator; RB: right border.

(Altmann *et al.*, 1995). The *Arabidopsis thaliana* HEXO family consists of three members (AtHEXO1–AtHEXO3) (Liebminger *et al.*, 2011; Strasser *et al.*, 2007). AtHEXO1 is found mainly in the vacuoles where it generates paucimannosidic *N*-glycans on vacuolar glycoproteins. AtHEXO2 activity and protein expression could not be detected in previous studies suggesting that AtHEXO2 represents an inactive or highly regulated enzyme that is expressed only in specific cell types. AtHEXO3, on the other hand, is an active β-hexosaminidase located at the plasma membrane and acts on secreted glycoproteins (Castilho *et al.*, 2014; Liebminger *et al.*, 2011). Although these enzymes have been well characterized from *A. thaliana* and their ability to act on *N*-glycans is well documented, their biological function is still unclear. Neither *hexo* single mutants, nor the *hexo1 hexo3* double mutant plants display any growth or developmental phenotype.

Expression of human α1-antitrypsin (A1AT) in the *A. thaliana hexo3* mutant resulted in the formation of the fully processed GnGnXF *N*-glycan instead of the paucimannosidic MMXF that was observed in wild-type (Castilho *et al.*, 2014). Consequently, HEXO3 activity is a major limitation for the production of distinct glycoproteins in plants as it generates unwanted truncated *N*-glycans and contributes to the overall *N*-glycan microheterogeneity. Here, we aimed to identify and inactivate the corresponding HEXOs from *N. benthamiana* plants. In addition, we performed *in planta* HEXO activity assays demonstrating that the attachment of core α1,3-fucose can influence the processing of *N*-glycans from a model glycoprotein. In summary, our study provides new insights into the specific function of plant HEXOs and paves the way for the efficient elimination of nonfavourable paucimannosidic *N*-glycan formation in plants to improve the quality of plant-made recombinant glycoproteins.

Results

A database search revealed several HEXO1 and HEXO3 candidates in *N. benthamiana*

The amino acid sequences from *A. thaliana* β-hexosaminidases (AtHEXO1 and AtHEXO3) were used to search in the *N. benthamiana* draft genome for genes encoding putative HEXO orthologs. As the expression, enzymatic activity and putative function of AtHEXO2 is unclear (Strasser *et al.*, 2007), no attempts were made to find orthologs from *N. benthamiana*. At least two different *N. benthamiana* sequences were identified which represent putative orthologs of AtHEXO1 and AtHEXO3. Based on this information, HEXO1 and HEXO3 coding regions were PCR amplified using cDNA derived from *N. benthamiana* leaf RNA. The obtained DNA sequence information suggested the amplification of a clone corresponding to a putative full-length HEXO1 homolog from *N. benthamiana* that was also annotated in different *N. benthamiana* sequence databases (Figure S2). Despite several attempts, no sequence corresponding to an additional HEXO1 candidate was obtained. RT-PCR from leaf RNA allowed the amplification of a complete open reading frame (ORF) corresponding to HEXO3. The cDNA sequence was slightly different from the annotation in the Sol Genomics Network database, but transcripts carrying the identified HEXO3 ORF were present in the *N. benthamiana* transcriptome database (Figure S3). RT-PCR analyses for other HEXO3 candidates resulted in the amplification of several clones harbouring incomplete or aberrant cDNA fragments. We fully sequenced 12 different clones from two independent RT-PCR amplification events, but we were

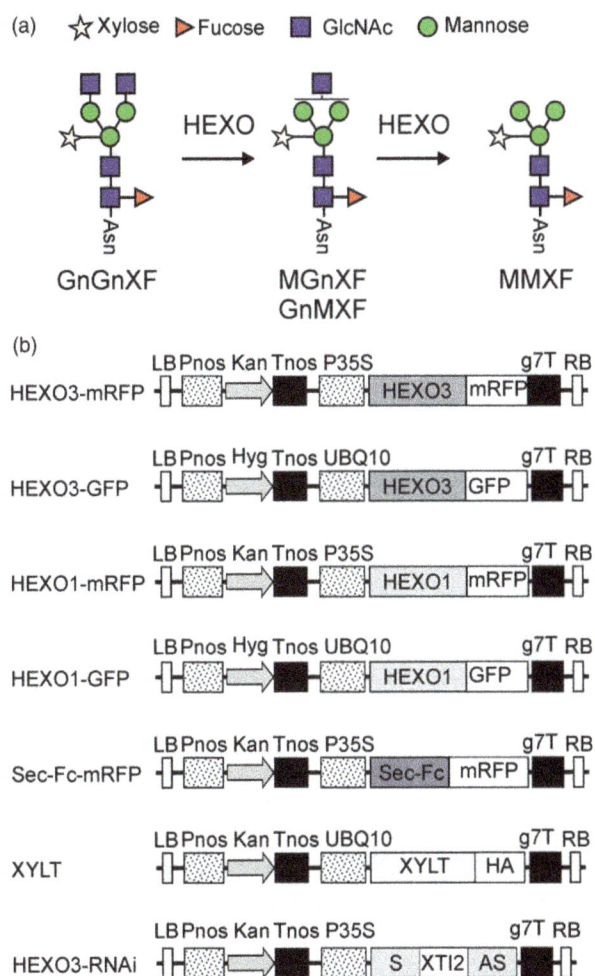

unable to confirm the presence of an additional HEXO3 ORF. All the sequenced clones were interrupted by small deletions or insertions resulting in frame shifts and the generation of premature stop codons. As a consequence of these screening and cloning results, all further studies were performed with the clearly identified candidates for HEXO1 and HEXO3.

The HEXO1 ORF codes for a 541 amino acid protein with a predicted N-terminal signal peptide sequence (amino acids 1–24). Seven potential *N*-glycosylation sites are assigned on the protein backbone. The HEXO3 ORF encodes a 530 amino acid protein with a hydrophobic N-terminal region that could either represent a signal peptide or a single transmembrane domain. The predictions for possible transmembrane domains are not consistent. While TMHMM, for example, does not predict a clear transmembrane domain, HMMTOP suggests the presence of an N-terminal transmembrane helix and type II membrane protein topology with a large luminal catalytic domain. HEXO3 contains six possible *N*-glycosylation sites whereby one carries a proline in +1 position (NPS site) and is therefore very likely not glycosylated. HEXO1 displays 71% identity to AtHEXO1 at the amino acid level and 51% identity to HEXO3. HEXO3 displays 71% identity to AtHEXO3.

HEXO1 and HEXO3 are glycosylated proteins located in the vacuole and at the plasma membrane, respectively

To characterize the identified HEXO candidates, we cloned cDNA corresponding to the ORFs of HEXO1 and HEXO3 into plant expression vectors carrying the sequences for green and red fluorescent protein tags (Figure 1b). The constructs were transformed into agrobacterium and transiently expressed in *N. benthamiana* leaves by agroinfiltration. Leaf material was harvested 24, 48 and 72 h postinfiltration and analysed by immunoblotting (Figure 2a). HEXO1-mRFP was clearly detectable at all three time points. However, the migration position of bands changed considerably from 24 to 48 h suggesting that HEXO1 is subjected to additional processing or post-translational modifications. Likewise, immunoblot analysis of HEXO1-GFP revealed the presence of two bands when samples were harvested 48 h after infiltration (Figure 2b). HEXO3-mRFP was highly expressed at 24 h postinfiltration and showed like HEXO3-GFP predominately a single band of the expected size of approximately 90–95 kDa (Figure 2a,b).

To investigate the *N*-glycosylation state of the expressed HEXO variants, Endo H and PNGase F digestions were carried out. Both proteins, HEXO1-mRFP and HEXO3-mRFP, were fully resistant to Endo H and sensitive to PNGase F when expressed in ΔXT/FT plants that generate negligible amounts of PNGase F-resistant core α1,3-fucosylated *N*-glycans. These results show that HEXO1 and HEXO3 are glycosylated with Golgi-processed *N*-glycans (Figure 2c). The absence of Endo H-sensitive ER-derived oligomannosidic *N*-glycans indicates efficient exit from the ER and trafficking trough the Golgi to their final destination. To examine the subcellular location of the two HEXO proteins, we analysed the expression of fluorescently tagged variants by confocal laser scanning microscopy. HEXO1-mRFP displayed a clear vacuolar distribution (Figure 3a) that was also confirmed by colocalization with the vacuolar marker aleu-GFP (Humair et al., 2001) (Figure 3b). These data suggest that HEXO1 resides in the large central vacuole in *N. benthamiana* leaf epidermal cells. By contrast, HEXO3-mRFP labelled the outline of the epidermal cells indicating targeting to the plasma membrane and/or apoplast (Figure 3c). Co-expression of HEXO3-mRFP with the plasma

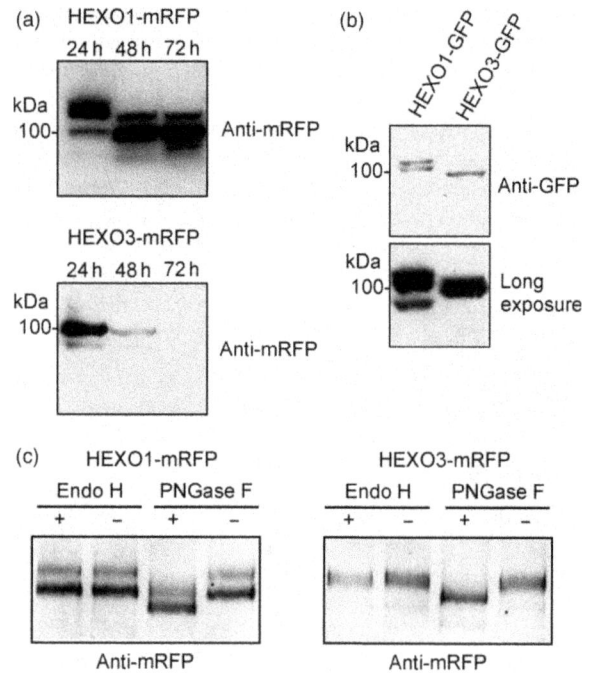

Figure 2 HEXO1 and HEXO3 are glycosylated with complex *N*-glycans. Immunoblot analysis of transiently expressed HEXO variants: (a) HEXO1-mRFP and HEXO3-mRFP protein expression analysed at three different time points (24, 48 and 72 h postinfiltration). (b) HEXO1-GFP and HEXO3-GFP expression 48 h postinfiltration. (c) Endo H and PNGase F digestion of HEXO1-mRFP and HEXO3-mRFP.

membrane marker EGFP-LTI6b (Kurup et al., 2005) (Figure 3d) or co-expression of HEXO3-GFP with the apoplast-targeted glycoprotein Sec-Fc-mRFP (Figure 3e,f) revealed colocalization. In summary, the identified cellular sites are consistent with the ones shown for the corresponding AtHEXOs further suggesting that they are functional orthologs (Liebminger et al., 2011; Strasser et al., 2007). For many recombinant glycoproteins, secretion to the extracellular space is the preferred subcellular site for their accumulation in plants. Our subcellular localization studies suggest that HEXO3 is the candidate enzyme for trimming of terminal GlcNAc residues from secreted glycoproteins.

HEXO1 and HEXO3 can complement A. thaliana hexo mutants

To monitor functional activities of HEXOs, *A. thaliana* mutants that lack either endogenous AtHEXO1 (*hexo1*) or AtHEXO3 activities (*hexo3*) were used in complementation experiments (Liebminger et al., 2011). Transgenic *hexo1* and *hexo3* mutants that overexpress HEXO1-GFP and HEXO3-GFP, respectively, were generated, and total *N*-glycans were analysed by MALDI mass spectrometry (Figure 4). Expression of HEXO1-GFP resulted in a clear increase of paucimannosidic MMXF *N*-glycans and a decrease of the complex *N*-glycan GnGnXF. Similarly, HEXO3-GFP could complement the HEXO deficiency of *hexo3* plants. The *A. thaliana hexo1 hexo3* double mutant lacks any detectable HEXO acting on *N*-glycans and therefore does not produce paucimannosidic *N*-glycans at all (Liebminger et al., 2011). To further confirm the enzymatic activity of HEXO3, complementation of *hexo1 hexo3* was analysed. Transgenic expression of HEXO3-GFP resulted in the formation of substantial amounts of

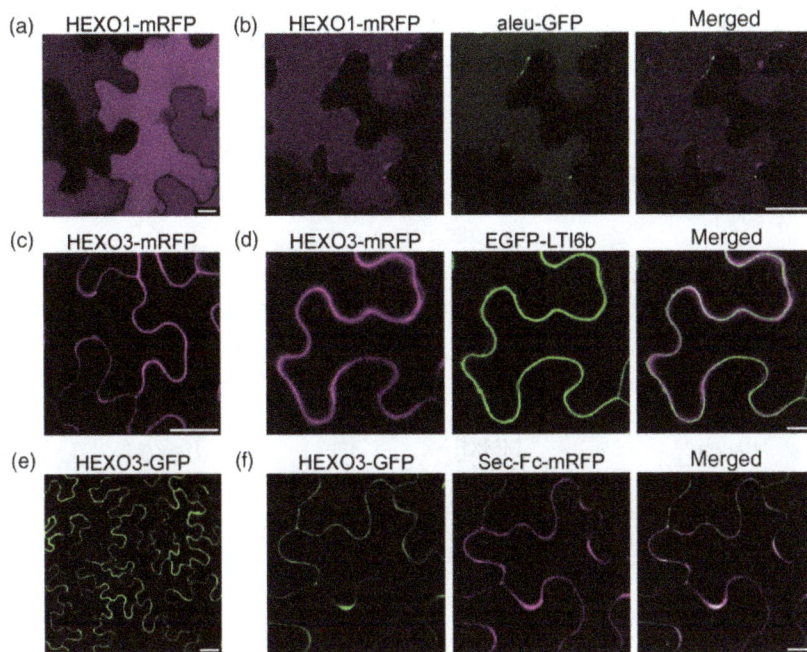

Figure 3 HEXO1 and HEXO3 are located in different subcellular compartments. Transient expression of HEXO1 and HEXO3 with or without different subcellular markers. (a) Confocal microscopy of HEXO1-mRFP, scale bar = 10 μm. (b) Colocalization of HEXO1-mRFP with aleu-GFP, scale bar = 25 μm. (c) HEXO3-mRFP, scale bar = 25 μm. (d) Colocalization of HEXO3-mRFP with EGFP-LTI6b, scale bar = 10 μm. (e) HEXO3-GFP, scale bar = 25 μm. (f) Colocalization of HEXO3-GFP with Sec-Fc-mRFP, scale bar = 10 μm.

N-glycans corresponding to the paucimannosidic MMXF structure in *hexo1 hexo3* that were not present in the double mutant (Figure 4). Collectively, the functional complementation and subcellular localization data demonstrate that HEXO1 and HEXO3 are functional orthologs of AtHEXO1 and AtHEXO3, respectively.

Core α1,3-fucose promotes HEXO3 activity

The subcellular localization analysis and data from complementation experiments indicate that HEXO3 is responsible for the cleavage of GlcNAc residues in the apoplast or during the trafficking of proteins from the Golgi to the extracellular space. To obtain more evidence that HEXO3 can act on recombinant glycoproteins, we transiently co-expressed HEXO3-mRFP with a glycoprotein in *N. benthamiana* leaves and analysed whether the overexpression increases the GlcNAc trimming from N-glycans. First, we have chosen a model glycoprotein (Sec-Fc-mRFP) containing a signal peptide, the Fc part from human IgG1, carrying a single N-glycosylation site, and mRFP (Figure 1b). As confirmed by confocal laser scanning microscopy, Sec-Fc-mRFP is targeted to the secretory pathway and accumulates in the apoplast (Figure 3f). When expressed in *N. benthamiana* wild-type, predominately GnGnXF structures were detected on the purified protein (Figure 5a). Likewise, when expressed in ΔXT/FT, the Fc N-glycosylation site carried mainly GnGn and virtually no paucimannosidic or incompletely processed complex N-glycans (Figure 5b). Co-expression of Sec-Fc-mRFP and HEXO3-mRFP in ΔXT/FT did not significantly alter the overall N-glycan profile. However, in *N. benthamiana* wild-type, the co-expression of HEXO3-mRFP resulted in a marked increase of paucimannosidic MMXF structures showing that HEXO3-mRFP is active when transiently co-expressed with a glycoprotein (Figure 5a).

The observed difference between wild-type and ΔXT/FT plants was unexpected and suggested that the presence of either β1,2-xylose or core α1,3-fucose promote trimming by HEXO3. To test this assumption, we transiently co-expressed Sec-Fc-mRFP and HEXO3-mRFP with the *A. thaliana* β1,2-xylosyltransferase (XYLT), *A. thaliana* core α1,3-fucosyltransferase (FUT11) or mouse core α1,6-fucosyltransferase (FUT8) and analysed the effect on GlcNAc

removal by LC-ESI-MS. While neither the presence of β1,2-xylose nor the presence of core α1,6-fucose caused an increase in terminal GlcNAc processing from the Fc N-glycosylation site, the presence of core α1,3-fucose resulted in the formation of considerable amounts of paucimannosidic N-glycans (Figure 6, S4). Consequently, our data show that the presence of an additional sugar in a specific linkage (core α1,3-fucose) causes alterations in N-glycan processing by HEXO3.

Knock-down of HEXO3 leads to increased amounts of complex N-glycans on recombinant A1AT

Based on the subcellular localization and data from *in planta* activity of co-expressed HEXO3, we hypothesized that reduction or complete elimination of HEXO3 activity will reduce the amounts of paucimannosidic N-glycans on secreted recombinant glycoproteins. To investigate whether the formation of pauci-mannosidic N-glycans can be blocked, we designed an RNAi construct for silencing of HEXO3. The HEXO3-RNAi construct was infiltrated into *N. benthamiana* wild-type and ΔXT/FT leaves. N-glycans from leaf extracts and isolated intercellular fluid (IF) were analysed 3 days postinfiltration by mass spectrometry. The complex N-glycans from total soluble proteins of wild-type and ΔXT/FT were increased in the presence of HEXO3-RNAi (Figure S5). Likewise, the transient expression of HEXO3-RNAi resulted in a decrease of paucimannosidic N-glycans on IF-derived proteins (Figure S6). To test the approach on a therapeutically interesting recombinant glycoprotein, we co-expressed human A1AT (Castilho *et al.*, 2014) together with the HEXO3-RNAi construct. N-glycans of recombinant A1AT extracted from the IF were analysed by LC-ESI-MS. In the absence of HEXO3-RNAi, considerable amounts of paucimannosidic MMXF and MM structures were found in wild-type (Figure 7a) or ΔXT/FT plants (Figure 7b and Table 1). Co-expression of HEXO3-RNAi led to a profound increase of GnGnXF (Figure 7c) and GnGn N-glycans (Figure 7d,S7). In summary, these results demonstrate that *N. benthamiana* HEXO3 activity is a critical factor that generates truncated N-glycans on secreted recombinant glycoproteins in plants.

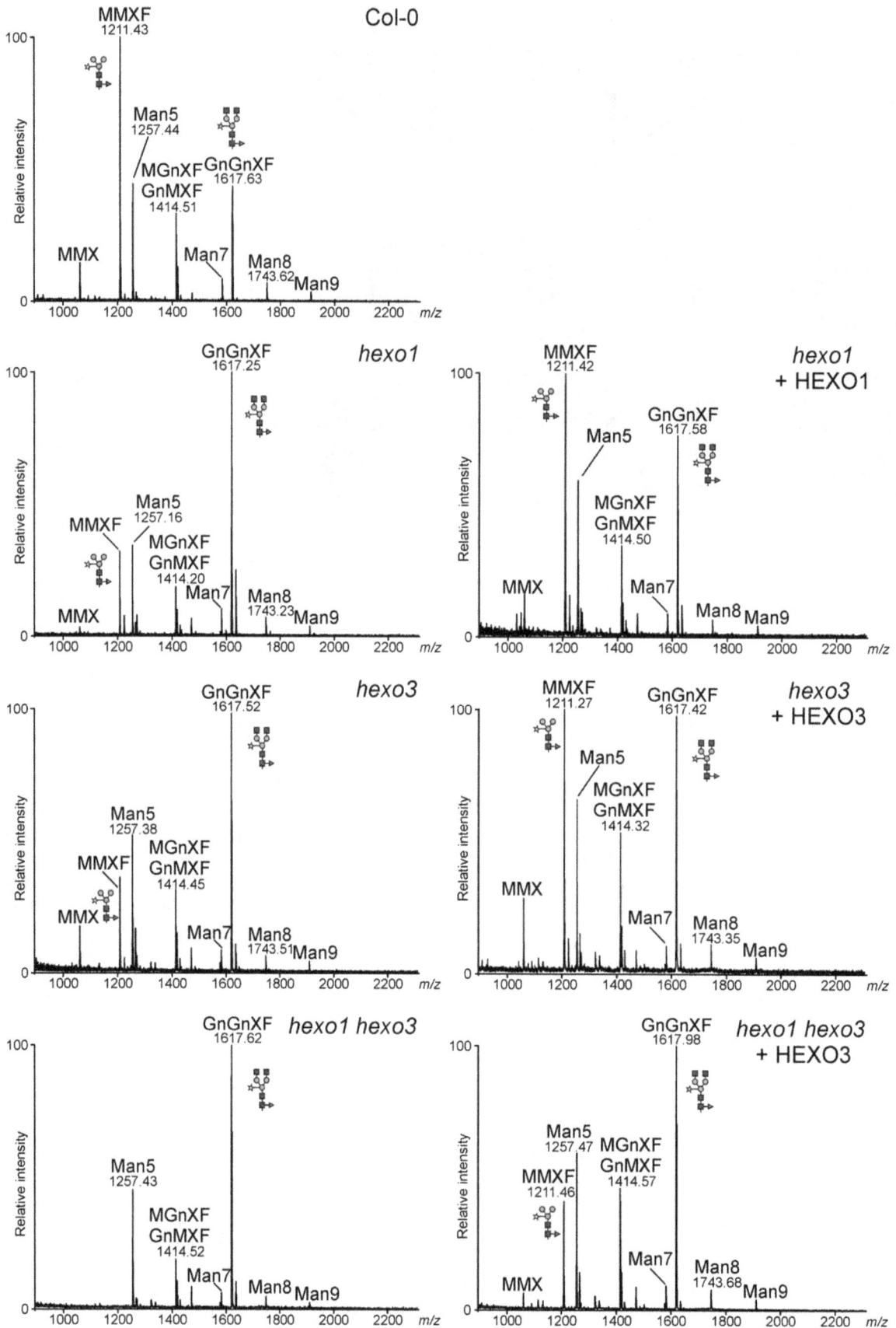

Figure 4 *Nicotiana benthamiana* HEXOs can complement the HEXO deficiency of *Arabidopsis thaliana hexo* mutants. Total *N*-glycan analysis by MALDI-MS of Col-0 wild-type, *hexo1*, *hexo1* expressing HEXO1-GFP, *hexo3*, *hexo3* expressing HEXO3-GFP, *hexo1 hexo3* and *hexo1 hexo3* expressing HEXO3-GFP. Peaks were labelled according to the ProGlycAn system (www.proglycan.com).

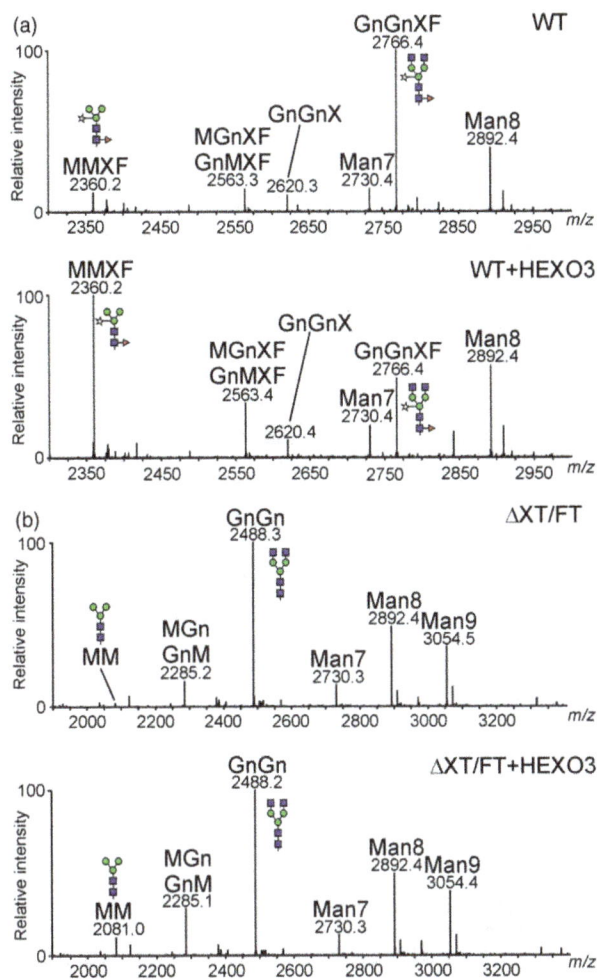

Figure 5 Overexpressed HEXO3 preferentially cleaves off GlcNAc residues from complex *N*-glycans with β1,2-xylose and core α1,3-fucose. Transient co-expression of HEXO3-mRFP with Sec-Fc-mRFP (a) in WT or (b) in ΔXT/FT plants. Sec-Fc-mRFP was purified, digested with trypsin and the glycosylated peptide EEQYNSTYR was analysed by LC-ESI-MS.

Figure 6 The presence of core α1,3-fucose enhances the trimming of GlcNAc from the Fc *N*-glycan. Relative abundance of complex and paucimannosidic *N*-glycans upon co-expression of Sec-Fc-mRFP, HEXO3-mRFP and different *N*-glycan processing enzymes (FUT11: *Arabidopsis thaliana* core α1,3-fucosyltransferase; FUT8: mouse core α1,6-fucosyltransferase; XylT: *A. thaliana* β1,2-xylosyltransferase). Tryptic digested glycopeptides from the Fc domain (EEQYNSTYR) were analysed by LC-ESI-MS. Different shades of blue, red and green represent percentages of paucimannosidic structures, complex *N*-glycans with one terminal GlcNAc residue and complex *N*-glycans with two terminal GlcNAc residues, respectively. Mean values from two to three biological replicates are shown. The corresponding *N*-glycan structures are indicated.

Discussion

N. benthamiana plants are a key expression platform for the production of recombinant proteins (Qiu *et al.*, 2014; Stoger *et al.*, 2014; Strasser *et al.*, 2014). In this study, we identified and characterized HEXOs that trim terminal GlcNAc residues from vacuolar or secreted glycoproteins in *N. benthamiana*. For the production of distinct recombinant glycoproteins, active HEXOs are a severe limitation because these enzymes generate truncated *N*-glycans that are not common on mammalian glycoproteins. Moreover, as HEXO enzymes are trafficking through the Golgi on their journey to their final destination and display β-hexosaminidase activity in the pH-milieu of the Golgi (Strasser *et al.*, 2007), it is possible that HEXOs cleave GlcNAc residues already in one of the Golgi subcompartments. Notably, insect cells that generate similar paucimannosidic *N*-glycans have a processing β-hexosaminidase that is found in the Golgi apparatus (Altmann *et al.*, 1995; Léonard *et al.*, 2006). GlcNAc removal in the Golgi interferes with other *N*-glycan processing reactions leading to increased *N*-glycan heterogeneity and may prevent further elongations with β1,4-galactose or sialic acid.

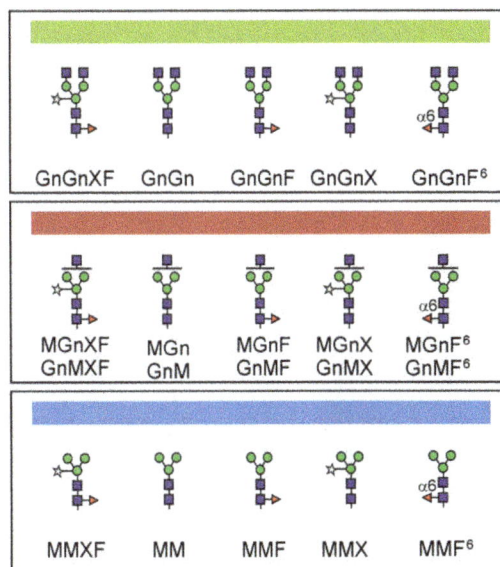

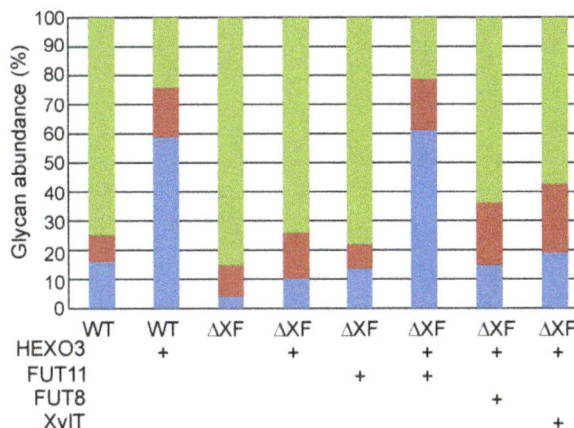

Our data confirm previous results indicating the existence of an active HEXO in the apoplast or plasma membrane that acts in a protein-specific manner (Castilho *et al.*, 2014; Liebminger *et al.*, 2011; Strasser *et al.*, 2007). Several recombinant human glycoproteins have been expressed in *N. benthamiana* leaves, and it seems that only some glycoproteins are substrates for HEXOs suggesting that the glycoprotein conformation or the interaction between the protein backbone and the attached glycan prevent processing. Due to the special structural features of the Fc homodimer, the single Fc *N*-glycan of human IgG1 appears, for example, quite resistant to processing by HEXOs. The same has been described for human EPO-Fc and transferrin when transiently expressed in *N. benthamiana* (Castilho *et al.*, 2011). These proteins carry mainly fully processed GlcNAc terminating complex

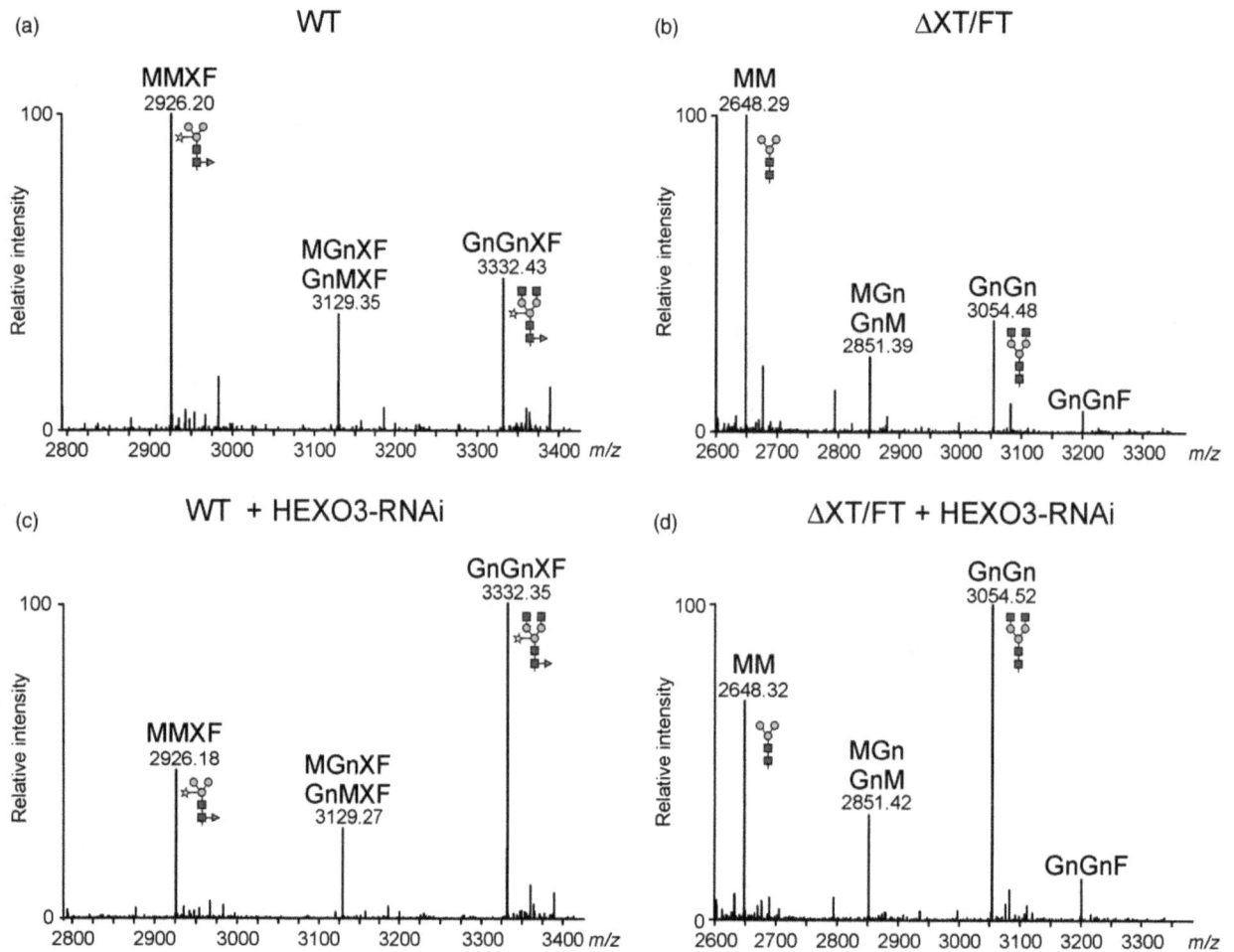

Figure 7 Transient co-expression of the HEXO3-RNAi construct leads to enhanced complex *N*-glycan formation on secreted A1AT. Human A1AT was transiently expressed in (a, c) WT or (b, d) ΔXT/FT plants, in the absence (a, b) or presence (c, d) of the HEXO3 silencing construct. LC-ESI-MS of trypsin-digested A1AT collected from the IF 3 days postinfiltration. The *N*-glycosylation profile of glycopeptide 3 ([243]YLGNATAIFFLPDEGK[259]) is shown.

Table 1 Relative amounts (%) of paucimannosidic (MM, MMXF), complex with one (GnM/MGn, GnMXF/MGnXF) or complex with two (GnGn, GnGnXF) GlcNAc residues on glycopeptide 2 (GP2) or 3 (GP3) from recombinant A1AT

Glycan	GP2	GP3	GP2	GP3
	ΔXTFT*		ΔXTFT + HEXO3 RNAi*	
MM	57.9	53.3	36.6	34.5
GnM/MGn	16.2	15.9	16.7	14.4
GnGn	25.9	30.8	46.7	51.0
	WT[†]		WT + HEXO3 RNAi[†]	
MMXF	39.8	45.7	37.3	26.3
GnMXF/MGnXF	27.7	17.3	21.9	16.6
GnGnXF	32.5	37	40.8	57.1

*Mean values from two independent biological replicates are shown.
[†]Amounts for WT are derived from a single analysis.

N-glycans. By contrast, human A1AT harbours considerably amounts of paucimannosidic structures on all three *N*-glycosylation sites (Castilho *et al.*, 2014). A similar result was observed for

the *N*-glycans of recombinantly expressed follicle stimulating hormone, recombinant glucocerebrosidase targeted to the apoplast or human lactoferrin (Dirnberger *et al.*, 2001; He *et al.*, 2012; Limkul *et al.*, 2016; Samyn-Petit *et al.*, 2003). Further studies are needed to identify the protein intrinsic features that lead to efficient processing by plant HEXOs.

Interestingly, the impact of core α1,3-fucose on GlcNAc trimming shows that specific *N*-glycans on proteins like the Fc domain-containing Sec-Fc-mRFP can be converted into HEXO3 substrates. The impact of core α1,3-fucose on *N*-glycan modifications has been described recently for the *N*-glycan from the Fc domain of the monoclonal antibody cetuximab (Castilho *et al.*, 2015). In addition to the conserved Fc *N*-glycan, cetuximab has a second *N*-glycan in the variable domain of the heavy chain. These two *N*-glycosylation sites on the heavy chain allow the comparison of *N*-glycan processing in a given cell in the presence or absence of an additional glycan modification. As a result of this study, it was shown that core α1,3-fucosylation increases branching, bisecting GlcNAc formation and in particular sialylation of the Fc *N*-glycan presumably by alleviating structural constraints between the Fc *N*-glycan and the IgG1 CH2 domains (Castilho *et al.*, 2015).

Co-expression of the HEXO3-RNAi construct resulted in a reduction of paucimannosidic structures, while complex *N*-

glycans increased. However, some paucimannosidic *N*-glycans were still detectable. One reason could be that the used transient silencing approach is not very efficient. It is plausible that residual HEXO3 protein is present in the apoplast that has been made before the HEXO3 silencing was established. An optimization of the infiltration procedure like pre-infiltration with the RNAi construct at an earlier time point and subsequent infiltration with the recombinant glycoprotein could significantly improve the down-regulation of the unwanted HEXO activity. Moreover, the production of stable HEXO3 silencing lines is ongoing and genome editing will be applied to completely inactivate HEXO3 in *N. benthamiana* in the future. The complete HEXO3 knockout will finally show whether the identified HEXO3 candidate is the only one acting on complex *N*-glycans or whether other HEXO candidates are also functional and contribute to the formation of paucimannosidic *N*-glycans on secreted glycoproteins.

For some therapeutic applications, it is beneficial to produce *N*-glycans with increased amounts of paucimannosidic *N*-glycans. This is, for example, the case for the carrot cell-based production of the recombinant glucocerebrosidase taliglucerase alfa that is used for treatment of Gaucher's disease (Shaaltiel *et al.*, 2007). The major *N*-glycan structures found on taliglucerase alfa are paucimannosidic MMXF with β1,2-xylose and core α1,3-fucose (Tekoah *et al.*, 2013). The exposed terminal mannose residues from recombinant glucocerebrosidase are essential for the internalization of the enzyme by macrophages. The *N*-glycans on mammalian cell-derived recombinant lysosomal enzymes have to be remodelled postproduction with glycosidases to obtain exposed mannose residues (Grabowski *et al.*, 1995). By contrast, for taliglucerase alfa, the efficient production of *N*-glycans with terminal mannose was achieved by targeting to the vacuole using a seven amino acid long vacuolar targeting signal fused to the recombinant glucocerebrosidase (Shaaltiel *et al.*, 2007). Using HEXO3 overexpression, the attachment of non-native targeting signals can be avoided and exposed mannose residues can be generated on secreted recombinant glycoproteins (Shen *et al.*, 2016).

In conclusion, we have identified active HEXOs from *N. benthamiana* and found that HEXO3 contributes to the formation of paucimannosidic *N*-glycans on secreted recombinant glycoproteins. Furthermore, we provide novel glyco-engineering tools that can be applied to eliminate paucimannosidic *N*-glycan formation or alternatively, generate mannose-terminated *N*-glycans on secreted glycoproteins. Implementation of these tools into the existing *N. benthamiana* expression platforms will increase the capacity of this plant for the expression of therapeutic glycoproteins for many different applications.

Experimental procedures

Cloning of N. benthamiana HEXO candidates

The *A. thaliana* HEXO1 (At3g55260) and HEXO3 (At1g65590) amino acid sequences were used to search in different *N. benthamiana* genome databases using tBLASTN (https://solgenomics.net/organism/Nicotiana_benthamiana/genome; http://benthgenome.qut.edu.au/). Based on the retrieved sequences, primers were designed and used for RT-PCR to amplify full-length ORFs coding for HEXO proteins. For this purpose, RNA was extracted from leaves of 4- to 5-week-old *N. benthamiana* using the SV Total RNA Isolation System (Promega, Mannheim Germany) and the iScript cDNA Synthesis Kit (Bio-Rad, Vienna Austria). An aliquot of the cDNA was used for the PCR amplification with

Phusion High-Fidelity DNA Polymerase (Biozym, Hessisch Oldendorf, Germany) and different primer combinations (Table S1). PCR products were ligated into a cloning vector using the Zero Blunt Topo PCR Cloning Kit (Thermo Fisher Scientific, Vienna, Austria) and fully sequenced. Sequence alignments were made using SeqMan II software (DNASTAR Lasergene, Madison, WI, USA).

Transient expression and immunoblots

For expression of *N. benthamiana* HEXO1 in plants, the corresponding ORF was amplified by PCR from the cloning vector using primers Nb-Hexo1-F4 and Nb-Hexo1-R4. The PCR product was *Spel*/*Bam*HI digested and cloned into *Xbal*/*Bam*HI digested plant expression vectors p31 (Schoberer *et al.*, 2013) and p47 (Hüttner *et al.*, 2014) to generate p31-NbHEXO1 and p47-NbHEXO1. In p31 expression is under the control of the CaMV 35S promoter and the recombinant protein is C-terminally fused to mRFP, whereas in p47 expression is under the control of the *A. thaliana* ubiquitin 10 promoter and the recombinant protein is C-terminally fused to GFP (Figure 1b). For expression of HEXO3, the corresponding ORF was amplified by PCR from the cloning vector with Nb-Hexo3-F6 and Nb-Hexo3-R5. The PCR product was *Xbal*/*Bam*HI digested and cloned into *Xbal*/*Bam*HI digested vectors p31 and p47 to generate p31-NbHEXO3 and p47-NbHEXO3. The plant expression vectors were transformed into *Agrobacterium tumefaciens* (strain UIA143 was used for all constructs). Syringe-mediated agroinfiltration was used for transient expression in leaves of 4- to 5-week-old *N. benthamiana* plants. At the indicated time points, leaf pieces were harvested from infiltrated plants and total protein extracts were prepared by grinding of frozen leaves with a mixer mill and steel balls. The ground leaves were dissolved in RIPA buffer (Sigma-Aldrich, Vienna, Austria) followed by centrifugation at 16 000 **g** for 10 min. An aliquot of the supernatant was mixed with SDS-PAGE loading buffer, denatured at 95 °C for 5 min and subjected to SDS-PAGE under reducing conditions. Protein gel blots were blocked in PBS containing 0.1% (v/v) Tween 20 and 3% (w/v) BSA. The membranes were probed with anti-GFP-horseradish peroxidase (MACS Miltenyi Biotec, Bergisch Gladbach, Germany) or anti-mRFP (Chromotek, Planegg-Martinsried, Germany) antibodies. Endo H (New England Biolabs, Frankfurt am Main, Germany) and PNGase F (New England Biolabs) digestions were performed as described in detail recently (Hüttner *et al.*, 2014).

Confocal imaging of fluorescent protein fusions

Leaves of 4- to 5-week-old *N. benthamiana* plants were infiltrated with agrobacterium suspensions carrying binary plant expression vectors for GFP- or mRFP-tagged proteins with the following optical densities (OD600): 0.1 for p31-NbHEXO1 (HEXO1-mRFP), p31-NbHEXO3 (HEXO3-mRFP) and p47-HEXO3 (HEXO3-GFP). Constructs for aleu-GFP (infiltrated with OD600 = 0.01) and for EGFP-LTI6b (infiltrated with OD600 = 0.1) were available from a previous study (Strasser *et al.*, 2007). The p39-Sec-Fc-mRFP construct was generated as follows: the DNA coding for the Fc domain from human IgG1 was amplified from p20F-Fc (Schoberer *et al.*, 2009) with primers Fc-1F/Fc-2R, *Bam*HI/*Bgl*II digested and cloned into the *Bam*HI site of p31-Sec-mRFP. To generate the vector p31-Sec-mRFP, a DNA fragment derived from annealing of primers GCSII_SP_F and GCSII_SP_R was ligated into the *Xbal*/*Bam*HI sites of vector p31. The fragment derived from these primers encodes the signal peptide of *A. thaliana* α-glucosidase II (GCSII).

Complementation of A. thaliana hexo *mutants*

Arabidopsis thaliana hexo single and *hexo1 hexo3* double knockout plants (Liebminger *et al.*, 2011) were transformed with p47-NbHEXO1 or p47-NbHEXO3 by floral dipping, as described previously (Strasser *et al.*, 2004). Hygromycin-resistant plants were screened by PCR with HEXO1- and HEXO3-specific primers, respectively. Leaves from different PCR-positive plants were pooled, and 500 mg was used for total *N*-glycan analysis. Preparation of *N*-linked glycans and matrix-assisted laser desorption ionization (MALDI) mass spectrometry was carried out as described previously (Strasser *et al.*, 2004).

In planta N-*glycan processing*

The p39-Sec-Fc-mRFP vector expressing the glycoprotein reporter was either expressed alone in *N. benthamiana* leaves by agroinfiltration or in combination with p31-NbHEXO3 and additional constructs for expression of different glycosyltransferases. Vectors for plant expression of *A. thaliana* core α1,3-fucosyltransferase 11 (FUT11) and mouse core α1,6-fucosyltransferase (FUT8) were available from a previous study (Castilho *et al.*, 2015). The *A. thaliana* β1,2-xylosyltransferase (XYLT) was amplified from *A. thaliana* cDNA using primers ARA_XT27F and ARA_XT29R. The PCR product was digested with *SpeI/BamHI* and cloned into the *XbaI/BamHI* site of p41 (a derivative of pPT2 with the UBQ10 promoter instead of the CaMV 35S promoter and a C-terminal HA tag for monitoring of protein expression) to generate p41-XYLT (Figure 1b). Two days postinfiltration 500 mg leaves were harvested, frozen leaves were grinded using a mixer mill and proteins were extracted with RIPA buffer. The Fc domain glycoprotein reporter was purified from the extract by binding to rProtein A Sepharose™ Fast Flow (GE Healthcare Europe, Vienna, Austria) as described in detail recently (Schoberer *et al.*, 2014). Purified protein was subjected to SDS-PAGE and Coomassie blue staining. The corresponding protein band was excised from the gel, destained, carbamidomethylated, in-gel trypsin digested and analysed by liquid chromatography electrospray ionization mass spectrometry (LC-ESI-MS), as described in detail previously (Stadlmann *et al.*, 2008). A detailed explanation of *N*-glycan abbreviations can be found at http://www.proglycan.com.

Transient knockdown of HEXO3

A synthetic DNA fragment consisting of intron 2 from *A. thaliana* XYLT (Strasser *et al.*, 2004, 2008) and an antisense DNA fragment corresponding to the coding sequence for amino acids 136–208 of HEXO3 was obtained by GeneArt gene synthesis (Thermo Fisher Scientific) (Figure S1). The obtained vector with the synthetic DNA was used as a template for PCR with primers Nb-HEXO3-F8 and Nb-HEXO3-R8. The 'sense' PCR product was digested with *XbaI/KpnI* and cloned into the synthetic DNA containing vector to generate a sense–intron–antisense hairpin construct. The sense–intron–antisense sequence was subsequently excised by *XbaI/BamHI* digestion and ligated into *XbaI/BamHI* digested plant expression vector pPT2 (Strasser *et al.*, 2007). In this vector, the RNAi construct is expressed under the control of the CaMV 35S promoter. The pPT2-NbHEXO3-RNAi vector was transformed into agrobacteria and transiently expressed by agroinfiltration in *N. benthamiana* leaves. Total *N*-glycan analysis was performed 3 days after infiltration as described for *A. thaliana*. Expression of human A1AT, extraction from the IF and LC-ESI-MS analysis of glycopeptides was performed as described in detail recently (Castilho *et al.*, 2014).

Acknowledgements

We thank Christiane Veit for help with cloning, Michaela Bogner for technical support, Jennifer Schoberer for help with confocal microscopy, Thomas Hackl, Karin Polacsek, Javier Fernando Montero, Daniel Maresch and Markus Windwarder (all from BOKU) for help with *N*-glycan analysis. This work was supported by a grant from the Federal Ministry of Transport, Innovation and Technology (bmvit) and Austrian Science Fund (FWF): TRP 242-B20 and by the Austrian Research Promotion Agency (FFG) in the frame of Laura Bassi Centres of Expertise (Grant Number 822757).

References

Altmann, F., Schwihla, H., Staudacher, E., Glössl, J. and März, L. (1995) Insect cells contain an unusual, membrane-bound beta-N-acetylglucosaminidase probably involved in the processing of protein N-glycans. *J. Biol. Chem.* **270**, 17344–17349.

Castilho, A., Gattinger, P., Grass, J., Jez, J., Pabst, M., Altmann, F., Gorfer, M. et al. (2011) N-glycosylation engineering of plants for the biosynthesis of glycoproteins with bisected and branched complex N-glycans. *Glycobiology*, **21**, 813–823.

Castilho, A., Windwarder, M., Gattinger, P., Mach, L., Strasser, R., Altmann, F. and Steinkellner, H. (2014) Proteolytic and N-glycan processing of human α1-antitrypsin expressed in *Nicotiana benthamiana*. *Plant Physiol.* **166**, 1839–1851.

Castilho, A., Gruber, C., Thader, A., Oostenbrink, C., Pechlaner, M., Steinkellner, H. and Altmann, F. (2015) Processing of complex N-glycans in IgG Fc-region is affected by core fucosylation. *MAbs*, **7**, 863–870.

Dicker, M., Tschofen, M., Maresch, D., König, J., Juarez, P., Orzaez, D., Altmann, F. et al. (2016) Transient glyco-engineering to produce recombinant IgA1 with defined N- and O-glycans in plants. *Front. Plant Sci.* **7**, 18.

Dirnberger, D., Steinkellner, H., Abdennebi, L., Remy, J. and van de Wiel, D. (2001) Secretion of biologically active glycoforms of bovine follicle stimulating hormone in plants. *Eur. J. Biochem.* **268**, 4570–4579.

Grabowski, G.A., Barton, N.W., Pastores, G., Dambrosia, J.M., Banerjee, T.K., McKee, M.A., Parker, C. et al. (1995) Enzyme therapy in type 1 Gaucher disease: comparative efficacy of mannose-terminated glucocerebrosidase from natural and recombinant sources. *Ann. Intern. Med.* **122**, 33–39.

He, X., Galpin, J.D., Tropak, M.B., Mahuran, D., Haselhorst, T., von Itzstein, M., Kolarich, D. et al. (2012) Production of active human glucocerebrosidase in seeds of *Arabidopsis thaliana* complex-glycan-deficient (cgl) plants. *Glycobiology*, **22**, 492–503.

Humair, D., Hernández Felipe, D., Neuhaus, J. and Paris, N. (2001) Demonstration in yeast of the function of BP-80, a putative plant vacuolar sorting receptor. *Plant Cell*, **13**, 781–792.

Hüttner, S., Veit, C., Vavra, U., Schoberer, J., Liebminger, E., Maresch, D., Grass, J. et al. (2014) Arabidopsis class I α-mannosidases MNS4 and MNS5 are involved in endoplasmic reticulum-associated degradation of misfolded glycoproteins. *Plant Cell*, **26**, 1712–1728.

Jez, J., Castilho, A., Grass, J., Vorauer-Uhl, K., Sterovsky, T., Altmann, F. and Steinkellner, H. (2013) Expression of functionally active sialylated human erythropoietin in plants. *Biotechnol. J.* **8**, 371–382.

Kurup, S., Runions, J., Köhler, U., Laplaze, L., Hodge, S. and Haseloff, J. (2005) Marking cell lineages in living tissues. *Plant J.* **42**, 444–453.

Léonard, R., Rendic, D., Rabouille, C., Wilson, I., Préat, T. and Altmann, F. (2006) The Drosophila fused lobes gene encodes an N-acetylglucosaminidase involved in N-glycan processing. *J. Biol. Chem.* **281**, 4867–4875.

Liebminger, E., Veit, C., Pabst, M., Batoux, M., Zipfel, C., Altmann, F., Mach, L. et al. (2011) β-N-acetylhexosaminidases HEXO1 and HEXO3 are responsible for the formation of paucimannosidic N-glycans in *Arabidopsis thaliana*. *J. Biol. Chem.* **286**, 10793–10802.

Limkul, J., Iizuka, S., Sato, Y., Misaki, R., Ohashi, T. and Fujiyama, K. (2016) The production of human glucocerebrosidase in glyco-engineered *Nicotiana benthamiana* plants. *Plant Biotechnol. J.* **14**, 1682–1694.

Loos, A., Gruber, C., Altmann, F., Mehofer, U., Hensel, F., Grandits, M., Oostenbrink, C. *et al.* (2014) Expression and glycoengineering of functionally active heteromultimeric IgM in plants. *Proc. Natl. Acad. Sci. USA*, **111**, 6263–6268.

Loos, A., Gach, J.S., Hackl, T., Maresch, D., Henkel, T., Porodko, A., Bui-Minh, D. *et al.* (2015) Glycan modulation and sulfoengineering of anti-HIV-1 monoclonal antibody PG9 in plants. *Proc. Natl. Acad. Sci. USA*, **112**, 12675–12680.

Qiu, X., Wong, G., Audet, J., Bello, A., Fernando, L., Alimonti, J.B., Fausther-Bovendo, H. *et al.* (2014) Reversion of advanced Ebola virus disease in nonhuman primates with ZMapp. *Nature*, **514**, 47–53.

Reusch, D. and Tejada, M.L. (2015) Fc glycans of therapeutic antibodies as critical quality attributes. *Glycobiology*, **25**, 1325–1334.

Samyn-Petit, B., Wajda Dubos, J.P., Chirat, F., Coddeville, B., Demaizieres, G., Farrer, S., Slomianny, M.C. *et al.* (2003) Comparative analysis of the site-specific N-glycosylation of human lactoferrin produced in maize and tobacco plants. *Eur. J. Biochem.* **270**, 3235–3242.

Schachter, H. (2009) Paucimannose N-glycans in *Caenorhabditis elegans* and *Drosophila melanogaster*. *Carbohydr. Res.* **344**, 1391–1396.

Schneider, J.D., Castilho, A., Neumann, L., Altmann, F., Loos, A., Kannan, L., Mor, T.S. *et al.* (2014) Expression of human butyrylcholinesterase with an engineered glycosylation profile resembling the plasma-derived orthologue. *Biotechnol. J.* **9**, 501–510.

Schoberer, J., Vavra, U., Stadlmann, J., Hawes, C., Mach, L., Steinkellner, H. and Strasser, R. (2009) Arginine/lysine residues in the cytoplasmic tail promote ER export of plant glycosylation enzymes. *Traffic*, **10**, 101–115.

Schoberer, J., Liebminger, E., Botchway, S.W., Strasser, R. and Hawes, C. (2013) Time-resolved fluorescence imaging reveals differential interactions of N-glycan processing enzymes across the Golgi stack in planta. *Plant Physiol.* **161**, 1737–1754.

Schoberer, J., Liebminger, E., Vavra, U., Veit, C., Castilho, A., Dicker, M., Maresch, D. *et al.* (2014) The transmembrane domain of N-acetylglucosaminyltransferase I is the key determinant for its Golgi subcompartmentation. *Plant J.* **80**, 809–822.

Sethi, M.K., Kim, H., Park, C.K., Baker, M.S., Paik, Y.K., Packer, N.H., Hancock, W.S. *et al.* (2015) In-depth N-glycome profiling of paired colorectal cancer and non-tumorigenic tissues reveals cancer-, stage- and EGFR-specific protein N-glycosylation. *Glycobiology*, **25**, 1064–1078.

Shaaltiel, Y., Bartfeld, D., Hashmueli, S., Baum, G., Brill-Almon, E., Galili, G., Dym, O. *et al.* (2007) Production of glucocerebrosidase with terminal mannose glycans for enzyme replacement therapy of Gaucher's disease using a plant cell system. *Plant Biotechnol. J.* **5**, 579–590.

Shen, J.S., Busch, A., Day, T.S., Meng, X.L., Yu, C.I., Dabrowska-Schlepp, P., Fode, B. *et al.* (2016) Mannose receptor-mediated delivery of moss-made α-galactosidase A efficiently corrects enzyme deficiency in Fabry mice. *J. Inherit. Metab. Dis.* **39**, 293–303.

Stadlmann, J., Pabst, M., Kolarich, D., Kunert, R. and Altmann, F. (2008) Analysis of immunoglobulin glycosylation by LC-ESI-MS of glycopeptides and oligosaccharides. *Proteomics*, **8**, 2858–2871.

Steinkellner, H. and Castilho, A. (2015) N-glyco-engineering in plants: update on strategies and major achievements. *Methods Mol. Biol.* **1321**, 195–212.

Stoger, E., Fischer, R., Moloney, M. and Ma, J.K. (2014) Plant molecular pharming for the treatment of chronic and infectious diseases. *Annu. Rev. Plant Biol.* **65**, 743–768.

Strasser, R. (2016) Plant protein glycosylation. *Glycobiology*, doi:10.1093/glycob/cww023

Strasser, R., Mucha, J., Schwihla, H., Altmann, F., Glössl, J. and Steinkellner, H. (1999) Molecular cloning and characterization of cDNA coding for beta1,2N-acetylglucosaminyltransferase I (GlcNAc-TI) from *Nicotiana tabacum*. *Glycobiology*, **9**, 779–785.

Strasser, R., Altmann, F., Mach, L., Glössl, J. and Steinkellner, H. (2004) Generation of *Arabidopsis thaliana* plants with complex N-glycans lacking beta1,2-linked xylose and core alpha1,3-linked fucose. *FEBS Lett.* **561**, 132–136.

Strasser, R., Bondili, J., Schoberer, J., Svoboda, B., Liebminger, E., Glössl, J., Altmann, F. *et al.* (2007) Enzymatic properties and subcellular localization of Arabidopsis beta-N-acetylhexosaminidases. *Plant Physiol.* **145**, 5–16.

Strasser, R., Stadlmann, J., Schähs, M., Stiegler, G., Quendler, H., Mach, L., Glössl, J. *et al.* (2008) Generation of glyco-engineered *Nicotiana benthamiana* for the production of monoclonal antibodies with a homogeneous human-like N-glycan structure. *Plant Biotechnol. J.* **6**, 392–402.

Strasser, R., Castilho, A., Stadlmann, J., Kunert, R., Quendler, H., Gattinger, P., Jez, J. *et al.* (2009) Improved virus neutralization by plant-produced anti-HIV antibodies with a homogeneous {beta}1,4-galactosylated N-glycan profile. *J. Biol. Chem.* **284**, 20479–20485.

Strasser, R., Altmann, F. and Steinkellner, H. (2014) Controlled glycosylation of plant-produced recombinant proteins. *Curr. Opin. Biotechnol.* **30C**, 95–100.

Tekoah, Y., Tzaban, S., Kizhner, T., Hainrichson, M., Gantman, A., Golembo, M., Aviezer, D. *et al.* (2013) Glycosylation and functionality of recombinant β-glucocerebrosidase from various production systems. *Biosci. Rep.*, **33**, 771–781.

Wilbers, R.H., Westerhof, L.B., Reuter, L.J., Castilho, A., van Raaij, D.R., Nguyen, D.L., Lozano-Torres, J.L. *et al.* (2016) The N-glycan on Asn54 affects the atypical N-glycan composition of plant-produced interleukin-22, but does not influence its activity. *Plant Biotechnol. J.* **14**, 670–681.

Yang, Z., Wang, S., Halim, A., Schulz, M.A., Frodin, M., Rahman, S.H., Vester-Christensen, M.B. *et al.* (2015a) Engineered CHO cells for production of diverse, homogeneous glycoproteins. *Nat. Biotechnol.* **33**, 842–844.

Yang, W.H., Aziz, P.V., Heithoff, D.M., Mahan, M.J., Smith, J.W. and Marth, J.D. (2015b) An intrinsic mechanism of secreted protein aging and turnover. *Proc. Natl. Acad. Sci. USA*, **112**, 13657–13662.

Zeitlin, L., Geisbert, J.B., Deer, D.J., Fenton, K.A., Bohorov, O., Bohorova, N., Goodman, C. *et al.* (2016) Monoclonal antibody therapy for Junin virus infection. *Proc. Natl. Acad. Sci. USA*, **113**, 4458–4463.

14

Co-expression of the protease furin in *Nicotiana benthamiana* leads to efficient processing of latent transforming growth factor-β1 into a biologically active protein

Ruud H. P. Wilbers*, Lotte B. Westerhof, Debbie R. van Raaij, Marloes van Adrichem, Andreas D. Prakasa, Jose L. Lozano-Torres, Jaap Bakker, Geert Smant and Arjen Schots

Laboratory of Nematology, Plant Sciences Department, Wageningen University and Research Centre, Wageningen, The Netherlands

*Correspondence
email ruud.
wilbers@wur.nl

Summary

Transforming growth factor beta (TGF-β) is a signalling molecule that plays a key role in developmental and immunological processes in mammals. Three TGF-β isoforms exist in humans, and each isoform has unique therapeutic potential. Plants offer a platform for the production of recombinant proteins, which is cheap and easy to scale up and has a low risk of contamination with human pathogens. TGF-β3 has been produced in plants before using a chloroplast expression system. However, this strategy requires chemical refolding to obtain a biologically active protein. In this study, we investigated the possibility to transiently express active human TGF-β1 in *Nicotiana benthamiana* plants. We successfully expressed mature TGF-β1 in the absence of the latency-associated peptide (LAP) using different strategies, but the obtained proteins were inactive. Upon expression of LAP-TGF-β1, we were able to show that processing of the latent complex by a furin-like protease does not occur *in planta*. The use of a chitinase signal peptide enhanced the expression and secretion of LAP-TGF-β1, and co-expression of human furin enabled the proteolytic processing of latent TGF-β1. Engineering the plant post-translational machinery by co-expressing human furin also enhanced the accumulation of biologically active TGF-β1. This engineering step is quite remarkable, as furin requires multiple processing steps and correct localization within the secretory pathway to become active. Our data demonstrate that plants can be a suitable platform for the production of complex proteins that rely on specific proteolytic processing.

Keywords: *Nicotiana benthamiana*, transforming growth factor β1, signal peptide, furin, proteolytic processing, codon optimization.

Introduction

Transforming growth factor beta (TGF-β) is a signalling molecule with crucial roles during early development and the regulation of immune responses in mammals. TGF-β controls cellular processes, including cell proliferation, recognition, differentiation and apoptosis (Blobe *et al.*, 2000; Li *et al.*, 2006). Three different TGF-β isoforms (TGF-β1, TGF-β2 and TGF-β3) exist in humans, which are all encoded by a different gene (Govinden and Bhoola, 2003). The TGF-β genes encode prepro-TGF-β that consists of the signal peptide, latency-associated peptide (LAP) and mature TGF-β protein. The signal peptide targets the protein to the endoplasmic reticulum (ER) where it is cleaved off. Pro-TGF-β forms homodimers and is further processed in the trans-Golgi by a furin protease. Proteolytic cleavage by this enzyme separates mature TGF-β from LAP; however, both polypeptides stay noncovalently associated (Dubois *et al.*, 1995). This dimeric complex is called the small latency complex (SLC) and keeps mature TGF-β in an inactive form. Upon secretion of the SLC, LAP binds to latent transforming growth factor binding protein (LTBP) in the extracellular matrix, resulting in a membrane-bound complex known as the large latency complex (LLC) (Annes *et al.*, 2003). TGF-β is activated by either proteolytic degradation of LAP, reactive oxygen species (ROS), an acidic environment or interaction with

integrin receptors (Annes *et al.*, 2003). Active TGF-β is a dimeric protein of ~25 kDa wherein a disulphide bridge connects the two monomers. Each monomer also has 4 internal disulphide bridges that create a cysteine knot structure (Archer *et al.*, 1993; Daopin *et al.*, 1992; Lin *et al.*, 2006; Schlunegger and Grutter, 1992).

The three TGF-β isoforms share high homology on amino acid level (70–80%, considering mature TGF-β) and bind to the same receptors (TβRI and TβRII). However, the three isoforms have distinct functions. It is unclear whether these distinct functions are caused by differences in receptor binding affinity or because the isoforms are further regulated by controlled expression in time, place or cell type. Another possibility is that activity can be controlled by different activation mechanisms. For instance, TGF-β1 and TGF-β3 can be activated via the interaction with integrin receptors, while TGF-β2 cannot (Ludbrook *et al.*, 2003). TGF-β1 is best known for its potent immunoregulatory functions, whereas TGF-β2 and TGF-β3 play key roles during early embryonic development.

From a pharmaceutical point of view, all three TGF-β isoforms have potential therapeutic applications. TGF-β1 maintains immune homeostasis by controlling lymphocyte proliferation, differentiation and survival, but also inhibits the maturation and activation of antigen presenting cells (Geissmann *et al.*, 1999; Takeuchi *et al.*, 1998). Also, TGF-β1 plays a key role in the differentiation of regulatory T cells (Tregs), a population of T cells that strongly suppresses immune responses (Chen et al., 2003). Thus, TGF-β1 might be used as a remedy for patients with (chronic) inflammatory diseases, like arthritis, inflammatory bowel disease and multiple sclerosis. However, in the presence of the pro-inflammatory cytokines IL-4 or IL-6, TGF-β1 gives rise to populations of T helper cells that can promote inflammation (Th9 and Th17 cells, respectively) (Dardalhon et al., 2008; Veldhoen et al., 2006, 2008). TGF-β2 also plays a role in controlling the immune system by inducing oral tolerance and suppressing immune responses. Next to that, TGF-β2 inhibits growth of intestinal epithelial cells and promotes their differentiation. Therefore, TGF-β2 might be effective in treating inflammatory bowel diseases (Oz et al., 2004). A disadvantage of TGF-β1 and TGF-β2 is the strong fibrotic response they can elicit. TGF-β3 on the other hand does not induce fibrosis and therefore has a therapeutic application in scar-free wound healing (O'Kane and Ferguson, 1997).

For large-scale production of recombinant TGF-β, a suitable expression system is required. TGF-β has been heterologously expressed in bacterial, insect and mammalian expression systems, but expression of TGF-β in large quantities seems difficult (Zou and Sun, 2004). TGF-β is improperly folded when using bacterial expression systems and requires chemical refolding. Mammalian expression systems have a low yield and require multistep purification procedures (Zou and Sun, 2006). In the last two decades, plants have emerged as an alternative expression system for the production of recombinant proteins. As eukaryotes, plants are capable of correctly folding complex proteins and perform post-translational modifications, like N-glycosylation. Plants are relatively cheap compared with other expression systems and are easy to scale up. One additional advantage for the production of immunoregulatory proteins in plants is that plants do not harbour human pathogens and therefore could be regarded as more safe. Recently, TGF-β3 has been produced in chloroplasts of Nicotiana tabacum (Gisby et al., 2011). A synthetic gene was used for chloroplast transformation and yielded up to 3 mg TGF-β3 per 80

grams of fresh tobacco leaves (37.5 μg/g leaf). However, because TGF-β3 was expressed in chloroplasts, the protein was not folded correctly due to the many cysteine bridges required for the cysteine knot structure in TGF-β. Plant-produced TGF-β3 therefore had to be chemically refolded upon purification, thereby losing the plants' economical advantage over bacterial expression systems.

In this study, we investigated the possibility to express active TGF-β1 in Nicotiana benthamiana plants without the need for chemical refolding. In a first attempt, we expressed the latent LAP-TGF-β1 complex and show that a furin-like cleavage, which is required for the release of mature TGF-β1, does not occur in planta. We therefore continued employing a strategy wherein mature TGF-β1 was expressed in the absence of LAP, but unfortunately these proteins lacked activity. Accumulation of latent LAP-TGF-β1 was then enhanced to 140 ng/mg total soluble protein (~2.7 μg/g leaf) using a plant chitinase signal peptide and by co-expressing human furin. Co-expression of furin also enabled the correct post-translational processing of LAP-TGF-β1 and increased the accumulation of biologically active TGF-β1 in plant leaves.

Results

Furin cleavage does not occur in Nicotiana benthamiana

As LAP could play a role in the folding process of mature TGF-β, we first expressed the complete open reading frame of human TGF-β1 in plants. Expression was achieved by agro-infiltration of leaves of N. benthamiana plants, and the yield in crude extracts was determined on 2–5 dpi using a sandwich ELISA (Figure 1a). Crude extracts were analysed with and without acid activation, but this did not influence the detection of TGF-β1 (only acid-activated samples are shown in Figure 1a). Maximum yield was found on 3 dpi, reaching 20 ng TGF-β1/mg TSP.

To investigate the conformation of LAP-TGF-β1, Western blot analysis was performed to detect mature TGF-β1. Leaf extracts were prepared under neutral pH conditions or low pH condi-

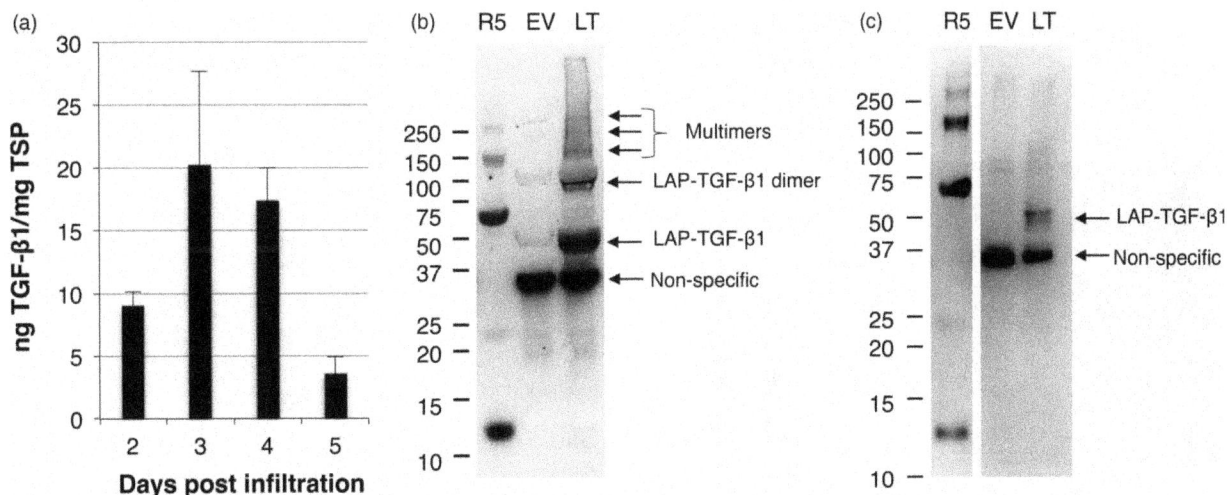

Figure 1 Analysis of LAP-TGF-β1 expression in Nicotiana benthamiana. Native human LAP-TGF-β1 is expressed in N. benthamiana leaves, but proteolytic cleavage between the latency-associated peptide (LAP) and TGF-β1 by a furin-like protease does not occur in plants. (a) Yield of human LAP-TGF-β1 in crude extracts at 2–5 days postinfiltration (dpi) as determined by ELISA after acid activation (n = 3, error bars indicate standard error). (b/c) Mature TGF-β-specific Western blot analysis with a polyclonal antibody under reducing conditions of plant-produced human LAP-TGF-β1 (LT). As controls, empty vector plant extract (EV) or 5 ng recombinant TGF-β1 (R5) was used. A molecular weight marker indicates protein size in kDa. Extracts were prepared at 3 dpi under neutral pH (b) and at a pH of 3 for acid activation (c).

tions (pH = 3) to mimic acid activation of the latent TGF-β1 complex (Figure 1b/c). As expected, recombinant TGF-β1 from mammalian cells (R5) was detected at ~12.5 kDa, but multimers were detected as well. LAP-TGF-β1 migrated at the expected size for latent LAP-TGF-β1, just below 50 kDa. When extracted under neutral conditions, also bands at 100, 150 and several bands >150 kDa were detected, which most likely represent dimers and multimers of LAP-TGF-β1. These multimers were not observed when LAP-TGF-β1 was extracted under acid conditions, which may be due to protein precipitation of large proteins under these conditions. Also a band just below 37 kDa was detected in both blots, but this band was also present in the empty vector (EV) sample. This band is likely to be the result of cross-reactivity of the polyclonal TGF-β antibody with a plant protein as this antibody is able to recognize multiple epitopes of TGF-β1. Our ELISA on the other hand is based on the capture and detection of TGF-β1 with two different monoclonal antibodies, which makes it very unlikely that our ELISA detects this ~37 kDa plant protein. Also, we have never observed cross-reactivity with EV control samples. Most importantly, under both neutral and acid conditions mature TGF-β1 with an expected size of ~12.5 kDa was not detected. This indicates that a furin-type cleavage between LAP and TGF-β1 does not occur *in planta*.

Mature TGF-β1 can be expressed without LAP, but results in necrosis

As a furin-like cleavage did not occur *in planta*, TGF-β1 was expressed in the absence of the LAP sequence. A construct was created whereby mature TGF-β1 was fused in frame with its native signal peptide (SP). To investigate whether the signal peptide resulted in adequate uptake into the endoplasmic reticulum (ER), we fused GFP C-terminally to SP-TGF-β1 and followed expression by confocal imaging (Figure 2a). Plants expressing SP-TGF-β1-GFP at 3 dpi showed fluorescence in the nuclear envelope and the endoplasmic reticulum (ER). This indicates that the signal peptide was recognized and that the fusion protein was translocated into the ER.

Next, we assessed the effect of different strategies to boost the yield of mature TGF-β1 in plant leaves. Thereto, a C-terminal KDEL sequence to facilitate ER retention was added or codon use

of TGF-β1 was optimized. The yield of mature TGF-β1 constructs in crude extracts was determined on 2–5 dpi using a sandwich ELISA (Figure 2b). Although the maximum yield of ER-retained TGF-β1 seems slightly higher (up to ~60 ng/mg TSP), no significant differences in maximum yield were observed between mature SP-TGF-β1 constructs or with LAP-TGF-β1. The yield of SP-TGF-β1 quickly drops after 3 dpi, which could be explained by the observation that mature TGF-β1 induces necrosis in plant leaves. Necrosis appeared as early as 3 dpi for the codon-optimized SP-TGF-β1 and reached its maximum for all constructs at 5 dpi (see Figure 3a and 4b). Necrosis was never observed upon expression of LAP-TGF-β1.

When analysing plant extracts by Western blot, several bands were detected for SP-TGF-β1 with and without KDEL. The bands of 12.5, 25 and >75 kDa corresponded with the bands detected in the sample with recombinant human TGF-β1 (Figure 2c) and most likely represent monomers, dimers and multimers of TGF-β1. However, a band at 16 kDa was also detected and may represent TGF-β1 with its signal peptide (~3 kDa) still attached. This would mean that although the signal peptide is recognized and enables protein uptake into the ER, it is not cleaved during translocation into the ER.

Fcα-fusion of TGF-β1 enhances yield and prevents necrosis

To increase yield and circumvent improper signal peptide processing, we fused TGF-β1 to the C-terminus of a stable partner. The Fc portion of immunoglobulin alpha 1 (Fcα), a natural dimer, was used as a fusion partner allowing dimerization of TGF-β1. This strategy was previously used to express human interleukin-10, which is also active as a dimer (Westerhof *et al.*, 2012). In contrast to SP-TGF-β1 constructs, expression of Fcα-TGF-β1 did not result in necrosis in leaves (Figure 3a), even though Fcα-TGF-β1 yield increased significantly as analysed in crude extracts at 2–5 dpi (Figure 3b). Fcα-TGF-β1 yield was ~16-fold higher compared with SP-TGF-β1, reaching up to 338 ng TGF-β1/mg TSP on 3 dpi.

Finally, biological activity of Fcα-TGF-β1 in crude plant extracts was assessed in a cell-based assay using mink lung epithelial cells carrying a TGF-β-responsive luciferase reporter gene that is activated upon binding to the TGF-β receptor. Unfortunately, Fcα-

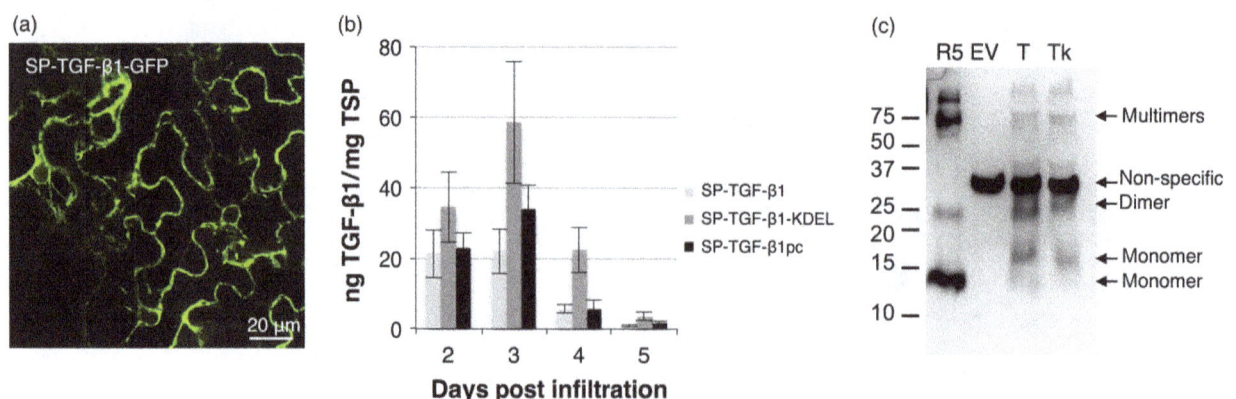

Figure 2 Analysis of mature TGF-β1 expression in *Nicotiana benthamiana*. (a) Whole mount confocal microscopy output of GFP fused C-terminally to TGF-β1 excluding LAP, but including its native signal peptide (SP) at 3 dpi. (b) Yield of SP-TGF-β1, SP-TGF-β1-KDEL and plant codon-optimized SP-TGF-β1pc in crude extracts at 2–5 dpi as determined by ELISA (*n* = 3, error bars indicate standard error). (c) Mature TGF-β-specific Western blot analysis with a polyclonal antibody under reducing conditions of plant-produced SP-TGF-β1 (T) and SP-TGF-β1-KDEL (Tk) at 3 dpi. As controls, empty vector plant extract (EV) and 5 ng recombinant TGF-β1 (R5) were used. A molecular weight marker is indicating size in kDa.

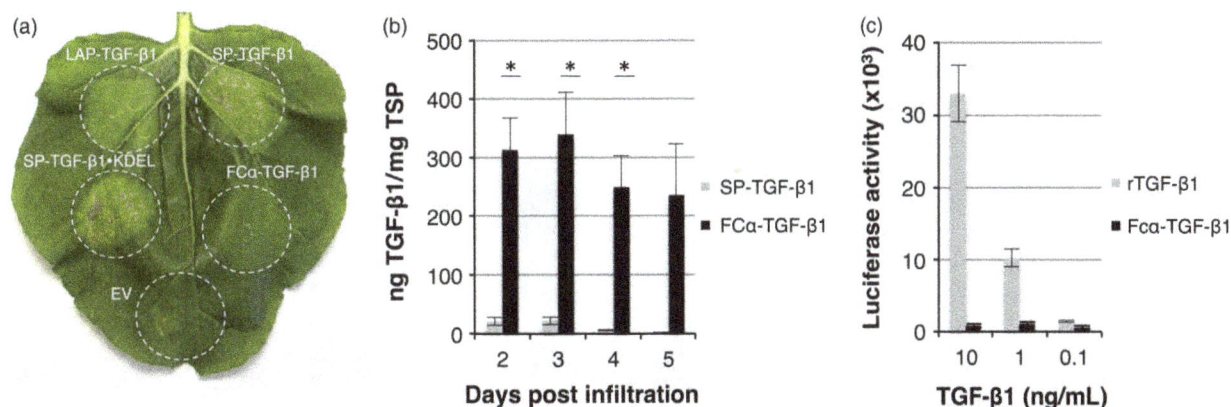

Figure 3 Analysis of the effect of Fcα-fusion on expression and activity. Necrosis is induced in *Nicotiana benthamiana* leaves upon expression of SP-TGF-β1 and SP-TGF-β1-KDEL, but not when fused to LAP or Fcα. (a) Leaf necrosis on 5 days postinfiltration of several TGF-β1 constructs. (b) Yield of SP-TGF-β1 and Fcα-TGF-β1 in crude extracts at 2–5 dpi as determined by ELISA ($n = 3$, error bars indicate standard error). Significant differences as determined by a Welch's t-test (P < 0.05) between samples are indicated with an asterisk. (c) Biological activity of Fcα-TGF-β1 was determined by using MLEC cells containing a TGF-β1-inducible luciferase reporter construct. Cells were treated with recombinant TGF-β1 (rTGF-β1 from mammalian cells) or plant-produced Fcα-TGF-β1 (from dpi 3), and luciferase expression was measured after overnight incubation.

TGF-β1 was not able to induce luciferase expression in the reporter cell line (Figure 3c). We therefore conclude that Fcα fusion of TGF-β1 aids in stability of the protein, but the fusion protein lacks activity.

Improper signal peptide cleavage of TGF-β1 induces necrosis

To investigate whether an alternative signal peptide would improve signal peptide processing and yield, a construct was created where mature TGF-β1 was fused in frame with the signal peptide of an *Arabidopsis thaliana* chitinase gene (cSP). The cSP-TGF-β1 construct was used for agro-infiltration and analysed by Western blot (Figure 4a). In extracts of transformed leaves with cSP-TGF-β1 (cTpc), a band was detected around ~12.5 kDa, which corresponds with the size of recombinant TGF-β1 from mammalian cells. A band of ~16 kDa, which resembles TGF-β1 with an uncleaved SP, was not detected this time. We therefore conclude that the *A. thaliana* chitinase signal peptide is properly cleaved. Besides the band for monomeric TGF-β1, several other bands representing dimeric and multimeric TGF-β1 were detected as well, whereas bands for recombinant TGF-β1 (R5) and SP-TGF-β1pc (Tpc) are very faint. Next to that, cSP-TGF-β1 does not induce necrosis in *N. benthamiana* leaves from 3 dpi onwards (Figure 4b). We therefore also conclude that the improper processing of the signal peptide was responsible for the necrotic symptoms.

As the cSP-TGF-β1 construct does not induce necrosis, we used the silencing inhibitor p19 to further boost TGF-β1 expression. The yield of cSP-TGF-β1 was determined in crude extracts at 5 dpi by sandwich ELISA. Figure 4c reveals that the yield of cSP-TGF-β1 was enhanced significantly by p19 when compared to SP-TGF-β1 (P = 0.004). This is most likely explained by the lack of necrosis due to the replacement of the signal peptide. However, the maximum yield of ~50 ng TGF-β1/mg TSP was not significantly higher as native SP-TGF-β1-KDEL at dpi 3 as shown in Figure 2b. With a yield of 50 ng/mg TSP, only 2.5 ng of TGF-β1 was loaded on gel for Western blot analysis (Figure 4a); however, the band intensity of cSP-TGF-β1 is much stronger than the 5 ng recombinant TGF-β1. Plant-expressed cSP-TGF-β1 could be folded

incorrectly, thereby explaining the difference in detection between ELISA and Western blot.

To test whether plant-produced mature TGF-β1 is active, crude extracts were applied in the TGF-β-responsive luciferase reporter assay (Figure 4d). Recombinant human TGF-β1 from mammalian cells was able to induce the expression of luciferase in the reporter cells, even in the presence of plant proteins. Unfortunately, mature TGF-β1 from plants was not active. Thus, replacing the signal peptide of TGF-β1 can prevent leaf necrosis, but this form of plant-produced mature TGF-β1 still lacks biological activity. It is therefore likely that mature TGF-β1 is not folded correctly in plants and might require LAP for proper folding.

Proteolytic processing of LAP-TGF-β1 in plants requires co-expression of furin

As mature TGF-β1 from plants lacks biological activity, we further investigated whether furin is required for the production of TGF-β1 in plants. We first replaced the native signal peptide of LAP-TGF-β1 with the *Arabidopsis* chitinase signal peptide (cSP) and analysed the yield for both LAP-TGF-β1 constructs at 3, 5 and 7 dpi while co-expressing the p19 silencing suppressor. The expression profile for both LAP-TGF-β1 constructs is given in Figure 5a. Replacing the native signal peptide with the chitinase signal peptide increased LAP-TGF-β1 yield 4.5-fold (P = 0.002). In a next experiment, we investigated the effect of co-expression of human furin. Figure 5b shows that co-expression of furin enhances the yield of both LAP-TGF-β1 constructs. Maximum yield of cSP-LAP-TGF-β1 was ~140 ng TGF-β1/mg TSP, which is 2.7-fold (P = 0.084) higher than without co-expression of furin.

Next, we performed a mature TGF-β1-specific and LAP-specific Western blot to analyse the post-translational processing of LAP-TGF-β1 by furin. Under nonreducing conditions, LAP-TGF-β1 is detected as a band just above 100 kDa with both antibodies, which is the expected size for the latent LAP-TGF-β1 homodimer (Figure 5c; first and fourth panel, respectively). Multimers were detected as well, but no differences were observed upon co-expression of furin. Under reducing conditions, LAP-TGF-β1 was detected as a monomeric band of approximately 50 kDa with the TGF-β antibody (Figure 5c, second panel). However, the intensity

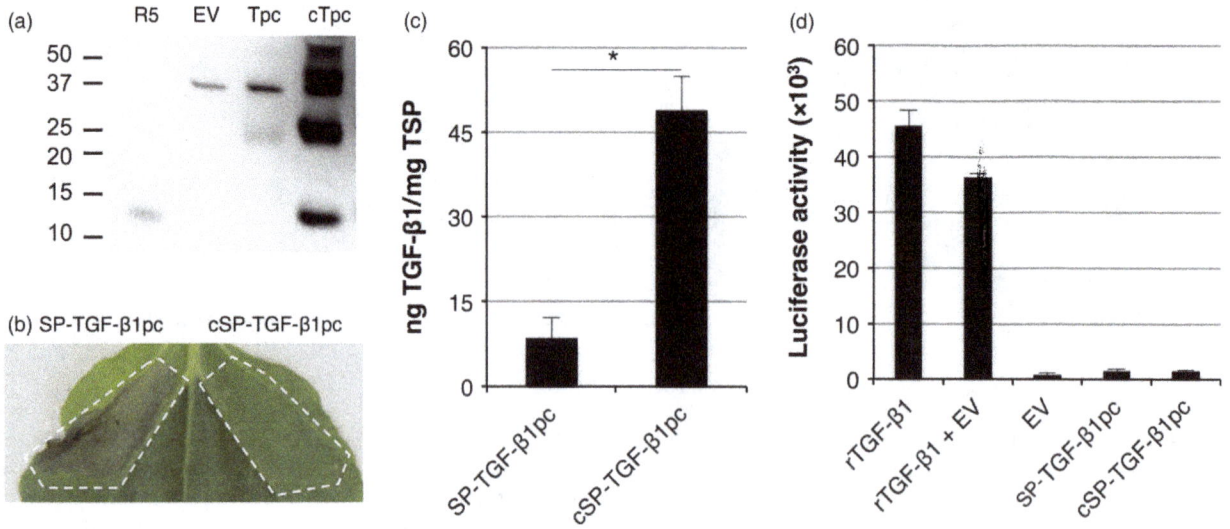

Figure 4 Analysis of the effect of a plant signal peptide on necrosis. Improper cleavage of the native signal peptide results in necrosis. (a) Mature TGF-β1-specific Western blot analysis with a polyclonal antibody under reducing conditions of plant-produced SP-TGF-β1pc (Tpc) containing the native signal peptide and cSP-TGF-β1pc (cTpc) containing the *Arabidopsis thaliana* chitinase signal peptide at 3 dpi. A molecular weight marker indicates protein size in kDa. As controls, empty vector plant extract (EV) and 5 ng recombinant TGF-β1 (R5) were used. (b) Leaf necrosis at 5 dpi of SP-TGF-β1pc and cSP-TGF-β1pc. (c) Yield of SP-TGF-β1 and cSP-TGF-β1 in crude extracts at 5 dpi as determined by ELISA (*n* = 3, error bars indicate standard error). Co-expression of p19 was used in this experiment to further enhance TGF-β1 expression. (d) Biological activity of 5 ng/mL recombinant TGF-β1 (rTGF-β1) and plant-produced TGF-β1 (from dpi 5 + p19) as determined by the induction of luciferase expression in MLEC reporter cells.

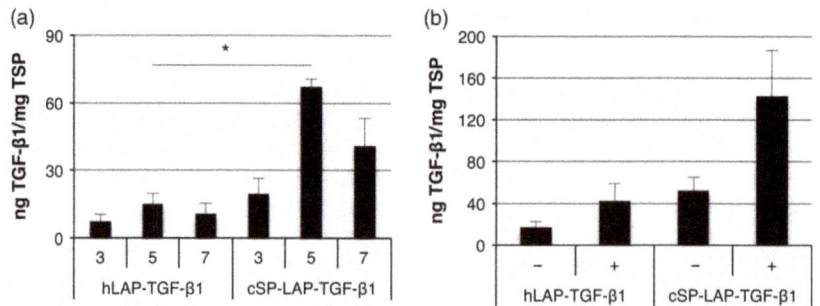

Figure 5 Co-expression of furin enables proteolytic processing of LAP-TGF-β1. (a) Yield of LAP-TGF-β1 and cSP-LAP-TGF-β1 in crude extracts at 3, 5 and 7 days postinfiltration (dpi) upon co-expression of p19 as determined by ELISA after acid activation (*n* = 3, error bars indicate standard error). Significant differences between samples as determined by a Welch's t-test (P < 0.05) are indicated with an asterisk. (b) Yield of LAP-TGF-β1 constructs in crude extracts at 5 dpi upon co-expression of furin (*n* = 3, error bars indicate standard error, +: furin co-expression). (c) Mature TGF-β1- and LAP-specific Western blot analysis with polyclonal antibodies under nonreducing and reducing conditions of cSP-LAP-TGF-β1 at dpi 5 (EV: empty vector control).

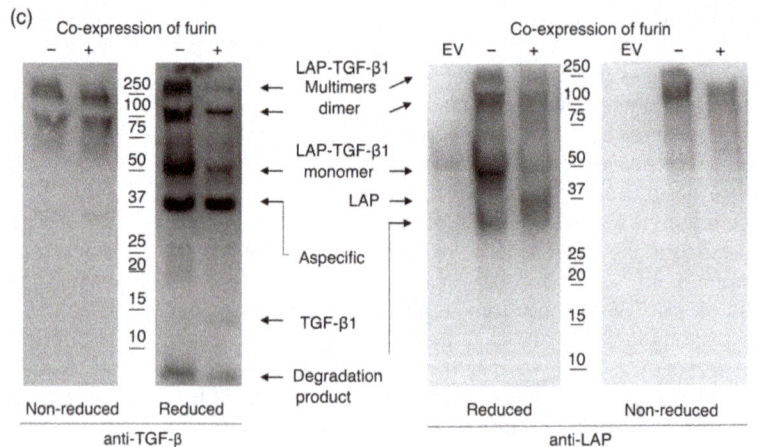

of this band is strongly reduced upon co-expression of furin. The Western blot probed with anti-LAP also shows this reduction in intensity for the LAP-TGF-β1 monomer (Figure 5c, third panel). Also, a faint band around 12.5 kDa (the size of monomeric TGF-β1) appears upon co-expression of furin (Figure 5c, second panel). Furthermore, we detected a band of approximately

37 kDa under reducing conditions when using LAP-specific antibodies. This likely represents properly processed LAP which should migrate around 37 kDa. However, the anti-LAP blot also reveals a degradation product of LAP of approximately 30 kDa in both samples. Therefore, LAP-TGF-β1 seems to be processed by endogenous proteases in plants as well. Altogether, we conclude

that co-expression of furin enables the correct post-translational processing of latent LAP-TGF-β1.

Co-expression of furin enhances the accumulation of active TGF-β1

To investigate the efficiency of furin cleavage, we isolated latent TGF-β1 from the apoplastic space of agro-infiltrated leaves at 5 dpi and analysed the conformation on Western blot. First, we determined the secretion efficiency of TGF-β1 into the apoplast. In Figure 6a, we demonstrate that co-expression of furin significantly increases the secretion efficiency of native cSP-LAP-TGF-β1 (P = 0.018). Quite unexpectedly, secretion efficiency of cSP-LAP-TGF-β1 was also significantly increased upon codon optimization (P = 0.017) and furin did not further increase secretion of codon-optimized cSP-LAP-TGF-β1.

Next, the efficiency of furin cleavage was evaluated by performing a TGF-β1-specific Western blot on both crude extracts and apoplastic fluids (Figure 6b). As described in the previous paragraph, the intensity of the 50 kDa band, which resembles the size of unprocessed LAP-TGF-β1, is reduced in both crude extracts and apoplast fluids upon furin co-expression. However, the processing of LAP-TGF-β1 is not complete as unprocessed LAP-TGF-β1 is still detected in the apoplast. We also observed clear bands around 12.5 kDa upon furin co-expression. The intensity of this monomeric mature TGF-β1 is stronger in crude extracts, which is striking as the absolute amount of TGF-β1 on gel should be higher for apoplast fluids according to the ELISA. Furthermore,

several degradation products were detected, which indicates that LAP-TGF-β1 is sensitive to the activity of endogenous proteases.

To evaluate the biological activity of cSP-LAP-TGF-β1 constructs, the effect of acid-activated crude extracts or apoplast fluids on mink lung epithelial cells carrying the TGF-β1-responsive luciferase reporter gene was assessed. Figure 6c shows that co-expression of furin enhances the biological activity of TGF-β1 in crude extracts to a similar level as recombinant TGF-β1. Unfortunately, cSP-LAP-TGF-β1 isolated from the apoplast was reduced in its activity when compared to crude extracts and recombinant TGF-β1. It is therefore likely that cSP-LAP-TGF-β1 is degraded by endogenous proteases in the apoplast and therefore loses activity.

Discussion

Here, we show that expression of biologically active TGF-β without the requirement of chemical refolding is feasible in plants. Previously, TGF-β3 was successfully produced in chloroplasts of Nicotiana tabacum plants (Gisby et al., 2011) and yielded considerable amounts of this protein (37.5 μg/g FW and 80% purity). However, this strategy required chemical refolding of purified TGF-β3 to obtain biological activity. Therefore, chloroplast expression, as a plant-based expression system, loses one of its major advantages over bacterial expression systems, namely cost-effectiveness. When we studied the transient expression of the whole open reading frame of the TGF-β1 gene

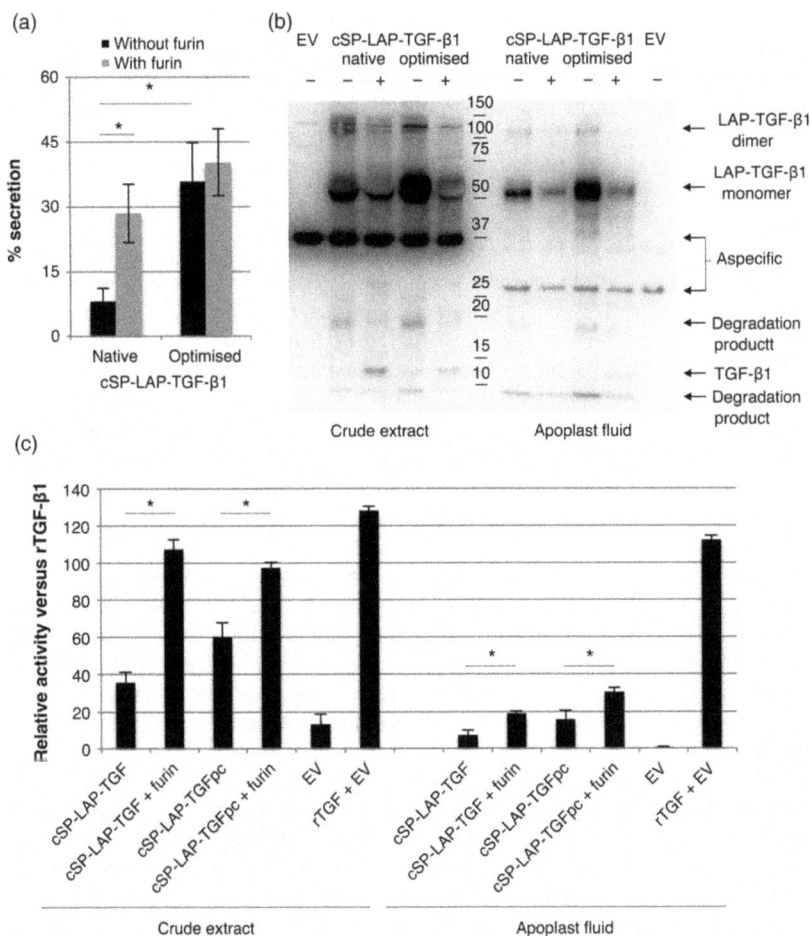

Figure 6 Furin enhances the accumulation of biologically active TGF-β1. (a) Native and codon-optimized cSP-LAP-TGF-β1 were transiently expressed in Nicotiana benthamiana leaves with or without co-expression of furin. The percentage of secreted cSP-LAP-TGF-β1 was determined at 5 dpi by determining the amount of cytokine in apoplast fluids versus crude extracts (n = 5, error bars indicate standard error). (cSP: Arabidopsis thaliana chitinase signal peptide; TSP: total soluble protein). Asterisk indicates significant differences as determined by a Welch's t-test (*P < 0.05). (b) Mature TGF-β1-specific Western blot analysis with a polyclonal antibody at 5 dpi under reducing conditions of apoplast fluids and crude extracts containing native or codon-optimized cSP-LAP-TGF-β1 with or without co-expression of furin (EV: empty vector control; pc: codon-optimized; +: furin co-expression). (c) Biological activity of 5 ng/mL recombinant TGF β1 (rTGF β1) and plant-produced TGF-β1 (from dpi 5) as determined by the induction of luciferase expression in MLEC cells. Relative activity of plant-produced TGF-β1 versus recombinant TGF-β1 (in %) is given (n = 3, error bars indicate standard error).

in *N. benthamiana* plants, the obtained yield of TGF-β1 was ~20 ng/mg TSP. This yield was considerably lower compared with many other proteins heterologously expressed in plants, like antibodies or other cytokines (Westerhof *et al.*, 2012, 2014). More importantly, we revealed that a furin-like cleavage between LAP and TGF-β1 does not occur in *N. benthamiana* plants. This furin-like cleavage is required to release active TGF-β1 from the latent complex. Furin is a mammalian subtilisin-/kex2p-like endoprotease. Although a kex2p-like pathway was shown to exist in *N. tabacum* (Kinal *et al.*, 1995), no furin cleavage of LAP-TGF-β1 was observed in our expression system. Hence, it is likely that a protease with the right specificity does not exist in *N. benthamiana*. The lack of furin cleavage might be a crucial factor that limits the yield of TGF-β1 in plants.

To circumvent the post-translational processing of LAP-TGF-β1, we expressed mature TGF-β1 as an in-frame fusion with its native signal peptide. We confirmed that the signal peptide of TGF-β1 is functional by monitoring localization of a GFP fusion protein with confocal imaging, as SP-TGF-β1 is taken up into the secretory pathway. However, after translocation into the ER, the signal peptide is not completely removed from the mature protein. When the signal peptide stays attached to mature TGF-β1, it could compromise the function of the plant ER. The unprocessed signal peptide could act as membrane anchor (High and Dobberstein, 1992), thereby facilitating accumulation of unprocessed TGF-β1 in the ER. On the other hand, the signal peptide could also affect the folding of TGF-β1 and trigger the unfolded protein response. Disturbance of ER function could ultimately lead to necrosis (Ye *et al.*, 2011). This disturbance of ER function may explain the necrosis that we observed upon expression of SP-TGF-β1.

Replacing the signal peptide with an *A. thaliana* chitinase signal peptide circumvents improper cleavage of the TGF-β1 signal peptide, prevented the induction of necrosis and increased yield of mature TGF-β1. Surprisingly, signal peptide replacement with the chitinase signal peptide also significantly increased the yield of LAP-TGF-β1. The chitinase signal peptide may also influence expression by changing the mRNA secondary structure at the 5'-end. The secondary structure of the 5' UTR sequence containing the *Alfalfa mosaic virus* RNA 4 (AlMV) leader and the first 40 nucleotides of LAP-TGF-β1 was predicted by the Vienna RNA fold software (Lorenz *et al.*, 2011). In Figure S1, drawings of the minimum free energy predictions for both LAP-TGF-β1 genes are given. This prediction reveals that the combination of the AlMV leader with the chitinase signal peptide increases the minimal free folding energy 6.5-fold, making the secondary structure less stable. Furthermore, strong GC-rich secondary structures are present downstream of the start codon in the native signal peptide. As low minimal free folding energy has been suggested to enhance translation initiation (Hall *et al.*, 1982; Tuller *et al.*, 2010), these observed differences could explain differences in yield.

Even though the yield was improved, a major bottleneck for the accumulation of mature TGF-β1 seems to be low stability or misfolding of the protein. Plants are rich in proteases, and a wide variety of proteases reside within the secretory pathway of plants as well as in the apoplast (Goulet *et al.*, 2012; van der Hoorn, 2008). It is therefore inevitable that recombinant secretory proteins encounter proteases on their way out. Proteolytic processing is regarded as one of the largest bottlenecks of plant expression systems. A classic example is the unintended processing of antibodies when expressed in plants (De Muynck *et al.*,

2010). A common strategy to increase recombinant protein yield and circumvent proteolytic degradation is to retain the protein in the ER by using a C-terminal KDEL targeting sequence (Schouten *et al.*, 1996). The ER is regarded as a protein friendly environment due to the absence of many proteases. We have used ER retention in order to increase the yield of TGF-β1, but this strategy did not result in significantly higher yields. As cytokines are inherently unstable, we also attempted to increase the yield of TGF-β1 by fusing them to a stable partner. In our study, fusion of TGF-β1 to Fcα increased the yield 16-fold, but led to the production of inactive TGF-β1. The observation that plant-produced mature TGF-β1 or Fcα-TGF-β1 is inactive is therefore most likely explained by misfolding of TGF-β1 in the absence of LAP. This idea is further supported by the fact that we observed a strong difference in the detection of mature TGF-β1 by ELISA and Western blot. Misfolded TGF-β1 is likely not detected properly in a sandwich ELISA, but might be detected by Western blot under reducing conditions as continuous epitopes become available. Proper folding of biologically active TGF-β1 likely requires expression of the full LAP-TGF-β1 gene.

Post-translational modification of LAP-TGF-β1 is a complex process, which is controlled at multiple levels. The secretion of biologically active TGF-β1 dimers first of all requires the dimerization of the LAP-TGF-β1 proprotein, which occurs in the endoplasmic reticulum (Gray and Mason, 1990). LAP-TGF-β1 is also N-glycosylated at three residues (82, 136 and 176) and the presence of the N-glycans on residues Asn82 and Asn136 was shown to be required for proper secretion of LAP-TGF-β1 (Brunner *et al.*, 1992; Sha *et al.*, 1989). Yet, secretion of latent TGF-β1 is a slow process as LAP-TGF-β1 is retained within the *cis*-Golgi in an Endo H-sensitive form (Miyazono *et al.*, 1992). *Cis*-Golgi retention of LAP-TGF-β1 is presumed to be caused by its capture by a sequestering protein, such as the chaperone GRP78 (Oida and Weiner, 2010). Retention of LAP-TGF-β1 in the ER/*cis*-Golgi is the limiting step for secretion and furin processing (Oida and Weiner, 2010). Only properly glycosylated LAP-TGF-β1 is transported to the *trans*-Golgi where it is processed by furin (Dubois *et al.*, 1995).

Within our study, we co-expressed human furin with LAP-TGF-β1 to investigate the proper proteolytic processing of LAP and TGF-β1 in plants. First of all, we observed that co-expression of furin slightly increased the yield of LAP-TGF-β1 by 2.7-fold ($P = 0.084$), but significantly enhanced the secretion of cSP-LAP-TGF-β1 into the apoplast ($P = 0.002$). We also observed that the majority of extracted LAP-TGF-β1 is properly processed by co-expressed furin (Figures 5c and 6b). This engineering strategy is quite remarkable as furin requires multiple autocatalytic processing steps and correct localization within the secretory pathway to become active (Vey *et al.*, 1994).

Co-expression of furin also enhanced the accumulation of biologically active TGF-β1 in *N. benthamiana* leaves, but activity of secreted LAP-TGF-β1 was reduced. This is most likely explained by the fact that LAP-TGF-β1 is sensitive to the activity of plant endogenous proteases that reside within the apoplast. A common activation mechanism for latent TGF-β1 in mammals is the proteolytic degradation of LAP by proteases, like plasmin, matrix metalloproteinases or thrombospondin (Boor *et al.*, 2010). It seems that endogenous proteases with similar activity exist in plants and therefore future work has to focus on how the stability of LAP-TGF-β1 in the apoplast can be enhanced. When undesired proteolytic processing of LAP-TGF-β1 can be prevented, co-expression of furin will enable the production of active TGF-β1 in plants without the need for chemical refolding.

Our study demonstrates that the production of biologically active TGF-β1 in plants is feasible without the need for chemical refolding. Thereby, we avoid difficult and expensive chemical refolding steps, which are required upon bacterial or chloroplast expression. We also show that the post-translational machinery of plants can be engineered to allow proteolytic processing of heterologous expressed mammalian proteins. Therefore, this study highlights the fact that plants are a suitable expression platform for the production of complex glycoproteins, like TGF-β1, that rely on correct post-translational processing for biological activity. This engineering strategy can further facilitate plant-based expression of other proproteins that rely on furin cleavage, like neural growth factor (NGF), bone-morphogenic proteins (BMPs) or even the Ebola Zaire envelope glycoprotein (Thomas, 2002). Furthermore, co-expression of other proteases can be explored as well to facilitate plant-based expression of a variety of complex proproteins.

Experimental procedures

Construct design

The complete native open reading frame (ORF) of human (h)LAP-TGF-β1 was amplified from the MegaMan™ Human Transcriptome cDNA library (Bio-Connect BV, Huissen, The Netherlands). Similarly, mouse LAP-TGF-β1 was amplified from the First-Choice™ PCR-Ready Mouse Spleen cDNA library (Life Technologies, Bleiswijk, The Netherlands). To remove LAP from the mature ORF, mature hTGF-β1 was re-amplified whereby a SacII restriction site was introduced at the 5′ end. This SacII was used to clone mature human TGF-β1 in frame with the mouse signal peptide, as SacII was uniquely present in the mouse (and not human) LAP-TGF-β1 signal peptide. Thereafter, SP-TGF-β1 was re-amplified to introduce the ER retention signal KDEL at the 3′ end of the gene. The complete ORF of human furin (NM_002569) was amplified from the MegaMan™ Human Transcriptome cDNA library. An internal BspHI site was removed by means of overlap-extension PCR to ensure subsequent cloning into pBIN+.

To create a GFP fusion of hTGF-β1, the enhanced green fluorescent protein (eGFP) gene was re-amplified and used to replace the mature hIL-10 sequence in the previously published Fcα-IL-10 construct (Westerhof et al., 2012) using SpeI/KpnI. Similarly, hTGF-β1 was re-amplified and used to replace Fcα in the same Fcα-hIL-10 construct using NcoI/SacI. In this way, a SP-TGF-β1-GFP construct was created where a glycine–serine linker separated hTGF-β1 and GFP. To create a Fcα-fusion of hTGF-β1, mature hTGF-β1 (SpeI/KpnI) was re-amplified and used to replace the mature hIL-10 sequence in the previously mentioned Fcα-IL-10 construct.

The mature TGF-β1 sequence was codon-optimized using an in-house optimization procedure and ordered at GeneArt (Life Technologies). The codon-optimized TGF-β1 gene was cloned in frame with the Arabidopsis chitinase signal peptide (cSP, AAM10081.1) by means of overlap-extension PCR. Similarly, the mature sequence for native LAP-TGF-β1 sequence was cloned in frame with cSP.

All construct sequences were confirmed by sequencing (Macrogen, Amsterdam, The Netherlands) in the expression vectors pBIN+ and pHYG (Westerhof et al., 2012) and an overview of all constructs is given in Figure S2. In pBIN+ and pHYG, expression is controlled by the 35S promoter of the Cauliflower mosaic virus with duplicated enhancer (d35S) and the Agrobacterium tumefaciens nopaline synthase transcription terminator (Tnos). A 5′ leader sequence of the Alfalfa mosaic virus RNA 4 (AlMV) is included between the promoter and construct to boost translation. The expression vectors were subsequently transformed to Agrobacterium tumefaciens strain MOG101 for plant expression. In some experiments, the silencing suppressor p19 from tomato bushy stunt virus was co-infiltrated to enhance heterologous expression. The vector pBIN61-p19 was obtained from Dr. D. Baulcombe (Voinnet et al., 2003).

Agro-infiltration of Nicotiana benthamiana

Agrobacterium tumefaciens clones were cultured for 16 h at 28 °C/250 rpm in LB medium (10 g/L pepton 140, 5 g/L yeast extract, 10 g/L NaCl with pH = 7.0) containing 50 μg/mL kanamycin, 20 μg/mL rifampicin and 20 μM acetosyringone. The bacteria were resuspended in MMA infiltration medium (20 g/L sucrose, 5 g/L MS salts, 1.95 g/L MES, pH5.6) containing 200 μM acetosyringone to obtain an OD (600 nm) of 1. For co-infiltration experiments, a final OD of 0.5 of each individual culture was used. After 1–2 h of incubation at room temperature, the two youngest fully expanded leaves of 5- to 6-week-old N. benthamiana plants were infiltrated completely.

Total soluble protein extraction

Leaves were immediately snap-frozen upon harvesting and homogenized in liquid nitrogen. Homogenized plant material was ground in ice-cold extraction buffer (50 mM phosphate-buffered saline (PBS) pH = 7.4, 100 mM NaCl, 0.1% v/v Tween-20 and 2% w/v immobilized polyvinylpolypyrrolidone (PVPP)) using 2 mL/g fresh weight. For extraction at low pH, a 30 mM sodium citrate buffer (30 mM sodium citrate buffer pH = 3, 100 mM NaCl, 0.1% v/v Tween-20 and 2% w/v PVPP) was used. Crude extracts were clarified by centrifugation at 16 100 g for 5 min at 4 °C, and supernatants were directly used in an ELISA and BCA protein assay.

Isolation of apoplast fluid

Leaves were vacuum-infiltrated with ice-cold extraction buffer (50 mM phosphate-buffered saline (pH = 8), 100 mM NaCl and 0.1% v/v Tween-20). Leaves were centrifuged for 20 min at 2000 g to isolate apoplast fluid. Apoplast fluids were clarified by centrifugation at 16 000 g at 4 °C. To determine the percentage of secretion, the remaining leaf material was processed as described above.

Quantification of human TGF-β1 protein levels

TGF-β1 protein concentration in crude plant extracts was determined by ELISA. Samples containing LAP-TGF-β1 were activated by the addition of 1M hydrochloric acid (10 μL acid/50 μL sample) and subsequently neutralized after 10 min by adding the same volume of 1M sodium hydroxide. The human TGF-β1 Ready-SET-Go!® ELISA kit (eBioscience, Vienna, Austria) were used according to the supplier's protocol. The ELISA is based on the capture of TGF-β1 with a monoclonal antibody and subsequent detection with a biotinylated monoclonal antibody and shows no cross-reactivity to plant proteins. For sample comparison, the total soluble protein (TSP) content was determined by the BCA method (Life Technologies) according to the supplier's protocol using bovine serum albumin (BSA) as a standard.

Protein analysis by Western blot

Soluble plant proteins (~50 μg for crude extract or ~20 μg for apoplast fluid) were separated under reducing or nonreducing conditions by SDS-PAGE on a 12% Bis-Tris gel. Recombinant human TGF-β1 (R&D Systems, Abingdon, UK) was used as a control. Proteins were transferred to a PVDF membrane (Life Technologies) by a wet blotting procedure. Thereafter, the membrane was blocked in PBST-BL (PBS containing 0.1% v/v Tween-20 and 5% w/v nonfat dry milk powder) for 1 h at room temperature, followed by overnight incubation with a rabbit anti-TGF-β polyclonal antibody (Bioké, Leiden, The Netherlands) or a goat anti-LAP polyclonal antibody (R&D Systems) in PBST (including 0.1% w/v bovine serum albumin) at 4 °C. HRP-conjugated secondary antibodies (Jackson ImmunoResearch, Suffolk, UK) were used for visualization. Finally, the SuperSignal West Femto substrate (Pierce) was used to detect the HRP-conjugated antibodies.

Confocal microscopy

Plants were agro-infiltrated with *A. tumefaciens* harbouring the pHYG vector containing the expression cassette encoding N-terminal fusions of human TGF-β1 to GFP as described previously. Leaves were taken from the plant at 3 days postinfiltration, and small sections were examined from the abaxial side using a Zeiss LSM510 (Zeiss, Oberkochen, Germany) confocal laser-scanning microscope in combination with an argon ion laser supplying a 488 nm wavelength.

Biological activity assay

The mink lung epithelial cell line with TGF-β-inducible luciferase reporter construct (Abe *et al.*, 1994) was kindly provided by Dr. C. Arancibia (Oxford University). MLEC cells were maintained at 37 °C with 5% CO_2 in RPMI-1640 medium containing 4 mM L-glutamine and 25 mM HEPES and supplemented with 10% foetal calf serum, 50 U/mL penicillin, 50 μg/mL streptomycin and 200 μg/mL G418. MLEC cells were harvested by trypsinization and seeded at a density of 1.5×10^6 cells/mL in 96-well plates and were allowed to rest for 4 h in medium without G418 prior to bioassays. For bioassays, cells were treated with 0.1–10 ng/mL plant-produced TGF-β1 (acid activated) or recombinant human TGF-β1 expressed in Chinese hamster ovary cells (R&D Systems). Recombinant TGF-β1 was supplemented with EV plant extract to keep plant protein levels equal. After overnight incubation, luciferase expression was analysed using the Bright-Glo™ Luciferase assay system (Promega, Leiden, The Netherlands) according to the supplier's protocol. Luminescence was measured using the FLUOstar OPTIMA microplate reader (BMG Labtech, Ortenberg, Germany).

Data analysis

All data shown in the figures (unless stated otherwise) indicate the average of at least three biological replicates (*n*). In the figure legends, *n* is indicated and error bars indicate standard error. Significant differences between samples were calculated using the Welch's t-test and regarded as significant when $P < 0.05$. Significant differences are indicated in the figures by an asterisk (*).

Acknowledgements

This study was financially supported in part by Synthon (Nijmegen, The Netherlands). We would like to thank Gerry Ariaans for all helpful discussions and Mohamed Abdi Hassan, Sophie van Gorkom, Kelly Heckman, Tram Hong, Xandra Schrama, Koen Verhees, Tineke Vliek and Michelle Yang for their input in the practical work of this study.

References

Abe, M., Harpel, J.G., Metz, C.N., Nunes, I., Loskutoff, D.J. and Rifkin, D.B. (1994) An assay for transforming growth-factor-beta using cells transfected with a plasminogen-activator inhibitor-1 promoter luciferase construct. *Anal. Biochem.* **216**, 276–284.

Annes, J.P., Munger, J.S. and Rifkin, D.B. (2003) Making sense of latent TGFbeta activation. *J. Cell Sci.* **116**, 217–224.

Archer, S.J., Bax, A., Roberts, A.B., Sporn, M.B., Ogawa, Y., Piez, K.A., Weatherbee, J.A. *et al.* (1993) Transforming growth factor-beta-1 - secondary structure as determined by heteronuclear magnetic-resonance spectroscopy. *Biochemistry*, **32**, 1164–1171.

Blobe, G.C., Schiemann, W.P. and Lodish, H.F. (2000) Mechanisms of disease: Role of transforming growth factor beta in human disease. *N. Engl. J. Med.* **342**, 1350–1358.

Boor, P., Ostendorf, T. and Floege, J. (2010) Renal fibrosis: novel insights into mechanisms and therapeutic targets. *Nat. Rev. Nephrol.* **6**, 643–656.

Brunner, A.M., Lioubin, M.N., Marquardt, H., Malacko, A.R., Wang, W.C., Shapiro, R.A., Neubauer, M. *et al.* (1992) Site-directed mutagenesis of glycosylation sites in the transforming growth factor-beta-1 (Tgf-Beta-1) and Tgf-Beta-2 (414) precursors and of cysteine residues within mature Tgf-Beta-1 - effects on secretion and bioactivity. *Mol. Endocrinol.* **6**, 1691–1700.

Chen, W.J., Jin, W.W., Hardegen, N., Lei, K.J., Li, L., Marinos, N., McGrady, G. *et al.* (2003) Conversion of peripheral CD4(+)CD25(-) naive T cells to CD4(+)CD25(+) regulatory T cells by TGF-beta induction of transcription factor Foxp3. *J. Exp. Med.* **198**, 1875–1886.

Daopin, S., Piez, K.A., Ogawa, Y. and Davies, D.R. (1992) Crystal-structure of transforming growth-factor-Beta-2 - an unusual fold for the superfamily. *Science*, **257**, 369–373.

Dardalhon, V., Awasthi, A., Kwon, H., Galileos, G., Gao, W., Sobel, R.A., Mitsdoerffer, M. *et al.* (2008) IL-4 inhibits TGF-beta-induced Foxp3(+) T cells and together with TGF-beta, generates IL-9(+) IL-10(+) Foxp3(-) effector T cells. *Nat. Immunol.* **9**, 1347–1355.

De Muynck, B., Navarre, C. and Boutry, M. (2010) Production of antibodies in plants: status after twenty years. *Plant Biotechnol. J.* **8**, 529–563.

Dubois, C.M., Laprise, M.H., Blanchette, F., Gentry, L.E. and Leduc, R. (1995) Processing of transforming growth-factor-beta-1 precursor by human furin convertase. *J. Biol. Chem.* **270**, 10618–10624.

Geissmann, F., Revy, P., Regnault, A., Lepelletier, Y., Dy, M., Brousse, N., Amigorena, S. *et al.* (1999) TGF-beta 1 prevents the noncognate maturation of human dendritic Langerhans cells. *J. Immunol.* **162**, 4567–4575.

Gisby, M.F., Mellors, P., Madesis, P., Ellin, M., Laverty, H., O'Kane, S., Ferguson, M.W.J. *et al.* (2011) A synthetic gene increases TGF beta 3 accumulation by 75-fold in tobacco chloroplasts enabling rapid purification and folding into a biologically active molecule. *Plant Biotechnol. J.* **9**, 618–628.

Goulet, C., Khalf, M., Sainsbury, F., D'Aoust, M.A. and Michaud, D. (2012) A protease activity-depleted environment for heterologous proteins migrating towards the leaf cell apoplast. *Plant Biotechnol. J.* **10**, 83–94.

Govinden, R. and Bhoola, K.D. (2003) Genealogy, expression, and cellular function of transforming growth factor-beta. *Pharmacol. Ther.* **98**, 257–265.

Gray, A.M. and Mason, A.J. (1990) Requirement for activin-a and transforming growth factor-beta-1 pro-regions in homodimer assembly. *Science*, **247**, 1328–1330.

Hall, M.N., Gabay, J., Debarbouille, M. and Schwartz, M. (1982) A role for mRNA secondary structure in the control of translation initiation. *Nature*, **295**, 616–618.

High, S. and Dobberstein, B. (1992) Mechanisms that determine the transmembrane disposition of proteins. *Curr. Opin. Cell Biol.* **4**, 581–586.

van der Hoorn, R.A. (2008) Plant proteases: from phenotypes to molecular mechanisms. *Annu. Rev. Plant Biol.* **59**, 191–223.

Kinal, H., Park, C.M., Berry, J.O., Koltin, Y. and Bruenn, J.A. (1995) Processing and secretion of a virally encoded antifungal toxin in transgenic tobacco plants - evidence for a Kex2p pathway in plants. *Plant Cell*, **7**, 677–688.

Li, M.O., Wan, Y.Y., Sanjabi, S., Robertson, A.K.L. and Flavell, R.A. (2006) Transforming growth factor-beta regulation of immune responses. *Annu. Rev. Immunol.* **24**, 99–146.

Lin, S.J., Lerch, T.F., Cook, R.W., Jardetzky, T.S. and Woodruff, T.K. (2006) The structural basis of TGF-beta, bone morphogenetic protein, and activin ligand binding. *Reproduction*, **132**, 179–190.

Lorenz, R., Bernhart, S.H., Honer Zu Siederdissen, C., Tafer, H., Flamm, C., Stadler, P.F. and Hofacker, I.L. (2011) ViennaRNA Package 2.0. *Algorithms Mol. Biol.* **6**, 26.

Ludbrook, S.B., Barry, S.T., Delves, C.J. and Horgan, C.M.T. (2003) The integrin alpha(v)beta(3) is a receptor for the latency-associated peptides of transforming growth factors beta(1) and beta(3). *Biochem J.* **369**, 311–318.

Miyazono, K., Thyberg, J. and Heldin, C.H. (1992) Retention of the transforming growth factor-beta 1 precursor in the Golgi complex in a latent endoglycosidase H-sensitive form. *J. Biol. Chem.* **267**, 5668–5675.

Oida, T. and Weiner, H.L. (2010) Overexpression of TGF-beta(1) Gene Induces Cell Surface Localized Glucose-Regulated Protein 78-Associated Latency-Associated Peptide/TGF-beta. *J. Immunol.* **185**, 3529–3535.

O'Kane, S. and Ferguson, M.W.J. (1997) Transforming Growth Factor beta s and wound healing. *Int. J. Biochem. Cell Biol.* **29**, 63–78.

Oz, H.S., Ray, M., Chen, T.S. and McClain, C.J. (2004) Efficacy of a transforming growth factor 82 containing nutritional support formula in a murine model of inflammatory bowel disease. *J. Am. Coll. Nutr.* **23**, 220–226.

Schlunegger, M.P. and Grutter, M.G. (1992) An unusual feature revealed by the crystal-structure at 2.2-angstrom resolution of human transforming growth-factor-beta-2. *Nature*, **358**, 430–434.

Schouten, A., Roosien, J., van Engelen, F.A., de Jong, G.A., Borst-Vrenssen, A.W., Zilverentant, J.F., Bosch, D. *et al.* (1996) The C-terminal KDEL sequence increases the expression level of a single-chain antibody designed to be targeted to both the cytosol and the secretory pathway in transgenic tobacco. *Plant Mol. Biol.* **30**, 781–793.

Sha, X., Brunner, A.M., Purchio, A.F. and Gentry, L.E. (1989) Transforming growth-factor beta-1 - importance of glycosylation and acidic proteases for processing and secretion. *Mol. Endocrinol.* **3**, 1090–1098.

Takeuchi, M., Alard, P. and Streilein, J.W. (1998) TGF-beta promotes immune deviation by altering accessory signals of antigen-presenting cells. *J. Immunol.* **160**, 1589–1597.

Thomas, G. (2002) Furin at the cutting edge: from protein traffic to embryogenesis and disease. *Nat. Rev. Mol. Cell Biol.* **3**, 753–766.

Tuller, T., Waldman, Y.Y., Kupiec, M. and Ruppin, E. (2010) Translation efficiency is determined by both codon bias and folding energy. *Proc. Natl Acad. Sci. USA*, **107**, 3645–3650.

Veldhoen, M., Hocking, R.J., Atkins, C.J., Locksley, R.M. and Stockinger, B. (2006) TGF beta in the context of an inflammatory cytokine milieu supports de novo differentiation of IL-17-producing T cells. *Immunity*, **24**, 179–189.

Veldhoen, M., Uyttenhove, C., van Snick, J., Helmby, H., Westendorf, A., Buer, J., Martin, B. *et al.* (2008) Transforming growth factor-beta 'reprograms' the differentiation of T helper 2 cells and promotes an interleukin 9-producing subset. *Nat. Immunol.* **9**, 1341–1346.

Vey, M., Schafer, W., Berghofer, S., Klenk, H.D. and Garten, W. (1994) Maturation of the trans-Golgi network protease furin: compartmentalization of propeptide removal, substrate cleavage, and COOH-terminal truncation. *J. Cell Biol.* **127**, 1829–1842.

Voinnet, O., Rivas, S., Mestre, P. and Baulcombe, D. (2003) An enhanced transient expression system in plants based on suppression of gene silencing by the p19 protein of tomato bushy stunt virus. *Plant J.* **33**, 949–956.

Westerhof, L.B., Wilbers, R.H.P., Roosien, J., van de Velde, J., Goverse, A., Bakker, J. and Schots, A. (2012) 3D domain swapping causes extensive multimerisation of human interleukin-10 when expressed in planta. *PLoS ONE*, **7**, e46460.

Westerhof, L.B., Wilbers, R.H., van Raaij, D.R., Nguyen, D.L., Goverse, A., Henquet, M.G., Hokke, C.H. *et al.* (2014) Monomeric IgA can be produced in planta as efficient as IgG, yet receives different N-glycans. *Plant Biotechnol. J.* **12**, 1333–1342.

Ye, C.M., Dickman, M.B., Whitham, S.A., Payton, M. and Verchot, J. (2011) The unfolded protein response is triggered by a plant viral movement protein. *Plant Physiol.* **156**, 741–755.

Zou, Z.C. and Sun, P.D. (2004) Overexpression of human transforming growth factor-beta 1 using a recombinant CHO cell expression system. *Protein Expre. Purif.* **37**, 265–272.

Zou, Z.C. and Sun, P.D. (2006) An improved recombinant mammalian cell expression system for human transforming growth factor-beta 2 and -beta 3 preparations. *Protein Expre. Purif.* **50**, 9–17.

15

Codon reassignment to facilitate genetic engineering and biocontainment in the chloroplast of *Chlamydomonas reinhardtii*

Rosanna E. B. Young* and Saul Purton

Algal Research Group, Institute of Structural and Molecular Biology, University College London, London, UK

*Correspondence
email r.young@ucl.ac.uk

Summary

There is a growing interest in the use of microalgae as low-cost hosts for the synthesis of recombinant products such as therapeutic proteins and bioactive metabolites. In particular, the chloroplast, with its small, genetically tractable genome (plastome) and elaborate metabolism, represents an attractive platform for genetic engineering. In *Chlamydomonas reinhardtii*, none of the 69 protein-coding genes in the plastome uses the stop codon UGA, therefore this spare codon can be exploited as a useful synthetic biology tool. Here, we report the assignment of the codon to one for tryptophan and show that this can be used as an effective strategy for addressing a key problem in chloroplast engineering: namely, the assembly of expression cassettes in *Escherichia coli* when the gene product is toxic to the bacterium. This problem arises because the prokaryotic nature of chloroplast promoters and ribosome-binding sites used in such cassettes often results in transgene expression in *E. coli*, and is a potential issue when cloning genes for metabolic enzymes, antibacterial proteins and integral membrane proteins. We show that replacement of tryptophan codons with the spare codon (UGG→UGA) within a transgene prevents functional expression in *E. coli* and in the chloroplast, and that co-introduction of a plastidial *trnW* gene carrying a modified anticodon restores function only in the latter by allowing UGA readthrough. We demonstrate the utility of this system by expressing two genes known to be highly toxic to *E. coli* and discuss its value in providing an enhanced level of biocontainment for transplastomic microalgae.

Keywords: *Chlamydomonas reinhardtii*, chloroplast, microalgae, non-sense suppression, transfer RNA.

Introduction

The microalgal chloroplast has many advantages as a production platform for recombinant proteins and small molecules including low culturing costs, lack of toxins and ease of genetic manipulation. The presence of multiple copies of the chloroplast genome per cell and lack of gene silencing give the chloroplast an advantage over nuclear-encoded transgene expression (Bock, 2015). The green alga *Chlamydomonas reinhardtii* is the most widely used for recombinant protein expression, with products such as vaccines, immunotoxins, therapeutics and industrial enzymes (reviewed by Rasala and Mayfield (2015) and Scaife *et al.* (2015)). Chloroplasts evolved from a cyanobacterial endosymbiont (Timmis *et al.*, 2004) and many chloroplast genes in *C. reinhardtii* have retained bacterial features such as −35 and/or −10 promoter elements and 70S ribosome-binding sequences. This is the case for the promoter and 5′ untranslated region (5′ UTR) of exon 1 of *psaA*, encoding a core subunit of photosystem I. The *psaA* promoter/ 5′ UTR is often used to drive robust expression of foreign genes in the *C. reinhardtii* chloroplast (Michelet *et al.*, 2011; Specht and Mayfield, 2013; Young and Purton, 2014), but it cannot be used for proteins that are detrimental to *Escherichia coli* as they will be expressed during cloning in this host and will prevent successful production of the plasmid vector for subsequent transfer to the microalga. For example the PanDaTox database lists over 40 000 microbial genes that are

predicted to be toxic to *E. coli* based on their failure to be propagated during genome sequencing projects (Amitai and Sorek, 2012). This lack of clonability can constrain the modification or introduction of biochemical pathways in *C. reinhardtii* for metabolic engineering due to alterations in carbon or nitrogen flux in *E. coli* or the generation of toxic intermediates, and may also prevent the cloning of genes for some antibacterial enzymes or integral membrane proteins.

Transfer RNAs (tRNAs) and their cognate aminoacyl-tRNA synthetases together determine the amino acid sequence that is encoded by messenger RNA, so manipulation of these components can alter the genetic code. In the standard genetic code, 61 of the 64 RNA triplet codons are translated as amino acids whereas the remaining three (UAA, UAG and UGA) are stop signals at which release factors aid the termination of translation. The *C. reinhardtii* chloroplast genome uses this standard genetic code; however, DNA sequencing revealed that there is a strong preference for UAA as the stop codon with 65 of the 69 protein-coding genes using this codon (Maul *et al.*, 2002). The remaining four genes use UAG, and UGA is not used at all, although early genetic evidence demonstrates that it can function as a stop codon in the *C. reinhardtii* chloroplast. For example non-photosynthetic mutants were isolated in which the chloroplast *rbcL* gene, encoding the large subunit of Rubisco, contained a TGG to TAG (amber) or TGA (opal) non-sense mutation (Spreitzer *et al.*, 1985). Chemical mutagenesis of the amber mutant, followed by selection for photosynthetic competence, produced a cell line in which the wild-type *trnW*$_{CCA}$ gene and a mutated version with an amber-specific CUA anticodon coexisted as a heteroplasmic mix in the polyploid plastome, thus allowing both UGG and UAG codons to be translated as tryptophan (Yu and Spreitzer, 1992). A similar experiment using the opal mutant also produced heteroplasmic non-sense suppressors but the genetic basis of the suppression was not characterized (Spreitzer *et al.*, 1984). These results suggested that it would be possible to genetically engineer the *C. reinhardtii* chloroplast *trnW* to recognize amber, and possibly opal, codons instead of UGG.

Here, we address the challenge of cloning genes whose products are toxic to *E. coli* by exploiting the unused UGA codon to create a genetic system in which the gene of interest (GOI) is modified to carry opal mutations at one or more tryptophan codons (i.e. UGG to UGA), thereby preventing synthesis of the full-length protein in either *E. coli* or the chloroplast. Translational read-through is restored in the chloroplast, but not in *E. coli*, by combining the GOI with a plastidial *trnW* gene encoding a tRNA with a modified anticodon.

The existence of an unused codon in the *C. reinhardtii* chloroplast genetic code, together with our demonstration that it can be integrated into coding sequences and translated without the need to eliminate any plastidial release factors, provides opportunities for future genetic engineering of the microalgal chloroplast involving canonical or non-canonical amino acids. The use of a non-sense codon to interrupt the coding sequence also reduces the risk of transgenes being translated into full-length proteins were they to spread to other organisms by horizontal gene transfer, thereby providing informational containment of the transgenes.

Results

Our scheme for cloning genes that are toxic to *E. coli* into a *C. reinhardtii* chloroplast expression vector requires firstly a version of the gene with one or more TGG codons modified to TGA 'stop' codons, and secondly a synthetic tRNA gene to read the TGA codon/s as tryptophan. Strategies for introducing these two elements into *C. reinhardtii* are outlined in Figure 1. Plasmids used in this work are detailed in Table 1.

Mutation of two TGG codons to TGA codons within a transgene (*crCD*) prevents accumulation of CrCD protein in *Chlamydomonas reinhardtii*

The first set of experiments was carried out using *crCD* as a test gene. This is an *E. coli* cytosine deaminase gene optimized for the *C. reinhardtii* chloroplast as a negative selectable marker (Young and Purton, 2014) and was chosen as a test gene due to the stability of CrCD protein in the chloroplast, ease of detection by Western blotting (via an added HA epitope) and clear phenotype of sensitivity to 5-fluorocytosine. CrCD is not toxic when expressed in *E. coli*, allowing appropriate control strains to be used.

A chloroplast expression vector containing *crCD* under the control of the *C. reinhardtii psaA* exon 1 promoter (plasmid pCD, previously called pRY127d; (Young and Purton, 2014)) was modified so that two of the TGG codons, encoding tryptophan, were altered to TGA (Table S1). The resulting plasmid, pCD**, was used to transform *C. reinhardtii* TN72 (a non-photosynthetic *psbH* mutant). The flanking region of pCD** contains an intact copy of *psbH*, so homologous recombination into the chloroplast genome restores phototrophic growth and allows the selection of transformants on minimal medium in the light. Transgene

Strategy 1: introduce *GOI*** and *trnW*$_{UCA}$ into *C. reinhardtii* using separate plasmids

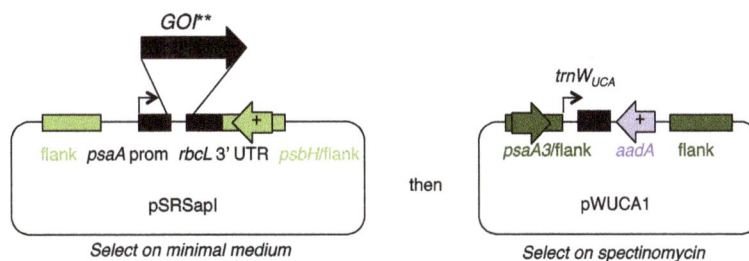

then

Strategy 2: introduce *GOI*** and *trnW*$_{UCA}$ into *C. reinhardtii* using one plasmid (*E. coli* cannot use *trnW*$_{UCA}$)

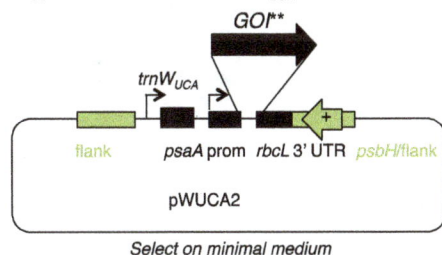

Figure 1 Strategy to clone genes into a chloroplast expression vector whilst preventing their expression in *Escherichia coli*. The gene of interest (GOI) is redesigned such that one or more tryptophan (TGG) codons are altered to TGA, indicated by asterisks. These changes can be integrated into the codon-optimized gene design prior to ordering the synthetic gene and integrating it into the *Chlamydomonas reinhardtii* chloroplast genome. A tRNA gene based on the *C. reinhardtii* plastidial *trnW*, but with the anticodon sequence altered to recognize UGA, is also introduced (*trnW*$_{UCA}$). This enables readthrough of the GOI in *C. reinhardtii*. Flanking regions amplified from chloroplast DNA allow targeted integration of the constructs into the chloroplast genome by homologous recombination; the target site is a neutral region downstream of either *psbH* or *psaA* exon 3, depending on the construct. The *psbH* gene can be used for selection in a *psbH* mutant recipient strain.

Table 1 Plasmids used in this work. All selection plates were incubated in the light except where pPsaA* was used. All synthetic protein-coding genes were codon-optimized for the *Chlamydomonas reinhardtii* chloroplast and encode a C-terminal HA tag

Plasmid	Synthetic gene	Protein encoded by gene	Expected protein size incl. HA tag (kDa)	Integration site in *C. reinhardtii* chloroplast	Selection
pCD	*crCD*	Cytosine deaminase (Young and Purton, 2014)	49	Downstream of *psbH*	Minimal medium (*psbH* restored)
pCD**	*crCD* with 2 internal TGA stop codons	Cytosine deaminase	49 (if pWUCA1 also present)	Downstream of *psbH*	Minimal medium (*psbH* restored)
pWUCA1	*trnW_{UCA}*	n/a	n/a	Downstream of *psaA* exon 3	Spectinomycin (*aadA* cassette)
pWUCA2	*trnW_{UCA}*	n/a	n/a	Downstream of *psbH*	Minimal medium (*psbH* restored or *psaA** translated, depending on recipient cell line)
pWUCA2-CD**	*crCD* with 2 internal TGA stop codons + *trnW_{UCA}*	Cytosine deaminase	49	Plasmid not used in *C. reinhardtii*	Plasmid not used in *C. reinhardtii*
pPsaA*	None; introduces TGG to TGA mutation at W693 in native *psaA* gene	Core protein of photosystem I	n/a	*psaA* exon 3 region	Spectinomycin (*aadA* cassette); incubate in dark
pSty	*Salmonella* Typhimurium phage gene SPN9CC_0043 with 1 internal TGA stop codon + *trnW_{UCA}*	Endolysin (lyses *Escherichia coli*; Lim et al., 2014)	19	downstream of *psbH*	Minimal medium (*psbH* restored)
pSde	*Shewanella denitrificans* Sden_1266 with 2 internal TGA stop codons + *trnW_{UCA}*	Hypothetical protein (very toxic to *E. coli*; Kimelman et al., 2012)	42	Downstream of *psbH*	Minimal medium (*psbH* restored)

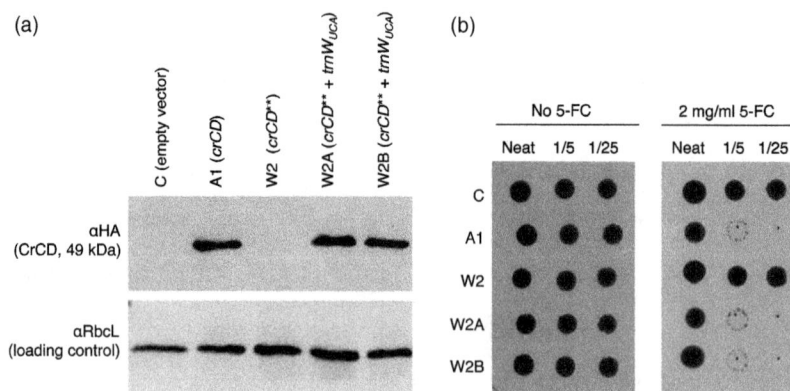

Figure 2 Introduction of a synthetic *trnW_{UCA}* gene into the *Chlamydomonas reinhardtii* chloroplast genome allows the expression of full-length, active CrCD protein from the *crCD** gene. (a) Western analysis. Equalized cell lysates were subjected to SDS-PAGE and two identical blots were probed with antibodies as indicated on the left. (b) Growth tests demonstrating that cell lines A1, W2A and W2B contain active CrCD protein that leads to cell death on media containing 5-fluorocytosine (5-FC). *C. reinhardtii* cultures were adjusted to equal optical densities and spotted onto TAP agar containing no drug (left panel) or 5-FC (right panel), with serial fivefold dilutions left to right. Plates were incubated under 50 µE/m²/s light for 10 days.

integration and homoplasmy across the approximately 80 copies of the chloroplast genome per cell was confirmed by PCR (Figure S1a and Table S2). As expected, the introduction of TGA codons into the *crCD* gene prevented the accumulation of full length CrCD protein as detected by immunoblotting with an antibody against the C-terminal HA tag (cell line W2 in Figure 2a). This suggests that no tRNA in the *C. reinhardtii* chloroplast is able to recognize the UGA codon and insert an amino acid at the corresponding position in the peptide chain, so the chain terminates prematurely.

Introduction of a synthetic *trnW*$_{UCA}$ gene into the chloroplast genome allows the accumulation of full length, active CrCD protein and does not affect growth

There is a single codon for tryptophan in the standard genetic code (UGG) and a single copy of the tryptophan tRNA gene, *trnW*$_{CCA}$, in the *C. reinhardtii* chloroplast genome (Cognat *et al.*, 2013; Maul *et al.*, 2002); see the PlantRNA database at http://plantrna.ibmp.cnrs.fr. A tRNA with a 5′-UCA-3′ anticodon would be expected to recognize UGA codons within the mRNA, such as those transcribed from the TGA codons that had been inserted into *crCD*. To test this, a version of the *C. reinhardtii* chloroplast *trnW* gene with 100 bp of its flanking sequence each side, which included its promoter, was synthesized with a mutated anticodon (CCA to UCA) and cloned into a chloroplast targeting vector to make plasmid pWUCA1 (Figure 1). The vector inserts the *trnW*$_{UCA}$ gene downstream of *psaA* exon 3 in the chloroplast genome and contains an *aadA* spectinomycin resistance cassette. Homoplasmic transformants were recovered following selection on medium containing spectinomycin (Figure S1b and Table S2). The transformation of *C. reinhardtii* W2 with plasmid pWUCA1 was found to elicit the accumulation of full-length CrCD protein (Figure 2a, cell lines W2A and W2B), demonstrating that the synthetic *trnW*$_{UCA}$ gene is expressed and indeed recognizes the UGA codon. This also confirms that the 100 bp flanking sequences included around *trnW*$_{UCA}$ were sufficient for its transcription and any subsequent 5′ and 3′ processing by RNaseP and other RNases. There was no observable difference in CrCD protein yield between W2A and a control *C. reinhardtii* cell line that had been transformed with pCD (Figure S2).

Cytosine deaminase (CrCD) normally converts cytosine to uracil but can also convert the synthetic compound 5-fluorocytosine to a toxic product, 5-fluorouracil. The activity of the CrCD protein made using the synthetic tRNA was demonstrated by the reduced growth of cell lines W2A and W2B on medium containing 5-fluorocytosine (Figure 2b).

The expression of a synthetic tRNA that reassigns a stop codon sometimes slows the growth of the organism, presumably due to the lack of proper termination of endogenous proteins (Wang *et al.*, 2014). As UGA is not used as a stop or sense codon in the *C. reinhardtii* chloroplast, we did not expect to see a growth

defect in this case. Indeed, we found that the synthetic *trnW*$_{UCA}$ gene expressed in the chloroplast had no detrimental effect on the growth rate of *C. reinhardtii* when tested under mixotrophic or phototrophic conditions at 25 °C (Figure S3).

The tRNA and gene of interest can be introduced into the chloroplast in a single homologous recombination step

The production of *C. reinhardtii* cell lines that express transgenes using the synthetic tRNA could be streamlined if both the transgene and tRNA gene were combined into one vector. This would require a single algal transformation step and would minimize the use of drug selection cassettes, especially if the restoration of *psbH* (i.e. growth on minimal medium) was used as the selection method. However, for this strategy to be useful for transgene containment and cloning toxic genes, *E. coli* must be unable to use the synthetic tRNA to translate the foreign protein. This was shown to be the case by inserting the *trnW*$_{UCA}$ gene and its 100 bp flanks upstream of the *crCD* expression cassette in plasmid pCD**, generating pWUCA2-CD**. No CrCD protein was detected in *E. coli* pWUCA2-CD** lysates by immunoblotting with an anti-HA antibody (Figure 3a), indicating that *E. coli* cannot use the synthetic tRNA to read through the UGA codons within the *crCD*** mRNA. This is more likely to be due to a lack of tRNA function or to competition with other factors than to a lack of tRNA transcription, as the *trnW*$_{UCA}$ flank contains an exact bacterial consensus promoter (see Discussion and Appendix S1). The presence of a *trnW*$_{UCA}$ plasmid did not affect the growth rate of *E. coli* (Figure 3b).

A new *trnW*$_{UCA}$ chloroplast expression vector, pWUCA2, was then constructed into which a transgene can be cloned between the *psaA* exon 1 promoter/5′ UTR and the *rbcL* 3′ UTR using SapI and SphI restriction enzymes. The pWUCA2 plasmid carries the *trnW*$_{UCA}$ gene upstream of this expression cassette (see Strategy 2 in Figure 1).

The *trnW*$_{UCA}$ gene rescues the mutation of an essential tryptophan codon to TGA in *psaA*

Tryptophan is the largest canonical amino acid and is the only one to carry an indole side-chain. The double ring structure of its indole moiety is often involved in stacking interactions that are important for substrate binding and catalysis in some enzymes

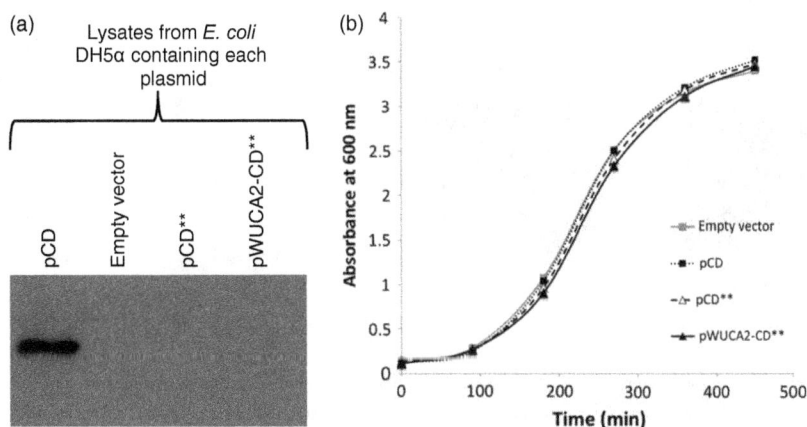

Figure 3 In *Escherichia coli*, the synthetic *Chlamydomonas reinhardtii* chloroplast *trnW*$_{UCA}$ gene does not permit full length CrCD protein expression from the *crCD*** gene. (a) Western analysis using equalized lysates of *E. coli* DH5α carrying four different plasmids. The blot was probed with an αHA antibody to detect HA-tagged CrCD protein. (b) Growth curve of *E. coli* DH5α carrying the four different plasmids.

(Nakamura et al., 2013; Zhang et al., 2004a). To demonstrate that the synthetic trnW$_{UCA}$ adds tryptophan rather than any other amino acid to the growing peptide chain, we carried out experiments on the C. reinhardtii chloroplast psaA gene, which encodes a core component of photosystem I (PSI).

The tryptophan at position 693 of PsaA is π-stacked through the indole moiety with the bound phylloquinone cofactor (Boudreaux et al., 2001; Jordan et al., 2001). Chlamydomonas reinhardtii strains in which W693 has been substituted for another amino acid have a functional PSI complex but are highly sensitive to oxygen during phototrophic growth, possibly from the formation of free radical species (Purton et al., 2001).

We substituted the W693 TGG codon in psaA with TGA by homologous recombination (plasmid pPsaA* in Table 1), using an aadA spectinomycin resistance cassette for the selection of C. reinhardtii transformants. Homoplasmic integration of aadA downstream of psaA exon 3 was confirmed by PCR (Figure 4a), and the introduction of the non-sense codon into psaA was confirmed by DNA sequencing (Figure S4). The resulting strain, cw15 + pPsaA*, shows the loss of phototrophy and the light sensitivity typical of PSI-deficient mutants (Figure 4c).

The cw15 + pPsaA* cell line was transformed with pWUCA2, plating on minimal medium in the light under aerobic conditions. Although the intact psbH gene in pWUCA2 can be used for selection in a psbH mutant line such as TN72, the recipient cell line used here already has an intact psbH so this gene was merely being used as a homologous flanking region for integration of trnW$_{UCA}$ into the chloroplast genome. Instead, selection was directly for trnW$_{UCA}$ to translate full-length, W693-containing PsaA protein for restored photosynthesis, effectively using the tRNA gene as a positive selectable marker. Four out of seven cw15 + pPsaA* + pWUCA2 colonies checked by PCR were homoplasmic for trnW$_{UCA}$ after a single round of streaking out on minimal medium, demonstrating that selection was successful. We continued with transformants 1 and 2 (Figure 4b and Table S2); DNA sequencing of part of psaA3 confirmed that they were not W693 TGA→TGG revertants (Figure S4). These cell lines can grow on minimal medium in the presence of oxygen

(Figure 4c), contrasting with the oxygen-sensitive phenotype of PsaA mutants that have other amino acids at position 693 (Purton et al., 2001).

The ability of trnW$_{UCA}$ to complement the psaA* mutation indicates that the single nucleotide change in the anticodon loop from the natural tryptophan tRNA, trnW$_{CCA}$, to the synthetic tRNA, trnW$_{UCA}$, does not prevent the tryptophanyl tRNA synthetase from recognizing this as a tRNA to be charged with tryptophan.

Genes whose products are toxic to Escherichia coli can be cloned using the synthetic tRNA system

The trnW$_{UCA}$ cloning scheme using the combined vector (Figure 1) was tested using two genes whose products are known to be toxic to E. coli. The SPN9CC endolysin, from a Salmonella Typhimurium bacteriophage, has previously been shown to lyse E. coli (Lim et al., 2014). A codon-optimized version of this endolysin gene was designed in which a single TGG codon was altered to TGA. This was cloned into pWUCA2 to make plasmid pSty. The second gene was a Shewanella denitrificans ORF identified from the PanDaTox database as unclonable in E. coli during genome sequencing. Further work by the compilers of the database showed that the ORF can be cloned under an inducible promoter in E. coli but that cells die upon induction of expression (Kimelman et al., 2012). A codon-optimized version of this ORF was designed with two TGG to TGA mutations, and cloned into pWUCA2 to give plasmid pSde. DNA sequences for the two genes are given in Appendix S1.

Chlamydomonas reinhardtii TN72 was transformed with pSty and pSde separately; homoplasmic integration of the transgenes and trnW$_{UCA}$ was demonstrated by PCR (Figure 5a and Table S2). The accumulation of the SPN9CC endolysin and the S. denitrificans protein was demonstrated by immunoblotting with an anti-HA antibody (Figure 5b).

Stop codon usage in other microalgae

The selection of microalgal species for industrial biotechnology depends on many factors including the ease of genetic manip-

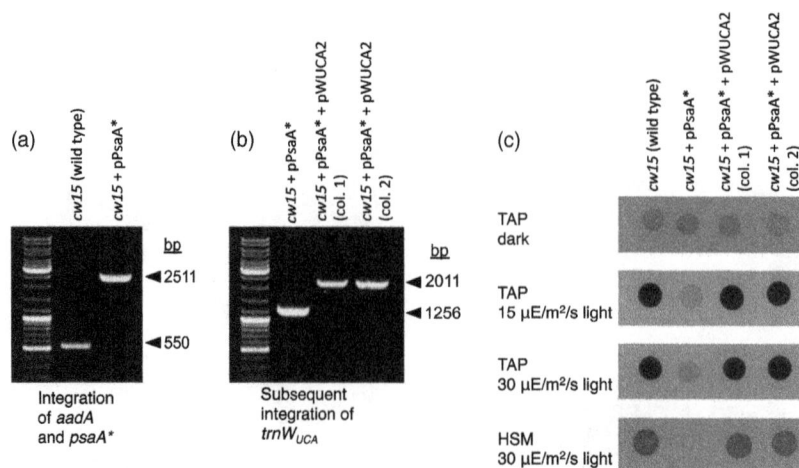

Figure 4 The mutation of a tryptophan codon that is essential for PsaA function in Chlamydomonas reinhardtii can be complemented by trnW$_{UCA}$. (a) PCR on C. reinhardtii cell lines showing the homoplasmic integration of aadA downstream of psaA in the chloroplast genome, with the concurrent mutation of the psaA W693 codon to TGA (confirmed by sequencing). (b) PCR showing the homoplasmic integration of pWUCA2 into the C. reinhardtii pPsaA* cell line. Selection was directly for trnW$_{UCA}$ to restore phototrophic growth by allowing the translation of psaA*. (c) Growth properties of the cell lines on media containing acetate (TAP) or minimal medium (HSM). Chlamydomonas reinhardtii cultures were adjusted to equal optical densities, spotted onto agar and incubated at 25 °C in the presence of oxygen.

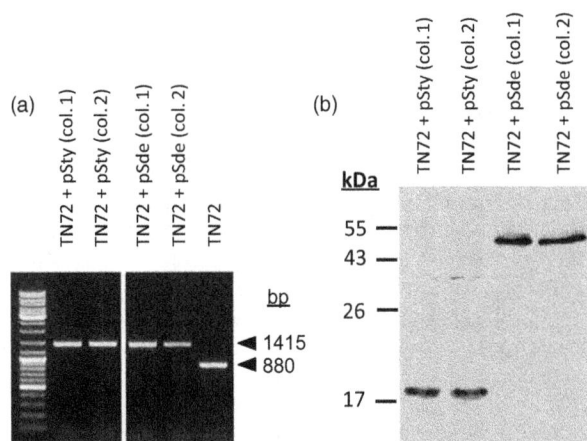

Figure 5 Use of the *trnW*_{UCA} system to clone and express genes in *C. reinhardtii* whose products are toxic to *Escherichia coli*. (a) PCR demonstrating homoplasmic integration of the transgenes into *C. reinhardtii* cell line TN72. (b) Western blot with anti-HA antibody, demonstrating accumulation of the foreign proteins.

ulation, growth rates, media costs, harvesting costs and, for some products, the lipid content. We surveyed the stop codon usage of 12 microalgal species for which chloroplast genome sequences were available to assess whether a similar approach to genetic code manipulation might be possible in species other than *C. reinhardtii* (Table S3). The chloroplast genomes ranged from 72 to 269 kb in size and were predicted to contain between 61 and 224 protein-coding genes according to the annotations in the NCBI database. For each of the 12 genomes, UAA was the most frequently used stop codon and UGA was the least frequently used. As noted by Robbens *et al.* (2007), the chloroplast genome of *Ostreococcus tauri* does not use the UGA codon. *Lobosphaera* (*Parietochloris*) *incisa* also lacks UGA stop codons, and only a single putative gene (encoding a 45 amino acid hypothetical protein) has this stop codon in *Chlorella sorokiniana* (Table S3). However, the chloroplast transformation of these three species has yet to be reported. *Dunaliella salina*, *Euglena gracilis* and *Phaeodactylum tricornutum* each use the UGA stop codon three to four times in the chloroplast genome, so may be amenable to emancipation or dual use of this codon. Species in each of these three genera have been shown to have transformable chloroplast genomes (Doetsch *et al.*, 2001; Georgianna *et al.*, 2013; Xie *et al.*, 2014).

Discussion

The UGA codon is one of three triplet nucleotide codons used as stop signals in the standard genetic code. However, some species are known to have reassigned UGA to a sense codon. In the *C. reinhardtii* nucleus, UGA encodes stop or selenocysteine depending on the RNA context (Novoselov *et al.*, 2002; Rao *et al.*, 2003). Gracilibacteria translate this codon as glycine (Rinke *et al.*, 2013), whilst UGA reassignments to tryptophan (known as Genetic Code 4) have been observed in mycoplasmas and their phages, in the bacterium *Candidatus* Hodgkinia cicadicola, and in some mitochondria (Ivanova *et al.*, 2014; McCutcheon *et al.*, 2009). The mitochondrion of the green alga *Pedinomonas minor* decodes both UGG and UGA as tryptophan using a single trnW (Turmel *et al.*, 1999). These examples set a precedent for the reassignment of UGA in the *C. reinhardtii* chloroplast.

The recognition of more than one codon sequence by an anticodon ('wobble') involves post-transcriptional modifications of the tRNA by one or more enzymes (Crick, 1966; El Yacoubi *et al.*, 2012). Unmodified uridine in the wobble position of an anticodon (U₃₄) binds only to adenosine in the third position of the codon (Agris *et al.*, 2007); in the case of the synthetic *C. reinhardtii* chloroplast trnW_{UCA} this would mean that only the codon UGA would be recognized, as desired. U₃₄ modifications in naturally occurring trnW_{UCA} include 5-taurinomethyluridine in *Bos taurus* mitochondria and 5-carboxymethylaminomethy-luridine in *Tetrahymena thermophila* mitochondria (see Modomics database at http://modomics.genesilico.pl). U₃₄ modifications can allow the anticodon to recognize certain other bases in the codon's wobble position, but the relationship is rather complex and not fully understood (Agris *et al.*, 2007). We have not determined whether U₃₄ in the synthetic tRNA is modified, but two findings indirectly suggest that wobble may not occur. First, the heteroplasmic nature of *C. reinhardtii* mutants that suppress a TGG→TGA mutation in *rbcL* (Spreitzer *et al.*, 1985) suggests that a homoplasmic CCA→UCA mutation in the endogenous trnW anticodon would be lethal due to the prevention of UGG translation. Second, pleiotropic effects might be expected if our synthetic tRNA could insert tryptophan in native proteins in response to UGU or UGC cysteine codons, but no change in growth rate was observed in *C. reinhardtii* cell lines containing the tRNA.

Peptide chain release factors specifically recognize stop codons and can antagonize attempts to reassign these to sense codons. However, we found no observable difference in the level of CrCD protein accumulation between *C. reinhardtii* cell lines with an intact *crCD* gene or *crCD** + *trnW*_{UCA} (Figure S2). This suggests that either there is no release factor that recognizes UGA codons in the *C. reinhardtii* chloroplast or such a factor is outcompeted successfully by trnW_{UCA}. In eubacteria and the *Arabidopsis thaliana* chloroplast, release factor PrfA (RF1) recognizes UAA and UAG codons whilst PrfB (RF2) recognizes UAA and UGA; both trigger peptidyl-tRNA hydrolysis. Since neither the chloroplast nor mitochondria in *C. reinhardtii* use UGA as a stop (or sense) codon, PrfB should not be required, but nevertheless a *prfB* orthologue (locus Cre01.g010864) is present in the nuclear genome. It is not clear from signal peptide analysis whether the resulting PrfB protein would be targeted to the chloroplast, mitochondria or both; the cytosol has its own eukaryotic release factor system. The motifs required for PrfB function and specificity are well studied in other organisms (Frolova *et al.*, 1999; Ito *et al.*, 2000; Johnson *et al.*, 2011; Wilson *et al.*, 2000). The stop codon recognition motif SPF, peptide release motif GGQ and essential Ser246 residue are all intact in the *C. reinhardtii* PrfB orthologue, although the predicted N-terminus is dissimilar to those of the *E. coli* and *A. thaliana* PrfB proteins. *Chlamydomonas reinhardtii* may eventually lose the *prfB* gene, as has happened in *Candidatus* Hodgkinia cicadicola (McCutcheon *et al.*, 2009). Alternatively it may be retained if it is required for efficient UAA termination or has been adapted and recruited to stabilize particular RNA transcripts, as is the case for PrfB3 in *A. thaliana* (Stoppel *et al.*, 2011).

The synthetic tRNA did not allow detectable readthrough of UGA codons in *E. coli*. This observation allows the gene of interest and *trnW*_{UCA} to be combined into a single vector and also reduces the risk that transgenes would be translatable if they spread to other organisms by horizontal gene transfer. The two main factors likely to contribute to this lack of readthrough in

E. coli are competition with the release factor PrfB and differences in the mechanism of specific tRNA function between *E. coli* and the *C. reinhardtii* chloroplast. Regarding the latter factor, the strongest recognition elements for the aminoacylation of *E. coli* trnW with tryptophan are the discriminator base G73, which is conserved across prokaryotic (but not eukaryotic) trnW, and the anticodon CCA (Himeno *et al.*, 1991; Hughes and Ellington, 2010). The *C. reinhardtii* chloroplast trnW and its synthetic trnW$_{UCA}$ counterpart do contain the G73 base. However, since the synthetic trnW$_{UCA}$ necessarily carries a mutated anticodon, the *E. coli* tryptophanyl-tRNA synthetase may not recognize it to be charged with tryptophan. Differences in the synthesis of the tRNA 3′ acceptor stem may also contribute to a lack of tRNA transferability between chloroplasts and *E. coli*: this trinucleotide sequence (also CCA) is included in tRNA genes in *E. coli*, whereas in the *C. reinhardtii* chloroplast and cyanobacteria it is added post-transcriptionally by a nucleotidyltransferase enzyme (Schmidt and Subramanian, 1993; Xiong and Steitz, 2006).

An alternative strategy for cloning genes that are toxic to *E. coli* into chloroplast expression vectors was demonstrated by Oey *et al.* (2009), who inserted bacterial transcription termination signals between a selectable marker gene and a downstream endolysin gene. After cloning in *E. coli* and transformation of tobacco, the selectable marker gene and termination signals were subsequently removed by Cre-*loxP* recombination to enhance endolysin expression. Whilst this strategy worked well for the gene tested and is a good compromise for genetic systems that lack a spare codon, there was a low level of leaky expression in *E. coli* so it would not be suitable for proteins that are highly toxic to this bacterium. In contrast, with the trnW$_{UCA}$ strategy we did not detect any CrCD protein in the *E. coli* pWUCA2-CD** cell line. If necessary, any leaky translation could be reduced further by increasing the number of TGA codons in the gene, with the maximum being the number of tryptophan residues in the protein.

Another approach to reduce the expression of bactericidal proteins during cloning is to culture the *E. coli* at a lower temperature, which increases plasmid supercoiling and reduces transcription of the transgene, at least when plant *rrn* and *psbA* promoters are used (Madesis *et al.*, 2010). Due to the differential sensitivity of plastid promoters to topology (Stirdivant *et al.*, 1985), the efficacy of this strategy is likely to vary between chloroplast expression vectors. Antimicrobial peptides, which would be lethal to *E. coli* if expressed on their own, are often expressed as fusion proteins to temporarily mask their function when this bacterium is used as an expression platform or cloning host (Lee *et al.*, 2011; Li, 2009). This is an effective strategy but requires the extra processing step of protease cleavage, adding to the time and cost of protein production and making it inappropriate for the manipulation of metabolic pathways.

Genetic code manipulation can be used to introduce non-canonical amino acids into certain positions within proteins *in vivo*, with applications including altering enzyme properties, enabling chemical modifications and providing trophic biocontainment by making an organism dependent on unnatural amino acids (Ravikumar and Liu, 2015). For example Zhang *et al.* (2004b) adapted the *Bacillus subtilis* tryptophanyl tRNA and its cognate synthetase to incorporate 5-hydroxytryptophan in response to the opal stop codon UGA in mammalian cells. This required altering the tRNA anticodon sequence and a single amino acid in the active site of the synthetase.

5-hydroxytryptophan has unique spectral properties and can be used to study protein structure and function. As long as non-canonical amino acids can be taken up into the chloroplast from the growth medium, the spare UGA codon in the *C. reinhardtii* chloroplast genetic code could be used to encode such an amino acid. According to the Codon Usage Database (www.kazusa.or.jp) there are no spare codons in the plastomes of the model higher plants *Arabidopsis thaliana*, *Nicotiana tabacum* or *Zea mays*, but the reassignment of existing codons such as UGA or the use of quadruplet codons may be possible, if less efficient. Indeed, non-canonical amino acids have been engineered into bacteria, yeast and mammalian cells despite the lack of spare codons in these organisms (Niu *et al.*, 2013; Wang and Wang, 2012).

A typical *E. coli* genome contains 2765 TAA (ochre), 321 TAG (amber) and 1249 TGA (opal) stop codons (www.kazusa.or.jp, Codon Usage Database). Amber is the rarest codon in *E. coli* and has been successfully assigned to encode non-canonical amino acids using transgenic orthogonal aminoacyl-tRNA synthetase/ tRNA pairs. In standard synthetic *E. coli* amber suppressor lines, UAG is still used for translation termination in many endogenous genes, and recent studies show that the cells evolve to counteract amber suppression by inserting transposons into the new aminoacyl-tRNA synthetase gene and decreasing plasmid copy number (Wang *et al.*, 2014). Growth rates are also decreased. These issues could hinder the development of stable, fast-growing *E. coli* amber suppressor lines for industrial use. The generation of an *E. coli* strain with a truly emancipated amber codon for genetic engineering purposes involved the replacement of all 321 TAG stop codons in the genome with TAA, then the deletion of *prfA* encoding release factor 1, to prevent competition with the transgenic tRNA; unfortunately this strain has a 60% increased doubling time (Johnson *et al.*, 2012; Lajoie *et al.*, 2013).

In contrast, the *C. reinhardtii* chloroplast genome uses ochre, amber and opal stop codons 65, four and zero times respectively. This reflects the considerably smaller gene content of the chloroplast genome: most proteins in *C. reinhardtii* are encoded by the nuclear genome, including many chloroplast-targeted proteins such as some components of the photosynthetic machinery (Harris *et al.*, 2009). The genetic isolation of the plastidial translation system should enable its manipulation independently of the nuclear and mitochondrial genomes. Although UGA is the only codon that is completely absent from protein-coding genes in the *C. reinhardtii* chloroplast genome, several other codons are rarely used and are potential targets for reassignment if several different non-canonical amino acids were required in a single cell line. As well as the low number of amber (UAG) codons mentioned above, there are only six instances of CGG and 14 of CUC. The finding that the *C. reinhardtii* chloroplast UGA codon can be efficiently assigned simply with the addition of a modified tRNA gene provides a starting point for more advanced genetic code manipulation in microalgae.

Experimental procedures

Chlamydomonas reinhardtii strains and growth conditions

For experiments using pPsaA*, the cell wall-deficient strain *C. reinhardtii* cw15 was used as the recipient. For all other experiments, the recipient strain was *C. reinhardtii* TN72 (Young and Purton, 2014), which is a *cw15 psbH*-deletion mutant. Cell lines were maintained on Tris-acetate phosphate (TAP) plates with 2% agar (Harris *et al.*, 2009) and were cultured for growth tests and Western analysis in flasks containing 20 mL TAP, shaking at 120 rpm and 25 °C. Where required, optical density was measured at 750 nm using a spectrophotometer.

Plasmid construction

A synthetic *trnW*$_{UCA}$ gene was designed by taking the *C. reinhardtii* chloroplast *trnW* gene sequence with 100 nt flanking DNA on each side (i.e. Genbank accession number BK000554.2, position 17481 to 17207; Appendix S1), altering the anticodon from CCA to TCA, and adding MluI restriction sites at both ends of the fragment for cloning. The DNA was synthesized as a linear fragment by Integrated DNA Technologies (Coralville, IA, USA) and cloned into the MluI site in pBev1 (Hallahan *et al.*, 1995) to make pWUCA1, or into the MluI site in pSRSapI (Young and Purton, 2014) to make pWUCA2. The sequences of pWUCA1 and pWUCA2 are given in Appendix S1. Restriction enzymes were purchased from New England Biolabs (Ipswich, MA). Plasmids were cloned by the transformation of chemically competent *E. coli* DH5α using ampicillin selection (Sambrook and Russell, 2001) and extracted by alkaline lysis (Sambrook and Russell, 2001) or with a QIAfilter Plasmid Midi kit (Qiagen, Venlo, The Netherlands).

To make the pCD** construct, two TGG→TGA mutations were introduced into plasmid pCD/pRY127d (Young and Purton, 2014) by one-step isothermal assembly using three PCR products (Gibson *et al.*, 2009); primers are listed in Table S1 and DNA was amplified using Phusion High-Fidelity DNA Polymerase (Thermo Scientific, Waltham, MA) according to the manufacturer's instructions. The mutations correspond to tryptophans W21 and W147 of the 436 aa CrCD protein. This is the first and sixth of the seven tryptophan residues in CrCD.

To make the pPsaA* construct, a TGG→TGA mutation at amino acid position W693 was introduced into the *psaA* exon 3 ORF in the plasmid pBev1 (Hallahan *et al.*, 1995) by one-step isothermal assembly of a single PCR product whose ends overlapped by 22 bp; primer sequences are given in Table S1. The DNA sequence of pPsaA* is given in Appendix S1.

Modified versions of the *Salmonella* Typhimurium bacteriophage SPN9CC endolysin and *Shewanella denitrificans* Sden_1266 genes were synthesized by Eurofins Genomics (Ebersberg, Germany) and Integrated DNA Technologies, respectively; see Appendix S1. They were each cloned into pWUCA2 using SapI and SphI restriction enzymes, placing them under a *psaA* exon 1 promoter. The resulting plasmids (pSty and pSde) are described in Table 1.

Chlamydomonas reinhardtii transformation

Transformation was carried out using the glass bead vortex method as described in Young and Purton (2014), with selection on high-salt minimal medium for constructs that restore *psbH* or PsaA accumulation and selection on TAP + 100 μg/mL spectinomycin for constructs containing an *aadA* spectinomycin resistance cassette. Colonies were checked for homoplasmy of the insertion by PCR using the primers shown in Table S2 and Phusion Polymerase (see above). PCR products were analysed on 1% agarose gels alongside GeneRuler DNA Ladder Mix (Thermo Scientific).

Western blot analysis

10 mL mid-log phase *C. reinhardtii* cultures grown under 90 μE/m²/s light were harvested for Western blot analysis of proteins. Preparation of lysates, SDS-PAGE gels, blotting onto a nitrocellulose membrane and incubation in the primary αHA or αRbcL antibody were performed as described previously (Young and Purton, 2014) except that αHA was prepared in TBS with 0.1% Tween (TBS-T) and 0.5% milk. Blots were then incubated for 1 h in the secondary antibody, goat αrabbit Dylight 800 (Thermo Scientific) diluted 1:25 000 in TBS-T and 0.5% milk, washed in TBS-T and imaged with an Odyssey Fc Imaging System (LI-COR, Lincoln, NE) at 800 nm.

For the *E. coli* Western blot, strains were grown in LB with 100 μg/mL ampicillin overnight at 37 °C. Optical densities were measured at 600 nm then cultures were pelleted and resuspended in sample buffer to equal densities as described in the Mini-PROTEAN Tetra Cell manual (Bio-Rad, Hercules, CA). The protocol was then continued exactly as for the *C. reinhardtii* Western blots described in Young and Purton (2014). Briefly, samples were boiled, loaded onto a 15% acrylamide gel, blotted onto a nitrocellulose membrane and probed with αHA primary antibody and ECL αrabbit IgG HRP-linked secondary antibody, with detection via chemiluminescence.

5-fluorocytosine sensitivity test

Liquid *C. reinhardtii* cultures grown in TAP medium for 48 h were adjusted to an optical density of 0.4 at 750 nm. 5 μL was spotted onto TAP plates containing 2% agar and either no drug or 2 mg/mL 5-fluorocytosine (Sigma-Aldrich, St. Louis, MO, USA). Plates were incubated under 50 μE/m²/s light at 25 °C for 10 days.

Escherichia coli growth curve

5 mL LB broths containing 100 μg/mL ampicillin were inoculated with overnight *E. coli* DH5α cultures containing each plasmid so that the starting absorbance at 600 nm was 0.1. Cultures were incubated at 37 °C with shaking, and the absorbance at 600 nm was measured every 90 min using a spectrophotometer.

Acknowledgements

This work was funded by the UK Biotechnology and Biological Sciences Research Council, grant BB/I007660/1. We thank UCL for covering the cost of open access publication. The authors have no conflicts of interest to declare.

References

Agris, P.F., Vendeix, F.A. and Graham, W.D. (2007) tRNA's wobble decoding of the genome: 40 years of modification. *J. Mol. Biol.* **366**, 1–13.

Amitai, G. and Sorek, R. (2012) PanDaTox: a tool for accelerated metabolic engineering. *Bioengineered* **3**, 218–221.

Bock, R. (2015) Engineering plastid genomes: methods, tools, and applications in basic research and biotechnology. *Annu. Rev. Plant Biol.* **66**, 211–241.

Boudreaux, B., MacMillan, F., Teutloff, C., Agalarov, R., Gu, F., Grimaldi, S., Bittl, R. *et al.* (2001) Mutations in both sides of the photosystem I reaction center identify the phylloquinone observed by electron paramagnetic resonance spectroscopy. *J. Biol. Chem.* **276**, 37299–37306.

Cognat, V., Pawlak, G., Duchene, A.M., Daujat, M., Gigant, A., Salinas, T., Michaud, M. *et al.* (2013) PlantRNA, a database for tRNAs of photosynthetic eukaryotes. *Nucleic Acids Res.* **41**, D273–D279.

Crick, F.H. (1966) Codon-anticodon pairing: the wobble hypothesis. *J. Mol. Biol.* **19**, 548–555.

Doetsch, N.A., Favreau, M.R., Kuscuoglu, N., Thompson, M.D. and Hallick, R.B. (2001) Chloroplast transformation in *Euglena gracilis*: splicing of a group III twintron transcribed from a transgenic *psbK* operon. *Curr. Genet.* **39**, 49–60.

El Yacoubi, B., Bailly, M. and de Crecy-Lagard, V. (2012) Biosynthesis and function of posttranscriptional modifications of transfer RNAs. *Annu. Rev. Genet.* **46**, 69–95.

Frolova, L.Y., Tsivkovskii, R.Y., Sivolobova, G.F., Oparina, N.Y., Serpinsky, O.I., Blinov, V.M., Tatkov, S.I. *et al.* (1999) Mutations in the highly conserved GGQ motif of class 1 polypeptide release factors abolish ability of human eRF1 to trigger peptidyl-tRNA hydrolysis. *RNA*, **5**, 1014–1020.

Georgianna, D.R., Hannon, M.J., Marcuschi, M., Wu, S., Botsch, K., Lewis, A.J., Hyun, J. *et al.* (2013) Production of recombinant enzymes in the marine alga *Dunaliella tertiolecta. Algal Res.* **2**, 2–9.

Gibson, D.G., Young, L., Chuang, R.Y., Venter, J.C., Hutchison, C.A. 3rd and Smith, H.O. (2009) Enzymatic assembly of DNA molecules up to several hundred kilobases. *Nat. Methods* **6**, 343–345.

Hallahan, B.J., Purton, S., Ivison, A., Wright, D. and Evans, M.C. (1995) Analysis of the proposed Fe-SX binding region of Photosystem 1 by site directed mutation of PsaA in *Chlamydomonas reinhardtii. Photosynth. Res.* **46**, 257–264.

Harris, E.H., Stern, D.B. and Witman, G.B. (2009) *The Chlamydomonas Sourcebook*, 2nd edn. Oxford: Elsevier Inc.

Himeno, H., Hasegawa, T., Asahara, H., Tamura, K. and Shimizu, M. (1991) Identity determinants of *E. coli* tryptophan tRNA. *Nucleic Acids Res.* **19**, 6379–6382.

Hughes, R.A. and Ellington, A.D. (2010) Rational design of an orthogonal tryptophanyl nonsense suppressor tRNA. *Nucleic Acids Res.* **38**, 6813–6830.

Ito, K., Uno, M. and Nakamura, Y. (2000) A tripeptide 'anticodon' deciphers stop codons in messenger RNA. *Nature*, **403**, 680–684.

Ivanova, N.N., Schwientek, P., Tripp, H.J., Rinke, C., Pati, A., Huntemann, M., Visel, A. *et al.* (2014) Stop codon reassignments in the wild. *Science*, **344**, 909–913.

Johnson, D.B., Xu, J., Shen, Z., Takimoto, J.K., Schultz, M.D., Schmitz, R.J., Xiang, Z. *et al.* (2011) RF1 knockout allows ribosomal incorporation of unnatural amino acids at multiple sites. *Nat. Chem. Biol.* **7**, 779–786.

Johnson, D.B., Wang, C., Xu, J., Schultz, M.D., Schmitz, R.J., Ecker, J.R. and Wang, L. (2012) Release factor one is nonessential in *Escherichia coli. ACS Chem. Biol.* **7**, 1337–1344.

Jordan, P., Fromme, P., Witt, H.T., Klukas, O., Saenger, W. and Krauss, N. (2001) Three-dimensional structure of cyanobacterial photosystem I at 2.5 A resolution. *Nature*, **411**, 909–917.

Kimelman, A., Levy, A., Sberro, H., Kidron, S., Leavitt, A., Amitai, G., Yoder-Himes, D.R. *et al.* (2012) A vast collection of microbial genes that are toxic to bacteria. *Genome Res.* **22**, 802–809.

Lajoie, M.J., Rovner, A.J., Goodman, D.B., Aerni, H.R., Haimovich, A.D., Kuznetsov, G., Mercer, J.A. *et al.* (2013) Genomically recoded organisms expand biological functions. *Science*, **342**, 357–360.

Lee, S.B., Li, B., Jin, S. and Daniell, H. (2011) Expression and characterization of antimicrobial peptides Retrocyclin-101 and Protegrin-1 in chloroplasts to control viral and bacterial infections. *Plant Biotech J.* **9**, 100–115.

Li, Y. (2009) Carrier proteins for fusion expression of antimicrobial peptides in *Escherichia coli. Biotechnol. Appl. Biochem.* **54**, 1–9.

Lim, J.A., Shin, H., Heu, S. and Ryu, S. (2014) Exogenous lytic activity of SPN9CC endolysin against gram-negative bacteria. *J. Microbiol. Biotechnol.* **24**, 803–811.

Madesis, P., Osathanunkul, M., Georgopoulou, U., Gisby, M.F., Mudd, E.A., Nianiou, I., Tsitoura, P. *et al.* (2010) A hepatitis C virus core polypeptide expressed in chloroplasts detects anti-core antibodies in infected human sera. *J. Biotech.* **145**, 377–386.

Maul, J.E., Lilly, J.W., Cui, L., dePamphilis, C.W., Miller, W., Harris, E.H. and Stern, D.B. (2002) The *Chlamydomonas reinhardtii* plastid chromosome: islands of genes in a sea of repeats. *Plant Cell* **14**, 2659–2679.

McCutcheon, J.P., McDonald, B.R. and Moran, N.A. (2009) Origin of an alternative genetic code in the extremely small and GC-rich genome of a bacterial symbiont. *PLoS Genet.* **5**, e1000565.

Michelet, L., Lefebvre-Legendre, L., Burr, S.E., Rochaix, J.D. and Goldschmidt-Clermont, M. (2011) Enhanced chloroplast transgene expression in a nuclear mutant of *Chlamydomonas. Plant Biotechnol. J.* **9**, 565–574.

Nakamura, A., Tsukada, T., Auer, S., Furuta, T., Wada, M., Koivula, A., Igarashi, K. *et al.* (2013) The tryptophan residue at the active site tunnel entrance of *Trichoderma reesei* cellobiohydrolase Cel7A is important for initiation of degradation of crystalline cellulose. *J. Biol. Chem.* **288**, 13503–13510.

Niu, W., Schultz, P.G. and Guo, J. (2013) An expanded genetic code in mammalian cells with a functional quadruplet codon. *ACS Chem. Biol.* **8**, 1640–1645.

Novoselov, S.V., Rao, M., Onoshko, N.V., Zhi, H., Kryukov, G.V., Xiang, Y., Weeks, D.P. *et al.* (2002) Selenoproteins and selenocysteine insertion system in the model plant cell system, *Chlamydomonas reinhardtii. EMBO J.* **21**, 3681–3693.

Oey, M., Lohse, M., Scharff, L.B., Kreikemeyer, B. and Bock, R. (2009) Plastid production of protein antibiotics against pneumonia via a new strategy for high-level expression of antimicrobial proteins. *Proc. Natl Acad. Sci. USA* **106**, 6579–6584.

Purton, S., Stevens, D.R., Muhiuddin, I.P., Evans, M.C., Carter, S., Rigby, S.E. and Heathcote, P. (2001) Site-directed mutagenesis of PsaA residue W693 affects phylloquinone binding and function in the photosystem I reaction center of *Chlamydomonas reinhardtii. Biochemistry*, **40**, 2167–2175.

Rao, M., Carlson, B.A., Novoselov, S.V., Weeks, D.P., Gladyshev, V.N. and Hatfield, D.L. (2003) *Chlamydomonas reinhardtii* selenocysteine tRNA[Ser] Sec. *RNA*, **9**, 923–930.

Rasala, B.A. and Mayfield, S.P. (2015) Photosynthetic biomanufacturing in green algae; production of recombinant proteins for industrial, nutritional, and medical uses. *Photosynth. Res.* **123**, 227–239.

Ravikumar, A. and Liu, C.C. (2015) Biocontainment through reengineered codes. *ChemBioChem* **16**, 1149–1151.

Rinke, C., Schwientek, P., Sczyrba, A., Ivanova, N.N., Anderson, I.J., Cheng, J.F., Darling, A. *et al.* (2013) Insights into the phylogeny and coding potential of microbial dark matter. *Nature*, **499**, 431–437.

Robbens, S., Derelle, E., Ferraz, C., Wuyts, J., Moreau, H. and Van de Peer, Y. (2007) The complete chloroplast and mitochondrial DNA sequence of *Ostreococcus tauri*: organelle genomes of the smallest eukaryote are examples of compaction. *Mol. Biol. Evol.* **24**, 956–968.

Sambrook, J. and Russell, D.W. (2001) *Molecular Cloning: A Laboratory Manual*. Cold Spring Harbor: CSHL Press.

Scaife, M.A., Nguyen, G.T., Rico, J., Lambert, D., Helliwell, K.E. and Smith, A.G. (2015) Establishing *Chlamydomonas reinhardtii* as an industrial biotechnology host. *Plant J.* **82**, 532–546.

Schmidt, J. and Subramanian, A.R. (1993) Sequence of the cyanobacterial tRNA (w) gene in *Synechocystis* PCC 6803: requirement of enzymatic 3' CCA attachment to the acceptor stem. *Nucleic Acids Res.* **21**, 2519.

Specht, E.A. and Mayfield, S.P. (2013) Synthetic oligonucleotide libraries reveal novel regulatory elements in *Chlamydomonas* chloroplast mRNAs. *ACS Synth. Biol.* **2**, 34–46.

Spreitzer, R.J., Chastain, C.J. and Ogren, W.L. (1984) Chloroplast gene suppression of defective ribulosebisphosphate carboxylase/oxygenase in *Chlamydomonas reinhardtii*: evidence for stable heteroplasmic genes. *Curr. Genet.* **9**, 83–89.

Spreitzer, R.J., Goldschmidt-Clermont, M., Rahire, M. and Rochaix, J.D. (1985) Nonsense mutations in the *Chlamydomonas* chloroplast gene that codes for the large subunit of ribulosebisphosphate carboxylase/oxygenase. *Proc. Natl Acad. Sci. USA* **82**, 5460–5464.

Stirdivant, S.M., Crossland, L.D. and Bogorad, L. (1985) DNA supercoiling affects in vitro transcription of two maize chloroplast genes differently. *Proc. Natl Acad. Sci. USA* **82**, 4886–4890.

Stoppel, R., Lezhneva, L., Schwenkert, S., Torabi, S., Felder, S., Meierhoff, K., Westhoff, P. *et al.* (2011) Recruitment of a ribosomal release factor for light- and stress-dependent regulation of *petB* transcript stability in *Arabidopsis* chloroplasts. *Plant Cell* **23**, 2680–2695.

Timmis, J.N., Ayliffe, M.A., Huang, C.Y. and Martin, W. (2004) Endosymbiotic gene transfer: organelle genomes forge eukaryotic chromosomes. *Nat. Rev. Genet.* **5**, 123–135.

Turmel, M., Lemieux, C., Burger, G., Lang, B.F., Otis, C., Plante, I. and Gray, M.W. (1999) The complete mitochondrial DNA sequences of *Nephroselmis olivacea* and *Pedinomonas minor*. Two radically different evolutionary patterns within green algae. *Plant Cell* **11**, 1717–1730.

Wang, Q. and Wang, L. (2012) Genetic incorporation of unnatural amino acids into proteins in yeast. *Methods Mol. Biol.* **794**, 199–213.

Wang, Q., Sun, T., Xu, J., Shen, Z., Briggs, S.P., Zhou, D. and Wang, L. (2014) Response and adaptation of *Escherichia coli* to suppression of the amber stop codon. *ChemBioChem* **15**, 1744–1749.

Wilson, D.N., Guevremont, D. and Tate, W.P. (2000) The ribosomal binding and peptidyl-tRNA hydrolysis functions of *Escherichia coli* release factor 2 are linked through residue 246. *RNA*, **6**, 1704–1713.

Xie, W.H., Zhu, C.C., Zhang, N.S., Li, D.W., Yang, W.D., Liu, J.S., Sathishkumar, R. *et al.* (2014) Construction of novel chloroplast expression vector and development of an efficient transformation system for the diatom *Phaeodactylum tricornutum*. *Mar. Biotechnol. (NY)* **16**, 538–546.

Xiong, Y. and Steitz, T.A. (2006) A story with a good ending: tRNA 3'-end maturation by CCA-adding enzymes. *Curr. Opin. Struct. Biol.* **16**, 12–17.

Young, R.E. and Purton, S. (2014) Cytosine deaminase as a negative selectable marker for the microalgal chloroplast: a strategy for the isolation of nuclear mutations that affect chloroplast gene expression. *Plant J.* **80**, 915–925.

Yu, W. and Spreitzer, R.J. (1992) Chloroplast heteroplasmicity is stabilized by an amber-suppressor tryptophan tRNA(CUA). *Proc. Natl Acad. Sci. USA* **89**, 3904–3907.

Zhang, Y., Deshpande, A., Xie, Z., Natesh, R., Acharya, K.R. and Brew, K. (2004a) Roles of active site tryptophans in substrate binding and catalysis by alpha-1,3 galactosyltransferase. *Glycobiology*, **14**, 1295–1302.

Zhang, Z., Alfonta, L., Tian, F., Bursulaya, B., Uryu, S., King, D.S. and Schultz, P.G. (2004b) Selective incorporation of 5-hydroxytryptophan into proteins in mammalian cells. *Proc. Natl Acad. Sci. USA* **101**, 8882–8887.

16

Cotton (*Gossypium hirsutum*) 14-3-3 proteins participate in regulation of fibre initiation and elongation by modulating brassinosteroid signalling

Ying Zhou[†], Ze-Ting Zhang, Mo Li, Xin-Zheng Wei, Xiao-Jie Li, Bing-Ying Li and Xue-Bao Li*

Hubei Key Laboratory of Genetic Regulation and Integrative Biology, School of Life Sciences, Central China Normal University, Wuhan, China

*Correspondence
e-mail xbli@mail.ccnu.edu.cn
[†]Present address: School of Basic Medical Sciences, Wuhan University, Wuhan 430071, China.

Keywords: cotton (*Gossypium hirsutum*), fibre development, brassinosteroid (BR) signalling, 14-3-3 protein, regulation of gene expression, protein–protein interaction.

Summary

Cotton (*Gossypium hirsutum*) fibre is an important natural raw material for textile industry in the world. Understanding the molecular mechanism of fibre development is important for the development of future cotton varieties with superior fibre quality. In this study, overexpression of *Gh14-3-3L* in cotton promoted fibre elongation, leading to an increase in mature fibre length. In contrast, suppression of expression of *Gh14-3-3L*, *Gh14-3-3e* and *Gh14-3-3h* in cotton slowed down fibre initiation and elongation. As a result, the mature fibres of the *Gh14-3-3* RNAi transgenic plants were significantly shorter than those of wild type. This 'short fibre' phenotype of the *14-3-3* RNAi cotton could be partially rescued by application of 2,4-epibrassinolide (BL). Expression levels of the BR-related and fibre-related genes were altered in the *Gh14-3-3* transgenic fibres. Furthermore, we identified Gh14-3-3 interacting proteins (including GhBZR1) in cotton. Site mutation assay revealed that Ser163 in GhBZR1 and Lys51/56/53 in Gh14-3-3L/e/h were required for Gh14-3-3-GhBZR1 interaction. Nuclear localization of GhBZR1 protein was induced by BR, and phosphorylation of GhBZR1 by GhBIN2 kinase was helpful for its binding to Gh14-3-3 proteins. Additionally, 14-3-3-regulated GhBZR1 protein may directly bind to *GhXTH1* and *GhEXP* promoters to regulate gene expression for responding rapid fibre elongation. These results suggested that Gh14-3-3 proteins may be involved in regulating fibre initiation and elongation through their interacting with GhBZR1 to modulate BR signalling. Thus, our study provides the candidate intrinsic genes for improving fibre yield and quality by genetic manipulation.

Introduction

14-3-3 proteins are about 27–32 kDa and acidic regulatory proteins found in all eukaryotes (Aitken, 2006). The proteins were first described in brain tissues of mammals and given the name 14-3-3 on the basis of their migration pattern on starch gel electrophoresis (Moore and Perez, 1967). They act as major regulators of primary metabolism and cellular signal transduction, and interact with large numbers of cellular proteins which include metabolic enzymes, signalling proteins and transcription factors (Igarashi *et al.*, 2001; Ishida *et al.*, 2004; Schoonheim *et al.*, 2007; Sehnke *et al.*, 2002). RSXpSXP (Mode 1) and RXXXpSXP (Mode 2) are two classical 14-3-3 protein binding motifs (Sehnke *et al.*, 2002).

Brassinosteroid (BR) hormone signalling is a well-described signalling pathway in *Arabidopsis* that plays a crucial role in plant growth and development. In the absence of BRs, the negative regulator BKI1 (BRI1 kinase inhibitor 1) binds to BRI1 (brassinosteroid-insensitive 1) and inhibits its function (Jaillais *et al.*, 2011; Wang and Chory, 2006), and BIN2 kinase phosphorylates BES1 (BRI1-EMS suppressor 1) and BZR1 (brassinazole-resistant 1) that are plant-specific transcription factors and inhibits their DNA-binding activity (He *et al.*, 2005; Yin *et al.*, 2005). 14-3-3 proteins interact with the phosphorylated BES1/BZR1 to keep them in the cytoplasm. In this signalling pathway, BRs are perceived by a plasma membrane-bound receptor BRI1 (Hothorn *et al.*, 2011; Li and Chory, 1997; She *et al.*, 2011). BR binding to BRI1 activates BRI1 kinase, which promotes BKI1 phosphorylation, leading to the release of BKI1. The 14-3-3 proteins interact with the phosphorylated BKI1 through a motif that contains the two phosphorylation sites to release inhibition of BRI1 by BKI1. Meanwhile, the cytosolic BKI1 antagonizes the 14-3-3s and enhances accumulation of BES1/BZR1 in the cell nucleus to regulate BR-response genes (Wang *et al.*, 2011).

14-3-3 proteins play the important roles in BR signalling pathway. BZR1 is a target for 14-3-3 proteins in Arabidopsis and rice (Bai *et al.*, 2007; Gampala *et al.*, 2007; Ryu *et al.*, 2007). Phosphorylation mediated BZR1/14-3-3 interaction causes cytoplasmic retention or nuclear export of the BZR1 (He *et al.*, 2005; Ryu *et al.*, 2007). BR-induced dephosphorylation and nuclear accumulation of BZR1 causes brassinosteroid-induced growth and feedback regulation of brassinosteroid biosynthesis (Wang *et al.*, 2002). Mutation on putative 14-3-3 interaction binding sites of BZR1 abolishes the 14-3-3 binding, which in turn increases the nuclear partition of BZR1 and BR-regulated plant growth (Bai *et al.*, 2007; Gampala *et al.*, 2007; Gendron and Wang, 2007). BES1 and BZR1 have atypical basic helix–loop–helix (bHLH) DNA-binding domain and bind to E-box (CANNTG) more predominant in BR-induced genes and/or BRRE (BR-response element, CGTGT/CG) preferred in BR-repressed genes to regulate expression of BR target genes (Guo *et al.*, 2010; Li *et al.*, 2009; Oh *et al.*, 2012; Wang *et al.*, 2009; Zhang *et al.*, 2009).

Cotton (*Gossypium hirsutum* L.) produces the most prevalent natural fibres for the textile industry in the world. Fibre development can be divided into four stages: initiation (from 2 days before anthesis to 5 days postanthesis, DPA), elongation (3–20

DPA), secondary cell wall deposition (16–40 DPA) and maturation (40–50 DPA) (Basra and Malik, 1984). Cotton fibres are single-celled trichomes differentiated from the ovule epidermis and are considered a model system for studying cell elongation and cell wall biogenesis (Kim and Triplett, 2001). Fibre initiation and elongation are regulated by endogenous brassinosteroids. In vitro application of BR promotes fibre elongation of cotton, whereas treating cotton floral buds with brassinazole (BRZ, a brassinos-teroid biosynthesis inhibitor) results in a complete absence of fibre cell differentiation (Sun et al., 2005). In our previous study, six fibre-preferential genes encoding 14-3-3 proteins (L, a, e, f, g and h) were identified in cotton (Zhang et al., 2010). In this study, we demonstrated that GhBIN2-mediated phosphorylation directly influenced the interaction between Gh14-3-3L and GhBZR1. Overexpression of Gh14-3-3L in cotton promoted fibre elonga-tion, whereas suppression of expression of Gh14-3-3L, Gh14-3-3e and Gh14-3-3h in cotton slowed down fibre initiation and elongation. Expression of the genes (such as GhBZR1, GhBIN2, GhXTH1, GhEXP, GhCesA5 and GhTUB1) related to BR signalling and fibre development was altered in the transgenic fibres. These data suggested that cotton 14-3-3 proteins were involved in regulating fibre initiation and elongation through BR signalling pathway.

Results

Gh14-3-3s function in fibre initiation and elongation of cotton

To investigate the role of Gh14-3-3 genes in fibre initiation and development, several fibre-preferential 14-3-3 genes were iden-tified in cotton (Zhang et al., 2010). By Agrobacterium-mediated DNA transformation, we obtained over 200 transgenic cotton plants (T0 generation), including Gh14-3-3L overexpression lines and Gh14-3-3L, Gh14-3-3e and Gh14-3-3h RNAi (RNA interfer-ence) lines. Most of T0 transgenic plants harbouring the 35S: Gh14-3-3s cassette (RNAi or overexpression) showed normal morphology at stages of vegetative and reproductive growth, except fibre development. The homozygous transgenic lines of T2–T4 generations were chosen for further study. As shown in Figure S1, Gh14-3-3L expression was significantly increased in fibres of the Gh14-3-3L overexpression cotton plants, whereas was decreased in fibres of the RNAi lines. Similarly, Gh14-3-3e and Gh14-3-3h RNAi transgenic lines also showed the declined expression activities of Gh14-3-3e and Gh14-3-3h, respectively. Furthermore, Gh14-3-3L protein amount in cotton fibres was examined by Western blot analysis. The 14-3-3L protein content were significantly decreased in fibres of Gh14-3-3L RNAi transgenic plants, but slightly increased in the Gh14-3-3L overexpression transgenic fibres, compared with that in wild type. The 0–1 DPA fibres cells which initiated and elongated rapidly on ovules of the transgenic cotton plants (T1–T4 generations) and wild type were examined by microscopy. As shown in Figure 1a–j, fibre cells initiated on the ovules of the Gh14-3-3L/e/h RNAi plants were smaller than those on the ovules of wild type plants. Fibre elongation was also inhibited in the RNAi transgenic plants. Fibre cells of wild type and Gh14-3-3l overexpression transgenic lines were rapidly elongated at 1 DPA, whereas those on the Gh14-3-3L/e/h RNAi ovules were severely stunted. Furthermore, on the day of anthesis, the density of fibre initials on the Gh14-3-3L RNAi ovules were less than those on wild type. Measurement and statistical analysis indicated that the length of fibre initials on Gh14-3-3 RNAi ovules was shorter than

that of wild type. On the contrary, fibre length was increased on the ovules of Gh14-3-3L overexpression transgenic lines (Fig-ure 1k).

An ovule culture method (Beasley and Ting, 1973) was used to examine the effect of 2,4-epibrassinolide (BL, a biologically active BR) on fibre elongation. Fibre cell elongation was significantly reduced on the Gh14-3-3L RNAi ovules, whereas increased on the Gh14-3-3L overexpression ovules cultured in vitro. Furthermore, fibre cell elongation was partially restored on Gh14-3-3L RNAi ovules by adding 100 nM BL to the culture medium, compared with that of wild type (Figure 2a). Measurement and statistical analysis indicated that length of the Gh14-3-3L RNAi fibres was 41.1% shorter than that of wild type, whereas length of Gh14-3-3L overexpression fibres was increased 37%, compared with that of wild type. There was significant difference in fibre length among 14-3-3L RNAi and overexpression transgenic lines and wild type in the presence or absence of BL (Figure 2b). Similarly, TFU (total fibre unit) of the 14-3-3L RNAi transgenic lines was remarkably decreased, compared with that of the 14-3-3L overexpression lines and wild type in the presence or absence of BL. After application of BL, TFU of the wild type, 14-3-3L overexpression and RNAi lines was increased by 45.3%, 42.1% and 62.2%, respectively, compared with that of themselves in the medium without BL, suggesting fibre growth of the 14-3-3L RNAi cotton was partially restored by BL (Figure 2c). Additionally, the expression levels of GhCPD and GhDWF4 in the 14-3-3L RNAi fibres were noticeably reduced, but expressions of these BR biosynthetic genes in the overexpression transgenic fibres were similar to those of wild type in the medium without BL. On the other hand, in wild type and 14-3-3L overexpression lines, GhCPD and GhDWF4 were feedback-inhibited by BR, and expressions of these genes were significantly declined in the medium supple-mented with BL. GhCPD and GhDWF4 transcripts in the Gh14-3-3L overexpression cotton was remarkably decreased by 98.5% and 79.5%. In contrast, GhCPD expression was little changed, and GhDWF4 expression was decreased a little in the Gh14-3-3L RNAi fibres treated with BL (Figure 2d). These results suggested that Gh14-3-3 proteins may participate in feedback regulation of BR biosynthesis and fibre growth response.

To examine the role of Gh14-3-3s in fibre development, the length of mature fibres in the transgenic cotton plants was measured and statistically analysed, using wild type as control. As shown in Figure 3, Gh14-3-3 RNAi fibres were obviously shorter than the wild type, whereas Gh14-3-3L-overexpressing fibres were longer than the control. Similarly, fibre length of the 14-3-3e and 14-3-3h RNAi lines (eR3C1, eR2B2, eR1A1, hR24A1 and hR16A1) was shorter than that of wild type. We measured fibre length of the transgenic cotton progenies (T1 – T4 generations), and demonstrated that the changes in fibre length were stably inherited in each of the generations tested. Also, fibre quality characteristics (such as fibre length, strength, micronaire etc.) in the 14-3-3 transgenic cotton progenies were determined by a fibre quality detection system (see Methods). The results indicated that fibres of the 14-3-3L/e/h RNAi transgenic cotton progenies was still shorter, whereas fibres of the 14-3-3L overexpression transgenic cotton progenies were longer than those of wild type. Furthermore, fibre strength and micronaire were also changed in a few transgenic cotton lines, compared with those of wild type (Table S1). The above results suggested that these 14-3-3 proteins may participate in regulation of fibre initiation and elongation, but slightly affect the other fibre quality characteristics (such as fibre strength and micronaire).

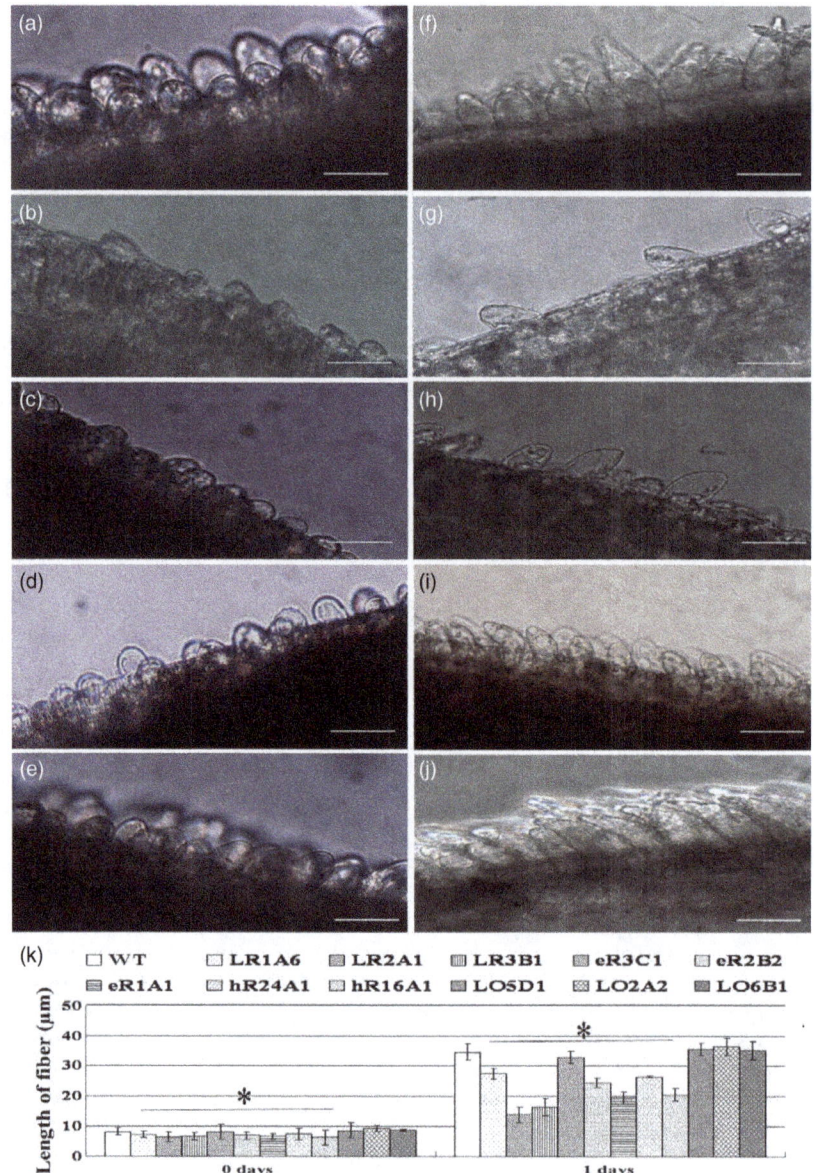

Figure 1 Comparison of fibre initiation and elongation between *Gh14-3-3* transgenic cotton plants and wild type. (a–e) Cross sections of the cotton ovules at the day of anthesis (0 DPA). (f–j) Cross sections of the cotton ovules at 1 DPA. (a and f) wild type; (b and g) *Gh14-3-3L* RNAi transgenic lines; (c and h) *Gh14-3-3e* RNAi transgenic lines; (d and i) *Gh14-3-3 h* RNAi transgenic lines; (e and j) *Gh14-3-3L* overexpression transgenic lines; (k) Measurement and statistical analysis of fibre length of the transgenic *Gh14-3-3* lines and wild type at 0 and 1 DPA. Ovules were sectioned and the length of 100 fibre cells was measured under a microscope for each transgenic line and wild type. Data were processed with Microsoft Excel, and error bars represent the standard deviations. Independent *t*-tests demonstrated that there was significant difference in length of early elongating fibres between the transgenic plants and wild type (*P value < 0.05). DPA, day postanthesis. WT, wild type; LR1A6, LR2A1 and LR3B1, *Gh14-3-3L* RNAi lines; LO5D1, LO2A2, LO6B1, *Gh14-3-3L* overexpression lines; eR3C1, eR2B2, eR1A1, *Gh14-3-3e* RNAi lines; hR24A1, hR16A1, *Gh14-3-3h* RNAi lines. Scale bar = 50 μm.

Expressions of the genes related to BR signalling and fibre development are altered in the Gh14-3-3 transgenic cotton

To investigate whether expressions of the genes related to BR signalling pathway and fibre development are altered in the 14-3-3 transgenic fibres, we analysed expression levels of the genes in 10 DPA fibres of the transgenic cotton progenies (T2–T4 generations). As shown in Figure 4, GhBZR1, GhXTH1 and GhCesA5 were down-regulated in the *14-3-3L/e/h* transgenic fibres, but their expression levels were slightly increased in the *Gh14-3-3L* overexpression plants, compared with those in wild type. GhEXP was up-regulated in the *Gh14-3-3L* overexpression fibres, but its expression was little changed in the 14-3-3 RNAi lines. GhTUB1 was up-regulated in the *Gh14-3-3L* overexpression lines and down-regulated in Gh14-3-3L/e/h RNAi transgenic fibres. On the other hand, GhBIN2 expression was little altered in both *Gh14-3-3L/e/h* RNAi fibres and *Gh14-3-3L* overexpression lines. The results suggested that Gh14-3-3s may be

involved in regulation of fibre elongation by modulating BR signalling.

The Ser163 in GhBZR1 and Lys51/56/53 in Gh14-3-3L/e/h are required for the protein–protein interaction

By yeast two-hybrid screening, 38 Gh14-3-3 interacting proteins were identified (Zhang *et al.*, 2010). Among them, one protein shares high similarity with *Arabidopsis* BZR1 and designated as GhBZR1 (accession number in GenBank: KM453728). Further study revealed that five 14-3-3 proteins (Gh14-3-3L/e/f/g/h) interacted with GhBZR1. Sequence analysis revealed that GhBZR1 protein shows high identity with *Arabidopsis* BZR1 (76%) and BZR2/BES1 (75%) and rice BZR1 (Os07g0580500) (64%). Highly conserved sequences among these proteins include the N-terminal DNA-binding domain, 22 GhBZR1's putative BIN2 phosphorylation sites (S/TXXXS/T), PEST domain, a potential 14-3-3 binding site (RXXXpSXP) and C-terminal region.

To further understand the Gh14-3-3s interaction with GhBZR1, a potential mode II type Gh14-3-3 binding site sequence was

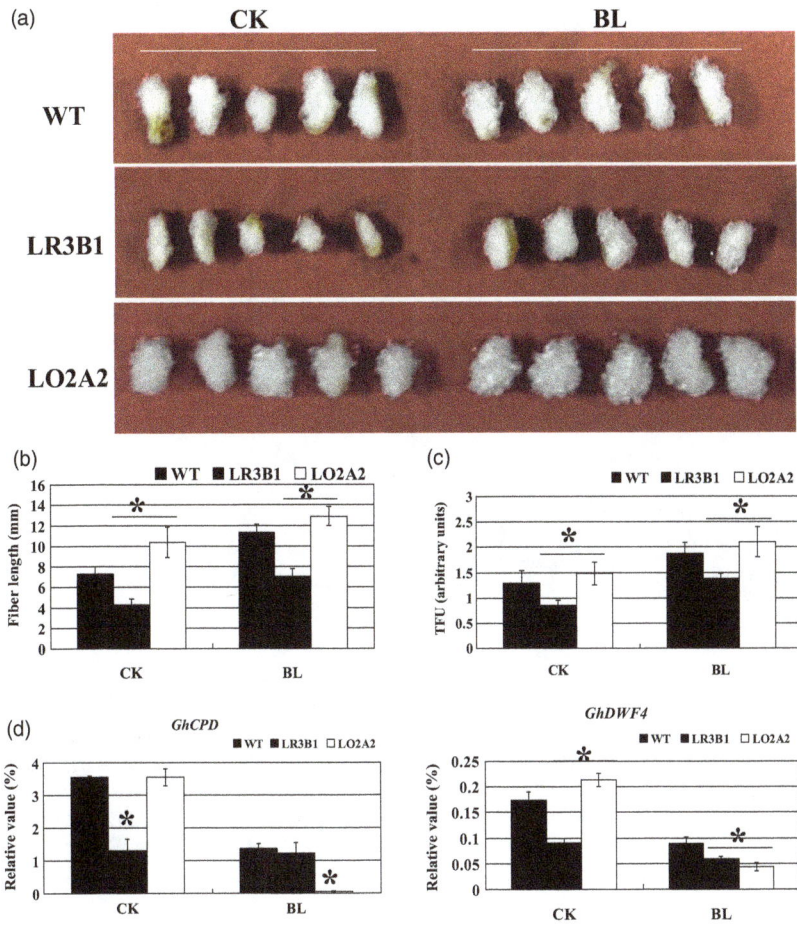

Figure 2 Effects of 2,4-epibrassinolide (BL) on fibre elongation of cotton. The cotton ovules (1 DPA) of wild type and *Gh14-3-3* RNAi and overexpression transgenic lines were *in vitro* cultured in liquid BT medium without (CK) or with 100 nM 2,4-epibrassinolide (BL) at 30 °C in darkness for 12 days. (a) Comparison of the phenotype of *in vitro* cultured ovules with fibres between the wild type and *Gh14-3-3L* transgenic lines. (b) Measurement and statistical analysis of fibre length on *in vitro* cultured ovules. (c) Fibre surface area was measured by dye binding in total fibre units (TFU) of *in vitro* cultured ovules. (d) Quantitative RT-PCR analysis of expression of *GhCPD* and *GhDWF4* genes in fibres of the *Gh14-3-3L* transgenic lines and wild type. Data were processed with Microsoft Excel, and error bars represent the standard deviations. Independent *t*-tests demonstrated that there was significant difference between the transgenic plants and wild type (*P value < 0.05). CK, without BL treatment; BL, treated with BL; WT, wild type; LR3B1, one of the Gh14-3-3L RNAi transgenic lines; LO2A2, one of the Gh14-3-3L overexpression transgenic lines. Accession numbers in GenBank are as follows: *GhCPD* (ACR20477) and *GhDWF4* (CO125422).

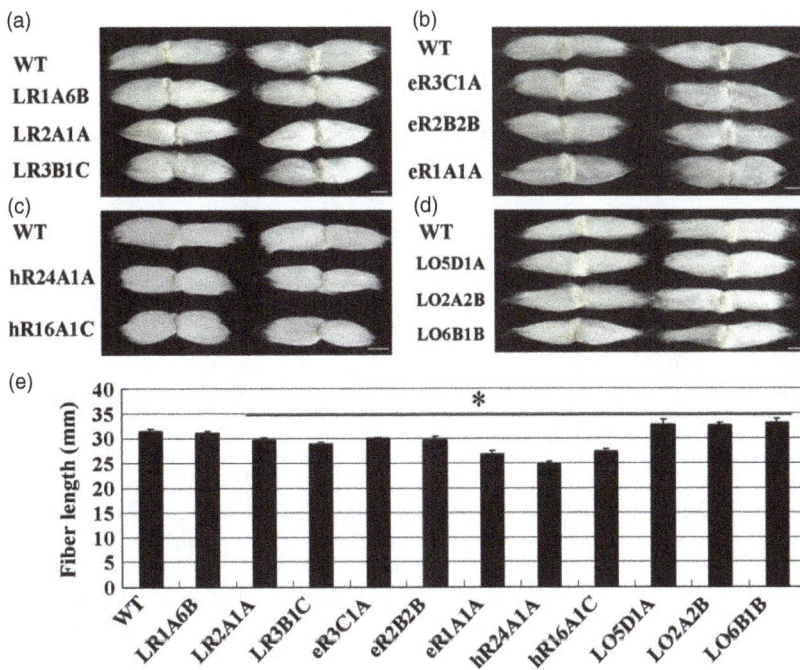

Figure 3 Comparison of mature fibre length between *Gh14-3-3* transgenic cotton plants and wild type. (a) Mature fibres of *Gh14-3-3L* RNAi lines and wild type. (b) Mature fibres of *Gh14-3-3e* RNAi lines and wild type. (c) Mature fibres of *Gh14-3-3 h* RNAi lines and wild type. (d) Mature fibres of *Gh14-3-3L* overexpression lines and wild type. (e) Measurement and statistical analysis of mature fibre length of *Gh14-3-3* transgenic lines and wild type ($n \geq 50$ cotton ovules per line). Data were processed with Microsoft Excel, and error bars represent the standard deviations. Independent *t*-tests demonstrated that there was significant difference between the transgenic plants and wild type (*P value < 0.05). WT, wild type; LR1A6, LR2A1 and LR3B1, *Gh14-3-3L* RNAi transgenic lines; eR3C1, eR2B2 and eR1A1, *Gh14-3-3e* RNAi transgenic lines; hR24A1 and hR16A1, *Gh14-3-3 h* RNAi transgenic lines; LO5D1, LO2A2 and LO6B1, *Gh14-3-3L* overexpression transgenic lines. Scale bar = 1 cm.

identified in GhBZR1 at 159–165th amino acids (RISNSAP), with the three required residues conserved also in GhBZR1 (Figure 5a). When Ser163 was mutated to Gly163 (GhBZR1S163G) or the binding site (RISNS, at position 159–163) was deleted (GhBZR1Δ), the GhBZR1 interaction with 14-3-3s was abolished in yeast cells. We also used flash-freezing filter assay to check the mating

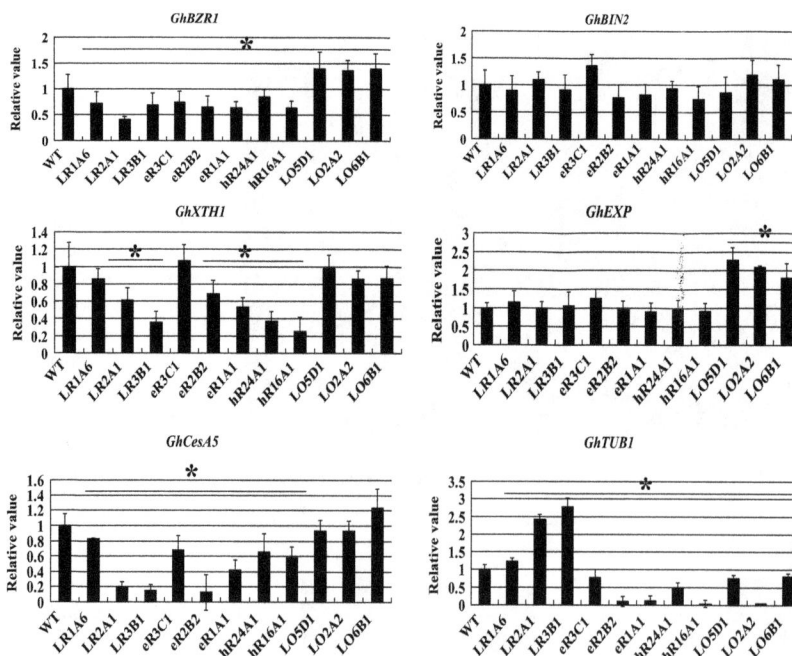

Figure 4 Quantitative RT-PCR analysis of expression of the related genes in fibres of the *Gh14-3-3* transgenic cotton plants. Total RNA was isolated from 10 DPA fibres of the *Gh14-3-3* transgenic lines and wild type. Expression levels of *GhBZR1*, *GhBIN2*, *GhXTH1*, *GhEXP*, *GhCesA5* and *GhTUB1* were determined by real-time quantitative RT-PCR, using *GhUBI1* as a quantification control. Data were processed with Microsoft Excel, and error bars represent the standard deviations. Independent t-tests demonstrated that there was significant difference (*P value < 0.05) in gene expression levels between the transgenic lines and wild type. DPA, day postanthesis. WT, wild type; LR1A6, LR2A1 and LR3B1, *Gh14-3-3L* RNAi lines (T2). LO5D1, LO2A2 and LO6B1, *Gh14-3-3L* overexpression lines. eR3C1, eR2B2 and eR1A1, *Gh14-3-3e* RNAi transgenic lines; hR24A1 and hR16A1, *Gh14-3-3h* RNAi transgenic lines. Accession numbers in GenBank are as follows: *GhBZR1* (KM453728), *GhBIN2* (KM453729), *GhXTH1* (HM749062), *GhEXP* (AY189969), *GhCesA5* (BM356396) and *GhTUB1* (BM356392).

diploids harbouring reporter gene *LacZ*. The yeast colony of pGBKT7-Gh14-3-3L/pGADT7-GhBZR1 turned blue, showing positive β-galactosidase activity. On the contrary, the pGADT7-GhBZR1S163G and pGADT7-GhBZR1Δ mating with pGBKT7-Gh14-3-3L were white, as did the empty vector pGADT7 (Figure 5b). These results suggested that the 14-3-3 proteins bind to GhBZR1 through the consensus binding site that requires phosphorylation of the conserved Ser163 residue.

In human, two of the charge-reversal mutations greatly (K49E) or partially (R56E) decreased the interaction of 14-3-3zeta with Raf-1 kinase. Lys49 is conserved among all 14-3-3 isoforms (Zhang *et al.*, 1997). Lys52 in a tobacco 14-3-3 protein (called D31) is identity with human Lys49. Using bioinformatics program blast, such potential Lys52 (K52) was also found in cotton 14-3-3L (K51), Gh14-3-3e (K56) and Gh14-3-3 h (K53) (Figure 5c). When Lys (Gh14-3-3LK51, Gh14-3-3eK56 and Gh14-3-3hK53) was mutated to Glu (Gh14-3-3LK51E, Gh14-3-3eK56E and Gh14-3-3hK53E), the interaction of the 14-3-3 proteins with GhBZR1 in yeast were abolished (Figure 5d), suggesting that Gh14-3-3L/e/h bind to GhBZR1 through the consensus binding site that requires Lys residue.

Brassinosteroid induces rapid nuclear localization of GhBZR1

To determine the subcellular localization of GhBZR1 protein in the presence of BR, the coding region of *GhBZR1* gene was fused with a green fluorescent protein (eGFP) gene under the control of CaMV 35S promoter (*35S:GhBZR1:GFP*) and transiently expressed in tobacco leaves by *Agrobacterium*-mediated DNA transfer.

Experimental results indicated that GFP fluorescent signals in the transformed tobacco leaf cells expressing *GhBZR1:eGFP* were localized in both cell nucleus and cytoplasm (Figure 6a–c). However, nuclear accumulation of GhBZR1:eGFP was increased rapidly after epibrassinolide (BL) treatment for 10 min (Figure 6d–e), suggesting that GhBZR1 as a BR signalling component may mediated BR signals into the nucleus by shuttling between the cytoplasm and the nucleus.

Phosphorylation of GhBZR1 promotes its binding to 14-3-3 proteins

To determine the interaction between GhBZR1 and GhBIN2 in BR signalling pathway during cotton fibre development, we identified a BIN2 gene (*GhBIN2*, accession number in GenBank: KM453729) in cotton. The isolated *GhBIN2* includes 1146 bp of open reading frame (ORF), and encodes a protein of 381 amino acids that shares 90% identity with *Arabidopsis* BIN2. We analysed the interaction between GhBIN2 and GhBZR1 by yeast two-hybrid and *in vitro* kinase assays. GhBIN2 that was fused to the GAL4-DNA-binding domain interacted with GhBZR1 that was fused to the GAL4-activation domain in yeast cells, leading to the cells grew well on SD/-Trp/-Leu/-His/-Ade nutritional selection medium (Figure 7a). Furthermore, *in vitro* kinase assay was also performed. His-GhBIN2 and MBP-GhBZR1 fusion proteins were expressed and purified from *Escherichia coli* cells and incubated together. His-GhBIN2 protein did not bind MBP, but did bind the MBP-GhBZR1 fusion proteins (Figure 7b). To determine if phosphorylation of GhBZR1 is required for Gh14-3-3 binding, the purified GhBZR1 protein was phosphorylated by GhBIN2 protein

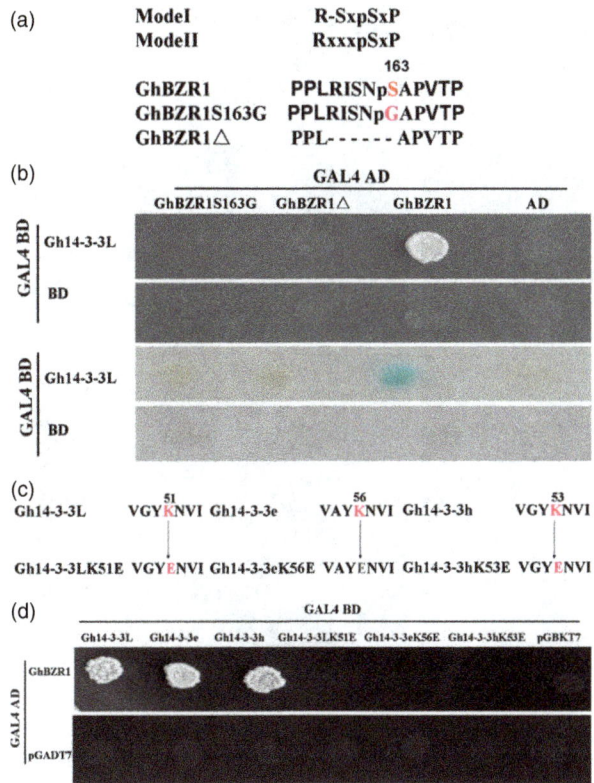

Figure 5 Gh14-3-3 proteins interact with GhBZR1 through consensus binding sites. (a) The motifs of GhBZR1 protein that Gh14-3-3s specifically interact with. Two known 14-3-3 binding site sequences (Mode I and Mode II) are aligned against GhBZR1 sequence. 'X' represents any given amino acid. Amino acids are presented as single letters namely: R, arginine; I, isoleucine; pS, phosphoserine; N, asparagine; C, cysteine; P, proline. Conserved serine residue that is crucial for 14-3-3 binding is numbered and marked in bold. Various mutations created in the 14-3-3 binding sites are also shown. Hyphen (-) represents deletion of an amino acid residue. GhBZR1S163G, the S_{163} of GhBZR1 was mutated to G_{163}. GhBZR1△, the RISNpS (159–163 amino acids) of GhBZR1 was deleted. (b) GhBZR1 interacts with Gh14-3-3L in yeast two-hybrid assay and flash-freezing filter assay of the β-galactosidase activity. Yeast transformants were streaked on SD/-Ade/-His/-Leu/-Trp/medium (SD minimal medium lacking Ade, His, Leu and Trp). Each yeast diploid contains the pGBKT7-14-3-3L prey construct and one of the following genes GhBZR1S163G, GhBZR1△, GhBZR1 in the pGADT7 vector, using pGADT7 empty vector (AD) as control. Growth of yeast cells and the yeast colonies including pGBKT7-Gh14-3-3L and pGADT7-GhBZR1 turned blue. (c) The motifs of Gh14-3-3L, Gh14-3-3e and Gh14-3-3 h that interact with GhBZR1. The K_{51} of Gh14-3-3L was mutated to E_{51}. The K_{56} of Gh14-3-3e was mutated to E_{56} and the K_{53} of Gh14-3-3h was mutated to E_{53}. K, lysine; E, glutamic acid. (d) Gh14-3-3L, Gh14-3-3e and Gh14-3-3h interact with GhBZR1 in yeast two-hybrid assay. Yeast transformants were streaked on SD/-Ade/-His/-Leu/-Trp/medium (SD minimal medium lacking Ade, His, Leu and Trp). Each yeast diploid contains the pGBKT7-Gh14-3-3L, pGBKT7-Gh14-3-3e, pGBKT7-Gh14-3-3h, pGBKT7-Gh14-3-3LK51E, pGBKT7-Gh14-3-3eK56E, pGBKT7-Gh14-3-3hK53E and pGBKT7 prey constructs, respectively, and the following pGADT7-GhBZR1 or pGADT7 vector.

in vitro. After phosphorylated or unphosphorylated GhBZR1 proteins were incubated with Gh14-3-3L proteins for 1 h, the GhBZR1 proteins were separated on a SDS-PAGE gel with Phos-tag™ Acrylamide AAL-107 Aqueous Solution and then blotted onto nitrocellulose membrane. The blotted membrane was incubated anti-GhBZR1 antibodies. As shown in Figure 7c, strong signals were detected in the lane with the phosphorylated GhBZR1, indicating that GhBZR1 phosphorylation by GhBIN2 facilitates GhBZR1 binding to Gh14-3-3 proteins.

As thus interaction in yeast is likely due to phosphorylation of GhBZR1 by yeast GSK3 kinase, *in vivo* interaction of GhBZR1 with Gh14-3-3L was determined by bi-molecular fluorescence complementation (BiFC) assay. The experimental results revealed that strong YFP fluorescence was observed in onion cells with co-expression of GhBZR1-cYFP and Gh14-3-3L-nYFP (Figure 7d–f), demonstrating that Gh14-3-3L can interact with GhBZR1 in plant cells.

In addition, we analysed *GhBZR1* and *GhBIN2* expression profiling in cotton tissues by quantitative RT-PCR. As shown in Figure S2, *GhBZR1* was preferentially expressed in elongating fibres of cotton. Likewise, *GhBIN2* gene was predominately expressed in elongating fibres. Furthermore, three 14-3-3 genes (*Gh14-3-3L/e/h*) were also preferentially expressed in elongating fibres (Zhang *et al.*, 2010). Therefore, the interaction among these proteins for BR signal transduction may be possible in elongating fibre cells of cotton.

GhBZR1 protein directly binds the *GhXTH1* and *GhEXP* promoter sequences

To further understand the molecular mechanism of the 14-3-3s-regulated BR signalling transduction in fibre development, we isolated the *GhXTH1* and *GhEXP* promoter fragments from cotton genome. Bioinformatics analysis revealed that the isolated *GhXTH1* promoter sequence contains putative BRRE (BR-response element, CGTGT/CG) and E-box (CANNTG). Two BRRE elements are located at −295 and −729 bp, while two E-box (CANNTG) elements are positioned at −688 and −755 bp upstream of the start codon (ATG), respectively. Likewise, the isolated *GhEXP* promoter sequence contains eight putative E-box (CANNTG), which are located at −210, −383, −485, −842, −1472, −1493, −1570 and −1647 bp upstream of the start codon (ATG), respectively (Figure 8a). To test GhBZR1 protein direct binding to GhXTH1 promoter, we performed yeast one-hybrid assay using pGADT7-GhBZR1 construct. The condition of Y1HGold yeast transformants of pGADT7-GhBZR1/GhXTH1-pAbAi on SD/-Leu nutritional selection medium was same as the Y1HGold yeast transformants of pGADT7/GhXTH1-pAbAi (negative control) and Y1HGold yeast transformants of pGADT7-Rec-p53/p53-AbAi (positive control). When transformants were assayed for growth on nutritional selection medium SD/-Leu with AbA (350 ng/ml), direct binding of GhBZR1 to *GhXTH1* promoter region was detected, as did the positive control (Figure 8b). Similarly, to test GhBZR1 protein direct binding to *GhEXP* promoter, yeast one-hybrid assay was also performed using pGADT7-GhBZR1 construct. We chose the *GhEXP* promoter sequence (−1832 to −1334 bp) containing possible four E-box (CANNTG) which are located at −1472 bp, −1493 bp, −1570 bp, −1647 bp. Y1HGold yeast transformants of pGADT7-GhBZR1/GhEXP-pAbAi were cultured on SD/-Leu nutritional selection medium as described above, and the direct binding of GhBZR1 to *GhEXP* promoter region was detected, as did the positive control (Figure 8c).

To confirm GhBZR1 as a transcription factor binding to E-box elements in *GhXTH1* and *GhEXP* promoters for regulating the gene expression, we employed electrophoretic mobility shift assay (EMSA) for protein-DNA-binding analysis. GhBZR1 protein was expressed and purified in *E. coli* cells. The oligonucleotide

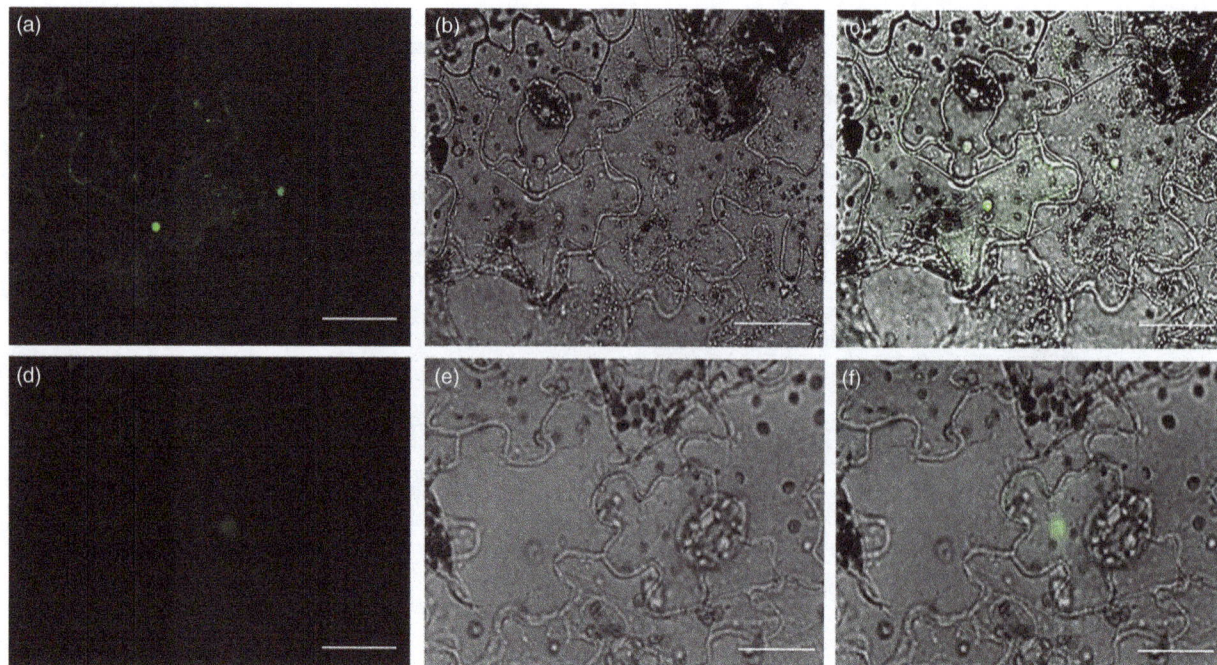

Figure 6 Subcellular localization of GhBZR1 protein in leaf cells of tobacco (*Nicotiana tabacum*). The GhBZR1:eGFP construct was transferred into tobacco leaf cells, and fluorescence images were obtained using confocal microscopy. Fluorescence signals of GFP were mainly detected in cell nucleus and cytoplasm of tobacco leaf cells harbouring the GhBZR1:eGFP construct. When treatment with 1 μM 2,4-epibrassinolide (BL), the fluorescence signals of GFP were detected only in cell nucleus. (a) Tobacco leaf cells expressing GhBZR1:eGFP. (b) The bright-field micrograph (transmission image) of the same cell. (c) Image A merged with its bright-field photograph. (d) Tobacco leaf cells expressing GhBZR1:eGFP were treated with 1 μM BL. (e) The bright-field micrograph (transmission image) of the same cell. (f) Image D merged with its bright-field photograph. Scale bar = 50 μm.

sequence containing two E-boxes and one BRRE of *GhXTH1* promoter was synthesized as DNA probe. Likewise, the oligonucleotide sequence containing two E-boxes of *GhEXP* promoter was also synthesized as DNA probe (see Methods). After stained DNA with SYBRH Green EMSA Nucleic Acid Gel Stain, one large molecular weight DNA band was presented in MBP-GhBZR1/*GhXTH1* DNA lane and MBP-GhBZR1/*GhEXP* DNA lane, respectively, whereas no signal was detected in the other lanes (MBP/DNA, MBP, MBP-GhBZR1 and DNA) (Figure 8d). These data suggested that 14-3-3-regulated GhBZR1 may directly bind *GhXTH1* and *GhEXP* promoter sequences to regulate expression of both genes for responding rapid fibre elongation of cotton.

Discussion

14-3-3 proteins are phosphopeptide-binding proteins that bind to phosphoserine/phosphothreonine motifs in a sequence-specific manner, and highly conserved in eukaryotes (Darling *et al.*, 2005). They interact with some metabolic enzymes and transcription factors in plants. In barley, about 150 proteins were identified to interact with 14-3-3 proteins (Schoonheim *et al.*, 2007). Our previous study revealed that five of the six fibre-preferential Gh14-3-3 proteins interacted with GhBZR1 protein (Zhang *et al.*, 2010). In this study, we further demonstrated the interaction of cotton 14-3-3s and BZR1 protein *in vitro* and *in vivo*.

Brassinosteroids (BR) participate in many developmental and physiological processes (Brosa, 1999). Previous studies revealed that 14-3-3 proteins mediate BR signal transduction by interacting with BZR1 to regulate the nuclear export of a transcriptional factor in *Arabidopsis* and rice (Bai *et al.*, 2007; Gampala *et al.*,

2007; Ryu *et al.*, 2007). Our data also demonstrated that Gh14-3-3 proteins function in BR signal transduction for cotton fibre development. Gh14-3-3 proteins bind to GhBZR1 through a consensus 14-3-3 binding site (RXXXpSXP). The S163G and RISNS mutation of GhBZR1 abolished its interaction with 14-3-3 proteins. On the other hand, all 14-3-3 isoforms contain a conserved Lys (K), like Lys49 of human Raf kinase and Lys52 of a tobacco 14-3-3 protein (D31). We found the mutation of K51E in Gh14-3-3L, K56E in 14-3-3e and K53E in 14-3-3h could be deprived of their ability interacted with GhBZR1, suggesting that the conserved site (Lys) in cotton 14-3-3 proteins is essential for Gh14-3-3-GhBZR1 interaction.

BZR1 is a highly phosphorylatable protein that is rapidly dephosphorylated in the presence of BR. The role of BZR1 phosphorylation has been proposed to regulate a variety of molecular events: protein stability, cytoplasmic retention and multimerization, and inhibition of the transactivation or repression of target genes via direct interaction with *cis* elements (Vert and Chory, 2006; Vert *et al.*, 2005). BR induces the nuclear translocation of the BZR1 protein. Our results are consistent with previous observations that the nuclear localization of BZR1 is enhanced by exogenous BR treatment (Wang *et al.*, 2002; Yin *et al.*, 2002). However, the other study suggested that BZR1 is constitutively localized in the cell nucleus (Vert and Chory, 2006). In this study, we showed GhBZR1 was phosphorylated by GhBIN2 to promote its binding to the Gh14-3-3 proteins, indicating that GhBIN2-mediated phosphorylation is required for GhBZR1 interacting with Gh14-3-3 proteins.

The log phase of cotton fibre elongation takes place during the first 10 DPA (DeLanghe, 1986). High expression of *Gh14-3-3L/e/h*, *GhBZR1* and *GhBIN2* in 3–10 DPA fibres correlates well with

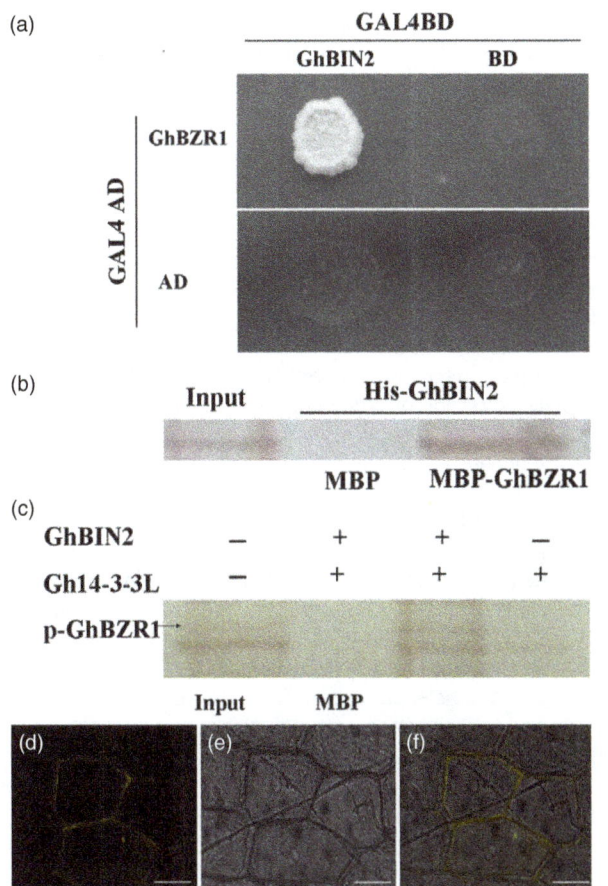

Figure 7 Assays of GhBZR1 phosphorylation by GhBIN2 and its binding with Gh14-3-3L. (a) GhBZR1 interacts with GhBIN2 in yeast two-hybrid assay. Yeast diploid contains the pGBKT7-GhBIN2 construct and pGADT7-GhBZR1 construct, using pGBKT7 (BD) and pGADT7 (AD) vectors as controls. (b) Interaction of GhBIN2 with GhBZR1 was detected by semi-*in vivo* pull-down assay. The coding sequence of GhBIN2 gene was cloned into pET32a vector and the coding sequence of GhBZR1 gene was cloned into pMAL vector. Pull-down assay was performed using His-GhBIN2, MBP, MBP-BZR1. GhBZR1 was detected by Western blotting with anti-GhBZR1 antibody. (c) phosphorylated GhBZR1 binds Gh14-3-3L. GhBZR1 proteins were phosphorylated by GhBIN2, and incubated with Gh14-3-3L proteins before pull-down assay (lanes 2 to 4), using MBP as negative control. Unphosphorylated GhBZR1 (lane 4) or phosphorylated GhBZR1 (lanes 1 and 3) were detected by Western blotting with anti-GhBZR1 antibody. (d) *In vivo* interaction between GhBZR1 and Gh14-3-3L was detected by bimolecular fluorescence complementation (BiFC). (e) Bright-field micrograph (transmission image) of the same cells. (f) Image D merged with its bright-field photograph. The N-terminal and C-terminal halves of YFP were fused to Gh14-3-3L and GhBZR1, respectively, and the constructs were co-transformed into onion epidermal cells and fluorescence images were observed using confocal microscopy. Scale bar = 10 μm.

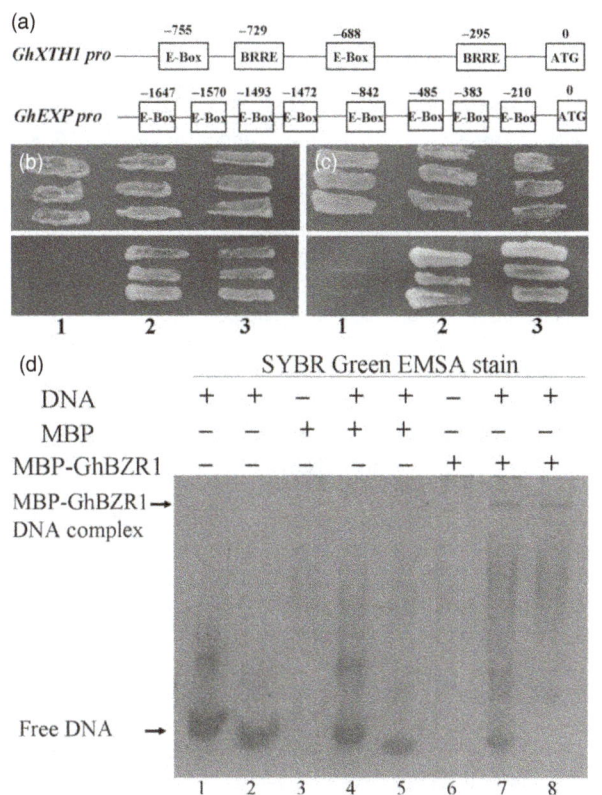

Figure 8 Interaction of GhBZR1 protein with *GhXTH1* and *GhEXP* promoter sequences. (a) The possible BRRE and E-box elements in GhXTH1 promoter sequence. (b) Yeast one-hybrid assay of GhBZR1 interacted with *GhXTH1* promoter sequence. Transformants grown on SD/-Leu nutritional selection medium with 350 ng/ml AbA (Aureobasidin A). 1, Y1HGold yeast transformants of pGADT7/GhXTH1-pAbAi used as the negative control; 2, Y1HGold yeast transformants of pGADT7-Rec-p53/p53-AbAi used as the positive control. 3, Y1HGold yeast transformants of pGADT7-GhBZR1/GhXTH1-pAbAi. (c) Yeast one-hybrid assay of GhBZR1 interacted with *GhEXP* promoter sequence. Transformants grown on SD/-Leu nutritional selection medium with 350 ng/ml AbA (Aureobasidin A). 1, Y1HGold yeast transformants of pGADT7/GhEXP-pAbAi used as the negative control; 2, Y1HGold yeast transformants of pGADT7-Rec-p53/p53-AbAi used as the positive control; 3, Y1HGold yeast transformants of pGADT7-GhBZR1/GhEXP-pAbAi. (d) Electrophoretic mobility shift assay (EMSA) of GhBZR1 binding DNA fragments from the promoters of *GhXTH1* and *GhEXP*. The gel was stained with SYBR® Green EMSA stain. 1, DNA fragment of the *GhXTH1* promoter; 2, DNA fragment of the *GhEXP* promoter; 3, Maltose-binding protein (MBP); 4, The binding reaction mixture of MBP and DNA fragment of the *GhXTH1* promoter; 5, The binding reaction mixture of MBP and DNA fragment of the *GhEXP* promoter; 6, The MBP-GhBZR1 fusion protein; 7, The binding reaction mixture of MBP-GhBZR1 fusion protein and DNA fragment of the *GhXTH1* promoter; 8, The binding reaction mixture of MBP-GhBZR1 fusion protein and DNA fragment of the *GhEXP* promoter.

rapid elongation of fibres. In *Gh14-3-3* RNAi transgenic cotton plants, fibre initiation and elongation were dramatically inhibited. When exogenous BL was supplied in the culture medium, fibre elongation and TFU of *Gh14-3-3L* RNAi lines were more increased than those of wild type and Gh14-3-3L overexpression lines. However, exogenous BL only partially restored the inhibition of fibre elongation of *Gh14-3-3L* RNAi ovules in *in vitro* culture (Figure 2).

The *bzr1-1D* mutant displays an increased feedback inhibition of BR biosynthesis, consistent with the enhanced BR signalling and the smaller stature of light-grown *bzr1-1D* plants (Wang et al., 2002). In *Arabidopsis*, BZR1 protein mediates feedback inhibition of the BR biosynthetic genes such as *CPD* and *DWF4* (He et al., 2005). Similarly, OsBZR1 is required for feedback regulation of BR biosynthetic genes in rice. The expression levels of BR biosynthetic genes *D2*, *D11*, and *BRD1* in *OsBZR1* RNAi plants were similar to the wild type without BR treatment but were not reduced by BR treatment (Bai *et al.*, 2007). In this study, expression of *GhCPD* and *GhDWF4* was strongly decreased in *Gh14-3-3* RNAi fibres. However, the expression levels of these genes were little changed in the *14-3-3* RNAi lines under BL treatment, compared with no BL treatment. In contrast, the expression of two BR biosynthetic genes were remarkably declined in the *Gh14-3-3L* overexpression cotton when the exogenous BL was applied. Furthermore, fibre quality is a key factor determining fibre price and quality of cotton textile products. Fibre quality is characterized mainly by fibre length (mm), strength (g tex-1) and micronaire (unitless), and the textile industry has a preference for long and strong fibres of moderate micronaire for producing high-quality yarns (Long *et al.*, 2010). In this study, we evaluated fibre quality characteristics of the 14-3-3 transgenic cotton progenies by HFI test. The data revealed that *Gh14-3-3* RNAi fibres were shorter, whereas *Gh14-3-3* overexpression fibres were longer than those of wild type. However, the fibre strength and micronaire were little altered in the 14-3-3 transgenic cotton (Figure 3 and Table S1). These results suggested that Gh14-3-3 proteins may be involved in modulating BR signalling for response to rapid fibre elongation.

BR promotes cell elongation by inducing the expression of the genes involved in cell wall-loosening or cell wall-modifying enzymes, such as xyloglucan endotransglycosylase/hydrolase (XTH), and pectin lyase, pectinesterase and expansins (EXP) (Li *et al.*, 2010). It has been supposed that BR regulates cotton fibre development, at least in part, by controlling the expression of genes involved in cell expansion. Expression of *GhXTH*, *GhEXP* and *GhTUB1* was up-regulated in the cultured cotton ovules with fibres by BL treatment, but was down-regulated in the cultured ovules with fibres by Brz (an inhibitor of BR) (Sun *et al.*, 2005). It was reported that mRNAs of cell wall-loosening enzymes (such as XTH and EXP) were preferentially expressed in elongating fibres of cotton (Ji *et al.*, 2003). EXP mediates cell wall extension in plants by breaking the hydrogen bonds between cellulose and hemicellulose, allowing these polymers to slip relative to each other (Mason and Cosgrove, 1995). Cosgrove (2000) asserted that EXPs are the most important wall-loosening factors in turgor-driven cell wall extension, while XTHs and other glucanases act secondarily in cell wall-loosening and reconstruction. In addition, a cotton β-tubulin gene, *GhTUB1*, was preferentially expressed in fibre and was associated with fibre initiation and elongation (Lee *et al.*, 2007; Li and Nam, 2002). Cortical microtubules are thought to be involved in determining the orientation of cell wall expansion and cellulose synthesis (Whittaker and Triplett, 1999). On the other hand, cellulose synthase catalytic subunit (*GhCesA5*) responsible for glucan chain elongation is believed to be a plasma membrane glycosyltransferase in cotton fibres. In this study, we found that expression of *GhBZR1*, *GhXTH1* and *GhCesA5* was down-regulated in *Gh14-3-3* RNAi transgenic fibres. *GhEXP* (alpha-expansin precursor) was down-regulated in *Gh14-3-3* RNAi transgenic fibres and up-regulated in the *Gh14-3-3* overexpression fibres (Figure 4). It was reported that *AtEXPA5* expression is reduced in a brassinosteroid-deficient mutant (*det2*) and a signalling mutant (*bri1-301*), while it is increased in dominant mutant *bzr1-1D*. The double mutant (*bzr1-1DXexpA5-1*) shows the reduced growth, compared with the *bzr1-1D* mutant, and the brassinazole resistance of *bzr1-1D* was impaired in the double mutant. These findings indicated that *AtEXPA5* is a

growth-regulating gene whose expression is controlled by BR signalling downstream of BZR1 in *Arabidopsis* (Park *et al.*, 2010). Through transcript profiling analysis and chromatin-immunoprecipitation microarray (ChIP-chip), 953 BR-regulated BZR1 target (BRBT) genes (including cell wall biosynthesis genes) were identified in *Arabidopsis*. A large number of BZR1 target genes were activated by BR and cell wall-related enzymes, such as cellulose synthase, pectinesterases, xyloglucosyl transferases and expansins, which were likely to mediate BR-promoted cell elongation and differentiation (Sun *et al.*, 2010). Similarly, our data suggested that 14-3-3 proteins may interact to regulate GhBZR1 activity in BR signalling pathway, and in turn, GhBZR1 protein may directly bind to *GhXTH1* and *GhEXP* promoters to regulate expression of both genes for promoting rapid fibre cell polar growth.

Cotton fibre is the main material in textile industry. During recent years, cotton textile of good quality is being highly demanded so that fibre modification becomes one of the main objectives for genetic breeding of cotton in the world. The data presented in this study revealed that up- and down-regulation of expressions of *Gh14-3-3* genes in fibre cells may alter the expression levels of BR-related and fibre growth-related genes, and consequently, affect fibre development of cotton. Thus, this study provides the evidence of a direct connection between 14-3-3-regulated BR signalling and fibre elongation through altering the expression of *14-3-3* genes in cotton and has significant implications for improving cotton fibre yield and quality by genetic manipulation.

Experimental procedures

Plant materials

Cotton (*Gossypium hirsutum* cv. Coker312) seeds were surface-sterilized with 10% H_2O_2 for 90 min, followed by washing with sterile water three to five times. The sterilized seeds were germinated on one-half strength Murashige and Skoog (MS) medium (12 h light/12 h dark cycle, 28 °C). Roots, cotyledons and hypocotyls were collected from these sterile cotton seedlings, and the other tissues were derived from cotton plants grown in field for DNA, RNA and protein extraction.

In vitro culture of cotton ovules with 2,4-epibrassinolide (BL) treatment

Bolls at 1 DPA (day postanthesis) from cotton plants were surface-sterilized with 70% ethanol for 1 min, followed by washing with sterile water. Ovules were picked out from these sterilized bolls, and cultured in BT liquid medium containing 5 μM NAA and 0.5 μM GA_3 (Beasley and Ting, 1974), supplemented with 0 (control) to 100 nM epibrassinolide (BL) at 30 °C in dark. Fibre length was measured after the ovules were cultured for 12 days. Experiments were repeated at least three times and run in three replicates each time. The total number of ovules counted in each group was 50, and *P* values were calculated according to the untreated series as controls.

Construction of Gh14-3-3 overexpression and RNA interference (RNAi) vectors and cotton transformation

For the overexpression construct, the coding sequence of *Gh14-3-3L* gene was cloned in pBI121 vector under control of the CaMV 35S promoter and *NOS* terminator. For the RNAi constructs, the specific 3'- UTR (untranslational region) sequences of *Gh14-3-3L*, *Gh14-3-3e* and *Gh14-3-3h* were cloned into pBluescript SK1 vector, respectively, to create the inverted repeat sequence of the transgene and then cloned into pBI121 vector. All primers used for these constructs are listed in Table S2. For introduced these constructs into cotton, hypocotyl explants from cotton (cv. Coker 312) seedlings were used for *Agrobacterium*-mediated transformation as described previously (Wang and Li, 2009). The homozygosity of transgenic plants was determined by segregation ratio of kanamycin selection marker and was further confirmed by PCR analysis.

Quantitative RT-PCR analysis

Total RNAs were extracted and purified from cotton tissues using the Qiagen RNeasy mini kit (Qiagen, Hilden, Germany). First-strand synthesis of cDNAs from these RNAs was performed using M-MLV reverse transcriptase (Promega, Madison, WI) according to the manufacturer's instructions. Gene expression were analysed by real-time quantitative PCR using the fluorescent intercalating dye SYBR-Green in a detection system (Opticon2; MJ Research) as described previously (Li *et al.*, 2005). A cotton ubiquitin gene (*GhUBI1*, access number in GenBank: EU604080) was used as a standard control in RT-PCR reactions. RT-PCR data are mean values and standard deviations (bar) of three independent experiments with three biological replicates. All primers used are listed in Table S3.

TFU analysis

For TFU (total fibre units) analysis, the cultured ovules of both wild type and transgenic plants were stained with 0.02% toluidine blue (Sigma) for 30 s, washed with running water for 1 min and dried. Then ovules (10 for each measurement) were eluted in the mixture (acetic acid:ethanol:water = 10 : 95 : 5) for 2 h, and the extracted solution was assayed by measuring absorbance at 624 nm using a diode array spectrophotometer (Beasley *et al.*, 1974; Gong *et al.*, 2014). Experimental data are mean values and standard deviations (bar) of three independent experiments with three biological replicates.

Fibre quality characteristic test

The mature fibres from the 14-3-3 overexpression/RNAi transgenic cotton plants and wild type grown in field at the same conditions were used to determine fibre quality (including fibre length, strength and micronaire etc.). The fibre samples (10 g/each sample) were prepared under the same humidity and temperature conditions before the measurements. Cotton fibre length, strength and micronaire etc. were determined with a high volume fibre test system (Premier HFT 9000, Premier Evolvics Pvt Ltd., Coimbatore, India). The fibre quality test was carried out in Institute of Agricultural Quality Standards and Testing Technology Research, Hubei Academy of Agricultural Sciences, Wuhan, China.

Subcellular localization

To construct GhBZR1:eGFP vector, the coding sequence of eGFP gene was cloned into pBluescript II SK+vector to obtain an intermediate construct pSK-eGFP. Subsequently, the coding sequence of GhBZR1 was cloned into the pSK-eGFP vector at a position upstream of eGFP gene. The constructed GhBZR1:eGFP fusion gene was cloned into pBI121 vector at *BamHI/SacI* sites, replacing *GUS* gene. All primers used are listed in Table S2.

Agrobacterium cells containing GhBZR1:eGFP construct were suspended in the induction medium (10 mM MgCl$_2$, 150 μM acetosyringone and 10 mM MES buffer, pH 5.6) and were infiltrated into young leaves of 4-week-old tobacco plants. After 36–48 h, GFP fluorescence in the leaf cells was detected by a SP5 Meta confocal laser microscope (Leica, Germany).

Bimolecular fluorescence complementation assay

To generate the vector system for Bimolecular fluorescence complementation (BiFC) analysis, fragments of *eYFP* gene encoding N-terminal 155 amino acids or C-terminal 84 amino acids were amplified by PCR to construct pSPYNE and pSPYCE vectors. Subsequently, the coding sequences of *Gh14-3-3L* and GhBZR1 genes were amplified by PCR using gene-specific primers (Table S2), and cloned into these vectors at a position upstream of the *eYFP* sequence, generating the recombinant constructs pSPYNE-Gh14-3-3L and pSPYCE-GhBZR1, respectively. The constructs were then introduced into onion epidermal cells by DNA particle bombardment according to the manufacturer's instructions (Biolistic PDS-1000/He Particle Delivery System; Bio-Rad, Hercules, CA 94547, USA). The YFP fluorescence was observed under a SP5 Meta confocal laser microscope (Leica, Wetzlar, Germany) after the epidermal layers of onions were cultured on MS medium at 25 °C in dark for 48 h (Li *et al.*, 2013).

Yeast two-hybrid assay

The coding sequences of *GhBIN2* and *Gh14-3-3* genes were cloned into pGBKT7 bait vector in frame with the GAL4-DNA-binding domain, respectively. The coding sequence of *GhBZR1* gene was cloned into the pGADT7 prey vector in frame with the GAL4-activation domain. The first 20 amino acids of GhBZR1 protein sequence was deleted in the construct. The constructs were transferred into yeast strain AH109, and interactions were tested by His selection and β-galactosidase assays (X-Gal filter lift assay) (Zhang *et al.*, 2010). All primers used are listed in Table S2.

Extraction of total proteins

Extraction of total proteins from cotton fibres was performed by the method described by Yao *et al.* (2006). Briefly, 2 g (fresh weight) of cotton fibres were ground to fine powders in liquid nitrogen with 10% SiO$_2$ and 10% PVP of sample weight, and then washed by 15 ml cold acetone. After lyophilized for 60 min, the powders were resuspended in 5 ml of extraction buffer (50 mM Tris-HCl, pH 8.65, 2% SDS, 30% sucrose and 2% 2-mercaptoethanol) and equal volume of Tris-saturated phenol (pH 7.8). After vortexed for 5 min, the samples were centrifuged for 10 min at 12 000 g. The phenolic phase was collected and precipitated with five volumes of 100 mM ammonium acetate in methanol at −20 °C for overnight. After centrifuging 15 min, the pellet was washed three times with cold 80% acetone in water, and then was lyophilized and stored at −80 °C. When the proteins were dissolved in WB buffer (7 M urea, 2 M thiourea and 4% CHAPS) and quantified using Bradford protein assay kit (Bio-Rad) for Western blot analysis.

SDS-PAGE and Western blot analysis

Proteins were quantified using a Bradford protein assay kit (Bio-Rad). Twenty micrograms of total protein extract was loaded per lane, separated by 10% SDS-PAGE, and electroblotted using semidry transfer method to polyvinylidene difluoridemembranes. The membranes were blocked in 5% nonfat dry milk in Trisbuffered saline (50 mM Tris-HCl, 150 mM NaCl, pH 7.5) overnight at 4 °C and then incubated with the specific anti-GhBZR1 antibodies (1 : 10 dilution) in Tris-buffered saline containing 0.1% Tween 20 (TBST) and 5% nonfat dry milk for 2 h at room temperature. Finally, the membranes were washed three times with TBST and incubated with goat anti-rabbit IgG secondary antibodies conjugated to horseradish peroxidase (Bio-Rad). Signals were visualized with a diaminobenzidine kit.

In vitro kinase assay

The coding sequence of *GhBIN2* gene was cloned into pET32a vector, and the coding sequence of *GhBZR1* gene was cloned into pMAL vector. His-GhBIN2 and MBP-BZR1 proteins were expressed in *Escherichia coli* cells and affinity-purified with the affinity tags, respectively. MBP-GhBZR1 is a MBP-GhBZR1 fusion protein containing amino acids 21–313 of GhBZR1. *In vitro* kinase reactions were performed with 10 mM ATP, 10 mM MgCl₂, 20 mM Tris-HCl (pH 7.5), and 100 mM NaCl in a total volume. The reaction was incubated at 30 °C for 1 h, or times indicated, and then stopped by adding SDS-PAGE sample buffer. After boiling for 5 min, proteins were separated on a SDS-PAGE gel with Phos-tag™ Acrylamide AAL-107 5 mM Aqueous Solution.

In vitro pull-down assay

Unphosphorylated and phosphorylated MBP-BZR1 proteins were mixed together and incubated with beads containing GST-Gh14-3-3L fusion protein in binding buffer (50 mM Tris-Cl, pH 7.5; 5 mM MgCl₂). After incubation for 1 h, the beads were washed with binding buffer, and Gh14-3-3L-binding proteins were eluted with elution buffer (50 mM Tris-Cl, pH 7.5, 5 mM Glutathione). Proteins were separated on a SDS-PAGE gel with Phos-tag™ Acrylamide AAL-107 5 mM Aqueous Solution and detected using a polyclonal anti-GhBZR1 antibody in Western blot analysis.

Yeast one-hybrid assay

GhXTH1 and *GhEXP* promoter fragments were integrated into the linearized pAbAi vector, respectively, generating recombinant vectors pAbAi-GhXTH1p and pAbAi-GhEXPp. The constructs were transferred into yeast strain Y1HGold and were tested by SD/-Ura with 350 ng/ml AbA (Aureobasidin A). The coding sequence of *GhBZR1* was cloned into pGADT7 vector and transferred into yeast strain Y1HGold which contained the pAbAi-GhXTH1p or pAbAi-GhEXPp. Interaction of GhXTH1p or pAbAi-GhEXPp and GhBZR1 were tested by SD/-Leu with 350 ng/ml AbA. All primers used are listed in Table S2.

Electrophoretic mobility shift assay

Electrophoretic mobility shift assay (EMSA) was carried out using a molecular probes' fluorescence-based EMSA Kit (Invitrogen, Life Technologies, Grand Island, NY 14072, USA). The coding sequence of *GhBZR1* was cloned into pMAL-c2 vector downstream the *malE* gene, which encodes maltose-binding protein (MBP), resulting in *MBP-GhBZR1* fusion gene. The MBP-GhBZR1 fusion proteins were expressed in *Escherichia coli* strain BL21 (DE3), and purified using MBP's affinity for maltose (NEW ENGLAND) for protein-DNA-binding analysis. A pair of 48-bp oligonucleotides from *GhXTH1* promoter was synthesized and annealed as DNA probes. Likewise, a pair of 39-bp oligonucleotides from *GhEXP* promoter was also synthesized and annealed as DNA probes. The binding reaction of MBP-GhBZR1 protein and DNA probe was performed in binding buffer (750 mM KCl, 0.5 mM dithiothreitol, 0.5 mM EDTA, 50 mM Tris, pH 7.4) at room temperature for 30 min. The reaction mixture was separated by nondenaturing polyacrylamide gel electrophoresis. The gel DNA was stained with SYBR® Green EMSA Nucleic Acid Gel Stain

(Invitrogen) and imaged using 254 nm UV epi-illumination (Kodak, Gel logic 2200).

Acknowledgement

This work was supported by National Natural Sciences Foundation of China (Grant No. 31171174), and the project from the Ministry of Agriculture of China for transgenic research (Grant No. 2014ZX08009-027B).

References

Aitken, A. (2006) 14-3-3 proteins: a historic overview. *Semin. Cancer Biol.* **16**, 162–172.

Bai, M.Y., Zhang, L.Y., Gampala, S.S., Zhu, S.W., Song, W.Y., Chong, K. and Wang, Z.Y. (2007) Functions of OsBZR1 and 14-3-3 proteins in brassinosteroid signaling in rice. *Proc. Natl Acad. Sci. USA*, **104**, 13839–13844.

Basra, A.S. and Malik, C.P. (1984) Development of the cotton fiber. *Int. Rev. Cytol.* **89**, 65–113.

Beasley, C.A. and Ting, I.P. (1973) The effects of plant growth substances on *in vitro* fiber development from fertilized cotton ovules. *Am. J. Bot.* **60**, 130–139.

Beasley, C.A. and Ting, I.P. (1974) The effects of plant growth substances on *in vitro* fiber development from unfertilized cotton ovules. *Am. J. Bot.* **61**, 188–194.

Beasley, C.A., Birnbaum, E.H., Dugger, W.M. and Ting, I.P. (1974) A quantitative procedure for estimating cotton fiber growth. *Stain Technol.* **49**, 85–92.

Brosa, C. (1999) Biological effects of brassinosteroids. *Crit. Rev. Biochem. Mol. Biol.* **34**, 339–358.

Cosgrove, D.J. (2000) Loosening of plant cell walls by expansins. *Nature*, **407**, 321–326.

Darling, D.L., Yingling, J. and Wynshaw-Boris, A. (2005) Role of 14-3-3 proteins in eukaryotic signaling and development. *Curr. Top. Dev. Biol.* **68**, 281–315.

DeLanghe, E.A.L. (1986) Lint development. In *Cotton Physiology* (Mauney, J.R. and Stewart, J.M., eds). Memphis, TN: The Cotton Foundation, pp. 325–350.

Gampala, S.S., Kim, T.W., He, J.X., Tang, W., Deng, Z., Bai, M.Y., Guan, S., Lalonde, S., Sun, Y., Gendron, J.M., Chen, H., Shibagaki, N., Ferl, R.J., Ehrhardt, D., Chong, K., Burlingame, A.L. and Wang, Z.Y. (2007) An essential role for 14-3-3 proteins in brassinosteroid signal transduction in Arabidopsis. *Dev. Cell*, **13**, 177–189.

Gendron, J.M. and Wang, Z.Y. (2007) Multiple mechanisms modulate brassinosteroid signaling. *Curr. Opin. Plant Biol.* **10**, 436–441.

Gong, S.Y., Huang, G.Q., Sun, X., Qin, L.X., Li, Y., Zhou, L. and Li, X.B. (2014) Cotton *KNL1* encoding a class II KNOX transcription factor is involved in regulation of fiber development. *J. Exp. Bot.* **65**, 4133–4147.

Guo, Z., Fujioka, S., Blancaflor, E.B., Miao, S., Guo, X. and Li, J. (2010) TCP1 modulates brassinosteroid biosynthesis by regulating the expression of the key biosynthetic gene DWARF4 in *Arabidopsis thaliana*. *Plant Cell*, **22**, 1161–1173.

He, J.X., Gendron, J.M., Sun, Y., Gampala, S.S., Gendron, N., Sun, C.Q. and Wang, Z.Y. (2005) BZR1 is a transcriptional repressor with dual roles in brassinosteroid homeostasis and growth responses. *Science*, **307**, 1634–1638.

Hothorn, M., Belkhadir, Y., Dreux, M., Dabi, T., Noel, J.P., Wilson, I.A. and Chory, J. (2011) Structural basis of steroid hormone perception by the receptor kinase BRI1. *Nature*, **474**, 467–471.

Igarashi, D., Ishida, S., Fukazawa, J. and Takahashi, Y. (2001) 14-3-3 proteins regulate intracellular localization of the bZIP transcriptional activator RSG. *Plant Cell*, **13**, 2483–2497.

Ishida, S., Fukazawa, J., Yuasa, T. and Takahashi, Y. (2004) Involvement of 14-3-3 signaling protein binding in the functional regulation of the transcriptional activator REPRESSION OF SHOOT GROWTH by gibberellins. *Plant Cell*, **16**, 2641–2651.

Jaillais, Y., Hothorn, M., Belkhadir, Y., Dabi, T., Nimchuk, Z.L., Meyerowitz, E.M. and Chory, J. (2011) Tyrosine phosphorylation controls brassinosteroid receptor activation by triggering membrane release of its kinase inhibitor. *Genes Dev.* **25**, 232–237.

Ji, S.J., Lu, Y.C., Feng, J.X., Wei, G., Li, J., Shi, Y.H., Fu, Q., Liu, D., Luo, J.C. and Zhu, Y.X. (2003) Isolation and analyses of genes preferentially expressed during early cotton fiber development by subtractive PCR and cDNA array. *Nucleic Acids Res.* **31**, 2534–2543.

Kim, H.J. and Triplett, B.A. (2001) Cotton fiber growth in planta and *in vitro*. Models for plant cell elongation and cell wall biogenesis. *Plant Physiol.* **127**, 1361–1366.

Lee, J.J., Woodward, A.W. and Chen, Z.J. (2007) Gene expression changes and early events in cotton fibre development. *Ann. Bot.* **100**, 1391–1401.

Li, J. and Chory, J. (1997) A putative leucine-rich repeat receptor kinase involved in brassinosteroid signal transduction. *Cell*, **90**, 929–938.

Li, J. and Nam, K.H. (2002) Regulation of brassinosteroid signaling by a GSK3/SHAGGY-like kinase. *Science*, **295**, 1299–1301.

Li, X.B., Fan, X.P., Wang, X.L., Cai, L. and Yang, W.C. (2005) The cotton ACTIN1 gene is functionally expressed in fibers and participates in fiber elongation. *Plant Cell*, **17**, 859–875.

Li, L., Yu, X., Thompson, A., Guo, M., Yoshida, S., Asami, T., Chory, J. and Yin, Y. (2009) Arabidopsis MYB30 is a direct target of BES1 and cooperates with BES1 to regulate brassinosteroid-induced gene expression. *Plant J.* **58**, 275–286.

Li, L., Ye, H., Guo, H. and Yin, Y. (2010) Arabidopsis IWS1 interacts with transcription factor BES1 and is involved in plant steroid hormone brassinosteroid regulated gene expression. *Proc. Natl Acad. Sci. USA*, **107**, 3918–3923.

Li, D.D., Ruan, X.M., Zhang, J., Wu, Y.J., Wang, X.L. and Li, X.B. (2013) Cotton plasma membrane intrinsic protein 2s (PIP2s) selectively interact to regulate their water channel activities and are required for fibre development. *New Phytol.* **199**, 695–707.

Long, R.L., Bange, M.P., Gordon, S.G., van der Sluijs, M.H.J., Naylor, G.R.S. and Constable, G.A. (2010) Fiber quality and textile performance of some Australian cotton genotypes. *Crop Sci.* **4**, 1509–1518.

Mason, M.S. and Cosgrove, D.J. (1995) Expansin mode of action on cell walls: analysis of wall hydrolysis stress relaxation and binding. *Plant Physiol.* **107**, 87–100.

Moore, B.W. and Perez, V.J. (1967) Specific acidic proteins of the nervous system. In *Physiological and Biochemical Aspects of Nervous Integration* (Carlson, F., ed.), pp. 343–359. Woods Hole, MA: Prentice Hall.

Oh, E., Zhu, J.Y. and Wang, Z.Y. (2012) Interaction between BZR1 and PIF4 integrates brassinosteroid and environmental responses. *Nat. Cell Biol.* **14**, 802–809.

Park, C.H., Kim, T.W., Son, S.H., Hwang, J.Y., Lee, S.C., Chang, S.C., Kim, S.H., Kim, S.W. and Kim, S.K. (2010) Brassinosteroids control *AtEXPA5* gene expression in Arabidopsis thaliana. *Phytochemistry*, **71**, 380–387.

Ryu, H., Kim, K., Cho, H., Park, J., Choe, S. and Hwang, I. (2007) Nucleocytoplasmic shuttling of BZR1 mediated by phosphorylation is essential in Arabidopsis brassinosteroid signaling. *Plant Cell*, **19**, 2749–2762.

Schoonheim, P.J., Sinnige, M.P., Casaretto, J.A., Veiga, H., Bunney, T.D., Quatrano, R.S. and de Boer, A.H. (2007) 14-3-3 adaptor proteins are intermediates in ABA signal transduction during barley seed germination. *Plant J.*, **49**, 289–301.

Sehnke, P.C., DeLille, J.M. and Ferl, R.J. (2002) Consummating signal transduction: the role of 14-3-3 proteins in the completion of signal-induced transitions in protein activity. *Plant Cell*, **14**, S339–S354.

She, J., Han, Z., Kim, T.W., Wang, J., Cheng, W., Chang, J., Shi, S., Wang, J., Yang, M., Wang, Z.Y. and Chai, J. (2011) Structural insight into brassinosteroid perception by BRI1. *Nature*, **474**, 472–476.

Sun, Y., Veerabomma, S., Abdel-Mageed, H.A., Fokar, M., Asami, T., Yoshida, S. and Allen, R.D. (2005) Brassinosteroid regulates fiber development on cultured cotton ovules. *Plant Cell Physiol.* **46**, 1384–1391.

Sun, Y., Fan, X.Y., Cao, D.M., Tang, W., He, K., Zhu, J.Y., He, J.X., Bai, M.Y., Zhu, S., Oh, E., Patil, S., Kim, T.W., Ji, H., Wong, W.H., Rhee, S.Y. and Wang, Z.Y. (2010) Integration of brassinosteroid signal transduction with the

transcription network for plant growth regulation in Arabidopsis. *Dev. Cell*, **19**, 765–777.

Vert, G. and Chory, J. (2006) Downstream nuclear events in brassinosteroid signaling. *Nature*, **441**, 96–100.

Vert, G., Nemhauser, J.L., Geldner, N., Hong, F. and Chory, J. (2005) Molecular mechanisms of steroid hormone signaling in plants. *Annu. Rev. Cell Dev. Biol.* **21**, 177–201.

Wang, X. and Chory, J. (2006) Brassinosteroids regulate dissociation of BKI1, a negative regulator of BRI1 signaling, from the plasma membrane. *Science*, **313**, 1118–1122.

Wang, X.L. and Li, X.B. (2009) The GhACS1 gene encodes anacyl-CoA synthetase which is essential for normal microsporogenesis in early anther development of cotton. *Plant J.* **57**, 473–486.

Wang, Z.Y., Nakano, T., Gendron, J., He, J., Chen, M., Vafeados, D., Yang, Y., Fujioka, S., Yoshida, S. and Asami, T. (2002) Nuclear-localized BZR1 mediates brassinosteroid-induced growth and feedback suppression of brassinosteroid biosynthesis. *Dev. Cell*, **2**, 505–513.

Wang, H., Zhu, Y., Fujioka, S., Asami, T., Li, J. and Li, J. (2009) Regulation of Arabidopsis brassinosteroid signaling by atypical basic helix-loop-helix proteins. *Plant Cell*, **21**, 3781–3791.

Wang, H., Yang, C., Zhang, C., Wang, N., Lu, D., Wang, J., Zhang, S., Wang, Z.X., Ma, H. and Wang, X. (2011) Dual role of BKI1 and 14-3-3s in brassinosteroid signaling to link receptor with transcription factors. *Dev. Cell*, **21**, 825–834.

Whittaker, D.J. and Triplett, B.A. (1999) Gene-specific changes in α-tubulin transcript accumulation in developing cotton fibers. *Plant Physiol.* **121**, 181–188.

Yao, Y., Yang, Y.W. and Liu, J.Y. (2006) An efficient protein preparation for proteomic analysis of developing cotton fibers by 2-DE. *Electrophoresis*, **27**, 4559–4569.

Yin, Y., Wang, Z.Y., Mora-Garcia, S., Li, J., Yoshida, S., Asami, T. and Chory, J. (2002) BES1 accumulates in the nucleus in response to brassinosteroids to regulate gene expression and promote stem elongation. *Cell*, **109**, 181–191.

Yin, Y., Vafeados, D., Tao, Y., Yoshida, S., Asami, T. and Chory, J. (2005) A new class of transcription factors mediates brassinosteroid regulated gene expression in Arabidopsis. *Cell*, **120**, 249–259.

Zhang, L., Wang, H., Liu, D., Liddington, R. and Fu, H. (1997) Raf-1 kinase and exoenzyme S interact with 14-3-3zeta through a common site involving lysine 49. *J. Biol. Chem.* **272**, 13717–13724.

Zhang, L.Y., Bai, M.Y., Wu, J., Zhu, J.Y., Wang, H., Zhang, Z., Wang, W., Sun, Y., Zhao, J., Sun, X., Yang, H., Xu, Y., Kim, S.H., Fujioka, S., Lin, W.H., Chong, K., Lu, T. and Wang, Z.Y. (2009) Antagonistic HLH/bHLH transcription factors mediate brassinosteroid regulation of cell elongation and plant development in rice and Arabidopsis. *Plant Cell*, **21**, 3767–3780.

Zhang, Z.T., Zhou, Y., Li, Y., Shao, S.Q., Li, B.Y., Shi, H.Y. and Li, X.B. (2010) Interactome analysis of the six cotton 14-3-3s that are preferentially expressed in fibres and involved in cell elongation. *J. Exp. Bot.* **61**, 3331–3344.

17

Site-specific proteolytic degradation of IgG monoclonal antibodies expressed in tobacco plants

Verena K. Hehle[1,†], Raffaele Lombardi[2,†], Craig J. van Dolleweerd[1], Mathew J. Paul[1], Patrizio Di Micco[3], Veronica Morea[4], Eugenio Benvenuto[2], Marcello Donini[2,*] and Julian K-C. Ma[1,*]

[1]Molecular Immunology Unit, Division of Clinical Sciences, St. George's University of London, London, UK
[2]Biotechnology Laboratory, ENEA Casaccia, Rome, Italy
[3]Department of Biochemical Sciences 'A. Rossi Fanelli', 'Sapienza' University of Rome, Rome, Italy
[4]CNR-National Research Council of Italy, Institute of Molecular Biology and Pathology, 'Sapienza' University of Rome, Rome, Italy

*Correspondence (JKCM:

email jma@sgul.ac.uk; MD:

email marcello.donini@enea.it)
[†]These authors contributed equally to this work.

Keywords: molecular farming, monoclonal antibodies, proteolysis, N-terminal sequencing.

Summary

Plants are promising hosts for the production of monoclonal antibodies (mAbs). However, proteolytic degradation of antibodies produced both in stable transgenic plants and using transient expression systems is still a major issue for efficient high-yield recombinant protein accumulation. In this work, we have performed a detailed study of the degradation profiles of two human IgG1 mAbs produced in plants: an anti-HIV mAb 2G12 and a tumour-targeting mAb H10. Even though they use different light chains (κ and λ, respectively), the fragmentation pattern of both antibodies was similar. The majority of Ig fragments result from proteolytic degradation, but there are only a limited number of plant proteolytic cleavage events in the immunoglobulin light and heavy chains. All of the cleavage sites identified were in the proximity of interdomain regions and occurred at each interdomain site, with the exception of the V_L/C_L interface in mAb H10 λ light chain. Cleavage site sequences were analysed, and residue patterns characteristic of proteolytic enzymes substrates were identified. The results of this work help to define common degradation events in plant-produced mAbs and raise the possibility of predicting antibody degradation patterns 'a priori' and designing novel stabilization strategies by site-specific mutagenesis.

Introduction

The production of monoclonal antibodies (mAbs) by plant biotechnology has been carried out for several years (Hiatt, 1991). This can now be achieved both by stable integration of immunoglobulin (Ig) transgenes in the nuclear genome of a host plant and by using a more rapid strategy comprising transient expression of the Ig genes using vectors containing plant viral sequences to drive expression (Komarova et al., 2010). Significant advances have been achieved in the last few years. The original expression levels of ~1% of total tobacco soluble protein reported by Hiatt and colleagues have now been increased by up to 10-fold in transgenic plants (Ramessar et al., 2008) and 40-fold using transient plant expression systems (Komarova et al., 2010). Reaching these yields has allowed the field to progress, develop manufacturing processes and move closer towards the prospect of achieving commercial viability. Indeed, manufacturing approval for plant-derived mAbs has been granted in Germany, and recently, a first-in-human Phase I clinical trial for a plant-derived mAb was approved by the UK Medicines and Healthcare product Regulatory Agency (MHRA).

However, proteolytic degradation of antibodies produced in plants is still a major issue. Hundreds of plant genes encoding enzymes involved in proteolytic pathways have been identified, which are directly or indirectly involved in the hydrolysis of peptide bonds (Pesquet, 2012; Rawlings et al., 2010). Moreover, proteolytic enzymes are very abundant in some subcellular compartments, such as the vacuoles, tonoplast and apoplast, which is the final destination of antibodies targeted to the secretory pathway (De Wilde et al., 1996; van der Hoorn, 2008). Unintended proteolysis driven by the plant proteolytic machinery can dramatically affect the final yield of intact IgGs and even lead to almost complete (i.e. more than 90%) degradation of the final product (Villani et al., 2009). In several studies, SDS-PAGE analysis under non-reducing conditions has highlighted the presence of protein fragments between 20 and 150 kDa, indicating a significant degradation of antibody expressed in plants (De Muynck et al., 2010; De Neve et al., 1993; Lombardi et al., 2012; Ma et al., 1994; Sharp and Doran, 2001; Stevens et al., 2000). These patterns may be ascribed either to antibody assembly intermediates or to proteolytic processes taking place in planta, at some stage of protein synthesis/intracellular trafficking, and/or ex planta, during extraction and subsequent purification procedures (Rivard et al., 2006; Sharp and Doran, 2001). Even where the major proportion of the extracted antibody remains intact, the presence of mixtures of immunoglobulin fragments presents important challenges for downstream processing, often increasing the cost of manufacture.

Recent work has increased our comprehension of the mechanisms of antibody degradation in tobacco plants by demonstrating that formation and accumulation of IgG proteolytic fragments can occur at several points in the secretory pathway, including the apoplastic space (Hehle et al., 2011). For example, specific aspartic, cysteine and serine peptidases can exert strong proteolytic activity in the tobacco extracellular space (Delannoy et al., 2008). Furthermore, it has been shown that ex planta proteolytic activity due to the release of proteases from subcellular compartments during tissue disruption and extraction is not a major issue under most commonly used extraction conditions (Hehle *et al.*, 2011).

IgG degradation is generally studied by Western blot analysis using anti-heavy (HC) and/or anti-light (LC) chain antibodies. This does not ensure detection of antibody fragments with incomplete epitopes, leading to potential underestimation of the extent of antibody degradation in plant extracts. Different approaches have been followed to overcome unintended proteolysis in plants. Recombinant proteins have been redirected to specific tissues or cellular organelles (Stoger *et al.*, 2005; Vitale and Pedrazzini, 2005). Production of protease knockdown organisms, which had been successful in bacteria and yeast, cannot be extended to complex multicellular organisms such as higher plants without affecting their viability (Schaller, 2004). Transient expression of specific protease inhibitors was described as a promising approach to minimize *in planta* hydrolysis of recombinant antibodies (Benchabane *et al.*, 2008; Komarnytsky *et al.*, 2006). In a recent study, specific inhibitors of several plant protease classes (aspartic, serine and cysteine proteases) were transiently expressed in *Nicotiana benthamiana* to assess their ability to stabilize proteins targeted to the apoplast (Goulet *et al.*, 2012). This approach increased the accumulation of recombinant proteins along the cell secretory pathway. While some increase in product amount and quality has been achieved by the aforementioned procedures, their success has been limited.

Recent studies have attempted to characterize antibody degradation *in planta* by determining N-terminal sequences of the fragments (De Muynck *et al.*, 2009; Niemer *et al.*, 2014). Identification of sequences specifically prone to proteolytic cleavage in plants would allow protease classes involved in antibody degradation to be identified and new strategies to increase the yield of intact IgG molecules to be developed. In this work, we performed a detailed comparative analysis of the degradation products of two human IgG1 mAbs, named 2G12

and H10, which recognize the gp120 envelope of human immunodeficiency virus (Trkola *et al.*, 1996) and tenascin-C, a glycoprotein that is over-expressed in neoplasia (Borsi *et al.*, 1992), respectively. The two mAbs have identical heavy chain constant domains (i.e., C_H1–C_H3) and different light-chain isotypes (κ and λ in 2G12 and H10, respectively).

We used a range of complementary approaches to elucidate the determinants of Ig degradation in tobacco plants and identify specific protease sensitive regions in Ig sequences. The results of this work may ultimately help to maximize mAbs expression in plants and reduce downstream processing required to eliminate contaminating Ig fragments.

Results

Degradation of mAb 2G12 and H10 in plants

Initially, expression of mAb 2G12 in transgenic *Nicotiana tabacum* was compared with transient expression in *N. tabacum*. Plant extracts were passed through a protein A/G affinity column and bound antibody fragments were eluted, electrophoresed under non-reducing conditions and immunodetected by Western blotting using anti-human-γ and anti-human-κ chain antiserum (Figure 1a,b).

Fully assembled IgG was identified as the uppermost band with Mr ~150K in all lanes (labelled *). Overall, anti-κ antiserum detected more immunoreactive bands (labelled a–g) than anti-γ chain antiserum (labelled a–c). All the bands identified in the anti-γ chain blot were also detected with anti-κ chain antiserum, suggesting that these correspond to antibody fragments containing both chains. There was no significant difference between samples from transgenic and transient expression in the anti-κ chain blot (Panel b).

An extract from *N. benthamiana* transiently expressing mAb H10 was passed through a protein A/G affinity column and

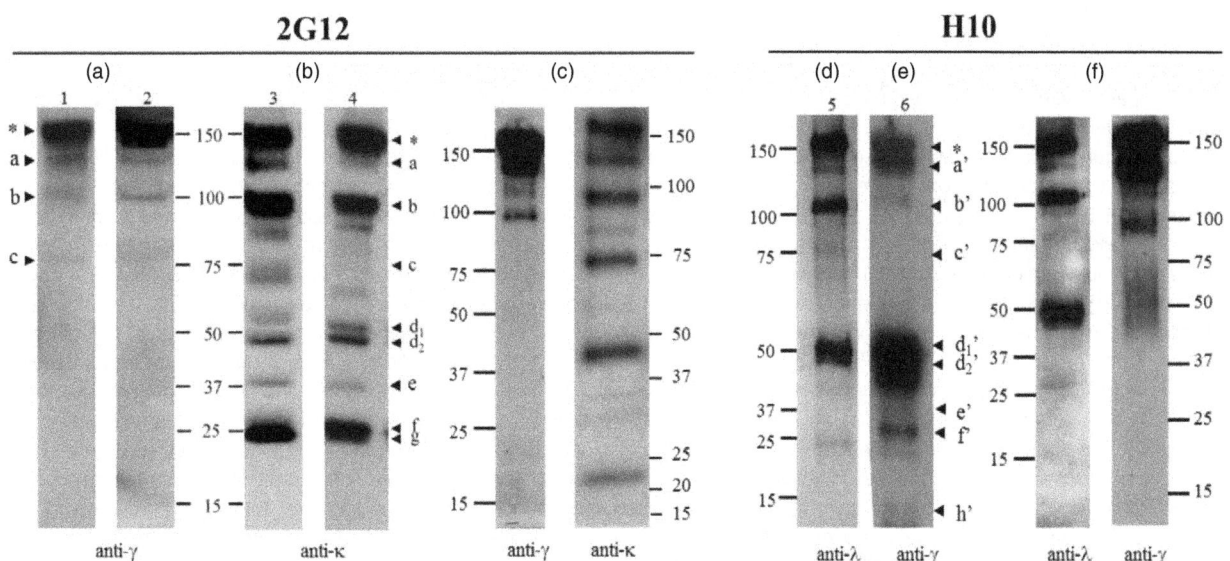

Figure 1 Analysis of plant-derived mAbs 2G12 and H10. SDS-4-15% (w/v) PAGE of agroinfiltrated leaf extracts or protein A/G purified antibodies was run under non-reducing conditions. For immunoblot analysis, (~100 ng of purified antibody was loaded on each lane) proteins were transferred onto nitrocellulose membrane. Detection was with anti-γ-chain (panels a and e), anti-κ-chain (panel b) or anti-λ-chain (panel d) antisera. Lanes 1 and 4: 2G12 from transgenic leaf extract. Lanes 2 and 3: 2G12 transiently expressed in *Nicotiana tabacum*. Lanes 5 and 6: H10 transiently expressed in *Nicotiana benthamiana*. (*) and lower case letters indicate fully assembled antibody and major antibody fragments bands, respectively. Plant extracts from transiently expressed 2G12 (c) and H10 (f) (~10 μg of total soluble proteins) were also detected by anti-γ-chain, anti-κ-chain or anti-λ-chain antisera.

bound antibody fragments were eluted electrophoresed under non-reducing conditions and detected by Western blotting and immunodetected using anti-λ and anti-γ chain antiserum (Sigma; Figure 1d,e). The number of fragments obtained for mAb H10 was similar to that observed for mAb 2G12, except that in the case of H10 the anti-γ chain antiserum detected more bands than the anti-λ one. H10 major bands with the same size as a, b, c, d_1, d_2, e and f in the mAb 2G12 blot were labelled a′, b′, c′, d_1′, d_2′, e′ and f′ accordingly. The H10 and 2G12 blots appeared to differ significantly only in the presence of low molecular weight H10 h′ band and 2G12 g band, and in the different intensities of some major bands (e.g. the signal of H10 band * and b′ is weaker, and that of bands d_1′ and d_2′ stronger, than the corresponding bands of 2G12).

Western blot analysis was also performed on extracts from tobacco plants transiently expressing mAb 2G12 and H10 (Figure 1c,f). Similar patterns in terms of number of bands were observed compared with Western blots deriving from purified antibodies. Some major differences in intensities were observed in the case of H10 bands b′ and d_2′ which were much more intense in the purified antibody (Figure 1d). This may be ascribed to the previously demonstrated protein A ability to specifically bind not only to the Fc but also to germ-line 3-derived V_H domain of H10 (Villani et al., 2009), which is present in both Fab and $(Fab)_2$ fragments. This result indicates that protein A/G affinity chromatography ensured the broadest enrichment of antibody fragments.

N-terminal sequencing of mAb fragments

N-terminal sequencing of antibody fragments was performed to identify major proteolytic cleavage sites. MAbs 2G12 and H10 were purified from plants by protein A/G affinity chromatography, separated by SDS-PAGE, blotted onto a PVDF membrane under non-reducing conditions and stained with Coomassie blue. The major bands a, b, d_2, f, g were clearly visible in the mAb 2G12 preparation (Figure 2a) as well as the two less intense bands c and e and band h. However, band d_1 was not detected. In the case of H10, the same bands revealed by Western blotting were visible, and the intensity of bands b′ and d_2′ was much higher compared with the corresponding 2G12 bands. The most intense and well-defined bands were individually cut from the membrane (Figure 2a,b left panels) and submitted for N-terminal sequencing by Edman degradation. The results of this analysis are summarized in Figure 2a,b right panels. As gel electrophoresis was run under non-reducing conditions, it was expected that some bands would comprise a mixture of heavy and light chains. In the case of 2G12 fragments, the N-terminal sequences of both light and heavy chains (DVVM and EVQL respectively) were identified from bands a, b and d_2 (Figure 2a). These were consistent with correct processing of the respective leader peptides. Analysis of fragment h revealed proteolytic cleavage of essentially the entire V_L and V_H as the light- and heavy-chain internal amino acids sequences (RVEI and TVSP) were identified. The light-chain cleavage site of fragment h was shared by fragment f, which had a different cleavage site in the heavy chain (leaving an N-terminal sequence of KVEP), resulting in the loss of the whole V_H and C_H1 domains. Band g appears to contain a mixture of light chain fragments as two N-terminal sequences were identified, one shared with fragments a, b and d_2 and an additional site near the junction between V_L and C_L (RVEI).

In the case of mAb H10, bands b′, c′, d_2′ and f′ all returned N-terminal sequences that were consistent with correct process-

ing of both light and heavy chain leader peptides (SELT and EVQL; Figure 2b). Within the heavy chain, three additional N-terminal sequences were identified: KKVE between C_H1 domain and hinge region, for band d_1′; and ISKA and KAKG at the end of the C_H2 domain, both for band h′, suggesting that this contains a mixture of heavy-chain fragments. No internal cleavage sites were identified within the H10 λ light chain.

To confirm this finding, extracts from mAb 2G12 and H10 expressing plants were prepared and electrophoresed under reducing conditions. Western blotting was then performed, using either anti-kappa or anti-lambda antisera for detection. Cleavage of the mAb 2G12 kappa chain was confirmed by the presence of a Mr~15K band, whereas no such cleavage product was observed for the H10 λ chain (Figure 2c).

Glycosylation and functional characterization of mAb fragments

Biotinylated concanavalin (con) A was used to identify glycosylated mAb 2G12 bands by Western blot (Figure 3a). Con-A is a lectin that recognizes high mannose glycans. Six major bands were identified, including the full-length antibody (*) and bands corresponding to a, b, c and d_2 in Figure 1. In addition, a further band between Mr 50–75K, which had previously been observed but not labelled, was detected. None of the fragments with a Mr <37K were detected (bands e, f, g and h). Individual protein bands were excised from an SDS-polyacrylamide gel and eluted. Samples from the eluted bands are shown in Figure 3b, in which the isolated fragments were electrophoresed, blotted onto nitrocellulose and probed with either anti-γ or anti-κ chain antisera. Western blot analysis of isolated fragment a shows two bands, one the expected size for band a and the other with a Mr similar to the intact antibody (band *) indicating that these two species cannot be completely separated. Each of the preparations of bands b, c, d_2, f, g and h contains a predominant species of the expected Mr. Western blotting with con-A on the purified Ig fragments confirmed the earlier results (not shown). Eluted proteins from bands b and d_2 were also subjected to SDS-PAGE under reducing conditions, transferred to nitrocellulose filters and probed with anti-human γ chain antiserum (Figure 3d). Three bands are visible, one at about Mr 50K corresponding to the size of the intact heavy chain (very faint in d_2), and two other bands at Mr 37K and 25K corresponding to HC fragments.

An HIV gp120–specific ELISA was performed to determine the activity of each fragment (Figure 3c). Bands were excised from Coomassie stained gel and all eluted using phosphate buffered saline (PBS). Due to the different intensity of the bands excised from the gel and the variation in the dimension the the of the manually excised gel slab, it was not possible to normalize the ELISA results and to perform a quantitative analysis of the binding activity. Antigen-binding activity was detected for bands a, d_2, f and g, as well as the full-length antibody (*). A weak signal was observed for bands b, c. However, antigen binding by these fragments cannot be excluded as excised bands (see Figure 2a) are much weaker compared with the other bands.

As for mAb 2G12, fragments of mAb H10 in bands *, a′, b′, c′, d_1′, d_2′ and h′ were excised from a Coomassie stained gel and eluted. The excised bands were visualized by Western blotting with anti-human γ (Figure 4a) and anti-human λ-chain (Figure 4b)–specific antisera. The purified bands were also subjected to SDS-PAGE under reducing conditions, transferred to

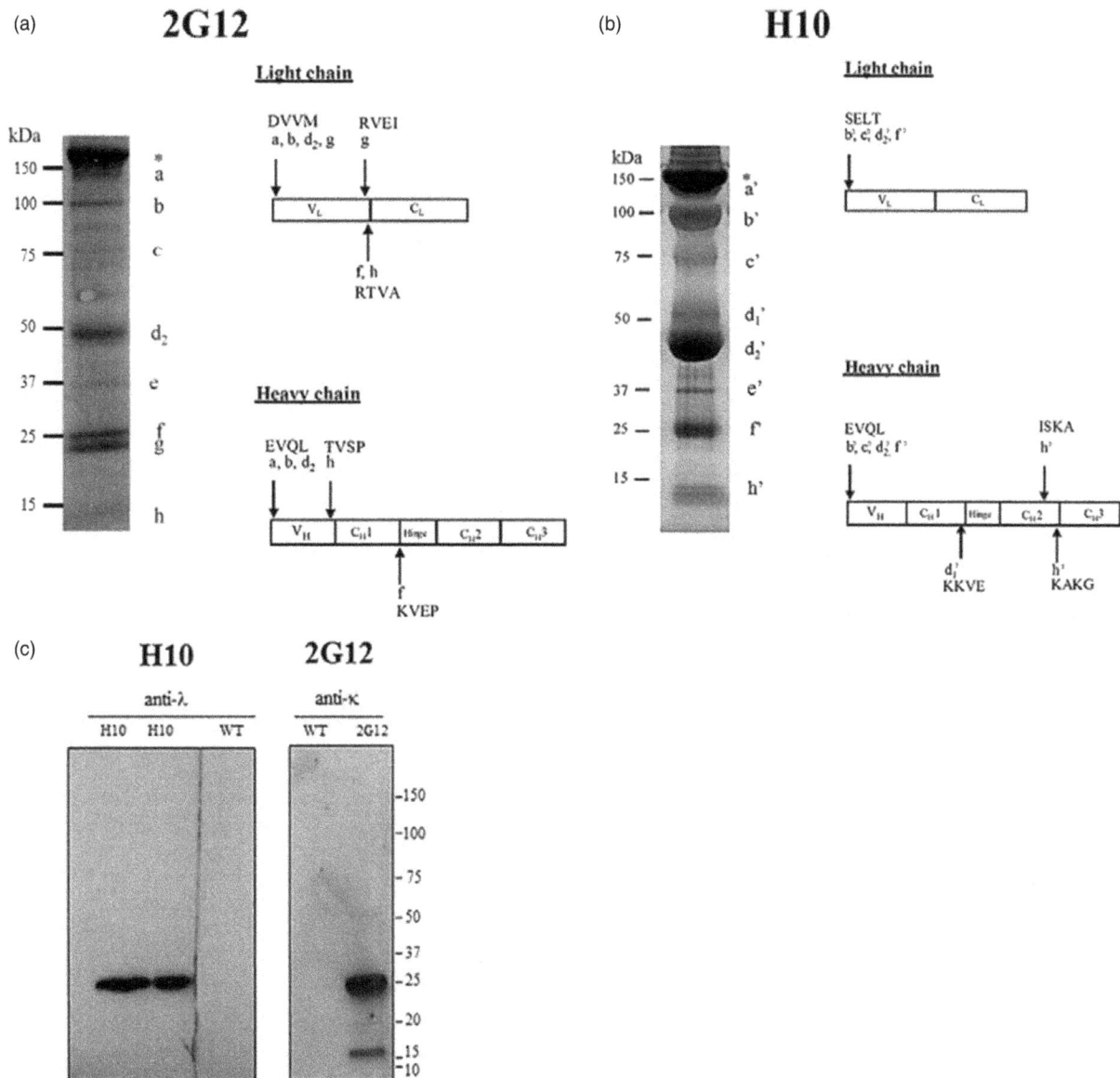

Figure 2 N-terminal sequencing results and cleavage sites for antibody fragments. Protein A/G purified mAb 2G12 (a, left panel) and H10 (b, left panel) were run on SDS-4-15% (w/v) PAGE under non-reducing conditions, transferred to PVDF membrane and stained with Coomassie blue. Protein bands were then excised from PVDF membranes and subjected N-terminal sequencing analysis. Right panels show diagrammatic representations of light and heavy chains. Amino acid sequences returned from N-terminal sequencing by Edman degradation are shown by upper case letters. Bands are labelled as in Figure 1. Panel (c) shows immunoblot analysis of light chain in plant extracts expressing mAb 2G12 and mAb H10. Tobacco leaf extracts transiently expressing mAb H10 and 2G12 (10 μg of total soluble proteins) were run on SDS-4-20% (w/v) PAGE under reducing conditions. Proteins were transferred onto nitrocellulose membrane and probed with horseradish peroxidase (HRP)-conjugated anti-human κ and λ-chain antiserum. WT represents the mock-infiltrated plants used as control.

nitrocellulose filters and probed with con-A or anti-human-γ chain antiserum (Figure 4c). In preparations of fragments *, a' and c' the relative molecular mass of the detected band was ~50K corresponding to the size of the intact HC. A clear signal with con-A was detected in these three samples, as well as for a smaller protein fragment in band sample h', approximate Mr 12K, proving that this fragment derives from a specific HC degradation. Analysis of bands b' and d₂' with anti-human γ-chain antiserum revealed specific bands at about Mr 25K (Figure 4c).

To detect antigen-binding activity of individual mAb H10 fragments, a whole-cell ELISA was performed using a human lung cancer cell line (A549) expressing the antigen tenascin-C (Fitch *et al.*, 2011). The results in Figure 4d indicate that the full-length antibody (band *) recognized the antigen as well as the positive control (+ve). Significant binding was also detected for fragments b' and d₂' and to some extent band c'. No antigen recognition was detected for band d₁'.

The results obtained for mAbs 2G12 and H10 are summarized in Table 1. From all data, it emerges that bands b and b' of both mAbs most likely comprise (Fab)₂ fragments while H10 bands d₁' and d₂' Fc and Fab fragments, respectively. Additionally, 2G12 bands b, d₂ and g likely contain different antibody fragments.

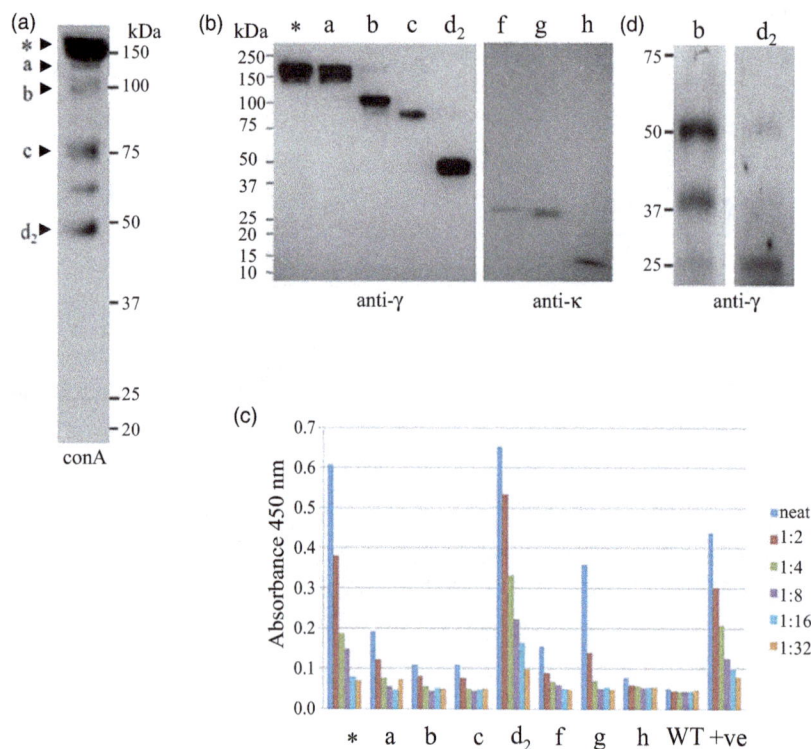

Figure 3 Immunoblot blot analysis of N-linked glycan structures for mAb 2G12 and antigen binding activity of purified fragments. (a) Affinity-purified mAb 2G12 (~500 ng) was separated by SDS-4-15% (w/v) PAGE under non-reducing conditions and transferred onto a nitrocellulose membrane. The membrane was blocked and probed with biotinylated con-A, followed by horseradish peroxidase (HRP)-conjugated streptavidin. (b) Affinity-purified mAb 2G12 was separated by SDS-4-15% (w/v) PAGE under non-reducing conditions and stained with Coomassie blue. Protein bands were excised individually from gel and extracted in 50 μL phosphate buffered saline (PBS). The bands were run again on SDS-4-15% (w/v) PAGE under non-reducing conditions, transferred onto nitrocellulose filters and probed with HRP-conjugated anti-human γ or κ-chain antiserum. (c) Antigen binding was assessed by ELISA using HIV gp120 coated on solid phase. Proteins extracted from gel were loaded at different dilutions as indicated. Detection of binding was with HRP-conjugated anti-human κ-chain-specific antiserum. WT indicates extract from non-transformed wild-type plant, +ve is CHO-derived 2G12 mAb. (d) Proteins eluted from bands b and d_2 were run on SDS-4-15% (w/v) PAGE under reducing conditions, transferred onto nitrocellulose filters and probed with HRP-conjugated anti-human γ.

Discussion

We have performed a detailed study of the degradation profiles of two human IgG1 mAbs of therapeutic interest produced in plants. Expression of IgG antibodies in plants has long been observed to be associated with a number of smaller Ig fragments, which have been ascribed to either assembly intermediates (e.g. free or aggregated light or heavy chains, or partially assembled IgG molecules) or undefined degradation products (Wongsamuth and Doran, 1997). Ig fragments have a highly consistent pattern between different murine antibodies and between different human antibodies, and preliminary findings suggest that they result predominantly from degradation rather than partial assembly (Hehle et al., 2011; Lombardi et al., 2010). Fragmentation of therapeutic recombinant proteins can also be a major problem for the biotechnology industry using CHO cells as a production platform. In some cases, proteolytic cleavage may result in reduced product yield, decreased product quality and diminished biological activity. For example, extensive proteolytic clipping of both recombinant interferon gamma and human nerve growth factor expressed in CHO cells has been documented (Rita Costa et al., 2010). Fragmentation of therapeutic antibodies has been well documented during storage of the final purified product (Vlasak and Ionescu, 2011) and preparation of a stabilizing formulation that prevents antibody deterioration

(degradation and aggregation) can be an important issue (Ishikawa et al., 2010). Interestingly, it has been shown that some residual CHO cell protease activity can still occur in highly purified monoclonal antibodies (Gao et al., 2010) and that human mAbs belonging to the same subclass (IgG1) share similar fragmentation patterns, with specific proteolytic cleavage mainly within the hinge region (Fan et al., 2012).

The fragmentation pattern of the two mAbs studied in this work was also highly similar, even though they contain different light chains. The majority of identified Ig fragments resulted from proteolytic cleavage, but a relatively small number of sites appeared to be involved. Results obtained from glycosylation analysis (conA) and antigen-binding experiments with gel excised purified fragments as well as N-terminal sequencing (summarized in Table 1) permitted us to identify the composition of several band complexes for both mAbs. H10 band b' (Mr 100K), most likely comprising tetrameric (Fab)$_2$ fragments, had, as expected, intact HC and LC N-termini, lacked glycan moiety and showed a clear antigen-binding recognition. Moreover, a band at about Mr 25K on reducing Western blot analysis using anti-γ-chain antibody was observed (Figure 4c). The corresponding 2G12 b band (Mr 100K) had also intact HC and LC N-termini but appeared glycosylated and had a very low signal on ELISA possibly due to the weak intensity of the gel-eluted band (Figure 2a). The presence of both intact HC and HC degradation products

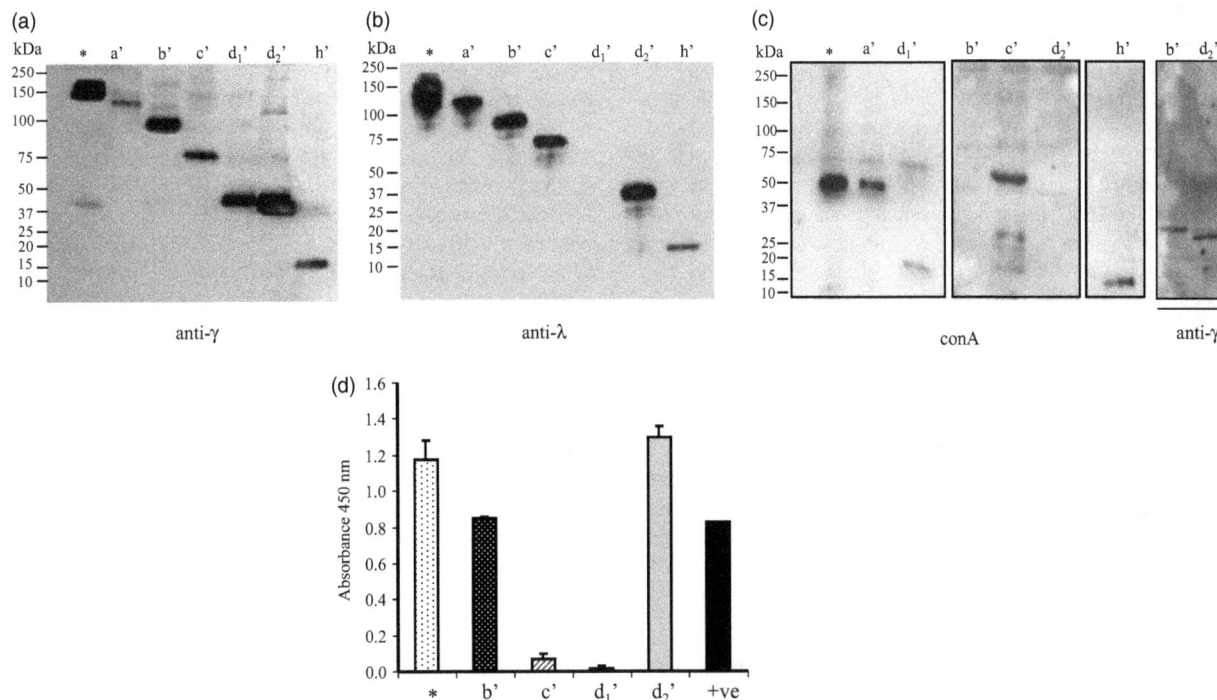

Figure 4 Immunoblot blot analysis of N-linked glycan structures for mAb H10 and antigen-binding activity of purified fragments. Affinity-purified mAb H10 was separated by SDS-4-15% (w/v) PAGE under non-reducing conditions and stained with Coomassie blue. Western blot analysis was then performed on individual protein bands that were excised from gel, homogenized, eluted and separated again by SDS-4-15% (w/v) PAGE under non-reducing or reducing conditions. Proteins were blotted to a nitrocellulose membrane and probed with: (a) horseradish peroxidase (HRP)-conjugated anti-human γ-chain antiserum (non-reducing conditions); or (b) HRP-conjugated anti-human λ-chain antiserum (non-reducing conditions); and (c) biotinylated concanavalin A followed by HRP-conjugated streptavidin (reducing conditions) and HRP-conjugated anti-human γ-chain antiserum (reducing conditions). Antigen binding was assessed by ELISA (d) using human lung cancer cells (A549) coated on the solid phase. Binding was detected with HRP-conjugated anti-human λ-chain-specific antiserum. H10 +ve is CHO-derived H10 mAb. Values are the mean of two independent experiments.

Table 1 Summary of the experimental results obtained from mAb 2G12 and mAb H10 bands using different strategies. HC, LC: heavy and light chain, respectively. (+) present (in the case of WB, the band is present in at least one experiment); (−) absent; n.a.: not assayed. N-V_L, N-V_H: N-terminus of V_L and V_H domain, respectively. V_L-C, V_H-C, C_H1-C, C_H2-C: C-terminal region of V_L, V_H, C_H1 and C_H2 domain, respectively. Binding activity of the fragments cannot be directly compared as it was determined using samples that could not be normalized for total soluble protein content (see Experimental procedures)

	WB HC band	WB LC band	Sugar chain	HC N-terminal sequence	LC N-terminal sequence	Binding	Mr (kDa) non-reducing conditions
2G12 band							
*	+	+	+	n.a.	n.a.	+	150
a	+	+	+	EVQL (N-V_H)	DVVM (N-V_L)	+/−	<150
b	+	+	+	EVQL (N-V_H)	DVVM (N-V_L)	+/−	100
c	+	+	+	n.a.	n.a.	+/−	75
d_1	−	+	−	n.a.	n.a.	n.a.	>50
d_2	+	+	+	EVQL (N-V_H)	DVVM (N-V_L)	+	50
f	−	+	−	KVEP (C_H1-C)	RTVA (VL-C)	+/−	>25
g	−	+	−	−	DVVM (N-V_L), RVEI (VL-C)	+	25
h	−	+	−	TVSP (V_H-C)	RTVA (VL-C)	−	<15
H10 band							
*	+	+	+	n.a.	n.a.	+	150
a'	+	+	+	n.a.	n.a.	n.a.	<150
b'	+	+	−	EVQL (N-V_H)	SELT (N-V_L)	+	100
c'	+	+	+	EVQL (N-V_H)	SELT (N-V_L)	+/−	75
d_1'	+	−	+	KKVE (C_H1-C)	−	−	>50
d_2'	+	+	−	EVQL (N-V_H)	SELT (N-V_L)	+	50
f'	+	−	n.a.	EVQL (N-V_H)	SELT (N-V_L)	n.a.	>25
h'	+	+	+	ISKA, KAKG (C_H2-C)	−	n.a.	<15

evidenced by reducing gel analysis (Figure 3d), indicates that this bands comprises different antibody fragments.

H10 band d_2' (about Mr 50K), most likely comprising Fab fragments, showed as expected, intact HC and LC N-termini, antigen-binding activity, lack of glycan moiety and a unique band at about Mr 25K on reducing Western blot analysis using anti-γ-chain antibody (Figure 4c). A similar pattern was observed for the corresponding 2G12 band d_2, which differed from d_2' for the presence of a sugar chain. This is consistent with the detection, along with a Mr 25K HC fragment, of an intact HC in the reducing gel analysis (Figure 3d). Therefore, this band could contain a mixture of Ig chains and Fab fragments. In the case of H10 band d_1' (>Mr 50K), the presence of a glycosylated band at about Mr 25K on reducing gel (Figure 4c), lack of antigen-binding activity and the KKVE HC N-terminal sequence indicates that it comprises Fc fragments.

Figure 5a shows the sequence alignment between the light chains of both antibodies and the heavy chains. The arrows indicate the cleavage sites identified in this study. Two cleavage sites (RVEI and RTVA), separated by five residues, were identified at the end of the VL domain of 2G12 κ chain, whereas none were

found in the λ light chain of H10 or in the CL domain of either mAb. In the heavy chain, one cleavage site was identified within the VH domain of 2G12 (TVSP); two at the end of the C_H1 domain of both mAbs (KVEP in 2G12 and KKVE in H10); and two at the end of the C_H2 domain of H10 (ISKA and KAKG). Interestingly, cleavage sites at the end of the C_H1 domain, just upstream of the hinge region, have been recently reported for an anti-HIV 2F5 antibody whose heavy chain constant domains are identical to H10 and 2G12, suggesting that this region represents a 'hot spot' for plant proteolysis (Niemer et al., 2014).

All of the detected cleavage sites are in the proximity of inter-domain regions, and each of these sites is involved, with the exception of the V_L/C_L interface in the λ light chain (Figure 5b,c). Interestingly, only two cleavages sites occurred within loop regions, whereas the others were in beta strands, and the residues in P4–P4' region of all cleavages were at least partially solvent accessible.

The comparison between the two mAbs helped to both strengthen and rationalize the obtained results. As an example, the RVEIKRTVA sequence at the end of the κ V_L domain of 2G12 (residues 105–113), containing two cleavage sites, is not present

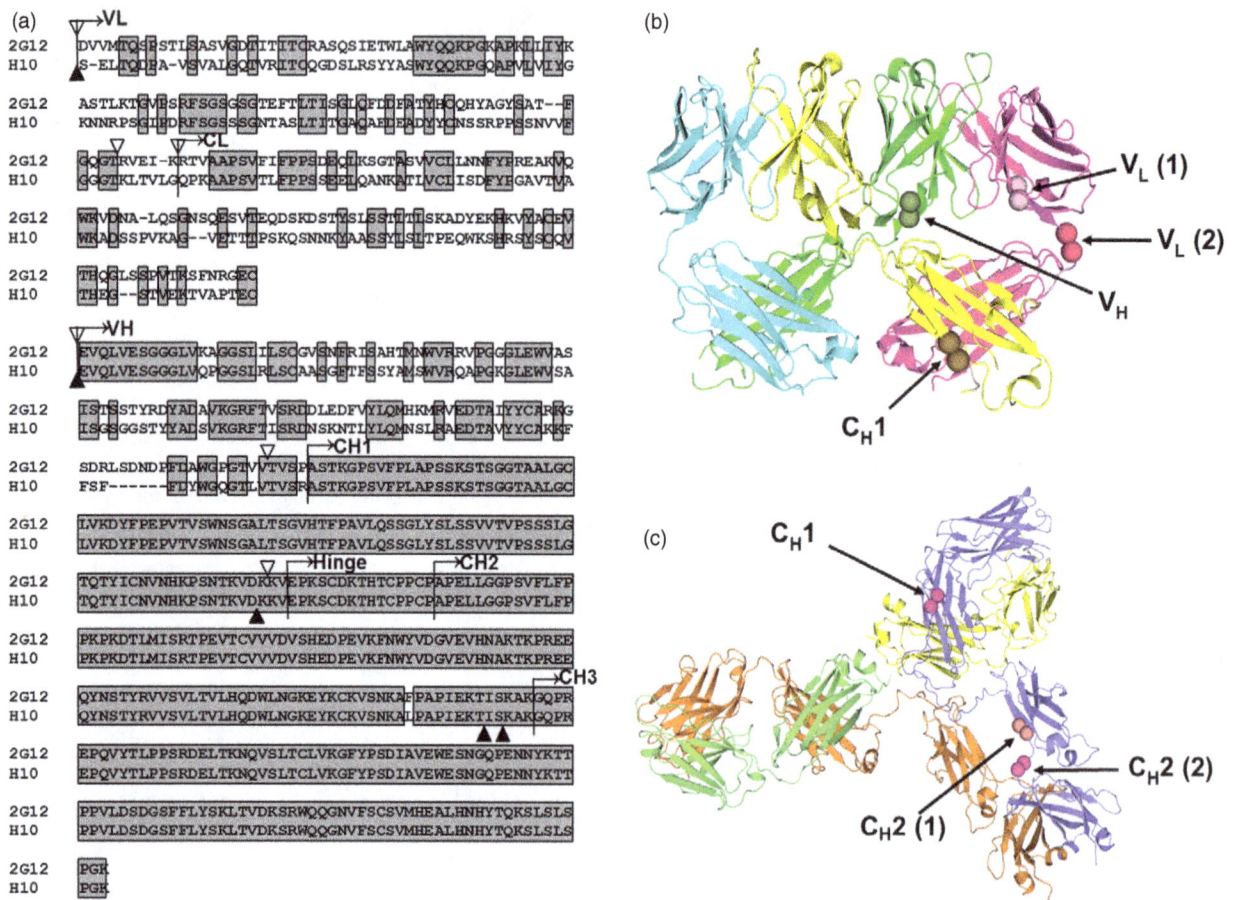

Figure 5 Cleavage sites of mAbs 2G12 and H10. (a) Amino acid light (top) and heavy (bottom) chain sequence alignments. Identical residues are shaded. The variable (V_L) and constant (C_L) domain of the light chain, and the variable (V_H) and constant (C_H1, C_H2, C_H3) domains and hinge region of the heavy chain are indicated. Open and closed arrows indicate cleavage sites for mAbs 2G12 and H10, respectively, in plants. (b) Ribbon representation of the three-dimensional structure of human 2G12 Fab homodimer (PDB ID: 1OP3) determined by X-ray crystallography. VL (1), V_L (2), V_H and C_H1 indicate cleavage sites detected in the V_L, V_H and C_H1 domains of 2G12, whose N-terminal sequences are: RVEI, RTVA, TVSP and KVEP, respectively. (c) Ribbon representation of the H10 homology model constructed using as templates the λ chain of CAT-2200 Fab fragment (PDB ID: 2VXS) and heavy chain of human IgG1 B12 (PDB ID: 1HZH). C_H1, C_H2 (1) and C_H2 (2) indicate cleavage sites detected in the C_H1 and C_H2 domains of H10, whose N-terminal sequences are: KKVE, ISKA and KAKG, respectively. (b, c) Cleavage sites identified by N-terminal sequencing are indicated by arrows. Residues involved in the cleaved peptide bond are shown as spheres.

in the λ chain of H10. As no evidence for λ-chain proteolysis was found, λ light chains may represent the isotype of choice for plant expressed antibodies.

In the heavy chain, a cleavage site at the interface of V_H and C_H1 was identified in mAb 2G12 but not H10. This region (residues 116-125) contains the sequence GT**V**VTVS**P**AS, the cleavage occurring between residues V119 and T120. Interestingly, a cleavage site between G116 and T117, that is, with a three residue shift towards the N-terminus, was found in 2G12 expressed in *N. benthamiana* (Niemer *et al.*, 2014). This suggests either that multiple cleavage events can occur in susceptible antibody regions, or that upon proteolytic cleavage, further trimming by exopeptidases may take place. The lack of H10 cleavage in the corresponding region might be ascribed to the substitution of V118 (P2 and P2' residue with respect to the cleavage sites at positions V119-T120 and G116-T117, respectively) with L, and of P123 (P4' residue with respect to the V119-T120 cleavage site) with R. Conversely, the cleavage sites identified at the end of the H10 C_H2 domain were not found in the corresponding 2G12 region, in spite of their sequence identity. It is possible that additional cleavage sites may be detected through a more extensive analysis of Ig fragments produced in tobacco leaves.

Three cleavage site pairs identified in this work are located in close proximity along mAb sequences at the end of 2G12 V_L, both 2G12 and H10 C_H1, and H10 C_H2 domains, and the cleavage site at the end of 2G12 V_H domain was very close to that identified in another work in the same antibody but also in the human anti-HIV mAb 2F5 (Niemer *et al.*, 2014). This peculiar location may be due to the presence of multiple and accessible endopeptidase cleavage sites within short sequence stretches, but also to exopeptidase trimming of N-terminal ends after initial endopeptidase cleavage. Such a phenomenon, similar to that previously described for recombinant aprotinin (Badri *et al.*, 2009) and equistatin (Outchkourov *et al.*, 2003), may mask the presence of initial endopeptidase cleavage. If this were the case, the process of protease degradation of antibodies expressed in tobacco could turn out to be very simple and involve a very limited number of proteolytic enzymes and mAb sites susceptible to degradation. Another interesting observation is that no proteolysis was found within the antigen-binding region of the V_H domain of either antibody. Niemer *et al.* (2014) showed that a major cleavage occurred in mAbs 2F5 and PG9 within the CDR3 loop, recognized as a major proteinase sensitive region, whereas no cleavage was observed in 2G12 or H10 heavy-chain CDR3 loops. Both our mAbs possess shorter heavy chain CDR3 loops (14 and 8 residues in 2G12 and H10, respectively, vs. 22 and 28 in 2F5 and PG9, respectively). We hypothesize that reduced length, and therefore, solvent accessibility of heavy chain CDR3 loops diminishes susceptibility to proteolytic cleavage. In the case of 2G12, it cannot be excluded that the peculiar three-dimensional arrangement consisting of a swap of the V_H domains of two adjacent Fabs (Calarese *et al.*, 2003) contributes to the proteolytic resistance of the V_H domain core.

Comparison of the substrate specificity of peptidases encoded by *Nicotiana* species provided by the MEROPS database (Rawlings *et al.*, 2010) with the amino acid sequences at the P4–P4' positions of the cleavage sites identified in the two mAbs allowed specific residues putatively responsible for proteolytic degradation to be identified. As an example, phytocalpains, which are associated with the plasma membrane and the endomembrane system in *N. benthamiana* (Ahn *et al.*, 2004;

Johnson *et al.*, 2008), cleave substrates with Val in P2 with high frequency, as in 2G12 V_H and H10 C_H1 cleavage sites, and might recognize Ile in P2, as in 2G12 V_L and the C-terminal H10 C_H2 cleavage site. Additionally, the *Nicotiana* genus encodes papain-like cysteine proteinases that have been previously demonstrated to contribute to mAb degradation in plants (Goulet *et al.*, 2012; Niemer *et al.*, 2014). Enzymes of this family recognize specifically bulky hydrophobic residues in position P2 of the substrate (Rawlings *et al.*, 2012) and might be responsible for the cleavage in the C-terminal region of 2G12 V_L and H10 C_H2 domain.

Various strategies have been attempted to reduce degradation of recombinant antibodies in plants. Extraction buffers commonly include a cocktail of protease inhibitors (Hellwig *et al.*, 2004; Rivard *et al.*, 2006), but these are expensive and probably not affordable at large scale. In any case, their effect appears to be limited, possibly because most degradation processes occur intracellularly, rather than during the extraction process (Hehle *et al.*, 2011). Co-expression of a recombinant protease inhibitor improved the yield of a rhizosecreted protein approximately threefold (Goulet *et al.*, 2012; Komarnytsky *et al.*, 2006). An alternative approach, which has been carried out in BY-2 cell cultures, consists of knockingout or knockingdown specific plant proteases (Mandal, 2011). However, the efficiency of silencing strategies is limited because plants typically possess multigene families (Pesquet, 2012). In addition, proteases often have important physiological functions in plant development, which would inevitably be affected.

By approaching the problem of degradation from the perspective of the target recombinant protein, we identified a relatively small number of specific protease susceptible sites, which were located in or close to interdomain mAbs regions. Except for the antigen-binding region, where no proteolytic cleavage was detected, antibodies have highly conserved structures and sequences. Therefore, it is not surprising that we obtained similar results for mAbs 2G12 and H10, and we expect similar results to be obtained for other human antibodies.

Taken together, the results presented in this work raise the possibility of stabilizing recombinant mAbs produced by tobacco plants by conservative site specific mutagenesis, by taking advantage of replacement amino acid sequences that are used by the human antibody repertoire.

Experimental procedures

Transgenic plant material

Experiments were performed using transgenic *Nicotiana tabacum* (*N. tabacum*, var. Petit Havana) lines, homozygous for both the γ1 heavy- and κ light-chain genes of the human IgG1 mAb 2G12 [kindly provided by Dr. Thomas Rademacher, University of Technology (RWTH), Aachen and Uni. Prof. Dr. Eva Stoeger University of Natural Resources and Applied Life Sciences (BOKU), Vienna]. The production of this plant line has been described (Rademacher *et al.*, 2008).

Transient expression of mAbs 2G12 in *Nicotiana tabacum* and H10 in *Nicotiana benthamiana* by agroinfiltration

For transient expression, the heavy- and light-chain genes of mAb 2G12 were cloned in the plant transformation vector pTRAk (courtesy of Dr. Thomas Rademacher, Fraunhofer IME Aachen). Here, each gene is under control of the duplicated *Cauliflower mosaic virus* (CaMV) 35S promoter, the *Tobacco etch virus* 5′ untranslated region and the CaMV 35S polyadenylation site/terminator.

Heavy- and light-chain genes of mAb H10 were obtained as previously described (Villani *et al.*, 2009) and cloned into the plant binary vector pBI-Ω (Marusic *et al.*, 2007) under the control of the CaMV 35S promoter and the omega translational enhancer sequence from tobacco mosaic virus.

Wild-type *N. tabacum* and *N. benthamiana* plants were cultivated for 10–12 weeks from seed. Recombinant *Agrobacterium tumefaciens* cultures EHA105 harbouring the light- and heavy-chain genes of mAbs 2G12 and H10 were grown overnight at 28 °C, with shaking at 250 r.p.m. in 10 mL of Luria Bertani medium supplemented with either carbenicillin (50 µg/mL) and rifampicin (100 µg/mL) for 2G12 or kanamycin (50 µg/mL) for H10. The cultures were centrifuged to form a pellet, which was resuspended in infiltration buffer (10 mM MES, 10 mM MgSO$_4$, pH 5.8). For the coinfiltration of heavy and light chains, aliquots of resuspended cell pellets were mixed together to reach 0.6 optical density (OD$_{600}$). Following careful penetration of the abaxial surface of a leaf with a pipette tip, the bacterial solution was injected directly using a syringe pressed against the leaf (Sainsbury and Lomonossoff, 2008). The plants were maintained for 5–7 days under normal greenhouse conditions (temperature 25 °C, 16/8 h light/dark cycle), and leaves were harvested for analysis of recombinant protein expression.

Extraction of mAbs from transgenic tobacco plants and agroinfiltrated tobacco plants

Mature leaves from transgenic tobacco plants expressing mAb 2G12 were homogenized with three volumes of PBS pH 7.4 at room temperature using a Waring blender. The plant extract was centrifuged at 17 000 *g* (Sorvall Instruments RC5C, Newtown, CT), for 30 min at 10 °C, and the supernatant was passed through Whatman #3 filter paper. The pH of the filtered plant juice was adjusted to pH 7.5–8.0 with 1 M NaOH and incubated for at least 30 min on ice, followed by re-centrifugation at 40 000 *g* for 20 min at 10 °C. The supernatant was filtered through a 0.22-µm Millex GP Filter (Millipore Express, Consett, Co Durham, UK). For agroinfiltrated plants, infiltrated leaves were sampled and homogenized for 5 min at 29 oscillation/s using a Mixer Mill MM400 (Retsch). Samples were centrifuged at 17 000 *g* (Sorvall Instruments RC5C) for 10 min at 10 °C. The supernatant was treated as described above. Clarified plant extracts were either used immediately or further purified.

ELISA

ELISA was performed to assess whether the H10 and 2G12 fragments in the bands separated by SDS-PAGE retained antigen binding ability. Bands with a molecular weight above 30 kDa were selected, ground in PBS 1× and centrifuged. The individual supernatants were loaded onto ELISA plate wells coated with either HIV gp120 (IIIB, Baculovirus; ImmunoDiagnostics Inc., Woburn, MA) or human lung cancer cells A549. A549 cells express on their surface, the large isoform of tenascin including the C domain specifically recognized by IgG H10 (Fitch *et al.*, 2011). Detection was performed with a peroxidase labelled anti-human κ-chain or anti-human λ-chain antiserum for mAb 2G12 or H10 respectively, followed by the addition of tetramethylbenzidine dihydrochloride (TMB; Sigma) substrate. The reaction was stopped with 2 M sulphuric acid, and the absorbance was detected at a wave length of 450 nm.

Affinity chromatography

Protein G-Sepharose® 4B resin (Sigma-Aldrich, Saint Louis, MO) and protein A-agarose (Sigma) (1 : 1 mix) were packed into a 5 cm × 1 cm glass chromatography column (Bio-Rad, Hemel Hempstead, Hertfordshire, UK) to give a final bed volume of ~1 mL. Filtered plant supernatant was applied to the column at a constant flow rate of 0.5–1 mL/min. The flow-through was re-applied to the column. The column was washed with ≥20 column volumes of PBS. MAbs and their fragments were eluted with 15 column volumes of 0.1 M glycine (pH 2.5) in 1 mL fractions. Each fraction was immediately neutralized by the addition of 50 µL of 1 M Tris base. To concentrate the samples, the pooled fractions were freeze-dried under vacuum overnight. Lyophilized samples were resuspended in 200 µL H$_2$O and dialysed overnight against PBS.

SDS-PAGE analysis

Purified plant antibody fragments or agroinfiltrated leaf extracts were separated by SDS-4-15% (w/v) PAGE Tris–HCl gradient gels (Bio-Rad). Samples were mixed with 5× SDS sample buffer in the absence of reducing agent, boiled for 3 min and loaded onto the gel. Samples were electrophoresed at 20 mA/gel. Separated proteins were visualized by Coomassie blue staining or Western blotting.

Purification of individual antibody bands

Purified plant-derived mAb 2G12 was subjected to electrophoresis on a SDS-4-15% (w/v) PAGE acrylamide gel under either reducing or non-reducing conditions. Proteins were stained with Instant Blue (Expedeon, Haston, Cambridgeshire, UK), and individual bands were excised from the gel using a sterile blade. Gel fragments were homogenized in 50 µL of PBS, incubated at room temperature for 30 min and centrifuged at 17 000 *g* for 10 min. The supernatant was collected for further analyses.

Western blot analysis

Protein transfer onto Hybond nitrocellulose (GE Healthcare, Little Chalfont, Buckinghamshire, UK) or poly(vinylidene difluoride) (PVDF) membrane (Millipore, Bedford, MA) was performed for 90 min at 0.4 mA/cm^2 and 50 V using a semi-dry blotting device (Bio-Rad). To block nonspecific sites, the membrane was incubated with 5% (w/v) non-fat milk powder in phosphate-buffered saline (PBS) for at least 30 min. Protein detection was performed with anti-human γ chain (8419; Sigma-Aldrich), anti-human λ chain (A5175; Sigma-Aldrich, Gillingham Dorset, UK) or anti-human κ chain (AU015, Binding Site) horseradish peroxidase-labelled antibodies for 1 h at room temperature in 2% (w/v) non-fat milk powder in PBS. Glycans were analysed using biotin-conjugated concanavalin A from *Canavalia ensiformis* (C2272; Sigma-Aldrich) in 1% BSA, 1× PBS and peroxidase-conjugated streptavidin (S5512; Sigma) 0.1% BSA, 1× PBS. The membrane was washed five times with 0.1% Tween-20 in PBS (5 min per wash) and developed using the ECL Plus Western blotting detection system (ECL, Plus; GE Healthcare).

Edman degradation

For Edman degradation analysis, purified mAbs 2G12 and H10 (typically 20–50 µg were loaded) were electrophoretically separated by SDS-4-15% (w/v) PAGE gels (Bio-Rad), blotted on Polyvinylidene fluoride (PVDF) membrane and stained with Coomassie suspension R250.

N-terminal sequencing was performed by Dr. Mike Weldon, University of Cambridge (http://www3.bioc.cam.ac.uk/pnac/index.html), on a Precise® Protein Sequencing System (Applied Biosystems, Foster City, CA).

Structure and sequence analysis

Basic Local Alignment Search Tool (BLAST; Altschul et al., 1997) was used to search the NCBI sequence database of proteins whose experimentally determined three-dimensional (3D) structures are available for Ab structures having high-sequence identity with H10. The following 3D structures were downloaded from the protein data bank (Rose et al., 2011): CAT-2200 Fab (Gerhardt et al., 2009; PDB identifier (ID): 2VXS) and B12 IgG (Saphire et al., 2001; PDB ID: 1HZH), which are closely related to H10 light and heavy chains, respectively; and 2G12 Fab (Calarese et al., 2003; PDB ID: 1OP3). Insight II (Accelrys Inc., San Diego, CA) and The PyMOL Molecular Graphics System, Version 1.5.0.4 (Schrödinger, LLC, New York) were used for structure visualization and mapping of the cleavage sites.

A list of proteolytic enzymes encoded by N. benthamiana (whose genome has been fully sequenced), N. tabacum and Nicotiana genus was obtained from the MEROPS peptidase database Release 9.10 (Rawlings et al., 2012). In the case of enzymes whose substrate specificity has been clearly established and provided by MEROPS, the pattern of residues characteristic of their substrates was compared with the P4–P4' sequence of the cleavage sites identified in mAbs 2G12 and H10.

Acknowledgements

We are grateful to Thomas Rademacher, Markus Sack and Eva Stoeger (Fraunhofer IME) who provided the transgenic plant seed for mAb 2G12 and the transient expression vector for mAb 2G12. Our thanks to Mike Weldon (University of Cambridge) for performing the N-terminal sequence analyses. We gratefully acknowledge the EU Framework 6 Pharma-Planta Project, ERC Future-Pharma Advanced Grant, the Molecular Farming COST Action (FA804) and The Hotung Trust for financial support for this research project. We also thank the Italian Ministry of Foreign Affairs, 'Direzione Generale per la Promozione del Sistema Paese' for the contribution. We thank Dr. Mariasole Di Carli for helpful discussion of the experiments.

References

Ahn, J.W., Kim, M., Lim, J.H., Kim, G.T. and Pai, H.S. (2004) Phytocalpain controls the proliferation and differentiation fates of cells in plant organ development. Plant J. **38**, 969–981.

Altschul, S.F., Madden, T.L., Schäffer, A.A., Zhang, J., Zhang, Z., Miller, W. and Lipman, D.J. (1997) Gapped BLAST and PSI-BLAST: a new generation of protein database search programs. Nucleic Acids Res. **25**, 3389–3402.

Badri, M.A., Rivard, D., Coenen, K. and Michaud, D. (2009) Unintended molecular interactions in transgenic plants expressing clinically useful proteins: the case of bovine aprotinin traveling the potato leaf cell secretory pathway. Proteomics, **9**, 746–756.

Benchabane, M., Goulet, C., Rivard, D., Faye, L., Gomord, V. and Michaud, D. (2008) Preventing unintended proteolysis in plant protein biofactories. Plant Biotechnol. J. **6**, 633–648.

Borsi, L., Carnemolla, B., Nicolò, G., Spina, B., Tanara, G. and Zardi, L. (1992) Expression of different tenascin isoforms in normal, hyperplastic and neoplastic human breast tissues. Int. J. Cancer, **52**, 688–692.

Calarese, D.A., Scanlan, C.N., Zwick, M.B., Deechongkit, S., Mimura, Y., Kunert, R., Zhu, P., Wormald, M.R., Stanfield, R.L., Roux, K.H., Kelly, J.W., Rudd, P.M., Dwek, R.A., Katinger, H., Burton, D.R. and Wilson, I.A. (2003) Antibody domain exchange is an immunological solution to carbohydrate cluster recognition. Science, **300**, 2065–2071.

De Muynck, B., Navarre, C., Nizet, Y., Stadlmann, J. and Boutry, M. (2009) Different subcellular localization and glycosylation for a functional antibody expressed in Nicotiana tabacum plants and suspension cells. Transgenic Res. **18**, 467–482.

De Muynck, B., Navarre, C. and Boutry, M. (2010) Production of antibodies in plants: status after twenty years. Plant Biotechnol. J. **8**, 529–563.

De Neve, M., De Loose, M., Jacobs, A., Van Houdt, H., Kaluza, B., Weidle, U., Van Montagu, M. and Depicker, A. (1993) Assembly of an antibody and its derived antibody fragment Nicotiana and Arabidopsis. Transgenic Res. **2**, 227–237.

De Wilde, C., De Neve, M., De Rycke, R., Bruyns, A.M., De Jaeger, G., Van Montagu, M., Depicker, A. and Engler, G. (1996) Intact antigen-binding MAK33 antibody and Fab fragment accumulate in intercellular spaces of Arabidopsis thaliana. Plant Sci. **114**, 233–241.

Delannoy, M., Alves, G., Vertommen, D., Ma, J., Boutry, M. and Navarre, C. (2008) Identification of peptidases in Nicotiana tabacum leaf intercellular fluid. Proteomics, **8**, 2285–2298.

Fan, X., Brezski, R.J., Fa, M., Deng, H., Oberholtzer, A., Gonzalez, A., Dubinsky, W.P., Strohl, W.R., Jordan, R.E., Zhang, N. and An, Z. (2012) A single proteolytic cleavage within the lower hinge of trastuzumab reduces immune effector function and in vivo efficacy. Breast Cancer Res. **14**, R116.

Fitch, P.M., Howie, S.E. and Wallace, W.A. (2011) Oxidative damage and TGF-β differentially induce lung epithelial cell sonic hedgehog and tenascin-C expression: implications for the regulation of lung remodelling in idiopathic interstitial lung disease. Int. J. Exp. Pathol. **92**, 8–17.

Gao, S.X., Zhang, Y., Stansberry-Perkins, K., Buko, A., Bai, S., Nguyen, V. and Brader, M.L. (2010) Fragmentation of a highly purified monoclonal antibody attributed to residual CHO cell protease activity. Biotechnol. Bioeng. **108**, 977–982.

Gerhardt, S., Abbott, W.M., Hargreaves, D., Pauptit, R.A., Davies, R.A., Needham, M.R., Langham, C., Barker, W., Aziz, A., Snow, M.J., Dawson, S., Welsh, F., Wilkinson, T., Vaugan, T., Beste, G., Bishop, S., Popovic, B., Rees, G., Sleeman, M., Tuske, S.J., Coales, S.J., Hamuro, Y. and Russell, C. (2009) Structure of IL-17A in complex with a potent, fully human neutralizing antibody. J. Mol. Biol. **394**, 905–921.

Goulet, C., Khalf, M., Sainsbury, F., D'Aoust, M.A. and Michaud, D.A. (2012) Protease activity–depleted environment for heterologous proteins migrating towards the leaf cell Apoplast. Plant Biotechnol. J. **1**, 1–12.

Hehle, V.K., Paul, M.J., Drake, P.M., Ma, J.K. and van Dolleweerd, C.J. (2011) Antibody degradation in tobacco plants: a predominantly apoplastic process. BMC Biotechnol. **11**, 128.

Hellwig, S., Drossard, J., Twyman, R.M. and Fischer, R. (2004) Plant cell cultures for the production of recombinant proteins. Nat. Biotechnol. **22**, 1415–1422.

Hiatt, A.C. (1991) Monoclonal antibodies, hybridoma technology and heterologous production systems. Curr. Opin. Immunol. **3**, 229–232.

van der Hoorn, R.A.L. (2008) Plant proteases: from phenotypes to molecular mechanisms. Annu. Rev. Plant Biol. **59**, 191–223.

Ishikawa, T., Ito, T., Endo, R., Nakagawa, K., Sawa, E. and Wakamatsu, K. (2010) Influence of pH on heat-induced aggregation and degradation of therapeutic monoclonal antibodies. Biol. Pharm. Bull. **33**, 1413–1417.

Johnson, K.L., Faulkner, C., Jeffree, C.E. and Ingram, G.C. (2008) The phytocalpain defective kernel 1 is a novel Arabidopsis growth regulator whose activity is regulated by proteolytic processing. Plant Cell, **20**, 2619–2930.

Komarnytsky, S., Borisjuk, N., Yakoby, N., Garvey, A. and Raskin, I. (2006) Cosecretion of protease inhibitor stabilizes antibodies produced by plant roots. Plant Physiol. **141**, 1185–1193.

Komarova, T.V., Baschieri, S., Donini, M., Marusic, C., Benvenuto, E. and Dorokhov, Y.L. (2010) Transient expression systems for plant-derived biopharmaceuticals. Expert Rev. Vaccines, **9**, 859–876.

Lombardi, R., Villani, M.E., Di Carli, M., Brunetti, P., Benvenuto, E. and Donini, M. (2010) Optimisation of the purification process of a tumour-targeting antibody produced in *N. benthamiana* using vacuum-agroinfiltration. *Transgenic Res.* **19**, 1083–1097.

Lombardi, R., Donini, M., Villani, M.E., Brunetti, P., Fujiyama, K., Kajiura, H., Paul, M., Ma, J.K. and Benvenuto, E. (2012) Production of different glycosylation variants of the tumour-targeting mAb H10 in *Nicotiana benthamiana*: influence on expression yield and antibody degradation. *Transgenic Res.* **21**, 1005–1021.

Ma, J.K., Lehner, T., Stabila, P., Fux, C.I. and Hiatt, A. (1994) Assembly of monoclonal antibodies with IgG1 and IgA heavy chain domains in transgenic tobacco plants. *Eur. J. Immunol.* **24**, 131–138.

Mandal, M.K. (2011) *Improving tobacco (Nicotiana tabacum) suspension cells as a molecular farming platform: a protease knock down approach.* Thesis Dissertation, http://publica.fraunhofer.de/documents/N-180476.html.

Marusic, C., Nuttall, J., Buriani, G., Lico, C., Lombardi, R., Baschieri, S., Benvenuto, E. and Frigerio, L. (2007) Expression, intracellular targeting and purification of HIV Nef variants in tobacco cells. *BMC Biotechnol.* **7**, e12.

Niemer, M., Mehofer, U., Torres Acosta, J.A., Verdianz, M., Henkel, T., Loos, A., Strasser, R., Maresch, D., Rademacher, T., Steinkellner, H. and Mach, L. (2014) The human anti-HIV antibodies 2F5, 2G12 and PG9 differ in their susceptibility to proteolytic degradation: down-regulation of endogenous serine and cysteine proteinase activities could improve antibody production in plant-based expression platforms. *Biotechnol. J.* **9**, 493–500.

Outchkourov, N.S., Rogelj, B., Strukelj, B. and Jongsma, M.A. (2003) Expression of sea anemone equistatin in potato. Effects of plant proteases on heterologous protein production. *Plant Physiol.* **133**, 379–390.

Pesquet, E. (2012) Plant proteases—from detection to function. *Physiol. Plant.* **145**, 1–4.

Rademacher, T., Sack, M., Arcalis, E., Stadlmann, J., Balzer, S., Altmann, F., Quendler, H., Stiegler, G., Kunert, R., Fischer, R. and Stoger, E. (2008) Recombinant antibody 2G12 produced in maize endosperm efficiently neutralizes HIV-1 and contains predominantly single-GlcNAc N-glycans. *Plant Biotechnol. J.* **6**, 189–201.

Ramessar, K., Rademacher, T., Sack, M., Stadlmann, J., Platis, D., Stiegler, G., Labrou, N., Altmann, F., Ma, J., Stöger, E., Capell, T. and Christou, P. (2008) Cost-effective production of a vaginal protein microbicide to prevent HIV transmission. *Proc. Natl Acad. Sci. USA*, **105**, 3727–3732.

Rawlings, N.D., Barrett, A.J. and Bateman, A. (2010) MEROPS: the peptidase database. *Nucleic Acids Res.* **38**, 227–233.

Rita Costa, A., Elisa Rodrigues, M., Henriques, M., Azeredo, J. and Oliveira, R. (2010) Guidelines to cell engineering for monoclonal antibody production. *Eur. J. Pharm. Biopharm.* **74**, 127–138.

Rivard, D., Anguenot, R., Brunelle, F., Le, V.Q., Vézina, L.P., Trépanier, S. and Michaud, D. (2006) An in-built proteinase inhibitor system for the protection of recombinant proteins recovered from transgenic plants. *Plant Biotechnol. J.* **4**, 359–368.

Rose, P.W., Beran, B., Bi, C., Bluhm, W.F., Dimitropoulos, D., Goodsell, D.S., Prlic, A., Quesada, M., Quinn, G.B., Westbrook, J.D., Young, J., Yukich, B., Zardecki, C., Berman, H.M. and Bourne, P.E. (2011) The RCSB Protein Data Bank: redesigned web site and web services. *Nucleic Acids Res.* **39**, D392–D401.

Sainsbury, F. and Lomonossoff, G.P. (2008) Extremely high-level and rapid transient protein production in plants without the use of viral replication. *Plant Physiol.* **148**, 1212–1218.

Saphire, E.O., Parren, P.W., Pantophlet, R., Zwick, M.B., Morris, G.M., Rudd, P.M., Dwek, R.A., Stanfield, R.L., Burton, D.R. and Wilson, I.A. (2001) Crystal structure of a neutralizing human IGG against HIV-1: a template for vaccine design. *Science*, **293**, 1155–1159.

Schaller, A. (2004) A cut above the rest: the regulatory function of plant proteases. *Planta*, **220**, 183–197.

Sharp, J.M. and Doran, P.M. (2001) Characterization of monoclonal antibody fragments produced by plant cells. *Biotechnol. Bioeng.* **73**, 338–346.

Stevens, L.H., Stoopen, G.M., Elbers, I.J.W., Molthoff, J.W., Bakker, H.A.C., Lommen, A., Bosch, D. and Jordi, W. (2000) Effect of climate conditions and plant developmental stage on the stability of antibodies expressed in transgenic tobacco. *Plant Physiol.* **124**, 173–182.

Stoger, E., Ma, J.K.C., Fischer, R. and Christou, P. (2005) Sowing the seeds of success: pharmaceutical proteins from plants. *Curr. Opin. Biotechnol.* **16**, 167–173.

Trkola, A., Purtscher, M., Muster, T., Ballaun, C., Buchacher, A., Sullivan, N., Srinivasan, K., Sodroski, J., Moore, J.P. and Katinger, H. (1996) Human monoclonal antibody 2G12 defines a distinctive neutralization epitope on the gp120 glycoprotein of human immunodeficiency virus type 1. *J. Virol.* **70**, 1100–1108.

Villani, M.E., Morgun, B., Brunetti, P., Marusic, C., Lombardi, R., Pisoni, I., Bacci, C., Desiderio, A., Benvenuto, E. and Donini, M. (2009) Plant pharming of a full-sized, tumour-targeting antibody using different expression strategies. *Plant Biotechnol. J.* **7**, 59–72.

Vitale, A. and Pedrazzini, E. (2005) Recombinant pharmaceuticals from plants: the plant endomembrane system as bioreactor. *Mol. Interv.* **5**, 216–225.

Vlasak, J. and Ionescu, R. (2011) Fragmentation of monoclonal antibodies. *mAbs*, **3**, 253–263.

Wongsamuth, R. and Doran, P.M. (1997) Production of monoclonal antibodies by tobacco hairy roots. *Biotechnol. Bioeng.* **54**, 401–415.

18

Feeding transgenic plants that express a tolerogenic fusion protein effectively protects against arthritis

Charlotta Hansson[1], Karin Schön[1], Irina Kalbina[2], Åke Strid[2], Sören Andersson[2,3], Maria I. Bokarewa[4] and Nils Y. Lycke[1,*]

[1]Department of Microbiology and Immunology, University of Gothenburg, Gothenburg, Sweden
[2]Örebro Life Science Center, School of Science and Technology, Örebro University, Örebro, Sweden
[3]Department of Laboratory Medicine, Örebro University hospital, Örebro, Sweden
[4]Department of Rheumatology and Inflammation Research, University of Gothenburg, Gothenburg, Sweden

*Correspondence
email
nils.lycke@microbio.gu.se

Summary

Although much explored, oral tolerance for treatment of autoimmune diseases still awaits the establishment of novel and effective vectors. We investigated whether the tolerogenic CTA1 (R7K)-COL-DD fusion protein can be expressed in edible plants, to induce oral tolerance and protect against arthritis. The fusion protein was recombinantly expressed in *Arabidopsis thaliana* plants, which were fed to H-2^q-restricted DBA/1 mice to assess the preventive effect on collagen-induced arthritis (CIA). The treatment resulted in fewer mice exhibiting disease and arthritis scores were significantly reduced. Immune suppression was evident in treated mice, and serum biomarkers for inflammation as well as anticollagen IgG responses were reduced. In spleen and draining lymph nodes, CD4$^+$ T-cell responses were reduced. Concomitant with a reduced effector T-cell activity with lower IFNγ, IL-13 and IL-17A production, we observed an increase in IL-10 production to recall antigen stimulation *in vitro*, suggesting reduced Th1, Th2 and Th17 activity subsequent to up-regulated IL-10 and regulatory T-cell (Treg) functions. This study shows that edible plants expressing a tolerogen were effective at stimulating CD4 T-cell tolerance and in protecting against CIA disease. Our study conveys optimism as to the potential of using edible plants for oral treatment of rheumatoid arthritis.

Keywords: autoimmunity, transgenic plants, edible plants, CIA, IL-10, FoxP3.

Introduction

Rheumatoid arthritis (RA) disease, as well as other autoimmune conditions, is treated with anti-inflammatory drugs that ameliorate symptoms and reduce tissue destruction, but do not offer cure to the disease (Ichim et al., 2008; Smolen and Aletaha, 2015). The exceptions to current therapies involve experimental attempts to cure disease by remodelling specific immune responses and, hence, controlling the attack on tissue functions (Ichim et al., 2008). Mucosal tolerance and, in particular, oral tolerance are well-established therapies to treat autoimmune diseases in experimental models, but few clinical trials have successfully shown a significant effect on human disease (Barnett et al., 1996, 1998; Cazzola et al., 2000; Choy et al., 2001; Trentham et al., 1993). The reason for these failures are thought to depend on the requirement for high and multiple doses of antigen or the fact that no clinically acceptable immunomodulation has been found that impose T-cell tolerance on the immune system (Faria and Weiner, 2006).

We have previously demonstrated that the patented CTA1 (R7K)-COL-DD fusion protein is an effective tolerance-inducing vector, preventing and greatly reducing collagen-induced arthritis (CIA) disease following intranasal administration in mice (Hasselberg et al., 2009). The mutant CTA1(R7K) moiety is derived from the A1 subunit of cholera toxin (CT), but lacks enzymatic activity due to a single amino acid substitution. The CTA1(R7K) molecule was genetically fused to a dimer of the D fragment of *Staphylococcus aureus* protein A, to target antigen-presenting cells (APC) (Hasselberg et al., 2010). These elements flank the immunodominant collagen type II (CII) peptide, amino acids

259–274 (COL), strongly associated with RA, and which effectively stimulates CD4$^+$ T cells in H-2^q DBA/1 mice (Trentham et al., 1977). Thus, the complex carries both APC targeting properties (DD) and immunomodulation through the mutant CTA1(R7K) moiety as well as the dominant species conserved peptide involved in RA disease development.

Most studies today ascribe the tolerogenic effects of mucosal antigen administration to the stimulation of regulatory CD4 T cells (Tregs) (Weiner et al., 2011). The generation of antigen-specific CD4 Tregs is a powerful strategy to reinstate tolerance in the immune system. The CTA1(R7K)-COL-DD tolerogen given intranasally stimulated Tregs and the mechanism of action appeared to be IL-10 dependent (Hasselberg et al., 2009, 2010). A major challenge to developers of mucosal tolerogenic therapies today is to identify a strategy for effective and sufficient delivery of antigen. Because nasal vaccine delivery has been met with scepticism by regulatory authorities, the use of other mucosal delivery routes have been favoured also for tolerance-inducing therapies (Jabbal-Gill, 2010). Whereas our fusion protein was ineffective when given orally in the CIA model and large quantities of soluble antigen are usually required for a tolerogenic effect, we explored alternative strategies to achieve an effective and highly compliant therapy for RA patients. Many investigators have turned to edible plants as a potential approach to achieve mucosal tolerance against autoimmune diseases, especially for type I diabetes and, to some extent, also for RA (Arakawa et al., 1998; Hashizume et al., 2008; Ma et al., 1997; Ruhlman et al., 2007). Plants are unlikely carriers of human pathogens and provide natural bioencapsulation of the antigen, which will prevent protein degradation in the gut and allow for effective uptake in the gut-associated lymphoid tissue (GALT) (Limaye et al., 2006). Although oral tolerance has proven difficult to achieve with plants expressing disease-relevant recombinant proteins or peptides only, the combination with mucosal targeting using expression systems exploiting the potential of cholera B-subunit (CTB) bound peptides have shown more promise. Examples of this strategy are lettuce or potatoes that recombinantly expressed human proinsulin-CTB, which had some reducing effects on disease in the experimental models (Arakawa et al., 1998; Ruhlman et al., 2007). This argues in favour of new and more effective tolerance-inducing components to be included in edible plants to make this approach clinically feasible.

RA is an autoimmune disease affecting approximately 1% of the population worldwide with symptoms that are chronic and incapacitating due to cartilage destruction and bone erosion in peripheral joints. It poses a significant health problem and economic challenge not only to Western societies but RA is also a growing problem in developing countries (Brooks, 2006; Okada et al., 2010). In RA, disease severity correlates with increased pro-inflammatory cytokines such as interferon-γ (IFN-γ), interleukin-6 (IL-6) and IL-17, as well as auto-aggressive CD4 T-cell infiltrates in the synovium (Firestein, 2003; McInnes and Schett, 2007). The broadly immunosuppressive biological agents used today are expensive, ineffective in a significant proportion of patients and often associated with side effects, such as increased risk of infections and cancer development (Beyaert et al., 2013; Bongartz et al., 2006; Salliot et al., 2009). This makes more effective and targeted therapies against RA with fewer side effects highly warranted.

This study was undertaken to investigate whether the tolerogenic CTA1(R7K)-COL-DD protein could be expressed in edible plants and enable the effective use of a targeted immunomodulating tolerogen for oral treatment of RA. We transformed Arabidopsis thaliana plants to express the tolerogenic CTA1(R7K)-COL-DD protein and fed mice that were immunized with CII to develop CIA disease. The results obtained clearly support the notion that edible plants can be used for tolerance treatment of autoimmune diseases, in general, and they offer promise as to the possibility to develop a disease-targeted oral therapy for RA, in particular.

Results

Arabidopsis thaliana plant expression of the CTA1(R7K)-COL-DD tolerogen

We genetically modified edible plants to express the tolerance-inducing gene fusion protein that we previously have demonstrated could protect against arthritis in the collagen-induced arthritis (CIA) model after intranasal administration (Hasselberg et al., 2009). Here, our aim was to develop an effective oral therapy for inducing tolerance and prevent CIA disease. Therefore, we incorporated the tolerogen encoding CTA1(R7K)-COL-DD genetic construct that carries the H-2^q-restricted collagen type II (CII) peptide 259–274 (COL) into the pGreen vector, which was then used for nuclear transformation of A. thaliana plants of the Col-0 ecotype (Figure 1a). After initial screening for successful transformation with BASTA (Clough and Bent, 1995), transgenic lines were identified and used for further analysis. Stable integration of the transgene (TG) and expression of the recombinant protein was monitored for up to six generations, and α-CTA1-DD antibodies were used to detect a protein of the correct size (~39 kDa) by Western blot (Figure 1b). A semiquantitative analysis

of the recombinant fusion protein content in the plant showed an expression of approximately 30 ng of CTA1(R7K)-COL-DD protein per mg of A. thaliana biomass, or 25 ng per μg of total soluble protein (TSP), corresponding to 2.5% of TSP (Figure 1c). The fusion protein CTA1(R7K)-COL-DD was detected in all plant organs with highest concentrations in the inflorescence and in the leaves (Figure 1d). A subfractionation analysis showed that as expected, no fusion protein was detected in the chloroplast fraction. Instead, it was mainly found in the cytosolic fraction, but with some protein retained in the insoluble fraction that could include cell debris and inclusion bodies (Figure 1e).

Therapeutic effects on CIA of feeding transgenic Arabidopsis thaliana plants

The CIA model is considered the best experimental model for RA, and it is frequently used to evaluate the efficacy of clinically relevant anti-inflammatory treatments (Brand et al., 2003). Therefore, DBA/1 mice were used to initiate CIA disease and a protocol for weekly feedings of either TG or control Arapidopsis thaliana plants was established. Control plants were either untransformed wild type (WT) or transformed with empty vector (pGreen) or peptide-free fusion protein (CTA1R7K-DD). In each cage, 10 mice were allowed to feed ad libitum on the plants, which were served 1 day of the week for 4 weeks (Figure 2a). Based on a semiquantitative analysis of protein expression (30 ng/mg of biomass), we estimated that each mouse would have sufficient amount of whole plant material to ingest approximately 200 μg of fusion protein and, thus, 10 μg of COL-peptide on each feeding occasion. Mice were monitored for CIA and, whereas 10–20% of the treated mice did not develop disease at all, the majority of mice feed TG plants exhibited significantly reduced CIA severity ($P < 0.05$), as opposed to mice fed WT plants, vector (pGreen) plants or controls plants CTA1(R7K)-DD, not expressing the COL-peptide (Figure 2b,c).

Protective mechanisms against CIA disease by feeding transgenic plants

Next, we investigated the immune mechanisms that were responsible for the protective effects of feeding TG plants on CIA disease development. To this end, we analysed humoral as well as cell-mediated immune responses and focused attention on evidence of CD4 T-cell tolerance, as our previous studies using the CTA1(R7K)-COL-DD tolerogen in soluble form indicated strong induction of peptide-specific regulatory CD4 T cells (Hasselberg et al., 2009, 2010). We found that TG plant-treated mice had lower levels of anti-CII protein IgG1 and IgG2a/b antibodies in serum than control mice fed WT plants, suggesting that both Th1 and Th2 responses were impaired in TG plant fed mice (Figure 3a). Furthermore, serum levels of MMP-3, reflecting cartilage and bone destruction, and which is strongly up-regulated in CIA disease (Seeuws et al., 2010), were significantly reduced in TG plant fed mice as compared to the levels observed in the control mice that developed more severe CIA. In fact, the level of serum MMP-3 was not significantly higher in TG plant fed mice than that found in normal naïve mice (Figure 3b). In addition, we found a significant reduction of the levels of serum IL-6 in TG plant fed mice as compared to those detected in control mice with CIA disease (Figure 3c), which is reminiscent of our previous observations with the CTA1(R7K)-COL-DD fusion protein. Taken together, these results indicated that TG plant feeding resulted in the suppression of systemic inflammatory parameters, indicating a preventive effect on CIA disease devel-

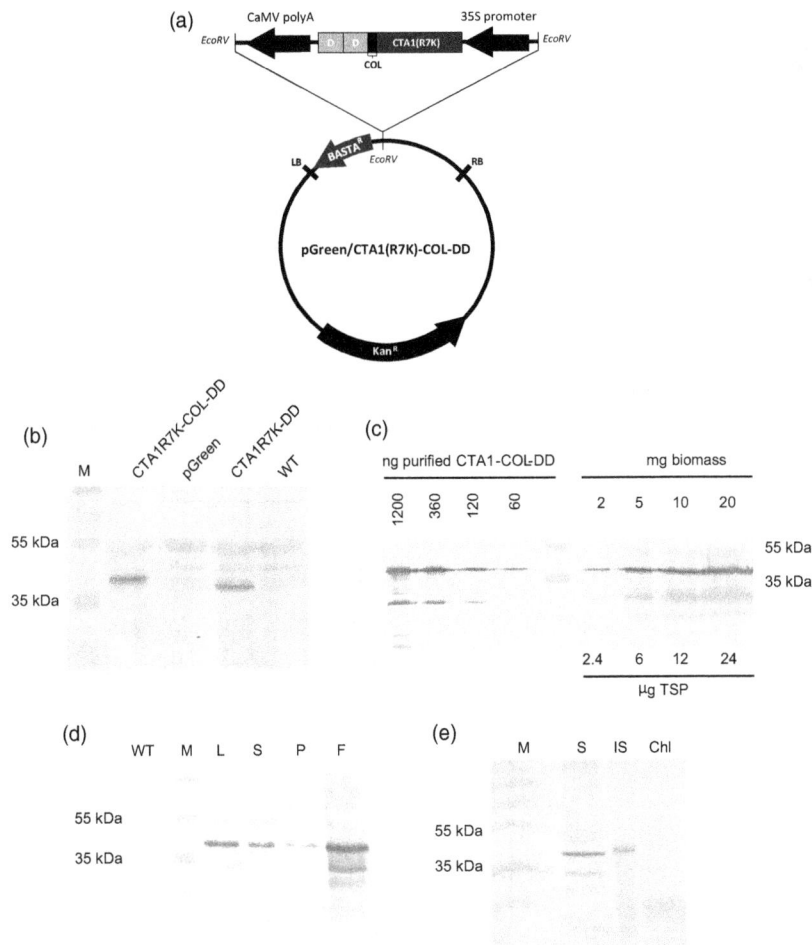

Figure 1 *Arabidopsis thaliana* plant expression of the CTA1(R7K)-COL-DD tolerogen. Schematic diagrams of the pGreen transformation vector and integration cassette/construct. Left border (LB) and right border (RB) flank the transfer DNA (t-DNA) integrated into the nuclear genome of *A. thaliana* (a). Western blot analysis of transgenic plants expressing CTA1(R7K)-COL-DD (1), empty vector pGreen (2), CTA1(R7K)-DD (3) or untransformed WT (4) using anti-CTA1-DD antibodies with positive bands at ~39 kDa (b). Expression levels of CTA1(R7K)-COL-DD in the TG plant were estimated by a semiquantitative analysis using recombinant CTA1(R7K)-COL-DD (c). Protein expression levels in different parts of TG plants were assessed in leaves (L), stems (S), siliques (P) and inflorescence (F). Protein was extracted with 50 mM Tris buffer (pH 7.5) containing 8 M urea. Extracts were 1 mg biomass/µL and 15 µL of each sample was loaded onto the gel. (d). Furthermore, the following subcellular fractions were examined for their content of CTA1(R7K)-COL-DD; soluble cytoplasmic (S), insoluble (IS) and chloroplast (Chl) fractions (e).

opment. Because such a global suppressive effect on inflammation could be ascribed to the stimulation of regulatory CD4 T cells, we analysed the recall response to COL-peptide in CD4 T cells from spleen or draining popliteal lymph nodes in TG plant fed mice. We observed significantly reduced cell proliferation and supernatants demonstrated strikingly lower levels of IL-1α, IL-6, IFNγ, IL-17 and GM-CSF as compared with CD4 T cells from control mice (Figure 4a). Of note, also IL-13 levels were reduced, supporting the notion that the suppressive effects we observed were not due to Th2-skewing of the COL-specific response. By contrast, IL-10 levels were enhanced in culture supernatants from CTA1(R7K)-COL-DD fed mice as compared to supernatants from control mice (Figure 4b). Hence, feeding tolerogenic TG plants stimulated regulatory CD4 T cells that produced IL-10, which strongly suppressed the expansion and/or function of COL-specific auto-aggressive T cells and significantly impaired disease progression in the CIA model. This interpretation was further supported by the increase in circulating FoxP3[+] Tregs in peripheral blood of TG plant fed mice, but which was not seen in mice fed WT plants (Figure 4c).

Feeding transgenic plants protects against disease and tissue destruction

Finally, we examined tissue sections of joints from CIA mice with or without feeding of the TG plants to assess the degree of tissue destruction. We found significantly less joint destruction and infiltration of inflammatory cells in the synovium of mice fed TG plants as compared to joints from mice in the control groups fed

WT plants (Figure 5a). The overall tissue destruction, as assessed by synovitis index and bone erosion index, was significantly lower in tolerogen fed mice compared to controls (Figure 5b). Hence, these histological data confirmed our clinical observations and demonstrated that feeding of TG plants expressing the CTA1(R7K)-COL-DD tolerogen had a strong protective effect on CIA disease progression.

Discussion

The present proof-of-principle study provides evidence that edible plants can be used to successfully treat autoimmune diseases and more specifically autoimmune arthritis. We report here that mice fed with TG *A. thaliana* expressing the CTA1(R7K)-COL-DD tolerogen prevented CIA in 10–20% of the mice and that those afflicted had significantly milder symptoms. Overall tissue destruction, as shown by inflammatory scores of the synovia as well as bone erosion index, was significantly lower in the treated mice. The results from our study shed light on the potential to treat patients directly by feeding edible plants as a salad or, perhaps, allowing the plants to be reformulated into tablets for oral treatment of disease. Nevertheless, we achieved this by combining immunomodulation with effective targeting of disease-associated peptides in a plant expression vector that provided shielding of the tolerogen from degradation. In this way, the combination represents a means to overcome the requirement for large amounts of peptides, which has hampered success and lowered enthusiasm for edible plants as treatment for

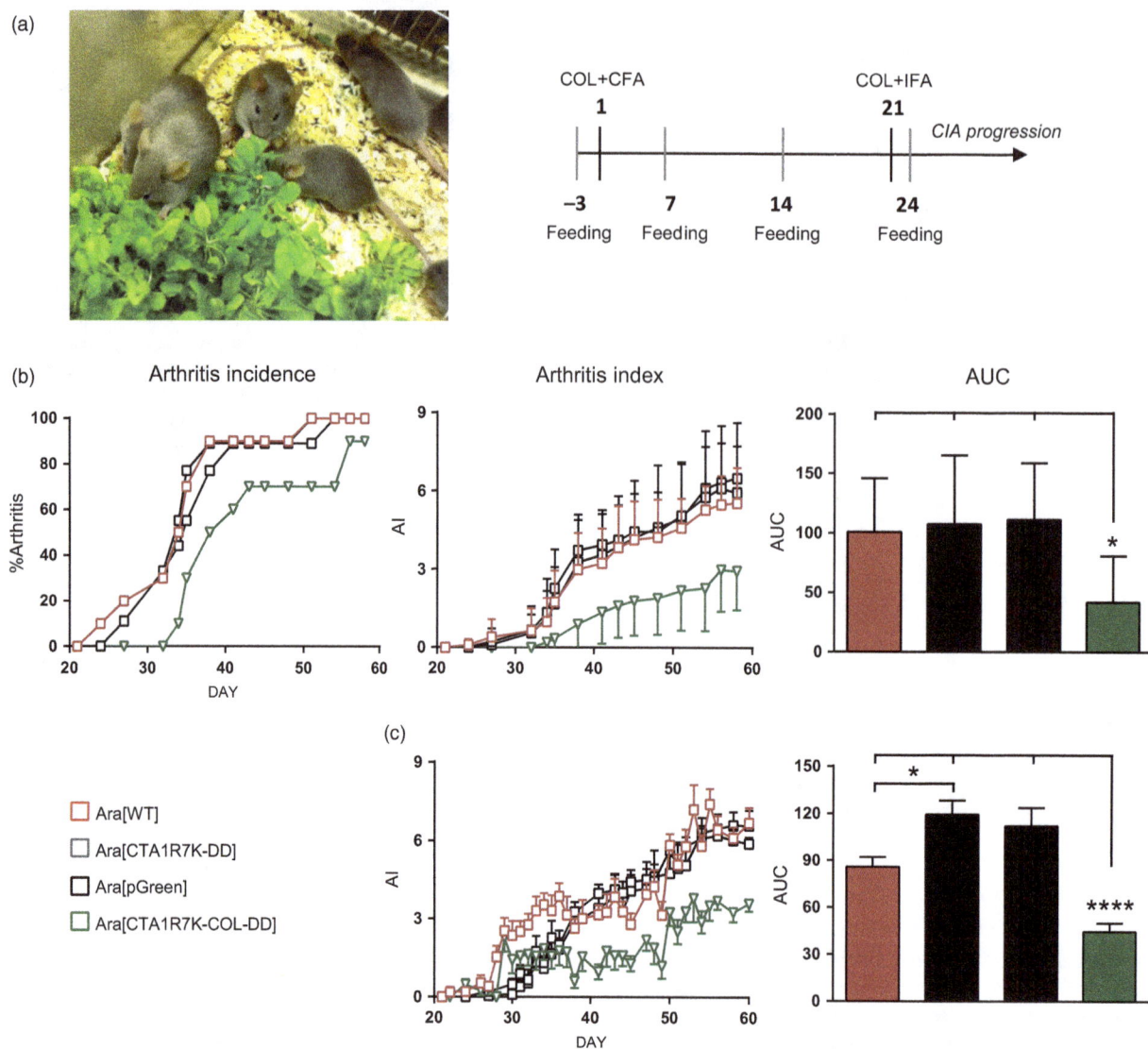

Figure 2 Feeding transgenic *Arabidopsis thaliana* plants to mice protects against CIA disease. DBA/1 mice were divided into groups of ten and fed approximately 70 g of WT or TG *A. thaliana* according to the experiment protocol (a). After the booster dose of CII in IFA on day 21, the mice were under observation for onset of CIA and scored regularly to determine the progression of disease. AUC valued were calculated for statistical analyses (b–c). Data are shown from one representative experiment and shown as mean ± CI (b) or from five pooled experiments ($n = 20$–50) and shown as mean ± SEM (c). Statistical significance was determined by a one-way ANOVA, where * $P < 0.05$ and **** $P < 0.0001$

autoimmune diseases in the past (Garcia *et al.*, 1999; Khare *et al.*, 1995; Park *et al.*, 2009; al-Sabbagh *et al.*, 1994; Thurau *et al.*, 1997; Zhu *et al.*, 2007).

In this study, we estimated that as little as 40 µg of the COL-peptide in total was sufficient to achieve significant clinical effects in the CIA mouse model. This was calculated on the basis of 2.5% of TSP of CTA1(R7K)-COL-DD fusion protein in TG *A. thaliana* plants. Compared to previous studies, this dose is low, especially considering that other studies have used 2 weeks of daily feeding of tolerogenic plants, amounting to, at least, a total of 350 µg of peptide (Hashizume *et al.*, 2008; Iizuka *et al.*, 2014). However, it should be noted that because mice were fed *ad libitum,* some animals may have ingested a higher dose and some a lower dose, which could have impacted on the outcome of our study. This is a frequently encountered problem with edible vaccines and has been discussed previously (Lal *et al.*, 2007). As a generalization, though, we can claim that the

individual mice ingested a dose that, at the group level, was sufficient to give a statistically significant reduction in disease score (IL-6) or tissue destruction (MMP-3). Individual variability is always a problem in any study like this, but given that we could undertake a controlled study with feeding exact doses of the plants, we may, in fact, have achieved an even better result. In our previous intranasal treatment protocols, a dose of 5 µg fusion protein corresponding to 0.2 µg of COL-peptide gave significant protective effects against CIA disease [6]. Although complete disease prevention was observed in more animals (60% reduction compared to 10–20%) following intranasal treatment, the effect on arthritic index and tissue destruction after feeding the tolerogen was roughly a 50% reduction and quite comparable to that seen following intranasal administration (Hasselberg *et al.*, 2009). This means that the fusion protein was well protected in the plant and effectively taken up by the GALT (Rigano *et al.*, 2003).

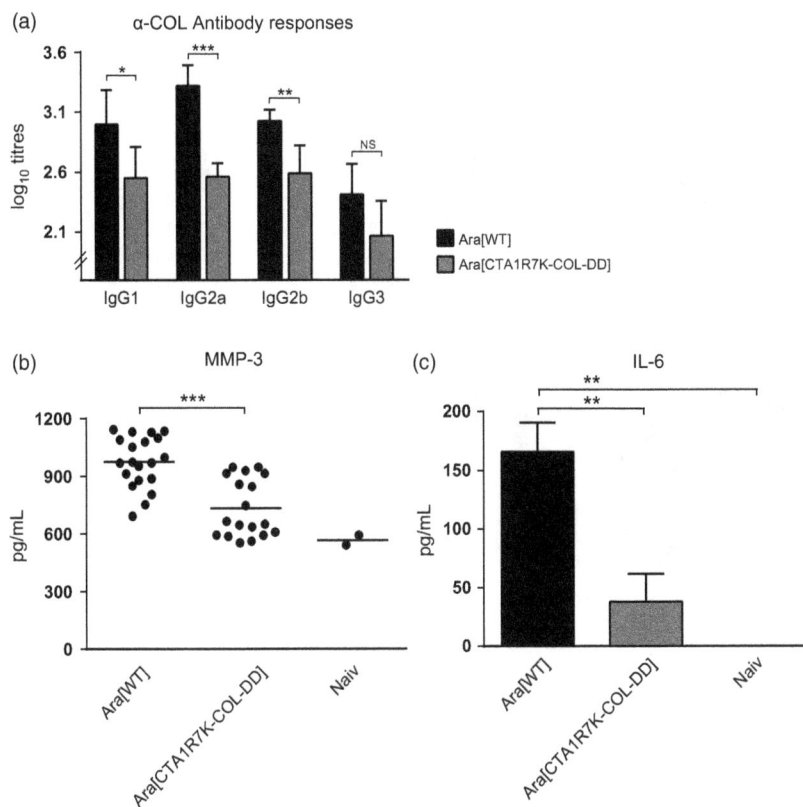

(a) α-COL Antibody responses

(b) MMP-3

(c) IL-6

Figure 3 Systemic responses to CIA are reduced in plant fed mice. Serum was collected from all remaining mice at the end of the experiment. CII-specific antibody levels of the different IgG subclasses were determined by ELISA, and values are shown as mean $\log_{10}$-titres $\pm$SD from one representative experiment ($n = 4$–5) (a). Levels of MMP-3 (b) and IL-6 (c) in serum are given as individual values and the mean $\pm$ SEM from pooled experiments (b–c). Statistical significance was calculated using an unpaired Student's t-test (a) or a one-way ANOVA (b–c), where * $P < 0.05$ and ** $P < 0.01$ and *** $P < 0.001$.

A predominant fraction of the fusion protein was found in the soluble fraction of the plant, but the insoluble fraction also hosted some fusion protein which could be trapped in inclusion bodies (IBs). This may, in fact, have contributed to shielding and effective uptake of our tolerogen in the intestine. (Howe et al., 2014; O'Hagan, 1996). Thus, although IBs traditionally represent a challenge in other biotechnological applications, in this case, it might be beneficial instead (Villaverde et al., 2012). Our study underscores the efficiency by which oral treatment with tolerogen-expressing edible plants can convey protection against autoimmune diseases. Furthermore, this study shows that DD-targeting and CTA1(R7K)-immunomodulation are important features of the tolerogen, also when given orally. Of note, in previous studies, we have shown that an equimolar dose of peptide alone given intranasally has no tolerance-inducing effect (Hasselberg et al., 2010). Moreover, Hashizume et al. expressed the CII250-270 peptide in glutelin as 4 tandem repeats in transgenic rice seeds and after feeding 25 µg of peptide daily for 2 weeks to mice they reported a suppression of collagen-specific IgG2a antibody levels, although these were not significant (Hashizume et al., 2008). This suggests that the peptide was taken up, but under suboptimal conditions and, thus, could not interfere effectively with ongoing immune responses. As our protein expression (2.5% of TSP) was comparable to that reported in many other studies, we believe the DD-targeting and CTA1(R7K) immunomodulation strongly contribute to explaining the difference between previous, less promising, attempts to tolerize against autoimmune diseases using edible plants and the success of this study (Avesani et al., 2010). The use of CTB as a delivery vehicle for protein or peptide for oral tolerization is well documented, for example, when expressed in potatoes or lettuce (Arakawa et al., 1998; Ruhlman et al., 2007). In support of the notion that adjuvants

greatly augment the tolerizing effect of edible plants, Ruhlmann et al. showed in the type 1 diabetes NOD mouse model that feeding CTB-proinsulin could lead to significant immunosuppression and less pancreatic insulitis, although this effect may be due to Th2-skewing rather than the induction of Tregs (Ruhlman et al., 2007).

Our clinical observations of milder arthritis following TG plant feeding were also validated by decreased levels of disease biomarkers in serum, indicative of a reduced severity of CIA. These biomarkers involved MMP-3 and IL-6. Whereas IL-6 is a general marker for inflammation and a reduction of IL-6 also reduces the differentiation of Th17 effector cells, MMP-3 levels in serum have been found to closely relate to RA disease intensity and joint destruction (Kolls and Linden, 2004; Seeuws et al., 2010). We monitored MMP-3 levels in TG plant fed mice, and the reduction was found to be almost to the level observed in healthy naïve mice. The low level of expression of the MMP-3 biomarker also correlated with significantly reduced synovitis and bone erosion scores, in agreement with our clinical scoring. We also observed a decreased serum level of specific total IgG antibodies. An extended analysis demonstrated that both IgG1- and IgG2a-specific antibody responses were reduced, suggesting the suppression of effector functions in both the Th1 and Th2 CD4 T-cell subsets. Moreover, we observed increased IL-10 production concomitant with decreased Th1 (IFNγ), Th2 (IL-13) and Th17 (IL-17 & GM-CSF) cytokine responses in supernatants from splenocytes exposed to recall antigen in vitro. This finding agrees well with our previous observation of induction of IL-10-producing systemic Tregs after intranasal treatment with the tolerogen (Hasselberg et al., 2010). Thus, we conclude that also after oral administration of plant-expressed tolerogen, we achieved induction of IL-10 producing Tregs. In our previous study, these were identified as Tr1 cells because they failed to express Foxp3, but

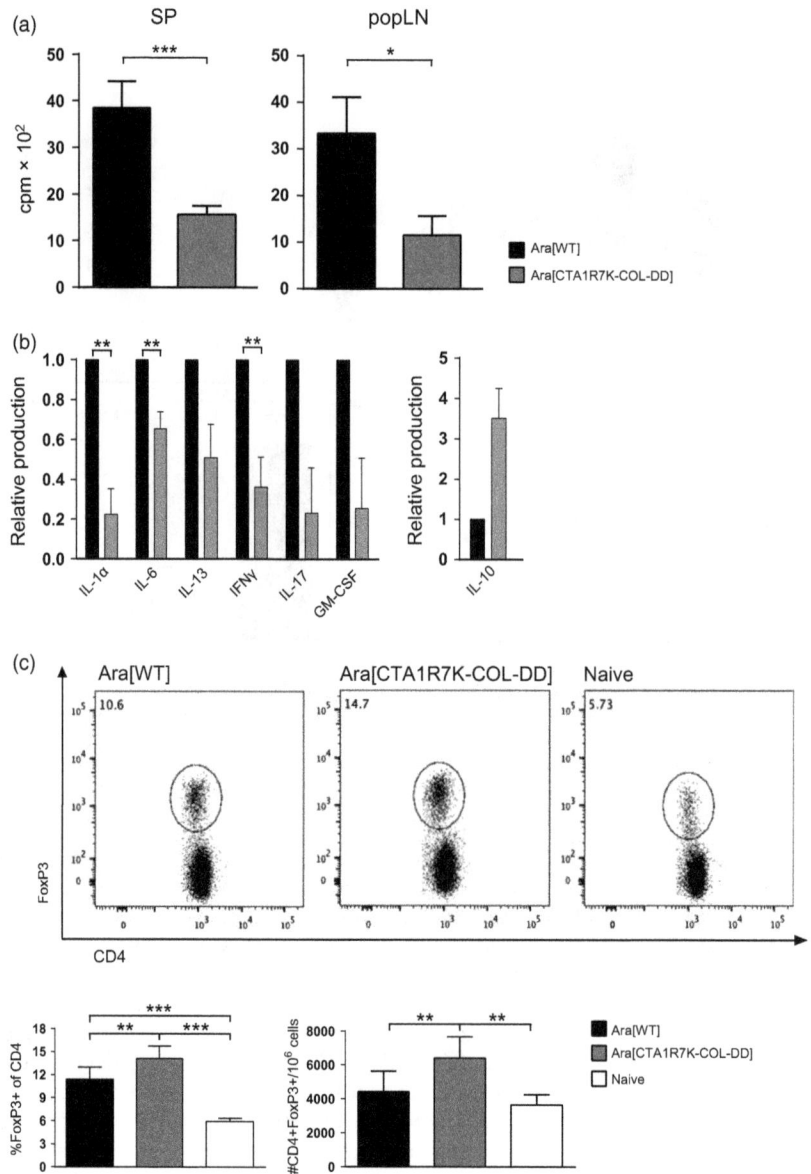

Figure 4 Suppressed CD4 T-cell responses to CIA in mice fed transgenic plants. At the end of the experiment, lymphocytes from spleen or popliteal lymphnodes were isolated from mice that had not yet been euthanized and restimulated with recall COL$_{259-274}$ peptide *in vitro*. Proliferation was assessed after 72 hours by thymidine incorporation (a), and supernatants were collected and analysed for cytokine content (b). Furthermore, FoxP3+ CD4 T cells in peripheral blood were quantified by FACS (c). Proliferation data (a) are summarized from one representative experiment of one (popLN) or four (SP) and shown as mean ± SEM. Relative cytokine production in TG plant fed mice vs WT plant fed mice (b) is pooled data from two to four experiments and shown as mean ± SEM. Representative FoxP3+ CD4 FACS plots (c) are shown and summarized as mean ± SD from one of two independent experiments with similar findings. Statistical significance was measured by Student's t-test (a–b) or a one-way ANOVA (c), where * $P < 0.05$ and ** $P < 0.01$ and *** $P < 0.001$.

several new phenotypic markers, such as CD49b and lymphocyte activation gene 3 (LAG-3) could have been used to better identify Tr1 cells (Gagliani *et al.*, 2013). Nevertheless, in the present study, we also observed an increase in circulating Foxp3$^+$ CD4 T cells after feeding tolerogenic plants. However, we have not yet investigated whether they were COL-specific, inducible or natural Tregs, and whether they were able to exert suppression. Interestingly, in a recent study, feeding of plants expressing haemophilia B factor IX and a Foxp3$^-$ Treg CD4 T-cell population expressing latency associated protein 3 (LAP3$^+$) were shown to exert oral tolerance (Sherman *et al.*, 2014; Wang *et al.*, 2015). Hence, the Treg subsets involved in oral tolerance after feeding recombinant edible plants may be complex and involve not only classical Foxp3$^+$ CD4 T cells, but also Foxp3$^-$ Tr1 cells and LAP$^+$ Tregs that produce not only IL-10, but also TGFβ.

In conclusion, we have greatly expanded on the clinical potential of our patented CTA1(R7K)-COL-DD fusion protein, which expressed in edible plants and fed to mice protected against CIA and ameliorated symptoms and tissue destruction. To our knowledge, this is the first study to clearly demonstrate how edible plants

can successfully be used to treat autoimmune arthritis. The study also serves as an example of how to combine fusion protein targeting and immunomodulation to achieve robust tolerance even with very low doses of the disease-relevant peptides. Our results convey optimism that edible plants, as for example, a salad, or a plant-derived tablet with the tolerogen, may be effective in treating RA and possibly also other autoimmune diseases. Needless to say, edible plants still represent a very cost-effective and safe treatment strategy for delivery of tolerogens for treatment of autoimmune diseases. Finally, it should be mentioned that although clinical use of this strategy is attractive several limitations need to be addressed. Patients will most likely not receive treatment until they present with significant symptoms and the disease is well underway. The therapeutic impact of edible plants expressing CTA1(R7K)-COL-DD in RA patients, therefore, awaits to be assessed. Moreover, as only one CII dominant peptide was used (aa 259–274), it is unknown whether a therapy based on this will be effective, given that epitope spreading is a well-known phenomenon in autoimmune diseases. However, it is well established that human HLA-DRB1*04 alleles can present this epitope in its glycosylated as well as nonglycosy-

(a)

(b)

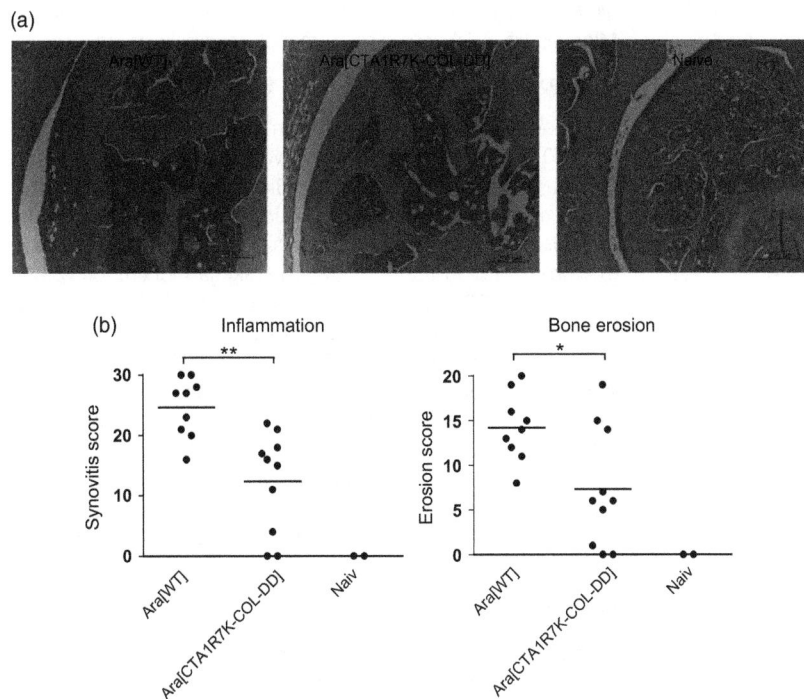

Figure 5 Feeding tolerance-inducing plants protects against CIA disease and tissue destruction. When mice were euthanized, before or at the termination of the experiment, joints were collected and stained for haematoxylin and eosin (a). Lymphocyte infiltrates and cartilage destruction were estimated to determine the severity of synovitis, inflammation and bone erosion (b). Data from one of two independent experiments is shown as a scatter dot plot, and statistical significance was determined by a one-way ANOVA, where * $P < 0.05$ and ** $P < 0.01$.

lated forms and that recognition was seen in 7 of 10 patients and that reactivity in the same individual over several years appeared to persist (Snir et al., 2012). These are all favourable elements that speak to the advantage of the CTA1R7K-COL-DD tolerogen in edible plants for treatment of RA patients.

Experimental procedures

Production of transgenic Arabidopsis thaliana

Transgenic plants were produced by a simplified Agrobacterium-mediated floral dip method as described by Clough and Bent (Clough and Bent, 1995). The transgene CTA1(R7K)-COL-DD was assembled in the pGreen vector (http://www.pgreen.ac.uk) (Hellens et al., 2000) kindly provided by Dr. P. Mullineaux, John Innes Centre and the Biotechnology and Biological Sciences Research Council (Norwich Research Part, UK). The expression cassette contained a constitutive plant promotor CaMV 35S and a CaMV polyA termination sequence, separated by a multicloning site. The vector was linearized at the multicloning site by SmaI restriction enzyme and used for cloning of the transgene (Figure 1a). The pGreen/CTA1(R7K)-COL-DD vector was used to transform Agrobacterium tumefaciens strain EHA105 by electroporation. Positive clones were selected on Luria Broth (LB) medium supplemented with kanamycin (50 μg/mL) and tetracyclin (5 μg/mL) and verified by PCR and sequencing (ABI PRISM 310 Genetic Analyser; Applied Biosystems, Waltham, MA). Four-week-old A. thaliana plants (ecotype Col-0; The European Arabidopsis Stock Centre, Loughborough, UK) were used for transformation of the nuclear genome with the Agrobacterium clone containing the transgene construct (Clough and Bent, 1995). Seeds were harvested from the plants that had undergone floral dip and spread onto agar plates with Murashige and Skoog medium supplemented with 10 μg/mL herbicide BASTA (Riedel-de Haën, Seelze, Germany) and 400 μg/mL cephotaxime (Sigma-Aldrich, Stockholm, Sweden). Resistant seedlings were transferred to potting mix (soil: perlite: vermiculite, 1:1:1) for analysis, self-

pollination and seed production. The seeds obtained from individual plants producing 100% BASTA-resistant progeny were used for feeding experiments after additional confirmation by PCR, sequencing and Western blot analysis.

Expression analysis of the transgenic plants

Cytoplasmic expression by the TG plants was analysed using immunoblotting. Arabidopsis tissues were ground in an extraction buffer containing 50 mM Tris (pH 7.5) with or without 8M urea containing 1 μM protease inhibitor PMSF (Sigma, Stockholm, Sweden). After centrifugation, the samples were separated by SDS-PAGE and blotted onto nitrocellulose membrane Hybond-C (GE Healthcare, Uppsala, Sweden). The membranes were blocked with 1% BSA in TBS-T (0.02 M Tris–HCl, 0.15 M NaCl, 0.1% Tween 20, pH 7.5) and hybridized with α-CTA1-DD (Agrisera, Vännäs, Sweden) antibodies. The protein: primary antibody complexes were then detected by alkaline phosphatase (AP)-conjugated anti-chicken or anti-rabbit antibodies (Promega, Madison, WI) and visualized with nitroblue tetrazolium chloride (NBT) and 5-bromo-4-chloro-3-indolyl phosphate (BCIP; Promega). Protein extract from nontransformed wild-type (WT) Arabidopsis was included in all runs as a negative control and recombinant CTA1-DD protein (Agren et al., 1997) was used as a positive control for semiquantitative analysis of the CTA1(R7K)-COL-DD expression in plants. Total soluble protein (TSP) in the TG plants was extracted using 50 mM Tris (pH 7.5) and quantified with the Bradford method/Bio-Rad protein assay (Bio-Rad Laboratories, Solna, Sweden) using bovine serum albumin (BSA) for construction of the standard curve.

Plant tissue subfractionation

Arabidopsis leaves were homogenized in ice-cold MES buffer (pH 6.5) using a prechilled blender, and the homogenate was filtered using two layers of Miracloth (Calbiochem, San Diego, CA). The released chloroplasts were pelleted at 3000 g, washed in 50 mM HEPES buffer (pH 8) and pelleted again at 5000 g three times. All steps were carried out at 4 °C. The debris retained in the Miracloth was returned to the blender and the homogenization and filtering steps were repeated twice. The remaining material in the Miracloth was an insoluble fraction composed of cellular debris and inclusion bodies. This fraction was further extracted with 50 mM TRIS buffer (pH 7.5) containing 8 M urea for analysis. After 3 washing steps and pelleting of the chloroplast fraction, the supernatants containing cytoplasmic soluble proteins were collected and concentrated using Amicon Ultra-10 filter device with a cut-off of 10 kDa (Merck Millipore, Billerica, MA). All three fractions (chloroplasts, cytoplasmic soluble proteins, and insoluble fraction) were analysed for the presence of the recombinant protein by Western blotting.

Mice, feeding and induction of CIA

Male DBA/1 (H-2^q) mice, ages 8–12 weeks (Harlan, The Netherlands), were kept under specific pathogen-free conditions at the department of experimental medicine (University of Gothenburg). Mice were fed transgenic or wild-type *A. thaliana* plants ad libitum once every week for 4 weeks, corresponding to 70 g of material, and conventional food was removed (<12 h) on these occasions (Figure 2a). CIA was induced by a 100 μL intradermal injection with 100 μg of collagen type II (CII) from bovine tracheal cartilage (MD Biosciences, St Paul, MN) emulsified in an equal volume of complete Freunds adjuvant (CFA; 4 mg/mL *Mycobacterium tuberculosis*; MD Biosciences). On day 21, an intradermal booster injection with 100 μL of 100 μg CII emulsified in Freunds incomplete adjuvant (IFA; Sigma) was given, and disease progression was monitored for 40–60 days. CIA severity was determined as previously described (Brand *et al.*, 2007; Tarkowski *et al.*, 1999). Briefly, limbs were scored 0–3, where 0: no inflammation, 0.5: toe or finger swelling, 1: mild swelling or redness, 2: swelling and redness, 3: marked swelling, redness and/or ankylosis. Mice with a score >6 during examination were euthanized. Studies were approved by the Ethics Committee for Animal Experimentation, University of Gothenburg.

Antibody determinations

96-well, transparent, flat-bottomed plates (Nunc, Roskilde, Denmark) were coated with 10 μg/mL rat CII (Sigma-Aldrich) and blocked with 1% BSA/PBS. Serial threefold dilutions of individual serum samples were incubated overnight at 4 °C. After washing with 0.1% BSA/PBS three times, either alkaline phosphatase-conjugated goat antimouse IgG1-, IgG2a-, IgG2b- or IgG3 antibodies (Southern Biotechnology, Birmingham, AL) were added. The 4-nitrophenyl phosphate disodium salt hexahydrate (NP; 1 mg/mL; Sigma) enzymatic reaction was analysed using a Tecan Sunrise ELISA Reader (Nordic Biolabs, Stockholm, Sweden). Anticollagen serum log$_{10}$-titres were defined as the reciprocal serum dilution, giving an absorbance of 0.4 nm above background levels.

Histological assessments

Paws were treated with paraformaldehyde (Histolab, Gothenburg, Sweden) and embedded in paraffin, as described previously (Tarkowski *et al.*, 1999). Tissue sections were stained with haematoxylin and eosin (Histolab) and an experienced pathologist performed a blinded examination of the tissues for the presence of synovial hypertrophy, pannus formation and bone erosion. Elbows, wrists, carpal joints, fingers, knees, ankles, tarsal joints and toes were scored for inflammation and erosion, where 1 = mild, 2 = moderate and 3 = severe synovitis. MMP3 levels

were assessed in serum by a quantitative ELISA kit (R&D Systems, Minneapolis, MN) according to manufacturer's instructions.

In vitro culturing and cytokine assays

Single-cell suspensions from spleen or popliteal lymph nodes (1 × 10^6/mL) were cultured in 96-well, round-bottomed plates (Nunc) using Iscove's medium (Biochrom, Berlin, Germany) supplemented with 10% heat-inactivated foetal calf serum (FCS; Biochrom), 50 μM 2-ME (Sigma-Aldrich), 1 mM 1-glutamine (Biochrom) and 50 μg/mL gentamicin (Sigma-Aldrich) for 72 h at 37 °C in 5% CO$_2$. Cells were incubated with or without 5 μM of the CII peptide COL (aa 259-274: GIAGFKGEQGPKGEPG; Agrisera) in triplicates, and after 72-h, T-cell cytokine production in supernatants or proliferation, the addition of 1 mCi [3H]-thymidine (PerkinElmer, Boston, MA) for 6 h, was assessed. A beta-scintillation counter (Beckman Coulter, Turku, Finland) was used to measure [3H]-thymidine incorporation. Cytokines in supernatants or serum were determined using a mouse Th1/Th2/Th17 cytometric bead array kit (eBioscience, Santa Clara, CA) according to the manufacturer's instructions.

FACS analysis

Peripheral blood was collected in di-potassium-EDTA-coated Microvette tubes (Sarstedt, Nümbrecht, Germany), and erythrocytes were lysed using BD Pharm Lyse™ lysing buffer (BD Biosciences, Stockholm, Sweden) and washed three times. Cells were resuspended in 0.1% BSA/PBS and incubated with the FcR blocking Ab (24G2) for 5 min before adding AlexaFluor 700-conjugated α-CD4 (BD Biosciences). Cells were then fixated and permeabilized using the FoxP3 staining buffer kit (eBioscience) and finally incubated with PE-conjugated α-FoxP3 (eBioscience) before analysis.

Statistical analysis

As specified in the figure legends, an unpaired Student's *t*-test or a one-way ANOVA was used to calculate statistical significance. The analyses were performed using the Prism software (GraphPad, La Jolla, CA), and *P*-values <0.05 were considered statistically significant.

Acknowledgements

This study was supported by grants from the Swedish Cancer-foundation, Vetenskapsrådet Medicin, Strategiska Stiftelserna, Knut & Alice Wallenbergs Stiftelse, AFA-försäkringar, EU projects in FP7 UniVacFlu and UNISEC.

References

Agren, L.C., Ekman, L., Lowenadler, B. and Lycke, N.Y. (1997) Genetically engineered nontoxic vaccine adjuvant that combines B cell targeting with immunomodulation by cholera toxin A1 subunit. *J. Immunol.* **158**, 3936–3946.

Arakawa, T., Yu, J., Chong, D.K., Hough, J., Engen, P.C. and Langridge, W.H. (1998) A plant-based cholera toxin B subunit-insulin fusion protein protects against the development of autoimmune diabetes. *Nat. Biotechnol.* **16**, 934–938.

Avesani, L., Bortesi, L., Santi, L., Falorni, A. and Pezzotti, M. (2010) Plant-made pharmaceuticals for the prevention and treatment of autoimmune diseases: where are we? *Expert Rev. Vaccines* **9**, 957–969.

Barnett, M.L., Combitchi, D. and Trentham, D.E. (1996) A pilot trial of oral type II collagen in the treatment of juvenile rheumatoid arthritis. *Arthritis Rheum.* **39**, 623–628.

Barnett, M.L., Kremer, J.M., St Clair, E.W., Clegg, D.O., Furst, D., Weisman, M., Fletcher, M.J., Chasan-Taber, S., Finger, E., Morales, A., Le, C.H. and Trentham, D.E. (1998) Treatment of rheumatoid arthritis with oral type II collagen. Results of a multicenter, double-blind, placebo-controlled trial. *Arthritis Rheum.* **41**, 290–297.

Beyaert, R., Beaugerie, L., Van Assche, G., Brochez, L., Renauld, J.C., Viguier, M., Cocquyt, V., Jerusalem, G., Machiels, J.P., Prenen, H., Masson, P., Louis, E. and De Keyser, F. (2013) Cancer risk in immune-mediated inflammatory diseases (IMID). *Mol. Cancer.* **12**, 98.

Bongartz, T., Sutton, A.J., Sweeting, M.J., Buchan, I., Matteson, E.L. and Montori, V. (2006) Anti-TNF antibody therapy in rheumatoid arthritis and the risk of serious infections and malignancies: systematic review and meta-analysis of rare harmful effects in randomized controlled trials. *JAMA,* **295**, 2275–2285.

Brand, D.D., Kang, A.H. and Rosloniec, E.F. (2003) Immunopathogenesis of collagen arthritis. *Springer Semin. Immunopathol.* **25**, 3–18.

Brand, D.D., Latham, K.A. and Rosloniec, E.F. (2007) Collagen-induced arthritis. *Nat. Protoc.* **2**, 1269–1275.

Brooks, P.M. (2006) The burden of musculoskeletal disease–a global perspective. *Clin. Rheumatol.* **25**, 778–781.

Cazzola, M., Antivalle, M., Sarzi-Puttini, P., Dell'Acqua, D., Panni, B. and Caruso, I. (2000) Oral type II collagen in the treatment of rheumatoid arthritis. A six-month double blind placebo-controlled study. *Clin. Exp. Rheumatol.* **18**, 571–577.

Choy, E.H., Scott, D.L., Kingsley, G.H., Thomas, S., Murphy, A.G., Staines, N. and Panayi, G.S. (2001) Control of rheumatoid arthritis by oral tolerance. *Arthritis Rheum.* **44**, 1993–1997.

Clough, S. and Bent, A. (1995) Floral dip: a simplified method for Agrobacterium-mediated transformation of *Arabidopsis thaliana. Plant J.* **16**, 735–743.

Faria, A.M. and Weiner, H.L. (2006) Oral tolerance: therapeutic implications for autoimmune diseases. *Clin. Dev. Immunol.* **13**, 143–157.

Firestein, G.S. (2003) Evolving concepts of rheumatoid arthritis. *Nature,* **423**, 356–361.

Gagliani, N., Magnani, C.F., Huber, S., Gianolini, M.E., Pala, M., Licona-Limon, P., Guo, B., Herbert, D.R., Bulfone, A., Trentini, F., Di Serio, C., Bacchetta, R., Andreani, M., Brockmann, L., Gregori, S., Flavell, R.A. and Roncarolo, M.G. (2013) Coexpression of CD49b and LAG-3 identifies human and mouse T regulatory type 1 cells. *Nat. Med.* **19**, 739–746.

Garcia, G., Komagata, Y., Slavin, A.J., Maron, R. and Weiner, H.L. (1999) Suppression of collagen-induced arthritis by oral or nasal administration of type II collagen. *J. Autoimmun.* **13**, 315–324.

Hashizume, F., Hino, S., Kakehashi, M., Okajima, T., Nadano, D., Aoki, N. and Matsuda, T. (2008) Development and evaluation of transgenic rice seeds accumulating a type II-collagen tolerogenic peptide. *Transgenic Res.* **17**, 1117–1129.

Hasselberg, A., Schon, K., Tarkowski, A. and Lycke, N. (2009) Role of CTA1R7K-COL-DD as a novel therapeutic mucosal tolerance-inducing vector for treatment of collagen-induced arthritis. *Arthritis Rheum.* **60**, 1672–1682.

Hasselberg, A., Ekman, L., Yrlid, L.F., Schon, K. and Lycke, N.Y. (2010) ADP-ribosylation controls the outcome of tolerance or enhanced priming following mucosal immunization. *J. Immunol.* **184**, 2776–2784.

Hellens, R.P., Edwards, E.A., Leyland, N.R., Bean, S. and Mullineaux, P.M. (2000) pGreen: a versatile and flexible binary Ti vector for Agrobacterium-mediated plant transformation. *Plant Mol. Biol.* **42**, 819–832.

Howe, S.E., Lickteig, D.J., Plunkett, K.N., Ryerse, J.S. and Konjufca, V. (2014) The uptake of soluble and particulate antigens by epithelial cells in the mouse small intestine. *PLoS ONE* **9**, e86656.

Ichim, T.E., Zheng, X., Suzuki, M., Kubo, N., Zhang, X., Min, L.R., Beduhn, M.E., Riordan, N.H., Inman, R.D. and Min, W.P. (2008) Antigen-specific therapy of rheumatoid arthritis. *Expert Opin. Biol. Ther.* **8**, 191–199.

Iizuka, M., Wakasa, Y., Tsuboi, H., Asashima, H., Hirota, T., Kondo, Y., Matsumoto, I., Takaiwa, F. and Sumida, T. (2014) Suppression of collagen-induced arthritis by oral administration of transgenic rice seeds expressing altered peptide ligands of type II collagen. *Plant Biotechnol. J.* **12**, 1143–1152.

Jabbal-Gill, I. (2010) Nasal vaccine innovation. *J. Drug Target.* **18**, 771–786.

Khare, S.D., Krco, C.J., Griffiths, M.M., Luthra, H.S. and David, C.S. (1995) Oral administration of an immunodominant human collagen peptide modulates collagen-induced arthritis. *J. Immunol.* **155**, 3653–3659.

Kolls, J.K. and Linden, A. (2004) Interleukin-17 family members and inflammation. *Immunity* **21**, 467–476.

Lal, P., Ramachandran, V.G., Goyal, R. and Sharma, R. (2007) Edible vaccines: current status and future. *Indian J. Med. Microbiol.* **25**, 93–102.

Limaye, A., Koya, V., Samsam, M. and Daniell, H. (2006) Receptor-mediated oral delivery of a bioencapsulated green fluorescent protein expressed in transgenic chloroplasts into the mouse circulatory system. *FASEB J.* **20**, 959–961.

Ma, S.W., Zhao, D.L., Yin, Z.Q., Mukherjee, R., Singh, B., Qin, H.Y., Stiller, C.R. and Jevnikar, A.M. (1997) Transgenic plants expressing autoantigens fed to mice to induce oral immune tolerance. *Nat. Med.* **3**, 793–796.

McInnes, I.B. and Schett, G. (2007) Cytokines in the pathogenesis of rheumatoid arthritis. *Nat. Rev. Immunol.* **7**, 429–442.

O'Hagan, D.T. (1996) The intestinal uptake of particles and the implications for drug and antigen delivery. *J. Anat.* **189**(Pt 3), 477–482.

Okada, H., Kuhn, C., Feillet, H. and Bach, J.F. (2010) The 'hygiene hypothesis' for autoimmune and allergic diseases: an update. *Clin. Exp. Immunol.* **160**, 1–9.

Park, K.S., Park, M.J., Cho, M.L., Kwok, S.K., Ju, J.H., Ko, H.J., Park, S.H. and Kim, H.Y. (2009) Type II collagen oral tolerance; mechanism and role in collagen-induced arthritis and rheumatoid arthritis. *Mod. Rheumatol.* **19**, 581–589.

Rigano, M.M., Sala, F., Arntzen, C.J. and Walmsley, A.M. (2003) Targeting of plant-derived vaccine antigens to immunoresponsive mucosal sites. *Vaccine* **21**, 809–811.

Ruhlman, T., Ahangari, R., Devine, A., Samsam, M. and Daniell, H. (2007) Expression of cholera toxin B-proinsulin fusion protein in lettuce and tobacco chloroplasts–oral administration protects against development of insulitis in non-obese diabetic mice. *Plant Biotechnol. J.* **5**, 495–510.

al-Sabbagh, A., Miller, A., Santos, L.M. and Weiner, H.L. (1994) Antigen-driven tissue-specific suppression following oral tolerance: orally administered myelin basic protein suppresses proteolipid protein-induced experimental autoimmune encephalomyelitis in the SJL mouse. *Eur. J. Immunol.* **24**, 2104–2109.

Salliot, C., Dougados, M. and Gossec, L. (2009) Risk of serious infections during rituximab, abatacept and anakinra treatments for rheumatoid arthritis: meta-analyses of randomised placebo-controlled trials. *Ann. Rheum. Dis.* **68**, 25–32.

Seeuws, S., Jacques, P., Van Praet, J., Drennan, M., Coudenys, J., Decruy, T., Deschepper, E., Lepescheux, L., Pujuguet, P., Oste, L., Vandeghinste, N., Brys, R., Verbruggen, G. and Elewaut, D. (2010) A multiparameter approach to monitor disease activity in collagen-induced arthritis. *Arthritis. Res. Ther.* **12**, R160.

Sherman, A., Su, J., Lin, S., Wang, X., Herzog, R.W. and Daniell, H. (2014) Suppression of inhibitor formation against FVIII in a murine model of hemophilia A by oral delivery of antigens bioencapsulated in plant cells. *Blood,* **124**, 1659–1668.

Smolen, J.S. and Aletaha, D. (2015) Rheumatoid arthritis therapy reappraisal: strategies, opportunities and challenges. *Nat. Rev. Rheumatol.* **11**, 276–289.

Snir, O., Backlund, J., Bostrom, J., Andersson, I., Kihlberg, J., Buckner, J.H., Klareskog, L., Holmdahl, R. and Malmstrom, V. (2012) Multifunctional T cell reactivity with native and glycosylated type II collagen in rheumatoid arthritis. *Arthritis Rheum.* **64**, 2482–2488.

Tarkowski, A., Sun, J.B., Holmdahl, R., Holmgren, J. and Czerkinsky, C. (1999) Treatment of experimental autoimmune arthritis by nasal administration of a type II collagen-cholera toxoid conjugate vaccine. *Arthritis Rheum.* **42**, 1628–1634.

Thurau, S.R., Chan, C.C., Nussenblatt, R.B. and Caspi, R.R. (1997) Oral tolerance in a murine model of relapsing experimental autoimmune uveoretinitis (EAU): induction of protective tolerance in primed animals. *Clin. Exp. Immunol.* **109**, 370–376.

Trentham, D.E., Townes, A.S. and Kang, A.H. (1977) Autoimmunity to type II collagen an experimental model of arthritis. *J. Exp. Med.* **146**, 857–868.

Trentham, D.E., Dynesius-Trentham, R.A., Orav, E.J., Combitchi, D., Lorenzo, C., Sewell, K.L., Hafler, D.A. and Weiner, H.L. (1993) Effects of oral administration of type II collagen on rheumatoid arthritis. *Science*, **261**, 1727–1730.

Villaverde, A., Garcia-Fruitos, E., Rinas, U., Seras-Franzoso, J., Kosoy, A., Corchero, J.L. and Vazquez, E. (2012) Packaging protein drugs as bacterial inclusion bodies for therapeutic applications. *Microb. Cell Fact.* **11**, 76.

Wang, X., Su, J., Sherman, A., Rogers, G.L., Liao, G., Hoffman, B.E., Leong, K.W., Terhorst, C., Daniell, H. and Herzog, R.W. (2015) Plant-based oral tolerance to hemophilia therapy employs a complex immune regulatory response including LAP+CD4+ T cells. *Blood*, **125**, 2418–2427.

Weiner, H.L., da Cunha, A.P., Quintana, F. and Wu, H. (2011) Oral tolerance. *Immunol. Rev.* **241**, 241–259.

Zhu, P., Li, X.Y., Wang, H.K., Jia, J.F., Zheng, Z.H., Ding, J. and Fan, C.M. (2007) Oral administration of type-II collagen peptide 250-270 suppresses specific cellular and humoral immune response in collagen-induced arthritis. *Clin. Immunol.* **122**, 75–84.

19

A novel and fully scalable *Agrobacterium* spray-based process for manufacturing cellulases and other cost-sensitive proteins in plants

Simone Hahn[‡], Anatoli Giritch[‡], Doreen Bartels, Luisa Bortesi[†] and Yuri Gleba[*]

Nomad Bioscience GmbH, Halle (Saale), Germany

*Correspondence
email
gleba@nomadbioscience.com
[†]Present address: Institute for Molecular Biotechnology, RWTH Aachen University, Worringerweg 1, 52074 Aachen, Germany.
[‡]These authors contributed equally to this work.

Keywords: *Agrobacterium*, bioethanol, cellulases, griffithsin, *Nicotiana*, thaumatin.

Summary

Transient transfection of plants by vacuum infiltration of *Agrobacterium* vectors represents the state of the art in plant-based protein manufacturing; however, the complexity and cost of this approach restrict it to pharmaceutical proteins. We demonstrated that simple spraying of *Nicotiana* plants with *Agrobacterium* vectors in the presence of a surfactant can substitute for vacuum inoculation. When the T-DNA of *Agrobacterium* encodes viral replicons capable of cell-to-cell movement, up to 90% of the leaf cells can be transfected and express a recombinant protein at levels up to 50% of total soluble protein. This simple, fast and indefinitely scalable process was successfully applied to produce cellulases, one of the most volume- and cost-sensitive biotechnology products. We demonstrate here for the first time that representatives of all hydrolase classes necessary for cellulosic biomass decomposition can be expressed at high levels, stored as silage without significant loss of activity and then used directly as enzyme additives. This process enables production of cellulases, and other potential high-volume products such as noncaloric sweetener thaumatin and antiviral protein griffithsin, at commodity agricultural prices and could find broad applicability in the large-scale production of many other cost-sensitive proteins.

Introduction

Transient expression of recombinant proteins in plants using *Agrobacterium* vectors has become a preferred biomanufacturing platform due to its speed, versatility, cost efficiency and industrially relevant scalability for many proteins (Giritch *et al.*, 2006; Marillonnet *et al.*, 2005; reviewed in Gleba and Giritch (2012) One major limitation of the current protocols, however, is their reliance on vacuum infiltration, a technique that requires mechanical manipulation of pot- or tray-grown plants for their submerged exposure to the inoculum in vacuum chambers (reviewed in Gleba and Giritch (2012). Such mechanical manipulation of the host plants, which is currently practised in larger scales via robotics and semi-automated systems, imparts cost and complexity to the inoculation process, and therefore, such systems lend themselves preferentially to the production of pharmaceutical proteins. Most industrial recombinant proteins such as industrial enzymes and biomaterials need to be manufactured inexpensively in high volumes; thus, the use of containerized greenhouse-grown plants is not practical or cost-effective for these applications. We therefore explored alternatives to vacuum infiltration that can be applied to plants regardless of how they are cultivated. We have found that the simple spraying of plant leaves with a dilute suspension of *Agrobacterium* vectors in the presence of a surfactant allows the effective transfection of 0.9–3.5% of leaf cells of most *Nicotiana* plants, and that the employment of T-DNA messages consisting of viral replicons capable of cell-to-cell movement increases the resultant transfection levels to up to 90% of leaf cells. This transfection procedure requires a slightly longer time to effect (10–14 days instead of the typical 4–7 days for vacuum infiltration), but the resultant expression levels are comparable to those of vacuum-infiltrated plants (Marillonnet *et al.*, 2005).

We evaluated this protocol in the manufacture of cellulases, one of today's most cost-sensitive biotechnology products that must ultimately be delivered in high volume to enable applications such as cellulosic ethanol production (Carroll and Somerville, 2009). Cellulases currently used in bioethanol production are all produced by microbial fermentation. Despite decades of research on lowering cellulase manufacturing costs, these enzymes still account for 20–40% of cellulosic ethanol production costs (Sainz, 2009). We expressed six cellulase genes of bacterial and fungal origin representing all four hydrolase classes needed for degradation of cellulose into glucose and identified several that are expressed at high levels and retain their expected enzymatic activity, even after the plant biomass was harvested and stored as silage without refrigeration for an extended period of time. The overall process is flexible and robust and consists of simple manipulations that are common to many agricultural practices. As the volumes of *Agrobacterium* inocula needed to induce cellulase biosynthesis are very small, the calculated costs of manufacturing and storing cellulases produced by this method are similar to the costs of other commodity agricultural products. The technology has been shown to work for other potential product candidates that would require high volume inexpensive production such as noncaloric sweetener protein thaumatin and griffithsin protein with broad antiviral activity.

Results

Plants can be transiently transfected by spraying with *Agrobacterium* vectors

We assessed the effect of direct spraying with diluted agrobacterial suspensions on transient transfection of *Nicotiana benthamiana* plants. Binary construct pICH18722 containing the TMV-based viral replicon with insertion of green fluorescent protein (GFP)-encoding gene (TMV(fsMP)-GFP) was selected for evaluating transfection efficiency (Figure S1). The viral replicon used was disabled for systemic movement by deletion of the coat protein (CP) gene; the vector also contained a frame-shift mutation in the movement protein (MP) coding sequence, resulting in loss of the cell-to-cell movement ability. Therefore, this vector expressed GFP only in cells directly transfected with T-DNA.

Plants were sprayed with agrobacterial suspensions diluted 10^{-2} and 10^{-3} relative to saturated overnight agrobacterial cultures (i.e. OD_{600} = 0.015, corresponding to approx. 10^7 CFU/mL and OD_{600} = 0.0015, corresponding to approx. 10^6 CFU/mL, respectively), supplemented with surfactant Silwet L-77® (GE Silicones, Inc., USA) at 0.1% (v/v), a concentration that is typically used in industrial agricultural applications (Harvey, 1998). Four days after spraying, separate foci of GFP fluorescence appeared in sprayed leaves and the same pattern was still visible for at least 2 weeks after spraying (Figure 1c). The proportion of GFP-expressing cells was counted after the isolation of leaf protoplasts. Depending on agrobacterial suspension concentration, 0.9–3.5% of total leaf cells was transfected as a result of *Agrobacterium*-mediated T-DNA transfer (Figure 1e).

Use of viral replicon-based vectors capable of cell-to-cell or systemic movement results in almost complete transfection of leaf cells

To increase the proportion of plant cells expressing the gene of interest, we then used T-DNA encoding viral replicon capable of movement in a plant. GFP-expressing TMV vector pICH18711 (TMV(MP)-GFP) is disabled for systemic movement due to the CP gene deletion, but retains cell-to-cell movement ability by the presence of MP (Figure S1). *Nicotiana benthamiana* plants were sprayed with 10^{-2} and 10^{-3} dilutions of agrobacterial cultures harbouring pICH18711 construct. Spots of GFP fluorescence appeared 3 days postspraying (dps) and enlarged over the time, merging together within approximately 12 days postinfection. Visually, complete transfection of sprayed leaves developed approximately 12 days postspraying without significant difference between 10^{-2} and 10^{-3} dilutions of agrobacterial cultures (Figure 1d). Counting of protoplasts prepared from three independent leaves sprayed with bacterial inocula revealed GFP expression in more than 90% of plant cells treated with 10^{-2} dilution of agrobacterial culture. Spraying with the 10^{-3} dilution of the inocula yielded the value of about 86% (Figure 1e). Spraying with the TMV-based vector resulted in abundant GFP expression in all sprayed leaves but not in the youngest ones newly developed after spraying (Figure 1b).

Multiple cellulases of microbial and fungal origin can be efficiently expressed in plants

We evaluated the expression of six cellulolytic enzymes of bacterial and fungal origin in *Nicotiana benthamiana* plants upon spraying agrobacterial inocula harbouring TMV viral vectors enabled for cell-to-cell movement. The enzymes included in this evaluation represent all three cellulase classes that are necessary for the conversion of cellulose to glucose: (i) endoglucanases (EC 3.2.1.4), which cleave cellulose chains internally, providing free reducing and nonreducing chain ends; (ii) cellobiohydrolases/exoglucanases (EC 3.2.1.91/EC 3.2.1.-), which operate on cellulose chain termini with specificity for reducing or nonreducing ends, respectively, releasing cellooligosaccharides, preferentially cellobiose; and iii) β-glucosidases which hydrolyse cellobiose to glucose (EC 3.2.1.21). The analysis provided here includes two

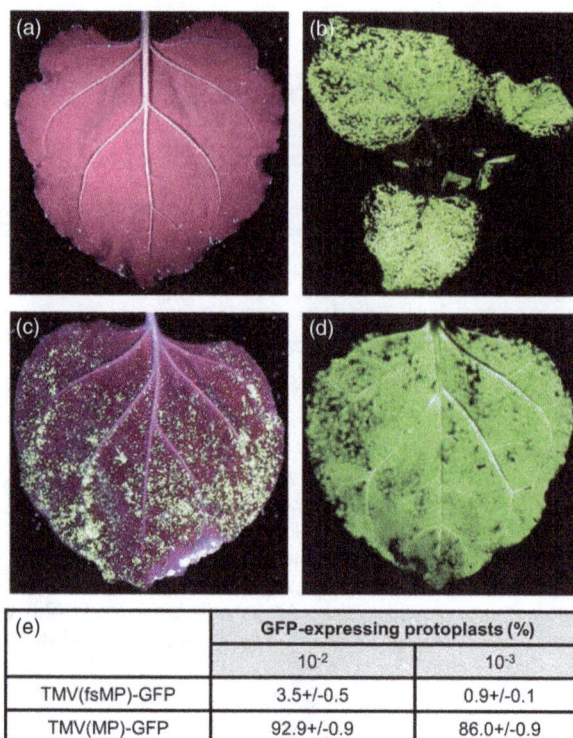

(e)	GFP-expressing protoplasts (%)	
	10^{-2}	10^{-3}
TMV(fsMP)-GFP	3.5+/-0.5	0.9+/-0.1
TMV(MP)-GFP	92.9+/-0.9	86.0+/-0.9

Figure 1 Transient transfection of *Nicotiana benthamiana* plants 12 days after spraying with suspension of agrobacterial cells harbouring TMV-based viral vectors with or without cell-to-cell movement ability. (a) Leaf of control nontransfected plant, (b) whole *Nicotiana benthamiana* plant transfected with TMV(MP)-GFP vector capable of cell-to-cell movement ability, (c) Leaf transfected with TMV(fsMP)-GFP vector lacking cell-to-cell movement ability, (d) Leaf transfected with TMV(MP)-GFP vector with cell-to-cell movement ability. (e) The percentage of GFP-expressing cells counted after the isolation of protoplasts from the whole leaf blade (the mean value and standard deviation (SD) of three different leaves of same plant. Numerals 10^{-2} and 10^{-3} show the dilution factor of the overnight agrobacterial cultures of OD=1.5 at 600 nm. In (b–d), 10^{-3} dilution was used.

endoglucanases, namely endoglucanase E1 from *Acidothermus cellulolyticus* 11B and endoglucanase Cel5A from *Thermotoga maritima*; one exoglucanase, β-1,4-exocellulase (E3) from *Thermobifida fusca* with a specificity for nonreducing chain ends; two cellobiohydrolases, exoglucanase 1 (CBHI) from *Trichoderma reesei/Hypocrea jecorina* and CBHI from *Humicola grisea*, both with reducing chain end specificity; and one β-glucosidase Bgl4 from *Humicola grisea* (Table 1).

Cellulases were screened for optimal expression using translational fusions with various signal peptides (SP) providing targeting into different cell compartments using assembly of 5′ and 3′ TMV pro-vector modules *in planta* (Marillonnet et al., 2004) (Figure S2). Selected fusions with the highest yield of active enzyme (listed in Table 2) were subcloned into assembled TMV-based viral vectors (Figure S3), and the time-course for expression was analysed. For this purpose, *Nicotiana benthamiana* plants were sprayed with *Agrobacterium* suspensions in dilutions ranging from 10^{-2} to 10^{-3}, and the plant material was inspected for recombinant protein expression from 4 to 18 dps using SDS-PAGE with Coomassie staining. The yields and timing of expression depended on the agrobacterial culture dilution and varied among the enzymes (Figure 2 and Table 1). In most cases, optimal yields

Table 1 Cellulolytic enzymes analysed for transient expression in *Nicotiana benthamiana*

No.	Cellulase	EC number	Source organism	Accession number	MW, kDa	pH optimum	t° optimum, °C	Yield*	Activity **
1	Endoglucanase E1	3.2.1.4	*Acidothermus cellulolyticus*	P54583.1	60.7	5–6 / Hood *et al.* (2007)	81 / Hood *et al.* (2007)	< 1% / < 0.03	17.83 ± 3.76 / 62.94 ± 4.8
2	Endoglucanase Cel5A	3.2.1.4	*Thermotoga maritima*	3MMW-D	37.4	6 / Kim *et al.* (2010)	80 / Kim *et al.* (2010)	12.5% / 0.68 ± 0.01	67.32 ± 15.6 / 363.22 ± 81.55
3	Exocellulase E3	3.2.1.-	*Thermobifida fusca*	AAA62211.1	63.5	7–8 / Zhang *et al.* (1995)	65 / Zhang *et al.* (1995)	25% / 1.6 ± 0.05	0.88 ± 0.46 / 5.58 ± 2.74
4	Exoglucanase (CBHI)	3.2.1.91	*Trichoderma reesei*	P62694.1	54.0	5 / Hood *et al.* (2007)	45–50 / Hood *et al.* (2007)	12.5% / 0.43 ± 0.09	8.06 ± 2.31 / 27.39 ± 6.94
5	Exoglucanase 1 (CBHI)	3.2.1.91	*Humicola grisea*	BAA09785.1	54.1	5 / Takashima *et al.* (1996)	60 / Takashima *et al.* (1996)	10% / 0.23 ± 0.07	6.9 ± 1.2 / 15.6 ± 1.82
6	β-glucosidase Bgl4	3.2.1.21	*Humicola grisea*	BAA74958.1	54.0	6 / Takashima *et al.* (1999)	55 / Takashima *et al.* (1999)	50% / 3.28 ± 0.4	16.27 ± 0.93 / 106.75 ± 14.44

*Yield of recombinant protein is expressed as a percentage of TSP (top) and mg recombinant protein/g fresh weight of plant biomass (bottom). **Enzymatic activity of recombinant protein is given in µmole pNP (T = 24 h) for Nr 1, 2, 4, and 5, in mg glucose (T = 46 h) for Nr 3 or IU for Nr 6 per mg of TSP (top) and per gram fresh weight of plant biomass (bottom). Measurements were taken on biological triplicates; the average and standard deviations are provided. Plant material was transfected for cellulase expression with 10^{-3} dilutions of *Agrobacterium* cultures and harvested 11 dpi.

were obtained for plant material sprayed with 10^{-3} dilutions at 11–12 dps (Figure 2). To standardize transfection and production, the same *Agrobacterium* inoculum density (10^{-3} dilution) and harvest time (11 dps) were used for all enzymes.

The yields of five of the six recombinant enzymes were in the range of 10% to 50% of TSP, or 0.23 to 3.28 mg/g of fresh weight (Table 1).

Plant-expressed cellulases maintain high enzymatic activity

To evaluate the enzymatic properties of expressed recombinant cellulases, plant total soluble protein (TSP) extracts were analysed for specific cellulolytic activities. Due to similar pH and temperature optima of selected enzymes and the proposed utilization of enzymes as a mixture in downstream processes, all cellulases were extracted with identical buffer, which enabled a simple pH-dependant removal of a considerable portion of nonrecombinant plant proteins by precipitation (Figure 3). All expressed cellulases were efficiently extracted in aqueous buffer and were active when incubated with the respective substrates under the same buffer and temperature conditions (Figure 3).

Plant biomass containing cellulases can be stored as silage without major loss of activity

As the harvest time of plant material expressing cellulases requires to be independent from the application of the enzymes in

Table 2 Storage stability of plant-made recombinant cellulases

Cellulase	Storage	Protein accumulation		Activity level	
		%age of TSP	mg recombinant protein/g fresh weight plant material	Per mg TSP	Per g fresh weight plant material
Rice amylase apoplast TP-E1	-80°C	<1%	< 0.03	17.83 ± 3.76	62.94 ± 4.8
storage 14 weeks		<1%	< 0.01	46.9 ± 1.83	46.9 ± 6.52
Cytosolic Cel5A	-80°C	12.5%	0.68 ± 0.01	67.32 ± 15.6	363.22 ± 81.55
storage 12 weeks		15%	0.16 ± 0.001	76.22 ± 11.29	81.41 ± 12.36
Chloroplast TP-His-EK-E3	-80°C	25%	1.6 ± 0.05	0.88 ± 0.46	5.58 ± 2.74
storage 15 weeks		15%	0.72 ± 0.04	3.59 ± 0.83	5.36 ± 1.51
Barley-α-amylase apoplast TP-CBHI (*Tr*)	-80°C	12.5%	0.43 ± 0.09	8.06 ± 2.31	27.39 ± 6.94
storage 14 weeks		35%	0.37 ± 0.03	17.68 ± 1.49	18.87 ± 2.89
Barley-α-amylase apoplast TP-CBHI (*Hg*)	-80°C	10%	0.23 ± 0.07	6.9 ± 1.2	15.6 ± 1.82
storage 12 weeks		20%	0.23 ± 0.02	17.08 ± 0.95	19.44 ± 1.14
Cytosolic Bgl4	-80°C	50%	3.28 ± 0.4	16.27 ± 0.93	106.75 ± 14.44
storage 16 weeks		90%	1.94 ± 0.23	26.03 ± 0.53	56.12 ± 6.65

Plant material was transfected for cellulase expression by spraying with 10^{-3} dilutions of *Agrobacterium* cultures admixed with 0.1% (w/w) Silwet L-77 surfactant, harvested 11 days postspraying and stored under different conditions. Activity level of recombinant protein is given in µmole pNP (T = 24 h) for E1, Cel5A, CBHI (*Tr*) and CBHI (*Hg*), in mg glucose (T = 46 h) for E3 or IU for Bgl4 per mg of TSP and per gram fresh weight of plant biomass. Measurements were taken on biological triplicates, with average and standard deviations provided.

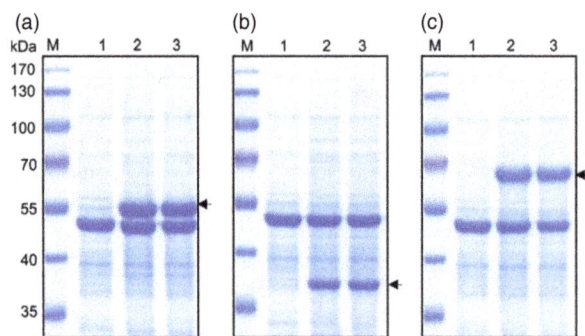

Figure 2 Expression of cellulases of microbial origin in *Nicotiana benthamiana* plants after the spraying with agrobacteria carrying TMV vectors. Coomassie-stained SDS protein gels loaded with crude extracts, corresponding to 1.0 mg fresh weight, from nontransfected leaf tissue (1) or plants transfected by *A. tumefaciens* carrying constructs encoding β-glucosidase Bgl4 (a), endoglucanase Cel5A (b) and exocellulase E3 (c) by spraying with dilutions of inoculation solutions of 10^{-2} (2), 10^{-3} (3). Harvesting time points of plant material were 7 (b, 2, 3), 11 (a, 2, 3; c, 2) and 14 (c, 3) days postinoculation. M, molecular weight marker with weights (kDa) shown on the left. Arrows indicate recombinant proteins.

downstream processes, there is a need for inexpensive and stable storage of cellulase-containing plant biomass. Consequently, protein stability and activity were analysed under different storage conditions, namely, at − 80°C or as silage stored at room temperature for 12 to 16 weeks (Figure 3, Table 2). Surprisingly, all cellulases retained their specific activities upon storage as silage. Two of the six cellulases analysed (Bgl4 and CBHI) were both very stable and had high enzymatic activity in ensiled plant material. Although some cleavage was observed for Cel5A and E3 proteins, they retained reasonable enzymatic activity (Figure 3, Table 2). Moreover, the majority of nonrecombinant plant proteins were largely degraded during the storage of plant material as silage, effectively resulting in an enrichment of some recombinant cellulases as a percentage of total soluble protein and achieving, at least in the case of Bgl4, levels of up to 90% of TSP (Table 2). Storage of plant-expressed cellulases in dried plant matter was also investigated, but the results were less favourable. Together, the data presented here show that representatives of all cellulase classes necessary to catalyse the complete hydrolysis of cellulosic biomass to glucose can be manufactured in plants using this simple transfection protocol. The protocol is robust, requires only simple agricultural manipulations to put into practice, and provides high expression levels of cellulases and high activity of these important enzymes at harvest and after extended storage as silage.

Recombinant griffithsin and thaumatin are stable upon the storage of plant biomass as silage

We tested our technology for two other cost-sensitive recombinant proteins: antiviral protein griffithsin from the red algae *Griffithsia* sp. (O'Keefe *et al.*, 2010) and the sweet protein thaumatin from the West African plant katempfe (*Thaumatococcus daniellii* Benth) (Edens *et al.*, 1982). We expressed griffithsin and thaumatin in *Nicotiana benthamiana* plants using spraying and vacuum inoculation with agrobacteria harbouring TMV-based viral vectors pNMD2971 and pICH95397, respectively (Figure S4). Recombinant protein stability was analysed upon the storage of plant biomass either at −80 °C or as silage at room temperature for up to 16 weeks. SDS-PAGE analysis revealed

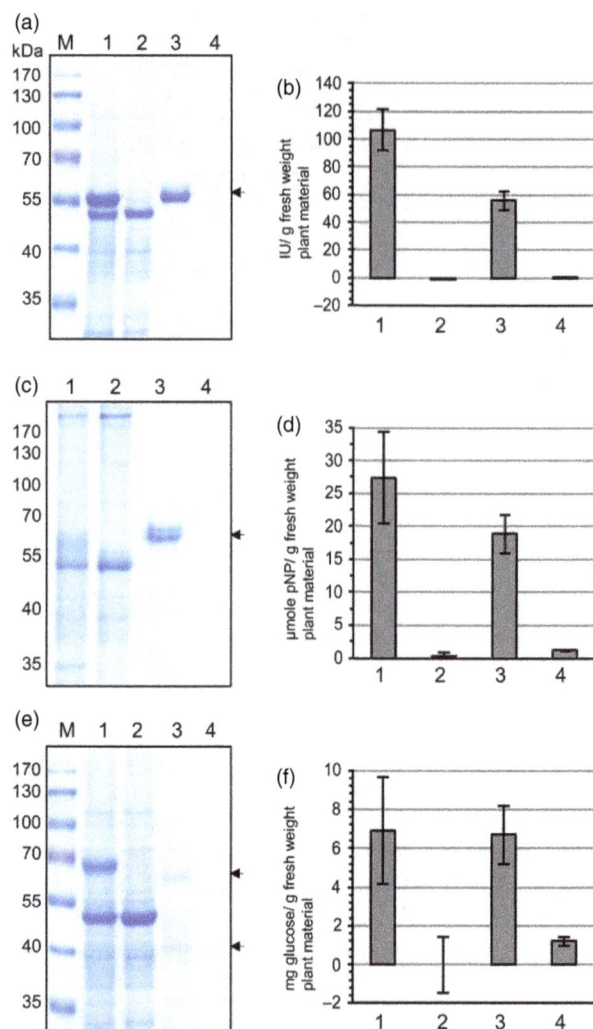

Figure 3 Storage stability of recombinant cellulases in the ensiled plant biomass: SDS-PAGE (a, c, e) and enzymatic activity analyses (b, d, f) for β-glucosidase Bgl4 (a, b), cellobiohydrolase CBHI (c, d) and exoglucanase E3 (e, f). Coomassie-stained SDS protein gels (a, c, e) are loaded with total soluble extracts (TSP, 5 μg) from *N. benthamiana* nontransfected leaf tissue (2, 4) or plants transfected by *Agrobacterium tumefaciens* carrying plasmid vectors (1, 3) pNMD1201 (a), pNMD1181 (c) or pNMD1229 (e) by spraying with 10^{-3} dilutions. Upon harvest 11 dpi, plant material was stored at −80 °C (1, 2) or as silage at ambient temperature (3, 4) for 16 weeks. M, molecular weight marker with weights (kDa) shown on the left. Arrows indicate recombinant proteins. Cellulase activities of TSP samples shown in b, d, f calculated per g of fresh weight plant material for frozen (1, 2) and ensiled (3, 4) biomass. Error bars indicate standard deviation of biological replicates, *n* = 3.

nearly complete degradation of native plant proteins in ensiled plant biomass already after 4 weeks of storage (Figure 4). In contrast, both recombinant griffithsin and thaumatin remained stable and did not show any significant degradation during whole analysed periods (16 weeks for griffithsin and 5 weeks for thaumatin).

Discussion

We provide here the example of robust industrially applicable technology for production of a desired recombinant protein product based on spraying of plants with *Agrobacterium* suspensions without the need of full virus. Earlier studies with *Agrobacterium*-mediated transfection employed either vacuum infiltration or alternative methods relying on the systemic viral infection. In the first case (Marillonnet *et al.*, 2005), bacteria were forced into plant leaves by flooding the intercellular leaf spaces upon release of a vacuum, resulting in direct transfection of the majority of leaf cells. However, this process requires a vacuum chamber to transfect pot- or tray-contained plants in batches and hence cannot be applied to field-grown plants. Several authors mention *Agrobacterium* application methods that do not rely on vacuum infiltration. In Ryu *et al.* (2004), and in Azhakanandam *et al.* (2007); 'agrodrench' (drenching the soil under plants) and 'wound-and-spray/agrospray' methods, respectively, have been described. However, as the successful delivery by *Agrobacterium* results in systemic viral infection when even very rare or single events of transfection could result in detectable symptoms, the efficiency of *Agrobacterium* delivery per se remains unknown. Moreover, the presented quantitative protein expression data in these studies are far below the vacuum infiltration (Marillonnet *et al.*, 2005). The spraying or dipping used for *Arabidopsis* transformation described by Chung *et al.* (2000) is an interesting solution that is practically restricted to *Arabidopsis* and does not provide any insight into efficiency of transient delivery step in connection with transformation. Our research clearly demonstrates that upon use of proper surfactants, between 0.9 and up to 3.5% of leaf cells can be directly transfected using spray application of bacterial suspensions containing 10^{6-7} CFU/ml as inoculum. In these studies, we used the organosilicon surfactant Silwet L-77®, but good results were also obtained with other adjuvants such as Tween®20 (ICI Americas, Inc., USA) and Triton X-100™ (Union Carbide Corporation, USA). We also show that a combination of this simple and indefinitely scalable DNA delivery procedure with deconstructed viral vectors capable of cell-to-cell movement results, within 10–14 days, in efficient transfection of most leaf cells, thus providing for expression of recombinant proteins at very high levels, often reaching over 50% of the total

soluble protein in leaves. Thus, even this early version of the process results in protein expression levels approaching the biological limits of plants. Although the results reported here are based on the use of *Nicotiana benthamiana* as a manufacturing host, we have good evidence that essentially the same protocol works with at least eight *Nicotiana* species, including *N. debneyi*, *N. excelsior*, *N. exigua*, *N. maritima*, *N. simulans*, *N. sylvestris* and some varieties of *N. tabacum* (e. g. Maryland Mammoth).

The process uses bacterial inocula efficiently such that treatment of 1 ton of host plant biomass could be effectively achieved with approximately 10 g of bacteria at the densities studied here. Fermentation costs for producing agrobacteria are estimated to be <5% of the total costs of the manufacturing and storage process. Undoubtedly, the procedure can be further optimized by improving the *Agrobacterium* vector and the plasmid constructs it carries, by additional selection of adjuvants and wetting agents and by screening and optimizing methods for spray delivery. In this study, an industrial *Agrobacterium tumefaciens* strain ICF320 developed by Icon Genetics GmbH (Halle, Germany) for magnI-CON® system (Bendandi *et al.*, 2010) was used because of biosafety reasons. ICF320 is a disarmed derivative of the strain C58 containing several auxotrophies (D*cysK*$_a$, D*cysK*$_b$, D*thiG*) that have been introduced to decrease the viability of bacteria in the environment. Several other *Agrobacterium* strains (*e.g.* standard laboratory strains AGL1, GV3101, EHA105) were also tested with our approach and provided good transfection results, too (data not shown).

Even before vector, host or procedural optimizations, we believe that this transient expression technology represents the most versatile, economical and efficient industrial process described to date for the plant-based manufacture of recombinant proteins in large scale.

The broad industrial use of this process would require release of genetically engineered agrobacteria which can be performed in a contained greenhouse environment even now. Open field application of the technology requires further studies and measures to assure the biological and environmental safety.

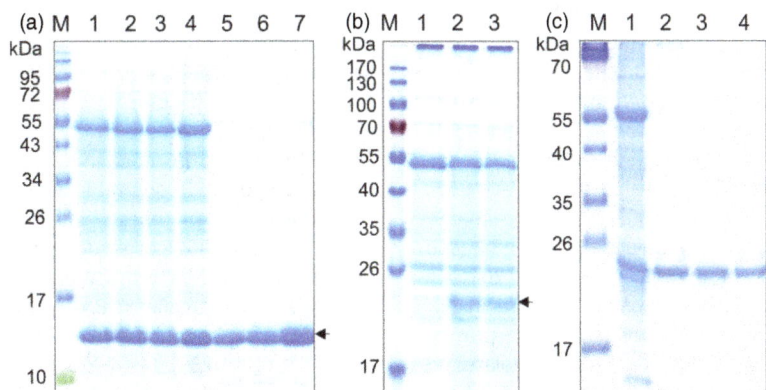

Figure 4 Expression levels and stability of recombinant griffithsin (a) and thaumatin (b, c) upon storage of plant biomass as silage. Extracts of frozen (−80 °C) and ensiled plant material resolved in a 12% (a, b) and 16.5% (c) polyacrylamide gel under reducing conditions. (a) Griffithsin-containing plant material extracted with 2 × Laemmli buffer. Lane 1, freshly harvested plant sample; lanes 2, 3 and 4, plant material after 4, 8 and 16 weeks storage at −80 °C; lanes 5, 6 and 7, ensiled plant material after 4, 8 and 16 weeks storage at ambient temperature. (b) Expression levels of thaumatin upon spray-transfection. Extracts with 2 × Laemmli buffer corresponding to 1.0 mg fresh weight. Lane 1, nontransfected plant material; lanes 2 and 3, plant material expressing thaumatin upon spraying with dilutions of inoculation solutions of 10^{-2} and 10^{-3} at 14 dpi. (c) Thaumatin-containing plant material after 5 weeks storage. Lanes 1 and 2, frozen samples extracted with 2 × Laemmli buffer and 100 mM NaCl, 50 mM acetic acid; lanes 3 and 4, ensiled plant material extracted with Laemmli buffer and 100 mM NaCl, 50 mM acetic acid, respectively. Loading corresponds to 2.0 mg fresh weight. M, molecular weight marker with weights (kDa) shown on the left. Arrows indicate recombinant proteins.

We chose to illustrate the potential power and utility of this transient expression-based protein manufacturing technology using cellulases, a broad class of enzymes needed in different processes but especially key to the economic conversion of cellulosic biomass to ethanol. One of the main reasons for the high cost of cellulosic ethanol today is the cost of cellulases, accounting for up to 20–40% of the cost of goods (Sainz, 2009). Relative to amylases used to produce ethanol from starch, cellulases would be needed in very large amounts because their specific activity is 10–100-times lower than that of amylases (Merino and Cherry, 2007); hence, their use rate is as high as 1% (w/w) of the cellulose treated (http://www.bioenergy.novozymes. com). As illustrated by the decades-long efforts of Novozymes, Genencor (DuPont-Danisco) and other companies and academic groups, even highly optimized microbial fermentation processes have been unable to solve the catalyst cost problem. The economics of cellulose bioconversion have also been unfavourable because of the high cost of transporting the cellulases from enzyme fermentation facilities to the ethanol manufacturing sites (Sticklen, 2008). The silage-based process described here offers a plausible lower cost option as the catalyst can be produced in immediate vicinity of feedstock production site or it could be co-transported with the feedstock.

Complete cellulose degradation requires three classes of glucanases, and in our work, we have tested six genes of bacterial and fungal origin encoding representative members all three classes. Although in these studies the expression of these enzymes was only partially optimized (e.g. codon-usage optimization and screening for best subcellular localization for each protein), we were able to express most proteins at high levels and demonstrate that all proteins expressed retained their expected enzymatic activity even when stored as silage, a traditional and inexpensive agricultural biomass preservation method. The production time and expression efficiencies of spray-based cellulase production surpass those achieved in plants stably transformed with nonreplicating expression vectors. For example, for endoglucanase Cel5A from T. maritima, expression levels of 4.5% and 5.2% were reported (Mahadevan et al., 2011) and (Kim et al., 2010), respectively) in stably transgenic tobacco plants compared to 12.5% of TSP in transiently transfected N. benthamiana. High accumulation of endoglucanase cel9A (about 40% of TSP) achieved by chloroplast transformation resulted in severe mutant phenotype of transgenic plants (Petersen and Bock, 2011). Expression of CBHI from T. reesei, E3 from T. fusca and Bgl4 from H. grisea in our system was as high as 12.5%, 25% and 50% of TSP, respectively (Table 2), compared to 4.1% for CBHI (Hood et al., 2007), 3–4% (Yu et al., 2007) and approximately 5% (Petersen and Bock, 2011) for exocellulase E3, or 5.8% and 9.6% for β-glucosidases (Gray et al., 2011; Jung et al., 2010) in stably transgenic plants. This is not surprising because in our system, cellulase production is induced at the moment of spraying with agrobacteria, and there is no negative effect of cellulases on plant growth. For all but one (E1 from A. cellulolyticus) of the modelled enzymes, expression levels exceeded 10% of TSP, the level which is thought to be sufficient for cell wall deconstruction without supplementation with microbially sourced enzymes (Sticklen, 2008).

The two other proteins expressed, thaumatin and griffithsin, are both good examples of cost-sensitive proteins. Thaumatin is an approved noncaloric sweet-tasting protein that is 3000 times sweeter than sucrose (on a weight basis). It can be extracted from the original source (Thaumatococcus daniellii) today inexpen-

sively, but the natural plant resources are very limited (Faus and Sisniega, 2003). Griffithsin is one of the most promising antivirals, and its potential use as a preventative microbicide against HIV (and other enveloped viruses) in a form of topical ointments or suppositories would require it to be manufactured very inexpensively (O'Keefe et al., 2009; Zeitlin et al., 2009). Our experiments demonstrate that both proteins can be easily and very efficiently manufactured by Agrobacterium spay-based transient expression and that the upstream process can be efficiently separated from the downstream process by storage of the biomass as silage.

Looking ahead, the process described will also benefit from identification of new or more efficient and stable enzymes. Based on the known costs of various commodity agricultural products, we estimate that the cost of cellulases produced by this method will be significantly less than $10 per kg of active protein, thus making the economics of this process highly competitive with microbial fermentation. We believe that the process should be easy to implement and scale because it relies on existing agricultural infrastructure and practices; hence, it constitutes a simple additional step that can be integrated at any site that generates cellulosic biomass. In addition to cellulases, we have been evaluating our process with multiple enzyme and protein candidates that need to be produced inexpensively and have observed expression levels that are comparable or surpass the levels seen in plants transfected using vacuum infiltration, which has been up to now the 'gold standard' for plant-based biomanufacturing.

Experimental procedures

Bacterial strains and growth conditions

Escherichia coli DH10B cells were cultivated at 37 °C in LB medium (lysogeny broth (Bertani, 1951)). Agrobacterium tumefaciens ICF320 (Bendandi et al., 2010) cells were cultivated at 28 °C in LBS medium (modified LB medium containing 1% soya peptone (Duchefa, Haarlem, Netherlands)).

Plasmid constructs

TMV-based assembled vectors pICH18711 and pICH18722 (Figure S1) were described in Marillonnet et al. (2005).

5'TMV-based pro-vector modules pICH22455, pICH20155, pICH20188, pICH22464, pICH20388, pICH20999, pICH20030 and pICH22474 (Figure S2a) used for the optimization of expression of cellulases by targeting into different subcellular compartments are described in Kalthoff et al. (2010).

For cloning of 3'-provector modules of TMV pro-vectors, coding sequences of genes of interest were cloned into the BsaI sites of the pICH28575 construct (Kalthoff et al., 2010). Genes of interest used in these constructs encoded β-glucosidase Bgl4 from Humicola grisea (GenBank: BAA74958.1; pNMD910), endoglucanase E1 from Acidothermus cellulolyticus 11B (Swiss-Prot: P54583.1; pNMD231), β-1,4-exocellulase (E3) from Thermobifida fusca (GenBank: AAA62211.1; pNMD251) and exoglucanase 1 (CBH1) from Trichoderma reesei/Hypocrea jecorina (Swiss-Prot: P62694.1, pNMD241) (Figure S2b). Coding sequences of all genes of interest were codon-optimized for Nicotiana benthamiana and synthesized by Entelechon GmbH (Bad Abbach, Germany).

The integrase vector module pICH14011 (Figure S2c) is described in Kalthoff et al. (2010).

Assembled TMV vectors for the expression of selected cellulase translational fusions were created on the basis of pICH18711 vector (Marillonnet et al., 2005). They contain actin 2-promoter-driven TMV RdRp with 14 introns, TMV MP with 2 introns, targeting presequence, and coding sequence of gene of interest, 3'TMV nontranslated region (3' NTR) as well as a nos terminator. The entire fragment is inserted between the T-DNA left and right borders of binary vector. pNMD1201 and pNMD3081 constructs contain coding sequences of β-glucosidase Bgl4 from Humicola grisea and endoglucanase Cel5A from Thermotoga maritima, respectively, without any fusion sequence. The pNMD1231 construct bears endoglucanase E1 from Acidothermus cellulolyticus fused with rice α-amylase 3A apoplast targeting presequence. pNMD1181 and pNMD3061 constructs contain exoglucanase 1 (CBHI) from Trichoderma reesei and exoglucanase 1 (CBHI) from Humicola grisea, respectively, fused with barley α-amylase apoplast targeting presequence (Figure S3).

Assembled TMV vectors for the expression of griffithsin and thaumatin were constructed similarly (Figure S4). pNMD2901 vector contained the coding sequence of griffithsin (GenBank: ACM42413.1) codon-optimized for Nicotiana benthamiana and synthesized by GeneArt (Life Technologies, Regensburg, Germany). pICH95397 construct contained the coding sequence of mature peptide of preprothaumatin 2 from Thaumatococcus daniellii (GenBank: AAA93095.1) fused with rice α-amylase 3A apoplast targeting presequence.

pICH18711, pICH18722, pICH22455, pICH20155, pICH20188, pICH22464, pICH20388, pICH20999, pICH20030, pICH22474, pICH28575, pICH95397 and pICH14011 constructs were kindly provided by Icon Genetics GmbH (Halle, Germany).

Plant material and inoculations

Nicotiana benthamiana plants were grown in the greenhouse (day and night temperatures of 19–23 °C and 17–20 °C, respectively, with 12 h light and 35–70% humidity). Six-week-old plants were used for inoculations.

For plant transfection, saturated Agrobacterium overnight cultures were adjusted to OD_{600} = 1.3 or 1.5 (approximately 1.2×10^9 cfu/mL) with Agrobacterium inoculation solution (10 mM MES pH 5.5, 10 mM $MgSO_4$), further diluted with same solution supplemented with 0.1% (v/v) Silwet L-77 (Kurt Obermeier GmbH & Co. KG, Bad Berleburg, Germany) and inoculation was carried out using a hand sprayer (Carl Roth GmbH + CO.KG, Karlsruhe, Germany).

Mesophyll protoplast isolation

Protoplasts were isolated as described in Giritch et al. (2006) for count of transfected GFP-expressing plant cells. For isolation of protoplasts, whole leaf blades of three different leaves of the same plant were used. For each leaf, approximately 8500 cells have been counted; the total number of cells for each treatment including three separate leaves was about 25 500. For the percentage of GFP-expressing cells, average and standard deviation (SD) values are provided.

Protein analysis

About 50 mg fresh weight N. benthamiana leaf material pooled from 3 leaves of different age were ground in liquid nitrogen, and crude protein extracts were prepared with 5 vol. 2× Laemmli buffer. Total soluble protein (TSP) was extracted from approximately 150 mg fresh weight plant material ground in liquid nitrogen with 5 vol. prechilled extraction buffer (50 mM sodium acetate pH 5.5, 100 mM NaCl, 10% (v/v) glycerol). The protein concentration of TSP extracts was determined by Bradford assay using Bio-Rad Protein Assay (Bio-Rad Laboratories GmbH, Munich, Germany) and BSA (Sigma-Aldrich Co., St-Louis, MO, USA) as a standard.

For analysis by 10%, 12% or 16.5% SDS-PAGE and Coomassie-staining using PageBlue™ Protein Staining Solution (Fermentas GmbH, St. Leon-Rot, Germany), protein extracts were denatured at 95 °C for 5 min. before loading. The estimation of the percentage of recombinant cellulase fusion of TSP was carried out by comparison of TSP extracts to known amounts of BSA (Sigma-Aldrich) on Coomassie-stained SDS-PAA gels.

Storage of plant material at −80 °C and as silage

Plant material sprayed with 10^{-3} dilutions of Agrobacterium cultures was harvested 11 dpi as a pool of plants. Midribs of leaves were removed, and leaf blades were chopped into pieces of about 5 cm^2 in size using a scalpel blade. Leaf pieces were spread out for dehydration at the room temperature for 22 h. Dehydrated leaf material (45–58% of fresh weight) was supplemented with 2% (w/dry weight) propionic acid, sodium salt and packed into plastic bags using a commercial vacuum food sealer (Food Saver V2040-I, Sunbeam Products, Inc., Boca Raton, FL) for conservation. This plant material was stored as silage at the room temperature in dark for 1–4 months. Aliquots corresponding to 150 mg fresh weight plant material were analysed in triplicates for enzymatic activity.

About 150 mg fresh weight aliquots (9 leaf discs of 1 cm in diameter pooled from 3 comparable leaves) were harvested in triplicates from identical plant material used for the preparation of silage. Samples were frozen in liquid nitrogen and stored at −80 °C.

Enzymatic activity measurements

Endoglucanase and cellobiohydrolase cellulase activity assay (Cel5A, CBHI, E1)

Enzyme activity was determined using p-nitrophenyl-β-D-cellobioside as substrate. 100 μg of plant TSP extracts was incubated in 50 mM NaAc pH 5.5, 100 mM NaCl, 5% (v/v) glycerol, supplemented with or without 5 mM p-nitrophenyl-β-D-cellobioside as enzyme blanks, respectively, in a final volume of 1 mL at 50 °C for 1–24 h. Aliquots of the reaction mixture were removed at different time points and diluted 1:10 with 0.15 M glycine pH 10.0 to terminate the reaction. The concentration of p-nitrophenol (pNP) released from p-nitrophenyl-β-D-cellobioside was determined by measurement of the absorbance at 405 nm. The commercial enzymes cellulase (endo-1,4-β-D-glucanase) from Trichoderma sp. (Megazyme, Bray, Ireland) and cellobiohydrolase (CBHI) from Trichoderma sp. (Megazyme) served as positive controls.

β-glucosidase cellulase activity assay (Bgl4)

For determination of β-glucosidase activity, cellobiose was used as a substrate. Different protein amounts of plant TSP extracts ranging from 1 to 40 μg were incubated in 50 mM NaAc pH 5.5, 100 mM NaCl, 5% (v/v) glycerol supplemented with or without 5 mM cellobiose as enzyme blanks, respectively, in a final volume of 1 mL at 50°C for 30 min. Concentration of released glucose in the samples were determined by D-glucose (GOPOD-Format) Kit (Megazyme). The commercial enzyme cellobiase from Aspergillus niger (synonym: Novozyme 188; Sigma-Aldrich) served as positive control. β-glucosidase activity was calculated in cellobiase units (CBU) according to IUPAC or international units (IU) like described in Ghose (1987) from samples releasing close to 1 mg glucose.

Exoglucanase cellulase activity assay (E3)

For determination of exoglucanase activity, the insoluble micro-crystalline cellulose substrate Avicel (Sigmacell cellulose type 20; Sigma-Aldrich) was used. 100 µg of plant TSP extracts was incubated in 50 mм NaAc pH 5.5, 50 mм NaCl, 5% (v/v) glycerol, 0.02% sodium azide, 1.2 mg/mL Novozyme 188 (β-glucosidase; Sigma-Aldrich), supplemented with or without 1% (w/v) Avicel as enzyme blanks, respectively, in a final volume of 3 mL at 50 °C and 90 rpm for 120 h. Aliquots of the reaction mixture were removed at different time points (2, 4, 24, 48, 72 and 120 h) for determination of the glucose concentration using D-glucose (GOPOD-Format) Kit (Megazyme). The commercial enzyme cel-lobiohydrolase (CBHI) from *Trichoderma* sp. (Megazyme) served as positive control.

Storage of griffithsin and thaumatin expressing plant material

Griffithsin-expressing plant material was generated, stored and analysed as described for cellulases. For thaumatin expression, plants were vacuum-inoculated with *Agrobacterium*; plant material was harvested at 9 dpi. Upon harvest, plant material (whole plants) was shredded in a table-top cutter and either frozen at −80 °C or supplemented with 2% (w/w) sodium propionate and compressed into a plastic beaker and weight down for storage as silage. For thaumatin stability analysis, crude extracts or TSP extracts were prepared from plant material corresponding to 10 g fresh weight with 3 vol. 2 × Laemmli buffer or 100 mм NaCl, 50 mм acetic acid, respectively.

Acknowledgements

We thank our colleagues from Icon Genetics GmbH (Halle, Germany) for providing us with magnICON® vectors. We thank Dr. Daniel Tusé, DT/Consulting Group, Sacramento, CA, for critical reading of manuscript.

The authors declare no conflict of interest.

References

Azhakanandam, K., Weissinger, S.M., Nicholson, J.S., Qu, R. and Weissinger, A.K. (2007) Amplicon-plus targeting technology (APTT) for rapid production of a highly unstable vaccine protein in tobacco plants. *Plant Mol. Biol.* **63**, 393–404.

Bendandi, M., Marillonnet, S., Kandzia, R., Thieme, F., Nickstadt, A., Herz, S., Fröde, R., Inogés, S., Lòpez-Dìaz de Cerio, A., Soria, E., Villanueva, H., Vancanneyt, G., McCormick, A., Tusé, D., Lenz, J., Butler-Ransohoff, J.E., Klimyuk, V. and Gleba, Y. (2010) Rapid, high-yield production in plants of individualized idiotype vaccines for non-Hodgkin's lymphoma. *Ann. Oncol.* **21**, 2420–2427.

Bertani, G. (1951) Studies on lysogenesis. I. The mode of phage liberation by lysogenic *Escherichia coli. J. Bacteriol.* **62**, 293–300.

Carroll, A. and Somerville, C. (2009) Cellulosic Biofuels. *Annu. Rev. Plant Biol.* **60**, 165–182.

Chung, M.H., Chen, M.K. and Pan, S.M. (2000) Floral spray transformation can efficiently generate *Arabidopsis* transgenic plants. *Transgenic Res.* **9**, 471–476.

Edens, L., Heslinga, L., Klok, R., Ledeboer, A.M., Maat, J., Toonen, M.Y., Visser, C. and Verrips, C.T. (1982) Cloning of cDNA encoding the sweet-tasting plant protein thaumatin and its expression in *Escherichia coli. Gene*, **18**, 1–12.

Faus, I. and Sisniega, H. (2003) Sweet-tasting proteins. In *Biopolymers, Volume 8, Polyamides and Complex Proteinaceous Materials II* (Fahnestock, S.R. and Steinbüchel, A., eds), pp. 203–222. Hoboken, NJ: Wiley-Blackwell.

Ghose, T.K. (1987) Measurement of cellulase activities. *Pure Appl. Chem.* **59**, 257–268.

Giritch, A., Marillonnet, S., Engler, C., van Eldik, G., Botterman, J., Klimyuk, V. and Gleba, Y. (2006) Rapid high-yield expression of full-size IgG antibodies in plants coinfected with noncompeting viral vectors. *Proc. Natl Acad. Sci. USA*, **103**, 14701–14706.

Gleba, Y.Y. and Giritch, A. (2012) Vaccines, antibodies, and pharmaceutical proteins. In *Plant Biotechnology and Agriculture. Prospects for the 21st century* (Altman, A. and Hasegawa, P.M., eds), pp. 465–479. San Diego, California, USA: Academic Press.

Gray, B.N., Yang, H., Ahner, B.A. and Hanson, M.R. (2011) An efficient downstream box fusion allows high-level accumulation of active bacterial beta-glucosidase in tobacco chloroplasts. *Plant Mol. Biol.* **76**, 345–355.

Harvey, L.T. (1998) *A Guide to Agricultural Spray Adjuvants Used in the United States.* Fresno, California, USA: Thompson publications.

Hood, E.E., Love, R., Lane, J., Bray, J., Clough, R., Pappu, K., Drees, C., Hood, K.R., Yoon, S., Ahmad, A. and Howard, J.A. (2007) Subcellular targeting is a key condition for high-level accumulation of cellulase protein in transgenic maize seed. *Plant Biotechnol. J.* **5**, 709–719.

Jung, S., Kim, S., Bae, H., Lim, H.S. and Bae, H.J. (2010) Expression of thermostable bacterial β-glucosidase (BglB) in transgenic tobacco plants. *Bioresour. Technol.* **101**, 7155–7161.

Kalthoff, D., Giritch, A., Geisler, K., Bettmann, U., Klimyuk, V., Hehnen, H.R., Gleba, Y. and Beer, M. (2010) Immunization with plant-expressed hemagglutinin protects chickens from lethal highly pathogenic avian influenza virus H5N1 challenge infection. *J. Virol.* **84**, 12002–12010.

Kim, S., Lee, D.S., Choi, I.S., Ahn, S.J., Kim, J.H. and Bae, H.J. (2010) *Arabidopsis thaliana* Rubisco small subunit transit peptide increases the accumulation of *Thermotoga maritima* endoglucanase Cel5A in chloroplasts of transgenic tobacco plants. *Transgenic Res.* **19**, 489–497.

Mahadevan, S.A., Wi, S.G., Kim, Y.O., Lee, K.H. and Bae, H.J. (2011) In planta differential targeting analysis of *Thermotoga maritima* Cel5A and CBM6-engineered Cel5A for autohydrolysis. *Transgenic Res.* **20**, 877–886.

Marillonnet, S., Giritch, A., Gils, M., Kandzia, R., Klimyuk, V. and Gleba, Y. (2004) *In planta* engineering of viral RNA replicons: efficient assembly by recombination of DNA modules delivered by *Agrobacterium. Proc. Natl Acad. Sci. USA*, **101**, 6852–6857.

Marillonnet, S., Thoeringer, C., Kandzia, R., Klimyuk, V. and Gleba, Y. (2005) Systemic *Agrobacterium tumefaciens*-mediated transfection of viral replicons for efficient transient expression in plants. *Nat. Biotechnol.* **23**, 718–723.

Merino, S.T. and Cherry, J. (2007) Progress and challenges in enzyme development for biomass utilization. *Adv. Biochem. Eng. Biotechnol.* **108**, 95–120.

O'Keefe, B.R., Vojdani, F., Buffa, V., Shattock, R.J., Montefiori, D.C., Bakke, J., Mirsalis, J., d'Andrea, A.L., Hume, S.D., Bratcher, B., Saucedo, C.J., McMahon, J.B., Pogue, G.P. and Palmer, K.E. (2009) Scaleable manufacture of HIV-1 entry inhibitor griffithsin and validation of its safety and efficacy as a topical microbicide component. *Proc. Natl Acad. Sci. USA*, **106**, 6099–6104.

O'Keefe, B.R., Giomarelli, B., Barnard, D.L., Shenoy, S.R., Chan, P.K., McMahon, J.B., Palmer, K.E., Barnett, B.W., Meyerholz, D.K., Wohlford-Lenane, C.L. and McCray, P.B.Jr.. (2010) Broad-spectrum *in vitro* activity and *in vivo* efficacy of the antiviral protein griffithsin against emerging viruses of the family *Coronaviridae. J. Virol.* **84**, 2511–2521.

Petersen, K. and Bock, R. (2011) High-level expression of a suite of thermostable cell wall-degrading enzymes from the chloroplast genome. *Plant Mol. Biol.* **76**, 311–321.

Ryu, C.M., Anand, A., Kang, L. and Mysore, K.S. (2004) Agrodrench: a novel and effective agroinoculation method for virus-induced gene silencing in roots and diverse Solanaceous species. *Plant J.* **40**, 322–331.

Sainz, M.B. (2009) Commercial cellulosic ethanol: The role of plant-expressed enzymes. *In Vitro Cell Dev. Biol. Plant.* **45**, 314–329.

Sticklen, M.B. (2008) Plant genetic engineering for biofuel production: towards affordable cellulosic ethanol. *Nat. Rev. Genet.* **9**, 433–443.

Takashima, S., Nakamura, A., Hidaka, M., Masaki, H. and Uozumi, T. (1996) Cloning, sequencing, and expression of the cellulase genes of *Humicola grisea* var. *thermoidea. J. Biotechnol.* **50**, 137–147.

Takashima, S., Nakamura, A., Hidaka, M., Masaki, H. and Uozumi, T. (1999) Molecular cloning and expression of the novel fungal β-Glucosidase genes from *Humicola grisea* and *Trichoderma reesei. J. Biochem.* **125**, 728–736.

Permissions

List of Contributors

Catherine Reinbold, Olivier Lemaire and Gérard Demangeat
Svqv, Inra, Université de Strasbourg, Colmar, France

Loréne Belval and Caroline Hemmer
Svqv, Inra, Université de Strasbourg, Colmar, France
Institut de Biologie Moléculaire des Plantes CNRS-UPR 2357, associée á l'Université de Strasbourg, CNRS, Strasbourg, France

Léa Ackerer
Svqv, Inra, Université de Strasbourg, Colmar, France
Institut de Biologie Moléculaire des Plantes CNRS-UPR 2357, associée á l'Université de Strasbourg, CNRS, Strasbourg, France
Institut Français de la Vigne et du Vin, Domaine de l'Espiguette, Le Grau-du-Roi, France

François Berthold, Corinne Schmitt-Keichinger and Christophe Ritzenthaler
Institut de Biologie Moléculaire des Plantes CNRS-UPR 2357, associée á l'Université de Strasbourg, CNRS, Strasbourg, France

Claude Sauter and Jean-Daniel Fauny
Institut de Biologie Moléculaire et Cellulaire du CNRS, UPR 9002, Architecture et Réactivité de l'ARN, Université de Strasbourg, Strasbourg, France

Zi Shi, Siela N. Maximova and Mark J. Guiltinan
Department of Plant Science and Huck Institute of Life Sciences, The Pennsylvania State University, University Park, PA, USA

Emily E. Helliwell
Department of Plant Science and Huck Institute of Life Sciences, The Pennsylvania State University, University Park, PA, USA
Center for Genome Research and Biocomputing and Department of Botany and Plant Pathology, Oregon State University, Corvallis, OR, USA

Brett M. Tyler
Center for Genome Research and Biocomputing and Department of Botany and Plant Pathology, Oregon State University, Corvallis, OR, USA
Virginia Bioinformatics Institute and Department of Plant Pathology, Physiology and Weed Science, Virginia Polytechnic Institute and State University, Blacksburg, VA, USA

Julio Vega-Arreguín
Virginia Bioinformatics Institute and Department of Plant Pathology, Physiology and Weed Science, Virginia Polytechnic Institute and State University, Blacksburg, VA, USA

Bryan Bailey
United States Department of Agriculture, Agricultural Research Service, Beltsville, MD, USA

Shunyuan Xiao
Institute for Bioscience and Biotechnology Research and Department of Plant Science and Landscape Architecture, University of Maryland, College Park, MD, USA

Danlong Jing, Jianwei Zhang, Yan Xia, Fangqun OuYang, Shougong Zhang and Junhui Wang
State Key Laboratory of Tree Genetics and Breeding, Key Laboratory of Tree Breeding and Cultivation of State Forestry Administration, Research Institute of Forestry, Chinese Academy of Forestry, Beijing, China

Lisheng Kong
Department of Biology, Centre for Forest Biology, University of Victoria, Victoria, BC, Canada

Hanguo Zhang
State Key Laboratory of Tree Genetics and Breeding, Northeast Forestry University, Harbin, China

Yashwant Kumar, Priyabrata Panigrahi, Bhushan B. Dholakia, Veena Dewangan, Sachin G. Chavan, Narendra Y. Kadoo, Ashok P. Giri and Vidya S. Gupta
Division of Biochemical Sciences, CSIR-National Chemical Laboratory, Pune, India

Limin Zhang, Xiangyu Wu and Ning Li
Key Laboratory of Magnetic Resonance in Biological Systems, National Centre for Magnetic Resonance in Wuhan, Wuhan Institute of Physics and Mathematics, Chinese Academy of Sciences, Wuhan, China

Huiru Tang
Key Laboratory of Magnetic Resonance in Biological Systems, National Centre for Magnetic Resonance in Wuhan, Wuhan Institute of Physics and Mathematics, Chinese Academy of Sciences, Wuhan, China
State Key Laboratory of Genetic Engineering, Metabolomics and Systems Biology Laboratory, School of Life Sciences, Fudan University, Shanghai, China

Shrikant M. Kunjir and Pattuparambil R. Rajmohanan
Central NMR Facility, CSIR-National Chemical Laboratory, Pune, India

Saravanan Kumar, Krishan Kumar, Sadhu Leelavathi and Vanga Siva Reddy
International Centre for Genetic Engineering and Biotechnology, New Delhi, India

Mogilicherla Kanakachari, Dhandapani Gurusamy, Prabhakaran Narayanasamy, Padmalatha Kethireddy Venkata, Amolkumar Solanke and Polumetla Ananda Kumar
National Research Centre on Plant Biotechnology, Indian Agricultural Research Institute (IARI), New Delhi, India

Savita Gamanagatti, Vamadevaiah Hiremath and Ishwarappa S. Katageri
University of Agricultural Sciences, Dharwad, India

Francesca De Marchis, Michele Bellucci and Andrea Pompa
Research Division of Perugia, Institute of Biosciences and Bioresources, National Research Council, Perugia, Italy

Katie L. Moore
School of Materials, University of Manchester, Manchester, UK

Paola Tosi
School of Agriculture Policy and Development, Reading University, Reading, UK

Richard Palmer, Malcolm J. Hawkesford and Peter R. Shewry
Rothamsted Research, Harpenden, UK

Chris R.M Grovenor
Department of Materials, University of Oxford, Oxford,UK

Ellen F. Mosleth
Nofima AS, Ås, Norway
Rothamsted Research, Harpenden, Hertfordshire, UK

Yongfang Wan, Artem Lysenko, Peter R. Shewry and Malcolm J. Hawkesford
Rothamsted Research, Harpenden, Hertfordshire, UK

Gemma A. Chope and Simon P. Penson
Cereals and Ingredients Processing, Campden BRI, Chipping Campden, Gloucestershire,

Pushkar Shrestha, Surinder P. Singh and Craig C. Wood
Csiro Agriculture, Canberra, ACT, Australia

Fatima Naim
Csiro Agriculture, Canberra, Act, Australia
School of Biological Sciences, The University of Sydney, Sydney, NSW, Australia

Peter M. Waterhouse
Csiro Agriculture, Canberra, Act, Australia
School of Biological Sciences, The University of Sydney, Sydney, NSW, Australia
School of Molecular Bioscience, The University of Sydney, Sydney, NSW, Australia

Stéphanie Robert, Marie-Claire Goulet and Dominique Michaud
Centre de recherche et d'innovation sur les végétaux, Pavillon Envirotron, Université Laval, Québec, QC, Canada

Marc-André D'Aoust
Medicago Inc., Québec, QC, Canada

Frank Sainsbury
Centre de recherche et d'innovation sur les végétaux, Pavillon Envirotron, Université Laval, Québec, QC, Canada
The University of Queensland, Australian Institute for Bioengineering and Nanotechnology, St Lucia, QLD, Australia

Reza Saberianfar and Rima Menassa
Department of Biology, Western University, London, on, Canada
Southern Crop Protection and Food Research Centre, Agriculture and Agri-Food Canada, London, on, Canada

Jussi J. Joensuu and Andrew J. Conley
VTT Technical Research Centre of Finland, Espoo, Finland

Yuan Shen and Dao-Xiu Zhou
Université Paris-sud 11, Institut de Biologie des Plantes, CNRS, UMR8618, Saclay Plant Science, Orsay, France

Martine Devic
Régulation Epigénétique et Développement de la Graine, Erl 3500 Cnrs-Ird Umr Diade, Ird centre de Montpellier, Montpellier, France

Loïc Lepiniec
Inra, Institut Jean-Pierre Bourgin, Saclay Plant Science, Versailles, France

Yun-Ji Shin, Alexandra Castilho, Martina Dicker, Flavio Sádio, Ulrike Vavra, Herta Steinkellner and Richard Strasser
Department of Applied Genetics and Cell Biology, University of Natural Resources and Life Sciences, Vienna, Austria

Clemens Grünwald-Gruber and Friedrich Altmann
Department of Chemistry, University of Natural Resources and Life Sciences, Vienna, Austria

Tae-Ho Kwon
NBM Inc., Wanju-gun, Jeollabuk-do, Korea

Ruud H. P. Wilbers, Lotte B. Westerhof, Debbie R. van Raaij, Marloes van Adrichem, Andreas D. Prakasa, Jose L. Lozano-Torres, Jaap Bakker, Geert Smant and Arjen Schots
Laboratory of Nematology, Plant Sciences Department, Wageningen University and Research Centre, Wageningen, The Netherlands

Rosanna E. B. Young and Saul Purton
Algal Research Group, Institute of Structural and Molecular Biology, University College London, London, UK

Ying Zhou, Ze-Ting Zhang, Mo Li, Xin-Zheng Wei, Xiao-Jie Li, Bing-Ying Li and Xue-Bao Li
Hubei Key Laboratory of Genetic Regulation and Integrative Biology, School of Life Sciences, Central China Normal University, Wuhan, China

Verena K. Hehle, Craig J. van Dolleweerd, Mathew J. Pau and Julian K-C. Ma
Molecular Immunology Unit, Division of Clinical Sciences, St. George's University of London, London, UK

Raffaele Lombardi, Eugenio Benvenuto and Marcello Donini
Biotechnology Laboratory, ENEA Casaccia, Rome, Italy

Patrizio Di Micco
Department of Biochemical Sciences 'A. Rossi Fanelli', 'Sapienza' University of Rome, Rome, Italy

Veronica Morea
CNR-National Research Council of Italy, Institute of Molecular Biology and Pathology, 'Sapienza' University of Rome, Rome, Italy

Charlotta Hansson Karin Schön and Nils Y. Lycke
Department of Microbiology and Immunology, University of Gothenburg, Gothenburg, Sweden

Irina Kalbina and Åke Strid
Örebro Life Science Center, School of Science and Technology, Örebro University, Örebro, Sweden

Sören Andersson
Örebro Life Science Center, School of Science and Technology, Örebro University, Örebro, Sweden
Department of Laboratory Medicine, Örebro University hospital, Örebro, Sweden

Maria I. Bokarewa
Department of Rheumatology and Inflammation Research, University of Gothenburg, Gothenburg, Sweden

Simone Hahn, Anatoli Giritch, Doreen Bartels, Luisa Bortesi and Yuri Gleba
Nomad Bioscience GmbH, Halle (Saale), Germany

Index

O
Oomycetes, 13, 24

P
Partial Desiccation Treatment, 25-27, 31-32

Phaseolin, 70-80

Picea Asperata, 25-26

Plant-pathogen Interaction, 37, 46

Polypeptides, 70-77, 119-120, 125, 128, 150

Protein Body, 82, 84-87, 119, 121-122, 124-125, 127-129

Protein Deposition, 82, 85, 114

Protein Folding, 47, 72, 76, 128

Protein Translation, 70

Proteolysis, 5, 8, 55, 116-117, 128, 182-183, 188-189, 191

Proteolytic Processing, 8, 11, 108, 112, 150, 153-154, 156-157, 191

Proteome, 23, 26, 29, 34-36, 38, 47, 49-54, 56, 61-63, 67-68, 70, 108-110, 112-117

Proteomics, 25-27, 35-38, 43, 46-51, 68-69, 114, 116, 118, 122, 125, 149, 191

R
Rheumatoid Arthritis, 193, 200-202

S
Seed Gene Expression, 130, 136

Signal Peptide, 15, 17, 71, 73-74, 76-77, 79, 114, 120, 141-143, 150, 152-157, 165

Soluble Protein, 71, 99, 119-121, 127, 151, 155, 157, 182, 187, 199, 203, 205, 207, 209

Somatic Embryo, 25-26, 35-36

Stress-related Protein, 25

Subcellular Localization, 122, 140, 142-143, 149, 176, 179, 191, 208

T
Thaumatin, 39, 63, 203, 206-210

Theobroma Cacao, 13-14, 23

Therapeutic Proteins, 71, 140, 160

Transcriptome, 49, 52-54, 61-63, 66-68, 89-90, 92-93, 95, 97-98, 141, 157

Transfer Rna, 160

Transgene Longevity, 99

Transgenic Plants, 2, 5, 10, 16-17, 20, 50, 79, 102, 106, 116-117, 131, 133, 170-173, 179, 182, 191-195, 198-199, 201, 208, 210

Transient Expression, 2-3, 9, 11, 13-14, 16-18, 20, 22, 107, 116-120, 124-126, 128, 140, 143, 147, 159, 182-183, 189, 191, 203, 205, 207-208, 210

V
Virus, 1-2, 5, 7-12, 68, 80, 99-100, 102, 106-108, 117-118, 120, 126, 128, 140-141, 149, 156-157, 159, 168, 183, 190, 192, 207, 210

Virus Like Particles, 1

W
Wheat, 43, 46, 49-51, 82-85, 87-91, 95-98